PHYSICS RESEARCH AND TECHNOLOGY

FRONTIER RESEARCH IN MICROSCALE AND NANOSCALE THERMAL AND FLUID SCIENCES

PHYSICS RESEARCH AND TECHNOLOGY

Additional books in this series can be found on Nova's website under the Series tab.

Additional E-books in this series can be found on Nova's website under the E-book tab.

NANOTECHNOLOGY SCIENCE AND TECHNOLOGY

Additional books in this series can be found on Nova's website under the Series tab.

Additional E-books in this series can be found on Nova's website under the E-book tab.

PHYSICS RESEARCH AND TECHNOLOGY

FRONTIER RESEARCH IN MICROSCALE AND NANOSCALE THERMAL AND FLUID SCIENCES

LIXIN CHENG
EDITOR

Nova Science Publishers, Inc.
New York

Library of Congress Cataloging-in-Publication Data

Frontier research in microscale and nanoscale thermal and fluid sciences / editor, Lixin Cheng.
p. cm.
Includes bibliographical references and index.
ISBN 978-1-61324-665-8 (hardcover)
1. Nanofluids--Thermal properties. 2. Microfluidics. I. Cheng, Lixin.
TJ853.4.M53F76 2011
620.1'06--dc23

2011015368

Published by Nova Science Publishers, Inc. † New York

CONTENTS

PREFACE

This book provides a vehicle for the exchange and dissemination of original research results, technical notes, and state-of-the-art reviews pertaining to thermal and fluid transport phenomena at microscales and nanoscales. It covers a wide range of topics on fundamentals and applications of microscale and nanoscale transfer processes of mass, momentum, and energy such as microscale and nanoscale heat transfer and fluid flow, nanofluid heat transfer and flow, microfluidics, nanofluidics, and technologies based on these transport processes such as various microscale and nanoscale thermal and fluid devices.

Chapter 1 - With the emergence of microscale and nanoscale thermal, fluidic and chemical systems, MEMS (Micro-Electo-Mechanical-Systems), NEMS (Nano-Electro-Mechanical-Systems), thermal management technologies for microelectronics, power electronics, aerospace and defence technologies etc., the development of ultra-compact heat exchangers, miniature and micro pumps, miniature compressors, micro-turbines, micro thermal systems for distributed power production, microfluidic and nanofludic device, lab-on-a-chip, microscale and nanoscale energy systems, nanofluid heat transfer technologies and others has become an important agenda of many researchers, research institutions and funding agencies.

Chapter 2 - Partly because of the technological challenge, partly because of stark necessity, there has been an increasing movement towards a miniaturization of appliances in the last decade. In all technical fields solutions are sought that encumber as little as possible without compromising on performance: in medical diagnostics, environmental sample analysis, military defence, consumer electronics, biomedical appliances, chemical reactors and heat management a constant research for quicker response times and portable devices has driven the field of microtechnology to impressive levels. In many applications it has been found that many small active components are more productive than few large ones, which is also in keeping with the growing trend towards modular design. Proper understanding of microscale transport phenomena is therefore fundamental for the designer of microfluidic devices. For this reason, many studies have been conducted to analyse the behaviour of convective flow through microchannels, both in single-phase and in two-phase flow. A first glance of the literature, especially for single-phase flow, leads to the conclusion that up to now the authors have had an agglomeration of disparate conclusions. In many cases the experimental data in microchannels disagree with the conventional theory and empirical correlations, but they also appear to be inconsistent with one another.

The present chapter is an attempt to critically analyse the available results for liquid single-phase and flow boiling heat transfer, trying to provide some sort of base note in the melisma of published data.

Chapter 3 - This chapter presents new experimental flow boiling heat transfer results in micro-scale tubes. The experimental data were obtained in a horizontal 2.32 mm I.D. stainless steel tube with heating length of 464 mm, R134a as working fluid, mass velocities ranging from 50 to 600 kg $m^{-2}s^{-1}$, heat flux from 5 to 55 kW m^{-2}, exit saturation temperatures of 22, 31 and 41 °C, and vapor qualities from 0.05 to 0.98. Flow pattern characterization was also performed from images obtained by high-speed filming. Heat transfer coefficient results from 2 to 14 kW m^{-2} K^{-1} were measured. It was found that the heat transfer coefficient is a strong function of the saturation pressure, heat flux, mass velocity and vapor quality. The experimental data were compared against the following flow boiling predictive methods from the literature: Liu and Winterton [3], Kandlikar and Balasubramanian [4], Zhang *et al.* [5], Saitoh *et al.* [6], Bertsch *et al.* [7] and Thome *et al.* [11]. Comparisons against these methods based on the data segregated according to flow patterns were also performed. Though not satisfactory, Saitoh *et al.* [6] worked the best and was able of capturing most of the experimental heat transfer trends.

Chapter 4 - In relation to the development of the interfacial-area transport equation, this study collected data sets on the axial development of area-averaged (one-dimensional) void fraction and interfacial-area concentration in vertical upward-developing bubbly flows using three mini pipes. The test pipe inner diameters were 0.55, 0.79 and 1.02 mm, and the lengths were 500 mm. Image processing was used to characterize the flow at five axial locations, z/D = 15.0, 75.0, 150, 250 and 450, where z and D are the axial distance from the inlet and inner pipe diameter, respectively. In the experiment, the superficial liquid velocity and the void fraction ranged from 0.395 m s^{-1} to 4.89 m s^{-1} and from 0.958 % to 28.6 %, respectively. The one-dimensional interfacial-area transport equation with a sink term due to wake entrainment was evaluated by using the newly obtained datasets. It was confirmed that the interfacial-area transport equation could reproduce the proper trend of the axial interfacial-area transport and predict the measured interfacial-area concentrations well. It was found that the present model shows promise for predicting the interfacial-area transport of mini-pipe bubbly flows in the test conditions of this study.

Chapter 5 - This study demonstrated the feasibility of simultaneously using a vibrating piezoelectric driver as an atomizer and condensing surface. The vibrations were of appropriate frequency and intensity to produce Vibration Induced Droplet Atomization (VIDA) in the condensate on the condenser surface.

The effect of VIDA was to disperse the vast majority of the condensate as a chaotic spray. The driver was a brass shim 21 mm in diameter and 0.1 mm thick. Driving signal was sinusoidal at 35 V_{p-p} and the frequency ranged from 2.0 to 4.0 kHz. The working fluid was water. Saturation temperature ranged from 23 to 32°C (0.028 to 0.032 bar). Condenser surface subcooling ranged from 0 to 17°C. The cold side of the driver was considered insulated. The piezoelectric driver/condenser surface was oriented in a vertical plane and was non-isothermal.

Surface temperature measurements revealed slight resistive heating in the piezoelectric element for the VIDA cases, which raised the temperature of over half the condenser surface above the saturation temperature of the vapor. This surface temperature change invalidated the theoretical model of the heat transfer process.

Despite this resistive heating, and resulting reduction in active condenser surface area, the VIDA cases still showed roughly equivalent overall heat transfer rate to the non-VIDA cases, indicating a more effective heat transfer process in the former.

Chapter 6 – The main fields of research at LTCM "Heat and Mass Transfer Laboratory" are: Two-phase flow and heat transfer (boiling and condensation) in microchannels, convective boiling of pure fluids and mixtures in plain and enhanced tubes, condensation of pure fluids and mixtures, falling film evaporation and condensation on enhanced tubes and tube arrays, boiling on tube bundles of plain and enhanced tubes, two-phase flows and modeling of flow regime transitions, flow visualization of two-phase phenomena and void fraction measurements with use of laser sheets, turbulence modeling of annular two-phase flows, and numerical modeling of adiabatic and condensing microscale two-phase flows. The LTCM laboratory was founded in 1998 by Prof. Thome upon his appointment at the EPFL. Currently, the laboratory has 12 Ph.D. students (plus one visiting) and 7 post-doctoral researchers.

Chapter 7 - An extensive review of flow boiling in microchannels has been made. Channel classification has been shown. Onset of nucleate boiling, liquid film thickness, flow pattern and bubble growth, heat transfer, pressure drop, critical heat flux and boiling instability with particular emphasis on different types of instabilities have been discussed. Direction of further research has been delineated in the conclusion.

Chapter 8 - In this chapter, a new microchannel heat sink with staggered structure was designed based on thermal boundary layer technology for enhancing heat transfer performance and reducing flow resistance. Comparative experiments were carried out between a test section having the new staggered structure microchannel and a conventional straight microchannel keeping the values of three control parameters, inlet temperature, heating power, and flow rate, very close. The results indicated that there was an obvious difference between the friction resistance factor from experimental data and calculated values from theoretical formula. The staggered structure increased the partial friction resistance to a certain extent, but the advantage for heat transfer was more evident. The temperature distribution of the wall was more uniform along the flow direction.

Chapter 9 - A two-phase closed thermosyphon (TPCT) is highly effective thermal spreader consisting of an evacuated closed tube filled with a certain amount of working fluid. In this study, nanofluids of aqueous CuO nanoparticles were prepared with volume concentrations varying from 1 to 3 percent. The effect of these different working fluids and the filling ratio on thermal performance of a TPCT was investigated and compared with pure water. Experimental results showed that the efficiency of TPCT significantly increases when the CuO/water nanofluid was used instead of pure water. Also, maximum efficiency of the TPCT was observed with filling ratio of 35%.

Chapter 10 - Using distilled water as the working fluid, the experiments with dual tube heating were made under two types of test section. The values of the onset of nucleate boiling Z_{ONB} could be measured in the experiments. And the values of the start of fully developed subcooled boiling Z_{FDB} were calculated by Bowring method. Differential equations for the relevant parameters such as true mass quality x, liquid subcooled temperature θ_l , distance along the heated channel measured from the inlet Z and void fraction α were developed by applying Zuber-Findlay correlation and improved Rouhani-Axelsson mechanistic model which considered the influence of mass flux. The fourth-order Runge-Kutta method was used to integrate the relevant equations simultaneously along the heating section while the experimental data were the initial conditions. The calculation results of the improved correlation were compared with the data of numerical calculation and agreed well with it while the deviation was less than ±12%

Chapter 11 – The Thermal-hydraulics and Reactor Safety Laboratory (TRSL) is a part of the Purdue University School of Nuclear Engineering and oversees several smaller laboratories. The research team at TRSL is led by two faculty members, Distinguished Professor Mamoru Ishii and Professor Takashi Hibiki. Working with these professors are a number of doctoral research associates and approximately 15 graduate research assistants working on specific research projects. Four full-time machinists work in the machine shop, which is attached to the laboratory.

Chapter 12 - Fluids with nano-sized particles have been proved that may effectively enhance the single-phase convective heat transfer performance. For pool boiling heat transfer, the published test results seems conflicted to each other. Some measured heat transfer coefficient decreased with increasing particle concentration but some showed no apparent difference.This study provides an experimental investigation on pool boiling heat transfer performance of refrigerants R-141b with and without nano-sized Au particles on horizontal plain tubes. The change of surface roughness and fluid particles condition were also observed for further understanding of the mechanisms that affect the pool boiling heat transfer performance.

The test results show that the boiling heat transfer coefficients increase with increasing nano-particles concentration. At particles concentration of 1.0 %, the heat transfer coefficients are more than twice higher than those without nano-particles. The SPM investigation shows that the test tube surface roughness decreased from 0.317 μm before boiling test to 0.162 μm after test. Further investigation by TEM and Dynamic Light Scattering particle analyzer shows that the nano-particles aggregated from the average size of 3 nm before test to 110 nm after test. This results show that the nano-sized Au particles are able to significantly increase pool boiling heat transfer of refrigerant R-141b on plain tube surface. The tube surface roughness and particle size changed after boiling test. Both of these effects degrade the boiling heat transfer coefficients.

Chapter 13 - The lattice Boltzmann model is a novel method to simulate gaseous or liquid flows. It successfully fills the gap between the macroscopic description, the Navier-Stokes equation, and particle based methods, such as the direct simulation of Monte-Carlo. In the last years, the lattice Boltzmann model has attracted increasing attention in modeling flows in microfluidic devices. This area has developed rapidly. Micro- electromechanical systems (MEMS), gaseous sensors, lab-on-chips produce signifi- cant grow in various areas of engineering and technology. In this chapter, the authors give an overview of the lattice Boltzmann method and its key concepts. They review recent developments related to the field of microflows and mi- crofluidics. Particular attention is paid for non-continuum behaviours observed at the micro-scale, such as the slip and jump of macroscopic variables in a micro-device. Finally, they provide further ideas and directionswhich can serve as targets of future developments. Beyond the lattice Boltzmannmodel, these ideas can stimulate progress and investment in applications and technology.

Chapter 14 - Heat transfer in electro-osmotic flow of power-law fluids in a slit microchannel is investigated for uniform wall heat flux thermal boundary condition. Energy equation is solved analytically for 0 _ _H _ 1 and numerically for _H > 1 using an implicit finite difference method. Previous analytical expressions for the velocity distribution are used to solve energy equation. The viscous dissipation effects are included in the analysis. The effects of electro kinetic parameter _H, flow behavior index n, and Brinkman number Br on the dimensionless local temperature and average heat transfer rate are examined and the results are presented graphically.

Chapter 15 - The modern tendency to improve the power sources efficiency is related to the fuel economy and nature protection. One of the promising solutions in the development of energy efficient technology of power sources is encompassed on the advanced solid sorption systems application (including co-generation and tri-generation, Photovoltaic/Thermal (PVT) solar collectors, fuel cells, Stierling and Diesel-generator) with heat pipe thermal management. Such technology provides the local heat, cold and electricity production. Some examples of heat pipes are shown oriented on its thermal control application in fuel cells and solid sorption heat pumps.

Chapter 16 – Dr. Chung's research activities for the past thirty years have been in the general areas of fluid mechanics and heat transfer with a special focus on phase change heat transfer and multiphase flows. Specifically, he has worked in the following areas : Numerical and analytical fluid mechanics and heat transfer associated with droplets and bubbles, boiling and condensation, heat and mass transfer in dehydration and porous media, two-phase flow and heat transfer, packed bed thermal storage system, particle and droplet dispersion in turbulent free shear flows, fluid mechanics and heat transfer in micro-gravity boiling, large-scale direct numerical simulation of transition to turbulence for heated flows by spectral methods, nuclear reactor thermal-hydraulic analysis, transport processes in manufacturing and materials synthesis for nano-cluster material. Recently his research is focused on nano to micro-scale transport phenomena in fundamental boiling, cryogenic two-phase flow and heat transfer, electro-thermal transport in fuel cell systems, and biomass and waste conversion to energy by thermal-chemical gasification.

Chapter 17 - This chapter reviews the current status of research on surface wettability, and classifies wettability into five categories: isotropic wettability, anisotropic wettability, interlaced wettability, gradient wettability, and tunable wettability. Examples from nature include lotus leaves (isotropic), bamboo leaves (anisotropic), Namibia desert beetles (interlaced), and butterfly wings (gradient wettabilty). The tunable wettability is primarily formed in laboratories. A droplet can move freely on a horizontal surface with wettability gradients due to differences in surface potential energies. As such, the wettability of a surface can be tuned using various methods, such as applying external electrical fields or mechanical stresses. This chapter also reviews the surface modification and fabrication methods currently used to achieve different wettability. Five main fabrication methods are considered in detail: etching, lithography, deposition, sol-gel processing, and templation. The transparency of the modified surface is also discussed since it is an important factor in some applications, such as solar cells. This review should be helpful in the design of surface modification, and future developments in this field.

Chapter 18 - An analysis is provided for the entropy generated in a capillary pumped device for the combined heat and mass transfer in a circular tube and micro/nano scale heat and mass transfer in a capillary tube in terms of the gradients of velocity, temperature and concentration as well as the physical properties of the fluid. In phase change transport devices, capillary forces drive the overall circulation of working fluid from an evaporator to a condenser section. The heat and mass transfer rates are assumed to be uniform on the surface of the tube. The optimum geometric configuration that corresponds to the minimization of entropy generated and minimization of the fluid flow resistances is identified. Capillary pumped heat transfer devices do not achieve the maximum predicted heat flux because of interfacial thermocapillary stresses arising from liquid vapor interfacial temperature gradients. This can be counteracted by introducing naturally occurring concentration gradients using binary fluid mixtures.

Chapter 19 - Due to environmental concerns, CO_2 has been reintroduced as a potential candidate to replace HFCs in refrigeration systems since the 1990s. In a refrigeration cycle, oil is utilized in lubricating a compressor. However, even though oil separators are installed after a compressor, oil is prone to leak to the whole system. In the case of the mixing of CO_2 and even a small amount of oil, the heat transfer performance in the heat exchanger is deteriorated and the pressure drop inside the tube increases. Therefore, to correctly design a CO_2 refrigeration system, it is necessary to precisely estimate the thermodynamic mixture properties of CO_2-lubricant oil. The commonly used method in estimating the mixture properties is the mole-based weighting model. However, the accuracy of the method cannot be assured for the CO_2-lubricant oil under the supercritical region. In the present study, CO_2-lubricant oil mixture properties, including density, viscosity, thermal conductivity, and heat capacity, were estimated by using the mixture models, based on the equation of state (EOS).

Chapter 20 - Copmarisons of 12 flow boiling heat transfer correlations to experimental data in microscale channels are presented. First, a database containing 1766 experimental data points of flow boiling heat transfer with R134a, R22, R410A and R245fa was compiled from the literature. The database covers a wide range of mass velocities from 20 to 860 kg/m^2s, heat fluxes from 5 to 55 kW/m^2, saturation temperatures from 0 to 50°C and tube diameters from 0.19 to 3 mm. Then, 12 flow boiling heat transfer correlations including 3 macroscale correlations and 9 microscale correlations have been collected from the literature.

The database was compared to these heat transfer prediction methods. Statistical analysis shows that no correlation predict the entire database well, with the best correlation only predicting 60% of the entire data within ±30%. Furthermore, some methods work well for some fluids while they do not work well for others.

In general, predictions for R134a and R410A data are not satisfactory while predictions for R245fa and R22 are acceptable and satisfactory. Effort should be made to develop a general correlation for flow boiling heat transfer in microscale channels according to heat transfer mechanisms and flow regimes which should also be fully understood in future.

Chapter 21 - MEMS thermal control Lab. has a research team that is consisted of two post-doctors, one research assistant, seven ph. D. students, and five master students. In this laboratory, the research team focuses on four research topics, a finger size smart sensor, innovative applications of microfluidic devices, surface modification of micro/nano hierarchical structures, and a disposable DNA sensor. Most fabrication process of devices or micro/nano structures are either out sourced to a fabrication company or performed in MEMS laboratory or Nanotechnology center in National Taiwan University. Our laboratory develops the testing instruments to measure the performance of developed sensors, devices, or micro/nano structures.

In: Frontier Research in Microscale and Nanoscale... ISBN 978-1-61324-665-8
Editor: Lixin Cheng

Chapter 1

MICROSCALE AND NANOSCALE THERMAL AND FLUID TRANSPORT PHENOMENA: RAPIDLY DEVELOPING RESEARCH FIELDS

Lixin Cheng[1]
School of Engineering, University of Aberdeen, King's College,
Aberdeen, AB24 3UE, Scotland, UK

With the emergence of microscale and nanoscale thermal, fluidic and chemical systems, MEMS (Micro-Electo-Mechanical-Systems), NEMS (Nano-Electro-Mechanical-Systems), thermal management technologies for microelectronics, power electronics, aerospace and defence technologies etc., the development of ultra-compact heat exchangers, miniature and micro pumps, miniature compressors, micro-turbines, micro thermal systems for distributed power production, microfluidic and nanofludic device, lab-on-a-chip, microscale and nanoscale energy systems, nanofluid heat transfer technologies and others has become an important agenda of many researchers, research institutions and funding agencies. Applications of microscale and nanoscale thermal and fluid transport phenomena involved in traditional industries and highly specialized fields such as bioengineering, micro-fabricated fluidic systems, microelectronics, aerospace technology, micro heat pipes, chips cooling etc. have been becoming especially important since the late 20th century [1-11]. However, microscale and nanoscale thermal and fluid transport phenomena are quite different from those at conventional scale or macroscale [12-14]. For example, gas liquid two-phase flow and flow boiling heat transfer characteristics in microcale channels are quite different from those in macroscale channels [1-3, 6-8, 12, 13]. Channel confinement also has a great effect on two-phase flow and flow boiling heat transfer characteristics [16, 17]. Furthermore, the available experimental results of single phase heat transfer and fluid flow in microscale channels are quite contradictory from one study to another [4, 5]. Studies of supercritical fluid flow and heat transfer in microscale channels have also exhibited contradictory results by various researchers [14]. Just to show several examples here. Therefore, there are many aspects to be clarified from both theoretical and applied aspects in microscale and nanoscale

[1] Founder and Editor-in-Chief of the IJMNTFTP, email: lixincheng@hotmail.com.

thermal and fluid transport phenomena. Furthermore, new research areas and technologies in the relevant fields have been emerging from time to time. For example, as a new research frontier of nanotechnology, the research of nanofluid two-phase flow and thermal physics is growing rapidly, however it has also posed new challenges as there are quite contradictory results in the available research [10]. For another example, advances in micro-electronics technology continue to develop with surprisingly rapidity and the thermal energy density of electronic devices to be dissipated is becoming much higher and higher. Therefore, it is essential to develop new high heat flux cooling technology to meet the challenging heat dissipation requirements [15]. All the newly emerging research areas require the understanding of fundamentals and applications of the microscale and nanoscale fluid and thermal transport phenomena.

Over the past years, a great amount of research work in the fields of microscale and nanoscale thermal and fluid transport phenomena has intensively been conducted. In response to the numerous evolving research topics, technologies and applications based on microscale and nanoscale thermal and fluid transport phenomena, it is highly the time to form a new journal to provide a high-quality forum specially for a wide range of papers dealing with original research results, technical notes and state-of-the-art reviews pertaining to thermal and fluid transport phenomena at microscale and nanoscale. Thus, this newly founded journal aims to meet such urgent needs and to bring worldwide these important frontier research works together. It covers a wide range of topics on fundamentals and applications of microscale and nanoscale transfer processes of mass, momentum and energy such as microscale and nanoscale heat transfer and fluid flow, nanofluid heat transfer and flow, microfluidcs, nanofludics and technologies based on these transport processes such as various microscale and nanoscale thermal and fluid devices, micro and nano energy systems, micro-cooling technology in the computer and electronics industries and information technologies etc., MEMS, NEMS and the interdisciplinary research related to microscale and nanoscale thermal and fluid transport phenomena in bio-engineering, medical engineering and life engineering etc. Furthermore, proposals for special issues on selected topics reflecting recent microscale and nanoscale thermal and fluid flow advances or on selected articles of conferences are also encouraged and may be submitted directly to the Editor-in-Chief. Reviews on important issues are invited by the Editor-in-Chief from tme to time and may also be submitted to the Editor-in-Chief in a normal way. One new feature of this journal is to present one leading research laboratory or group introduction in relevant fields in each issue. For example, a leading laboratory with interest in microscale two-phase flow and heat transfer research, led by Prof. John R. Thome, the Heat and Mass Transfer Laboratory (LTCM) at the Swiss Federal Institute of Technology Lausanne (EPFL), Switzerland is presented in this issue. If anyone is interested in presenting his/her lab or group in the journal, please contact the Editor-in-Chief directly for this.

As the founder and Editor-in-Chief, I would like to express my sincere thanks to all the editorial board members who have been giving their great support to this new journal. I welcome any suggestions and comments on improving this journal from our community. It is my greatest wish to make this journal a leading one in the microscale and nanoscale thermal and fluid transport phenomenon research fields in the near future. I am looking forward to receiving more support from our community, which I believe are invaluable.

REFERENCES

[1] L. Cheng, D. Mewes, Review of two-phase flow and flow boiling of mixtures in small and mini channels, *Int. J. Multiphase Flow* 32 (2006) 183-207.

[2] S.G. Kanlikar, Fundamental issues related to flow boiling in minichannels and microchannels, *Exp. Therm. Fluid Sci.* 26 (2002) 389-407.

[3] J.R. Thome, The new frontier in heat transfer: microscale and nanoscale technologies, *Heat Transfer Eng.* 27(9) (2006) 1-3.

[4] G.L. Morini, Single-phase convective heat transfer in microchannels: a review of experimental results, *Int. J. Therm. Sci.* 43 (2004) 631-651.

[5] G. Hetsroni, A. Mosyak, E. Pogrebnyak, L.P. Yarin, Fluid flow in micro-channels, Int. J. *Heat Mass Transfer* 48 (2005) 1982-1998.

[6] J.R. Thome, Boiling in microchannels: a review of experiment and theory, *Int. J. Heat Fluidflow* 25 (2004) 128-139.

[7] J.R. Thome, State-of-the art overview of boiling and two-phase flows in microschannels, *Heat Transfer Eng.* 27(9) (2006) 4-19.

[8] L. Cheng, G. Ribatski, J.R. Thome, Gas-liquid two-phase flow patterns and flow pattern maps: fundamentals and applications, *ASME Appl. Mech. Rev.* 61 (2008) 050802-1-050802-28.

[9] L. Cheng, D. Mewes, A. Luke, Boiling phenomena with surfactants and polymeric additives: a state-of-the-art review, *Int. J. Heat Mass Transfer* 50 (2007) 2744-2771.

[10] L. Cheng, E. P. Bandarra Filho, J.R. Thome, Nanofluid two-phase flow and thermal physics: a new research frontier of nanotechnology and its challenges, *J. Nanosci. Nanotech.* 8 (2008) 3315-3332.

[11] L. Cheng, Nanofluid heat transfer technologies, *Recent Patents on Engineering* 3(1) (2009) 1-7.

[12] G. Ribatski, L. Wojtan, J.R. Thome, An analysis of experimental data and prediction methods for two-phase frictional pressure drop and flow boiling heat transfer in micro-scale channels, exp. *Therm. Fluid Sci.* 31 (2006) 1-19.

[13] J.R. Thome, G. Ribatski, State-of-the-art of two-phase flow and flow boiling heat transfer and pressure drop of CO_2 in macro- and micro channels, *Int. J. Refrigeration* 28 (2005) 1149-1168.

[14] L. Cheng, G. Ribatski, J.R. Thome, Analysis of supercritical CO_2 cooling in macro- and micro channels, *In. J. Refrigeration.* 31 (2008) 1301-1316.

[15] L. Cheng, J.R. Thome, Cooling of microprocessors using flow boiling of CO_2 in a micro-evaporator: preliminary analysis and performance comparison, *Appl. Therm. Eng.* 29 (2009) 2426-2432.

[16] J. Moreno Quibén, L. Cheng, R.J. da Silva Lima, J.R. Thome, Flow boiling in horizontal flattened tubes: part I — two-phase frictional pressure drop results and model, *Int. J. Heat Mass Transfer* 52 (2009) 3634-3644.

[17] J. Moreno Quibén, L. Cheng, R.J. da Silva Lima, J.R. Thome, Flow boiling in horizontal flattened tubes: part II — flow boiling heat transfer results and model, *Int. J. Heat Mass Transfer* 52 (2009) 3645-3653.

In: Frontier Research in Microscale and Nanoscale... ISBN 978-1-61324-665-8
Editor: Lixin Cheng

Chapter 2

MICROSCALE HEAT TRANSFER IN SINGLE- AND TWO-PHASE FLOWS: SCALING, STABILITY, TRANSITION

G. P. Celata*

ENEA, Energy Department, Institute of Thermal-Fluid Dynamics
Via Anguillarese 301, 00123 S.M. Galeria, Rome, Italy

ABSTRACT

Partly because of the technological challenge, partly because of stark necessity, there has been an increasing movement towards a miniaturization of appliances in the last decade. In all technical fields solutions are sought that encumber as little as possible without compromising on performance: in medical diagnostics, environmental sample analysis, military defence, consumer electronics, biomedical appliances, chemical reactors and heat management a constant research for quicker response times and portable devices has driven the field of microtechnology to impressive levels. In many applications it has been found that many small active components are more productive than few large ones, which is also in keeping with the growing trend towards modular design. Proper understanding of microscale transport phenomena is therefore fundamental for the designer of microfluidic devices. For this reason, many studies have been conducted to analyse the behaviour of convective flow through microchannels, both in single-phase and in two-phase flow. A first glance of the literature, especially for single-phase flow, leads to the conclusion that up to now we have had an agglomeration of disparate conclusions. In many cases the experimental data in microchannels disagree with the conventional theory and empirical correlations, but they also appear to be inconsistent with one another.

The present paper is an attempt to critically analyse the available results for liquid single-phase and flow boiling heat transfer, trying to provide some sort of base note in the melisma of published data.

* E-mail: celata@enea.it

NOMENCLATURE

A	surface area, m^2
Co	confinement number, $=1/2\sqrt{2Eo}$, nondimensional
c_p	fluid specific heat, J/kgK
d, D	tube diameter, m
D_h	hydraulic duct diameter ($=4\Omega/P$), m
Ec	Eckert number, $W^2/2c_p\Delta\theta_{ref}$, nondimensional
Eo	Eötvös number, $g\Delta\rho D^2/8\sigma$, nondimensional
g	gravitational acceleration, m/s^2
G	mass flux, kg/m^2s
Gr	Grashof number, $gl^3\Delta\rho/\rho\nu^2$, nondimensional
Gz	Graetz number, RePr(d/z), nondimensional
h	heat transfer coefficient, W/m^2K
k	fluid thermal conductivity, W/mK
k	thermal conductivity, W/m K
L	channel length, m
l	characteristic length, m
M	axial conduction number, eq. (6) nondimensional
Nu	Nusselt number, hD/k, nondimensional
p	pressure of fluid in duct, Pa
P	wetted perimeter, m
Pe	Peclét number, RePr, nondimensional
Pr	Prandtl number, ν/α, nondimensional
q	linear heat flux, W/m
q”	heat flux, W/m^2
Re	Reynolds number, GD/μ, nondimensional
r	radius, m
T	dimensionless fluid temperature
u	axial fluid velocity, m/s
V	volume, m^3
W	fluid average velocity, m/s
x	vapour quality, nondimensional
x,y,z	dimensionless Cartesian co-ordinates
z	abscissa

Greek symbols

α	thermal diffusivity ($=k/\rho c_p$), m^2/s
δ	liquid thickness, m
$\Delta\theta$	temperature difference, K
$\Delta\rho$	liquid-vapour density difference, kg/m^3
θ	fluid temperature, °C
μ	dynamic viscosity, kg/m s

ν	kinematic viscosity, m^2/s
ξ,η,ζ	Cartesian co-ordinates, m
ρ	fluid density, kg/m^3
σ	surface tension, N/m
Ω	cross-sectional area, m^2

Subscripts

0	initial
b	bulk
exp	experimental
f	pertains to the fluid
i	internal
in	inlet
min	minimum
nucl	nucleation
o	external
ref	reference condition
TP	two-phase
w	pertains to the wall

1. Introduction

In recent years the research in the field of single- and two-phase flow heat transfer at a microscale level has been constantly increasing due to the rapid growth of the technology applications that require the transfer of high heat rates in a relatively small space and volume. Such applications span from compact heat exchangers to cooling systems for computer CPU to microfluidic devices.

Single- and two-phase flow heat transfer has been the subject of numerous researches for the last decades. A relevant knowledge has been constructed, which allows with more (single-phase flow) or less (two-phase flow) accuracy to cope with most heat transfer applications. Going down to microscale the first question which immediately pops up is: "Is the macroscale knowledge for single- and two-phase flow heat transfer applicable to microscale?" Over the last 10-15 years a good deal of publications have dealt with this matter producing a considerable amount of experimental data. Especially in the first years of publications, a large diversion of results to be discerned in the various experimental and numerical reports can be found. The diversion occurs in different contexts: fluid drag of laminar, transient and turbulent single-phase flows, heat transfer of liquid and gas flows, and two-phase flow in adiabatic and heated micropipes. On the one hand, it can be interpreted as displaying some new effects of heat transfer in flow in micro-channels. On the other hand, the phenomenon may be related to the discrepancy between the actual conditions of a given experiment, and theoretical or numerical solution obtained in the frame of conventional theory.

Looking at the most recent published works, at least in single-phase flow, a convergence of results would seem to be found, both in heat transfer and fluid flow. This can be explained by taking into account the significant improvement of the techniques of microfabrication with the consequent reduction of the issues of surface roughness of the microchannels and a more appropriate control of the channel cross sections. Besides, the improvement in the accuracy and the definition of a more standard measurement methodology have provided a more reliability of the more recent experimental data appeared in the literature.

The present lecture is an attempt to provide a critical evaluation of the current understanding on single (liquid) and two-phase (boiling) heat transfer in micropipes.

2. Single-Phase Flow Heat Transfer (Liquid)

Analyzing the first experimental results which have been published in liquid single-phase flow heat transfer the reader can realize that there are not convincing explanation of the difference between experimental and theoretical results obtained in laminar flow and between experimental and semi-empirical results obtained in turbulent flow. Even restricting the analysis to microtubes, early experimental results show differences between experimental outcome and macroscale knowledge which are not convincing. A thorough review has been published by Hetsroni et al. (2005), and the above conclusions are drawn from results plotted in figure 1, where the Nusselt number, $Nu = hD/k$, is plotted versus the Reynolds number, $Re = GD/\mu$, for laminar (left) and turbulent flow (right).

As already outlined in single-phase fluid flow, where the current knowledge shows a general agreement between microscale experimental data and conventional laws (friction factor in agreement with Hagen-Poiseuille law in laminar flow and Colebrook-White equation in turbulent flow), no apparent new mechanism should occur in single-phase flow heat transfer going down to microscale. However, on downscaling the system it might be that some of the assumptions are no longer valid, or that ulterior considerations have to be taken into account that might have been done away with as insignificant in the macro-scale account. What is it that might change physical conditions reducing the characteristic dimension below a certain limit?

Simple-mindedly, we can immediately gain an idea by looking at the definition of the characteristic dimension in terms of the volume, V, and surface area, A, of the system under consideration: $l = (V/A)$system. From this relation, it can be concluded that the volume-to-area ratio diminishes with smaller characteristic dimension (the pipe diameter is the obvious characteristic dimension in pipe flow), and with it, the importance of volume-tied with respect to surface-tied properties. Therefore, the more the system shrinks in size, the less important mass, heat capacity, charge, inertia etc. become in defining system behaviour, with respect to pressure, viscosity, hydrophobicity, contact heat transfer and other such properties tied to surface interactions.

A system can be described accurately even disregarding certain effects, if they have marginal weight on its behaviour, but on miniaturizing said system these disregarded effects might become influential because the characteristic dimension of the system has become of the same order of magnitude as the scale at which they operate.

This is what is understood as a scaling effect, and it is an important concept to keep in mind when one peruses the literature on the recent topic of microchannel flow: many times researchers claim to have unveiled a "new" micro-phenomenon, whereas they probably were thrown off-track because some peripheral effect that was not accounted for, turns out to have been influential in the measurements.

Very good and updated reviews of these aspects are given by Morini (2004, 2006). Scaling effects which play a relevant role in microscale single-phase flow heat transfer are viscous heating, thermal entrance length and axial conduction (conjugate heat transfer).

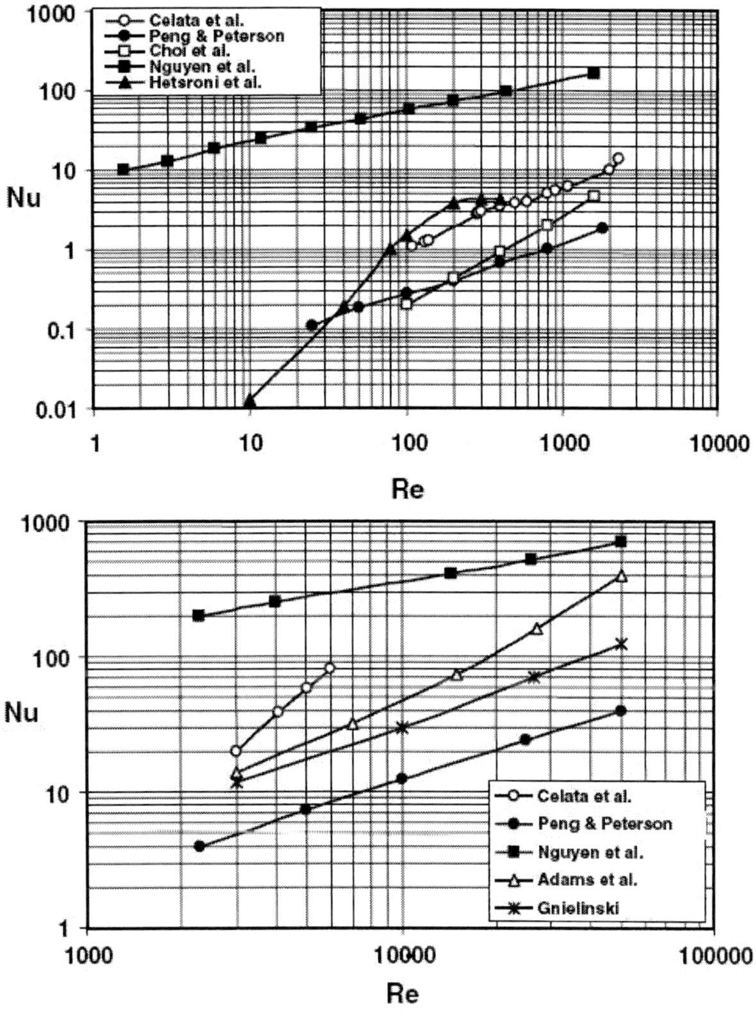

Figure 1. Comparison between the experimental correlations for the Nusselt number in micro-channels: (left) laminar flow and (right) turbulent flow.

2.1. Viscous Heating

For a given mass flow through a tube, reducing its diameter causes a squared increase in the fluid velocity. Thus, in microtubes, the fluid velocities achieved can be extremely high. Assuming, conventionally, that there is no slip at the wall, one can vividly imagine how exceptionally stretched-out the parabolic velocity profile must be to satisfy this condition. This very strong viscous deformation of the flow is what causes the exceedingly high pressure drop in a microchannel. From the point of view of the Second Law of Thermodynamics, this huge effort to drive the flow must cause a large irreversible loss somewhere. Indeed, just as a metal bar will heat up as it is drawn or extruded, so the fluid squeezed through a microscopic tube will generate entropy, and therefore heat. This phenomenon is known as viscous heating, or viscous dissipation, and is obviously alien to flow in large-diameter channels, in the sense that the viscous deformation of the velocity profile is quite reduced there, since a large area is available for the flow to pass through. Viscous heating effect has been theoretically treated by Morini (2005) and experimentally investigated by Celata et al. (2006a). Using Navier-Stokes and energy equations, based on the following simplifying assumptions: i) the fluid is Newtonian, incompressible and with a laminar fully developed profile of velocity and a uniform inlet temperature; ii) the transport processes are considered to be steady-state and bi-dimensional (fully developed velocity and temperature profiles); iii) thermal radiation is neglected; iv) all channel walls are rigid and non-porous; v) there is no fluid slip (zero velocity at the wall); vi) axial thermal conduction (Peclét number, $Pe >> 1$), natural convection ($Gr/Re^2 << 1$, where Gr is the Grashof number), and interior heat sources are neglected; vii) fluid thermophysical properties are assumed as constant; it is possible to get the fluid bulk temperature gradient along the microchannel:

$$\frac{\partial T_b}{\partial z} = \frac{q_w^* L^*}{Pe\Omega^*} + \frac{Ec}{\mathrm{Re}}\left[f \,\mathrm{Re}\, L^*\right] \tag{1}$$

being

$$z = \frac{\zeta}{L};\ \Omega^* = \frac{\Omega}{D_h^2};\ \nabla^* = D_h \nabla;\ T = \frac{(\theta - \theta_{in})}{\Delta\theta_{ref}};\ L^* = \frac{L}{D_h}$$

$$Ec = \frac{W^2}{2c_p \Delta\theta_{ref}};\ Pe = \frac{WD_h}{\alpha} = \mathrm{Re}\,\mathrm{Pr};\ q_w^* = \frac{q_w}{k\Delta\theta_{ref}} \tag{2}$$

In eq. (1) $\Delta\theta_{ref}$ is a reference temperature rise and its value can be fixed by considering the temperature sensitivity of fluid viscosity. For instance, it could be defined as the temperature rise for which the dynamic viscosity decreases, say, 2-3%. Besides, friction factor f is according to the definition given by Darcy-Weisbach. Thus, in order to neglect the effects of viscous dissipation in adiabatic flow through microchannels, the temperature rise so induced, between inlet and outlet, shall be less than the reference temperature rise. In heated or cooled macrosize channels, the first term is always predominant over the second term, especially in laminar regime. As an example, for a tube with inner diameter (ID) equal to 1 cm, and 1 m long, through which water flows in laminar regime ($Re = 1000$) the temperature rise due to viscous dissipation is of the order of 10^{-5} K; for an ID equal to 100 μm the temperature rise in same conditions becomes of the order of 10 K and thus significant. This is evidence that viscous heating can be considered a classic scaling effect for microchannels, as quantitatively verified in Celata et al. (2006a).

2.2. THERMAL ENTRANCE LENGTH

The thermal entrance length is a transitory region where flow at uniform temperature develops under influence of uniformly-heated walls to a steady-state profile of temperature that is dependent on fluid velocity and conductivity. Just like it is the case for the physics of flow in the hydrodynamic entrance region, the mechanism of heat transfer is rather different in the thermal entrance length – see figure 2. Fluid with uniform temperature enters the heated section of a pipe (let us assume that the velocity profile is already fully developed at this stage), and as the heat input is absorbed, the temperature of the fluid at the wall starts to rise. Through conduction across the fluid layers of the laminar flow, the temperature profile then develops over the entire cross-section. Superposition of the convective force then creates the thermal developing length (from $z = 0$ to $z = z_T$ in figure 2).

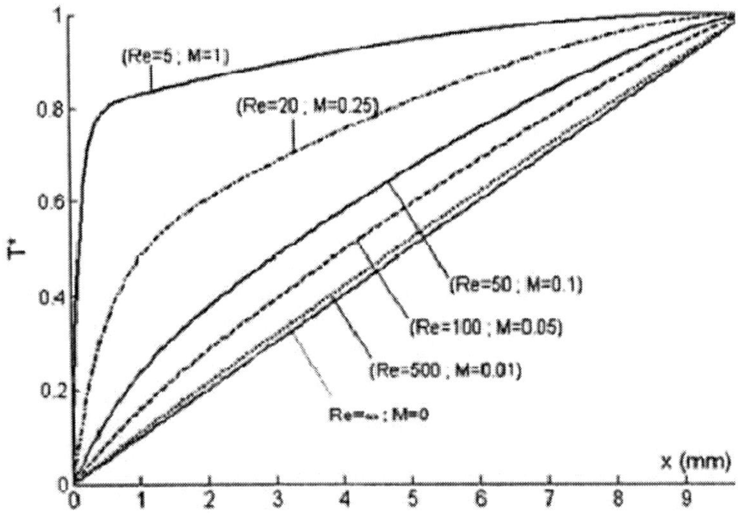

Figure 2. Thermal entrance length and developing temperature profile.

Whether a heated flow has reached this semi-steady thermal state (the bulk temperature will of course be increasing steadily), can be deduced from the value of the Graetz number:

$$Gz = \mathrm{Re}\,\mathrm{Pr}\frac{d}{z} \tag{3}$$

A thermally fully developed profile is achieved when $Gz < 10$, so the longer the distance z along the channel axis and the less force of convection *Re*, the more chance there is of obtaining a thermally fully developed profile.

The effect of a thermal entrance region is to increase the Nusselt number compared to that of the thermally fully developed region (see also Shah and London, 1978). The generalized Hausen correlation describes the mean Nusselt number as the sum of two terms: a fully developed value of the Nusselt number, Nu_∞, and a term taking into account the effects of the thermal entrance region:

$$< Nu >= Nu_{\infty} + K_1 \frac{Gz}{\left(1 + K_2 Gz^b\right)} \tag{4}$$

Here the Graetz number is taken over the entire length of channel, and in the case of constant wall heat flux and fully developed velocity profile, for circular ducts: $K_1 = 0.023$, $K_2 = 0.0012$ and $b = 1$, see Schlichting (1979). In case of rectangular channels we have: $K_1 = 0.14$, $K_2 = 0.05$ and $b = 2/3$. Typically, in laminar flow, the fully developed Nusselt number, Nu_{∞}, is constant, and equal to 4.36 for uniform heat flux, and equal to 3.66 for uniform temperature. Entrance effects for turbulent flow in tubes are more complicated than for laminar flow and cannot be expressed in terms of a simple function of the Graetz number. In general, the higher the Prandtl number and the lower the Reynolds number, the shorter the entry length. However, the thermal entry lengths are much shorter for turbulent flow than for their laminar counterpart, Holman (1992), and therefore less relevant for microflow considerations.

2.3. Axial Conduction (Conjugate Heat Transfer)

Generally, in macroscale the heat flow is assumed to take place only in radial direction. Obviously, if a fluid is heated in a pipe with constant heat flux, its temperature and the wall's will increase as it flows through the duct. This means that there will be a temperature difference between the outlet and the inlet of the heated tube. Temperature difference means heat conduction: thus, we have generated a potential for heat to flow axially along the wall (along the fluid is also possible, but as it possesses convective force in the opposite direction, backflow of heat through the working medium is suppressed), and the previous assumption will not hold any more. This superposition of convective and conductive heat fluxes (in perpendicular directions) is known as conjugate heat transfer. It has been studied in the researching past, but its practical importance was generally very limited, since the channel walls in conventionally-sized heat exchanging and fluid flow equipment are usually thin compared to the cross-sectional flow area, so that backflow of heat along the solid walls is always marginal compared to the main transport of heat inside the conduit. It can be felt coming: in microchannels this is not necessarily the case. Due to the exceedingly small diameter of a microtube, combined with the physical necessity of a certain thickness of the capillary for purely practical purposes of production and handling, the proportion of wall thickness and flow area is often inverted. This means that less resistance is put up to the backflow of heat along the walls, and axial conduction can become a competitive mechanism of heat transfer with respect to the internal convection. The consequence is that the wall heat flux density cannot stay uniform and that heat accumulates near the entrance of the microchannel, where more heat exchange will take place. Thus, also the bulk fluid temperature rise will not be linear any more, but rather convex as Maranzana et al. (2004) evidenced through numerical simulation, shown in figure 3, as, unfortunately, an analytical solution to the conjugate problem is generally not possible.

Therefore, the mean heat transfer coefficient, calculated with average temperatures, if one were to assume a linear fluid temperature distribution (the case of perfectly uniform wall heat flux):

$$< h > = \frac{q''}{\left(\overline{T}_w - \overline{T}_f\right)} \tag{5}$$

is underestimated then, because the mean value of the bulk temperature along the microchannel, $\overline{T}_f$, is underestimated. This is also an explanation for the dependence of the Nusselt number on the Reynolds number which is found in some experiments, as this effect of underestimation will become less as the convective term of heat transfer (~*Re*) increases with respect to the conductive term in the channel wall. The temperature distribution along the tube axis will then approach a linear tendency (although the effects of thermal development would increase, see Equation (5)).

A combination of physical parameters that describe the conjugate effects qualitatively has been made by Maranzana et al. (2004), though the exact same parameter was proposed by Chiou (1980) in describing the same situation, just not in a microchannel. Intuitively, the importance of axial conduction effects will be tied to: wall conductivity, the ratio of wall and flow passage cross-sectional area, channel length, the convective force and heat capacity of the fluid. The result is the definition of an axial conduction number, *M*, which for a circular tube it calculates as:

$$M = \left(\frac{k_w}{k_f}\right)\left(\frac{D_o^2 - D_i^2}{D_i L}\right)\frac{1}{\mathrm{Re}\,\mathrm{Pr}} \tag{6}$$

and allows the comparison of heat transfer by axial conduction in the wall with the convective heat transfer in the flow. If $M > 10^{2}$, it is assumed that axial conduction has to be accounted for in an appropriate evaluation of heat transfer rate.

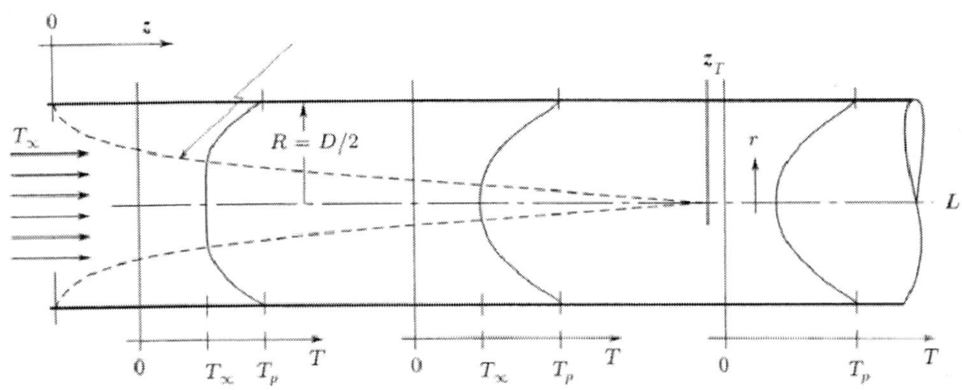

Figure 3. Normalized bulk temperature along the axis for uniform heat flux, Maranzana at al. (2004).

2.4. Definition of Microchannel in Liquid Single-Phase Flow Heat Transfer

It is now commonly acknowledged that the incipient relevance of scaling effects can be used as a criterion for identifying the threshold between the micro- and the macroscale, rather than barely using the channel diameter. This threshold is of course not unique and depends on thermal-fluid dynamic conditions (mainly velocity of the fluid), physical properties of the fluid (fluid viscosity and thermal conductivity), and size and material of the channel (thickness, diameter, thermal conductivity) as typically shown in figure 4, where the Nusselt number is plotted versus Reynolds number for water and iso-propanol, Morini (2006), and ranges of occurrence of different scaling effects are drawn.

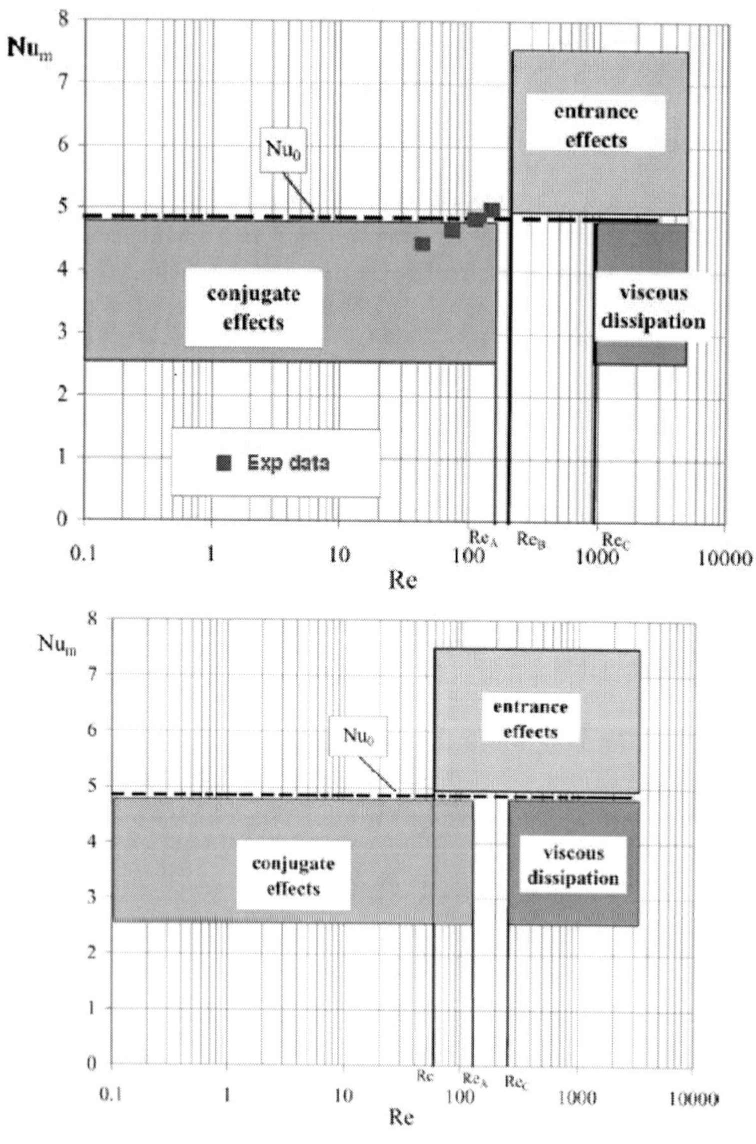

Figure 4. Scaling effects on the mean value of the Nusselt number for (left) water and (right) iso-propanol, Morini (2006).

In general, summarizing scaling effects occurrence we have:

- for low Reynolds numbers ($Re < Re_A$) the heat conduction along the solid wall is coupled to the internal convection: this effect tends to reduce the mean Nusselt number;
- entrance effects become important at high values of the Reynolds numbers ($Re > Re_B$) and tend to increase the mean value of the Nusselt number;
- for high Reynolds numbers ($Re > Re_C$) the effects related to viscous dissipation play up, tending to reduce the mean value of the Nusselt number and contrasting the effects of the entrance region.

2.5. Main Findings in Liquid Single-Phase Flow Heat Transfer

In the numerous experimental and theoretical investigations performed over the past years, a number of contradictory conclusions have been drawn. Most inconsistencies seem proven to originate from the difficulty of experimental conditions and high sensitivity of models to measurement errors, especially for the oldest researches. Among the published works we can quote in chronological order: Choi et al. (1991), Yu et al. (1995), Peng and Peterson (1996), Adams et al. (1998), Celata et al. (2002), Bucci et al. (2003), Kandlikar et al. (2003), Hetsroni et al. (2004), Lelea et al. (2004), Owhaib and Palm (2005), Celata et al. (2006b, 2007a).

The most recent works, where scaling effects above described in detail are properly accounted for, have shown how microscale heat transfer in liquid single-phase flow, both for laminar and turbulent flow, can be well predicted using what is known from the knowledge of heat transfer in macroscale, i.e., $Nu = 4.36$ for laminar flow and Gnielinski (1976) correlation for turbulent flow. A brief summary of most recent works is presented hereunder.

Lelea et al. (2004), have found a good agreement between water experimental data and conventional theory for laminar flow (Nusselt number, $Nu = 4.36$), including the entrance thermal effects (Shah and London, 1978), for pipe diameter ranging from 100 to 500 μm, as testified in figure 5, where typical results are reported in terms of the Nusselt versus the Reynolds number.

Hetsroni et al. (2004), as reported in figure 6, have observed a sharp decrease in the fully developed Nusselt number at Reynolds number below 200, doing experiments with water inside a 1.07 mm smooth tube. For $Re > 400$ the heat transfer coefficient tends towards the theoretical value, $Nu = 4.36$; this evidenced a clear effect of axial conduction along the channel wall, which is amplified at lower fluid mass fluxes, as discussed above. The axial conduction effect has been also studied by Tiselj et al. (2005) using water in a triangular silicon microchannel with hydraulic diameter of 160 μm in the range of Reynolds number $Re = 3.2\text{-}64$, and typical results are plotted in figure 7. The numerical simulation of wall (silicon) temperature is in good agreement with experimental data, and the non-linear behaviour of the wall and fluid temperature inside the microchannel, due to axial conduction, is clearly evidenced. Apparently, an inversion of wall-fluid temperature is observed which is of course to be avoided in a heat exchanger design.

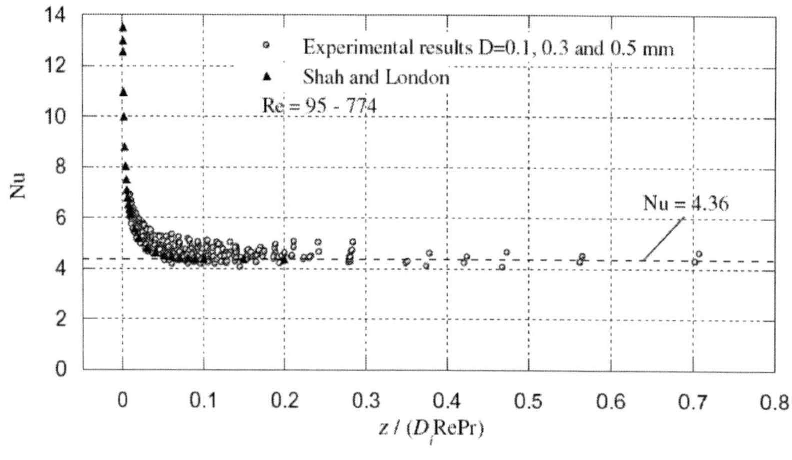

Figure 5. Relation of local value of Nu to non-dimensional axial distance for microtubes of D_i = 125.4, 300 and 500 μm, Lelea et al. (2004).

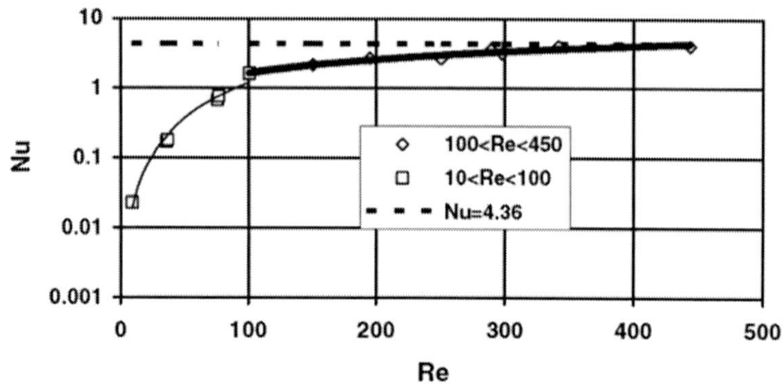

Figure 6. Experimental results for 1070 μm smooth circular tubes. Dependence of average Nusselt number on Reynolds number, Hetsroni et al. (2004).

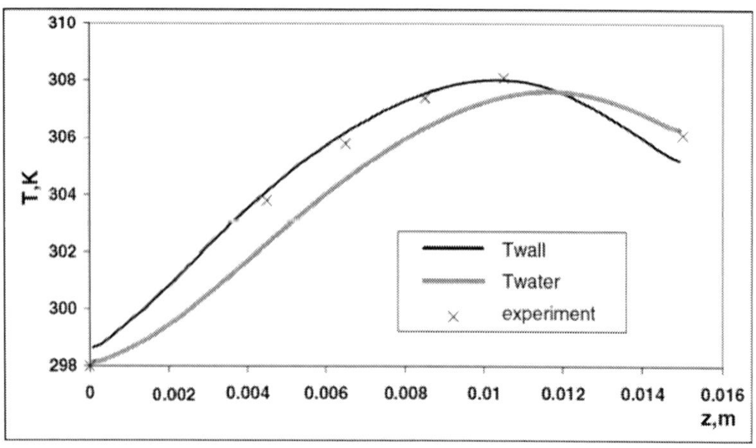

Figure 7. Water and silicon temperature distributions in 160 μm triangular microchannel, Tiselj et al. (2005).

Experiments of single-phase heat transfer with water in micropipes, both in laminar and turbulent flow, have been recently carried out by Celata et al. (2006, 2007), with pipe diameter in the range 50-528 μm. In the laminar flow regime the experimental Nusselt number, Nu, has been found in good agreement with the classical theory, i.e., $Nu = 4.36$, as shown in figure 8. In the turbulent flow regime, as shown in figure 9, experimental data are well predicted using the Gnielinski (1976) correlation, taking into account the experimental uncertainty. From figure 9 we can argue that experimental uncertainty is larger as the pipe diameter decreases.

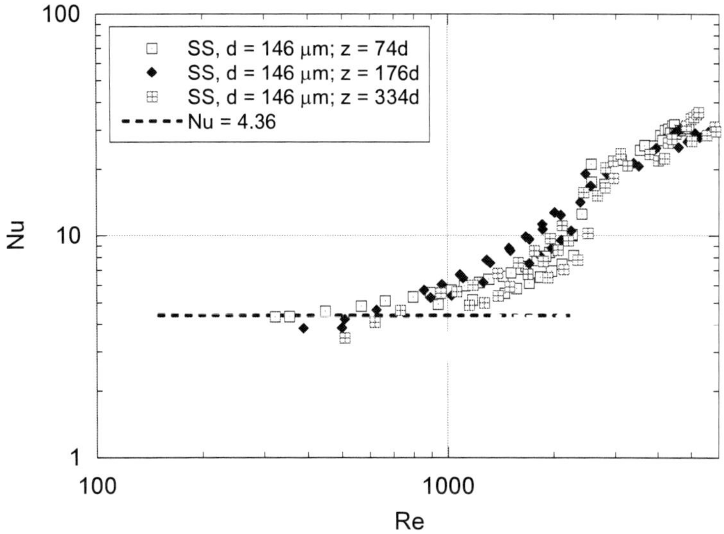

Figure 8. Local Nusselt versus Reynolds number, at three axial locations, in laminar and turbulent flow, for d = 146 μm pipe, Celata et al. (2007a).

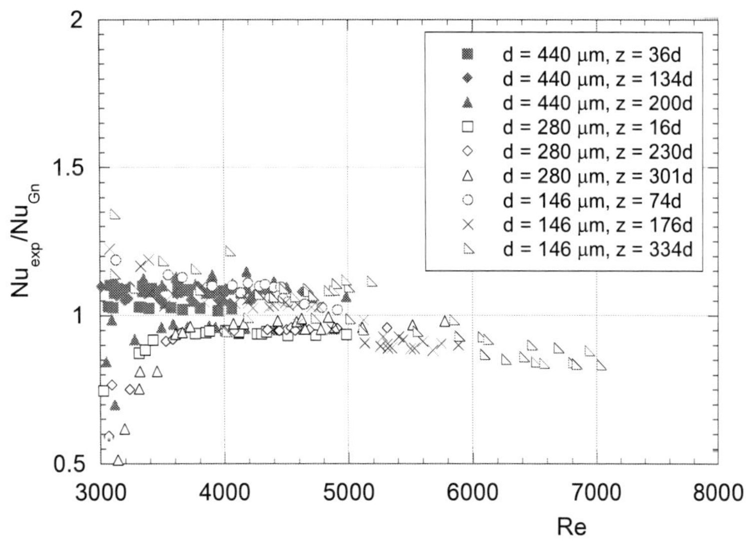

Figure 9. Ratio of experimental to calculated Nusselt number using the Gnielinski correlation, in turbulent flow, for d = 146-440 μm pipes, Celata et al. (2007a).

3. Two-Phase Flow Heat Transfer (Boiling)

If single-phase (liquid) flow heat transfer in microscale looks to be reasonably well understood, at least as a general trend, this is not the case of boiling heat transfer (as well as condensation heat transfer which will not be dealt with here).

3.1. Transition between Micro- and Macroscale

As a first consideration we can say that scaling effects analyzed in detail for single-phase flow generally loose their importance when two-phase flow takes place and can be again considered negligible: i) axial conduction is negligible as wall temperature along the pipe axis is generally flat in flow boiling; ii) thermal entrance length is negligible as much shorter than in single-phase flow; iii) viscous dissipation can be neglected as heat transfer in flow boiling is much more vigorous than in single-phase flow. Though this turns out to be a simplification in data analysis and experiment design, however it cannot be used to identify the threshold between the micro- and macroscale as in single-phase flow.

As a matter of fact, the transition between micro- and macroscale has been neither very well defined nor seriously experimentally investigated. As a consequence some people talk about microchannels some others refer to minichannels when dealing with 'small' channels, often referring to pipe diameter, such as Mehendale et al. (2000), and Kandlikar (2001). The most recent, and reasonable, criterion to identify this threshold has been recently proposed by Ullmann and Brauner (2007) on the basis of flow pattern maps, using the Eötvös number, Eo, proposing that we may talk about microscale when $Eo \leq 1.6$. It is evident that this criterion is correlated with the bubble confinement inside the channel. And is more than reasonable to assume that something is changing when reducing the channel diameter we reach a value which is of the same order of magnitude of the bubble size. From this point on, the bubble cannot grow and move as in an infinite medium and the two-phase flow dynamics is definitely altered by the geometry of the conduit. On the other hand, though starting from a less physical viewpoint, Kew and Cornwell (1997) have proposed the confinement number, $Co = 1/2\sqrt{2Eo}$ as reference parameter, assuming microscale flow when $Co > 0.5$. For the same fluid, i.e., R134a, at the saturation temperature of 0°C, the two criteria yield the transition between micro- and macroscale of 1.21 (Ullmann and Brauner, 2007) and 1.92 mm (Kew and Cornwell, 1997), respectively. Independent of which of the two criteria is better, we may say that microscale in two-phase flow might be of the order of the mm instead of microns. Although the bubble confinement looks an appropriate criterion for identifying the threshold between micro- and macroscale, we have to consider that inertial forces are not accounted for in the two criteria, while affecting bubble size, especially if buoyancy is missing (microgravity applications). This makes bubble confinement criterion reasonable enough if we restrict our analysis to terrestrial gravity applications. If we go to microgravity conditions, when gravity, g, tends to zero Eo (where g is at the numerator) is always less than 1.6 and Co (where g is at the denominator) is always larger than 0.5. Have we therefore to conclude that boiling heat transfer in microgravity is always a microscale phenomenon? Recent experiments of flow boiling in microgravity by Celata et al. (2007b) have evidenced that this is not the case. Bubble size in microgravity depends very much on fluid velocity and is also interrelated with vapour quality. If the fluid velocity is low generally the size of the bubble is large (much larger than at terrestrial gravity) and the bubble confinement criterion as suggested by Ullmann and Brauner, (2007) and by Kew and Cornwell (1997) gives correct prediction of the transition. Conversely, at high fluid velocity the bubble size is almost independent of the gravity level thus making the use of the bubble confinement criterion inappropriate. Maybe a further parameter, accounting for inertial forces, such as drag force, should be taken into account for extending the validity of threshold identification to space applications, where use of small channels is quite on demand for future applications.

3.2. Boiling Heat Transfer Mechanisms

Under these premises, in spite of the significant number of experimental data gathered so far (see for example Thome, 2004, 2006 and Kandlikar, 2002 for very good reviews on flow boiling heat transfer in microchannels without reporting here a long list of works) there is still a debate about the possible governing heat transfer mechanism, i.e., nucleate boiling or forced convective boiling. Typically, in nucleate boiling regime the heat transfer coefficient, *h*, is a function of heat flux and system pressure but is independent of vapour quality and mass flux. In convective boiling regime the heat transfer coefficient depends on vapour quality and mass flux but is not a function of heat flux. On this basis, many researchers have addressed their experimental results in microscale as governed by the nucleate boiling or the convective boiling regime depending on the heat transfer coefficient trend as a function of thermal-hydraulic parameters only.

Bao et al. (2000), among many others, have found the heat transfer coefficient behaving as the nucleate boiling regime up to vapour quality of 70%, as shown in figure 10, where *h* is plotted versus vapour quality, *x*, for different mass flux (left), and different heat flux (right), using R-11 and R-123 in a 1.95 mm pipe. Similar results have been found by Lazarek and Black (1982) using R-113 in a 3.15 mm pipe, by Tran et al. (1996) using R-12 in a 2.46 mm pipe, and by Owhaib et al. (2004) using R-134a in 0.826, 1.224, and 1.7 mm diameter pipe.

Huo et al. (2004), instead, using R-134a in 2.01 and 4.26 mm pipes, have found a more complex behaviour especially for vapour quality, *x*, larger than 20%, as outlined in figure 11, where *h* is plotted versus vapour quality for different heat flux. Similar results have been found by Yan and Lin (1998) using R-134a in a 2.0 mm 28-tubes channel, and by Lin et al. (2001) using R-141b in a 1.1 mm pipe. Even more recently, Hu and Kim (2006), although the heat transfer coefficient behaviour would suggest the nucleate boiling regime, as shown in figure 12, have argued that major flow pattern is similar with annular flow, which would not support the nucleate mechanism associated with an independency of *h* on *G* or *x*, which is instead the outcome of the heat transfer experimental data trend.

Apart from Hu and Kim (2006) considerations about the major flow pattern type encountered in their experiment, however not documented in the paper, there are other visual investigations which testify the occurrence of the annular flow regime, which would not support the nucleate boiling mechanism, even for low values of the vapour quality (this being interrelated with the size of the diameter), such as Brutin et al. (2003) who run boiling heat transfer visualization using n-pentane in a 0.889 mm hydraulic diameter rectangular channel, see figure 13, and Revellin et al. (2006) who conducted experiments of flow visualization of R-134a and R-245fa inside 0.5 and 0.8 mm diameter pipe, see figure 14.

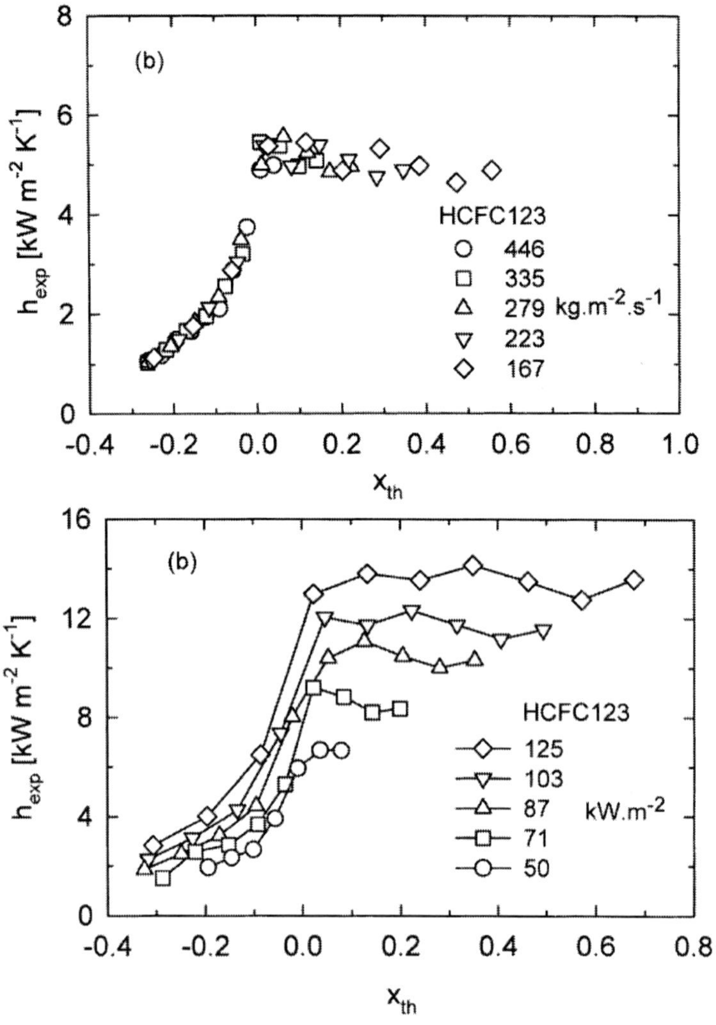

Figure 10. Flow boiling data for R-11 and R-123 in 1.95 mm tube plotted versus vapour quality, for different mass flux (left) and heat flux (right), Bao et al. (2000).

3.3. Flow boiling instability

Consolini and Thome (2009) have recently investigated on micro-channel flow boiling pointing out the sensitivity of small systems to oscillatory instabilities, as also pointed out, among others, by Qu and Mudawar (2003), Balasubramanian and Kandlikar (2005), Huh et al. (2007), Chang and Pan (2007), Qi et al. (2007). In broad terms, the oscillations that have been reported refer to either (1) parallel channel instabilities, in the case of parallel channel evaporators with common inlet headers, or (2) to compressible volume instabilities, which may develop even in single-channel systems (see Bergles and Kandlikar, 2005). While parallel channel instabilities are essentially due to the interaction of the two-phase flow among adjacent channels, compressible volume instabilities may propagate in the presence of a degree of compressibility (a body of vapour, flexible hosing, etc.) upstream, or within, the heated length.

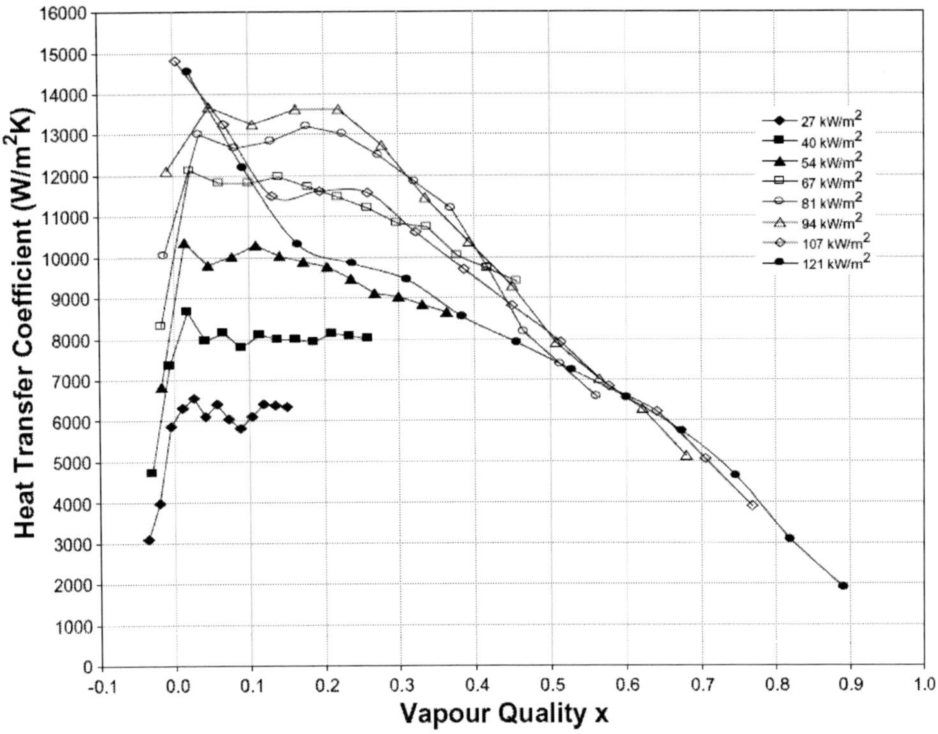

Figure 11. Flow boiling data plotted versus vapour quality, Huo et al. (2004).

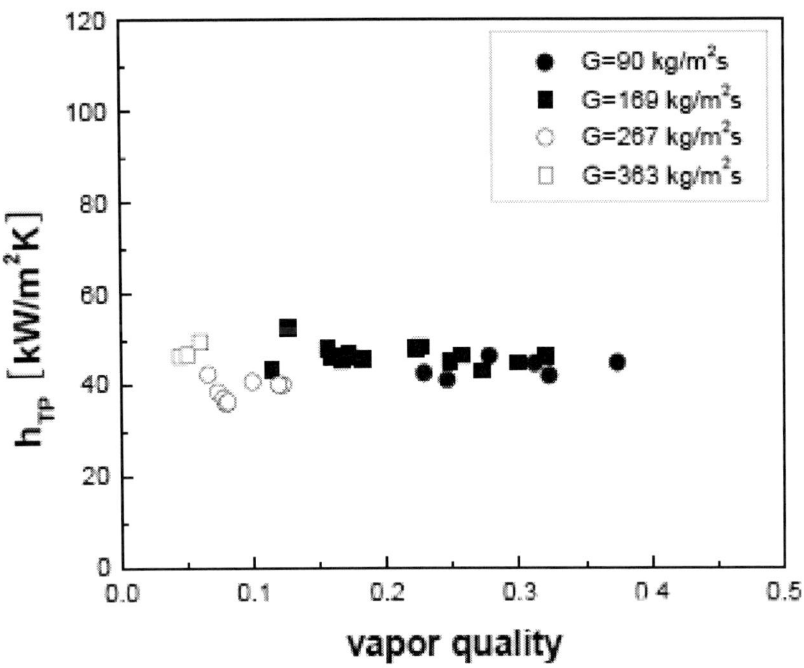

Figure 12. Flow boiling data versus vapour quality, Huh and Kim (2006).

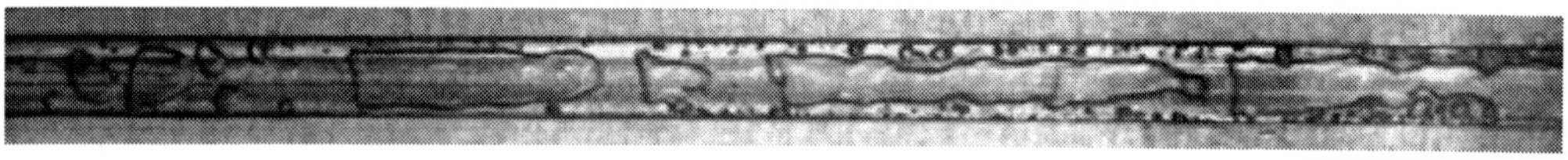

Figure 13. Typical steady boiling with two characteristic bubble diameters in a 0.5 x 4 x 200 mm^3 minichannel, Brutin et al. (2003).

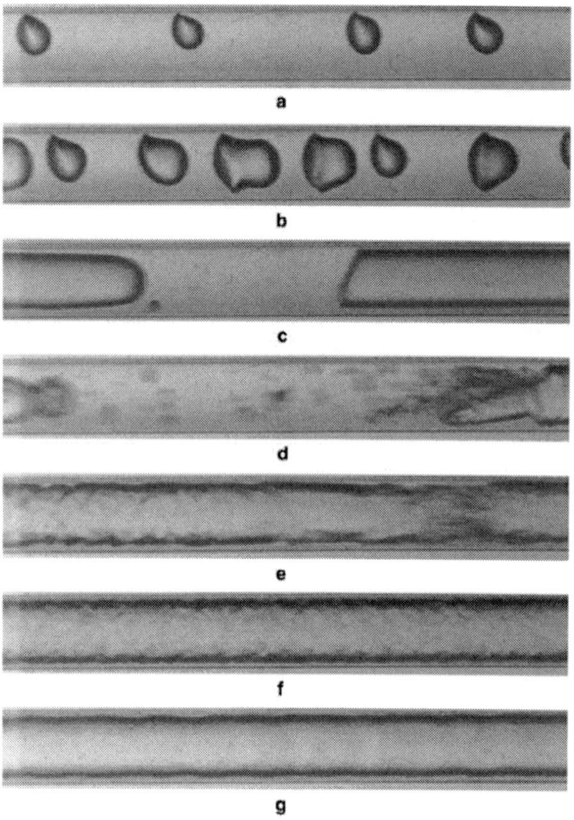

Figure 14. Flow patterns and transitions for D = 0.5 mm, L = 70.70 mm, G = 500 kg/m^2s, T_{sat} = 30 °C and ΔT_{sub} = 3 °C at entrance to heater: (a) bubbly flow at x = 2%; (b) bubbly/slug flow at x = 4%; (c) slug flow at x = 11%; (d) slug/semi-annular flow at x = 19%; (e) semi-annular flow at x = 40%; (f) wavy annular flow at x = 82%; (g) smooth annular flow at x = 82%, Revellin et al. (2006).

As the oscillation in the flow takes place, the evaporator experiences a cyclical boiling behaviour, which, in the most general case, may be described as the periodic development of five distinct stages: (1) single-phase liquid filling of the channel, (2) bubble nucleation over the channel length, (3) growth and coalescence of bubbles to form a vapour pouch over almost the entire cross-section, (4) a rapid expansion of the vapour that may involve the entire channel length, and (5) evaporation of the quasi-static liquid film left at the channel wall after the bubble expansion. The bubble in stage (4) may induce blockage of the flow in the heated section, or, if it grows towards the inlet, will provoke backflow of liquid into an upstream compressible volume or into adjacent channels. Oscillations in pressure drop, flow, and wall temperatures are the typical symptoms of these instabilities, and are often found in the micro-channel literature.

As observed by Peles et al. (2001) for flow boiling in a single capillary tube, see figure 15, growth of a large bubble leads to a blockage of the micro-channel, to a sharp change of the hydraulic resistance and ultimately to an unsteady gas–liquid flow. Consolini and Thome (2009) have shown the comparison between a 'stable' temporal profile for the outer wall temperature with the fluctuations generally within the uncertainty band, and that of an unstable flow, as reported in figure 16.

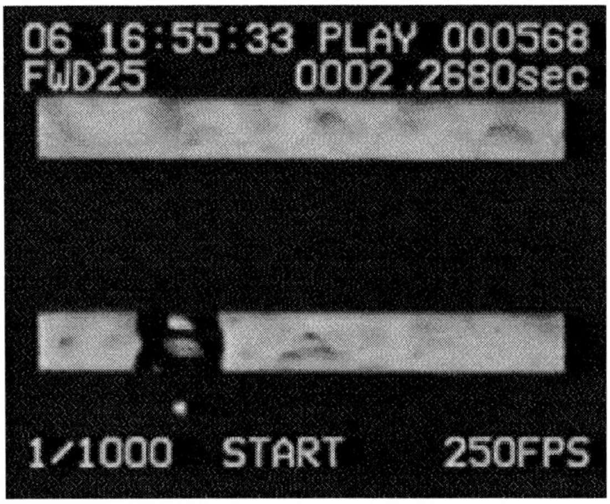

Figure 15. Bubble formation in a capillary pipe for water in a triangular shape microchannel, D_h = 50-200 μm, Peles et al. (2001).

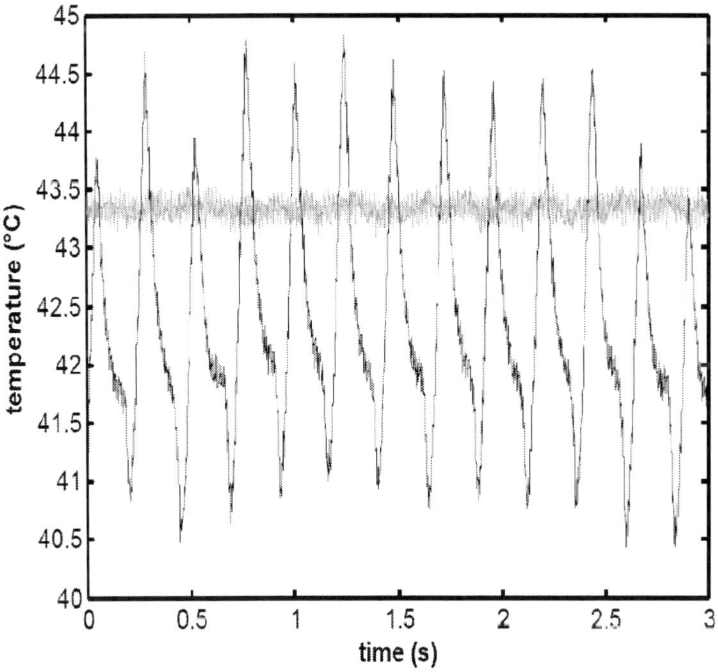

Figure 16. Steady outer wall temperature, and wall temperature during flow instability for R-245fa in a 510 μm tube, Consolini and Thome (2009).

They have also found that the amplitude of the oscillations increase when increasing heat flux, while maintaining its periodicity. As far as the heat transfer coefficient is concerned, large difference in the heat transfer rate has been observed between stable (higher values) and unsteady flow, as shown in figure 17. Flow stability can be regulated by the valve at the entrance to the test section which "isolates" the boiling flow in the evaporator from the upstream compressibility. Authors claim these results show quite conclusively that published data must be segregated into stable and unstable categories in order to be useful for development of prediction methods. Many results from the literature show the heat transfer coefficient to be independent of the vapour quality, as reported in figures 10 and 12, but no detail about the flow stability is provided. Visual studies of boiling in microchannels have been conducted, among others, by Hetsroni et al. (2004) and by Wang et al. (2008). Instability in parallel microchannels are associated with non uniform size of parallel channels and different bubble distribution/pressure drop in channels.

Figure 18 illustrates two-phase flow, driven by a pump in the microchannels of the test module in the work by Hetsroni et al. (2004). The vapour is generated only in part of the microchannels, and in different parts of the parallel microchannels. In the microchannels where vapour is generated two regions could be considered: one region, located close to the inlet collector, and the region located downstream from the inlet collector. The instability is clearly evident in the multichannel system. Wang et al. (2008) have recently performed experiments of flow boiling in parallel microchannels (D_h = 186 μm) using water experiencing significant flow instability. They have also verified that using inlet orifices upstream compressible volume instability and excursive instability are both suppressed thus obtaining stable flow conditions, as shown in figure 19.

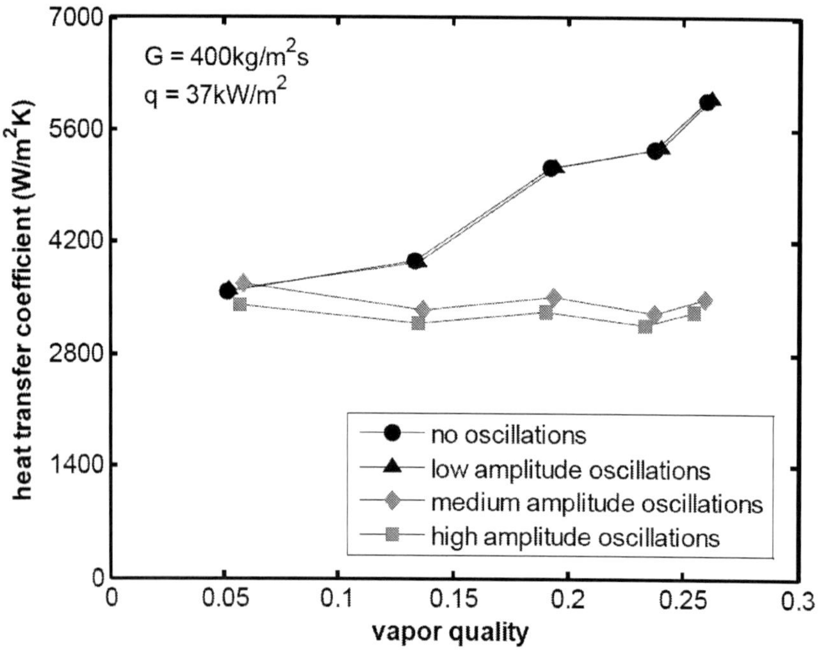

Figure 17. Stable and unstable heat transfer coefficients for R-245fa with D = 510 μm, Consolini and Thome (2009).

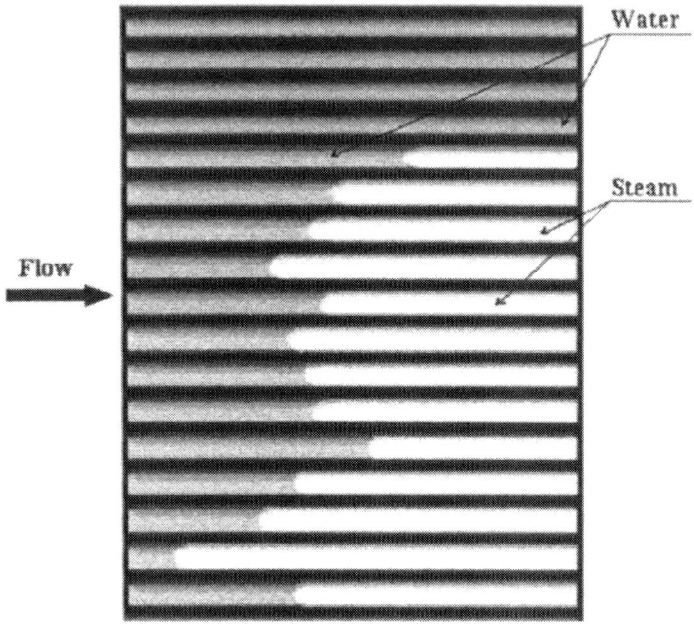

Figure 18: Boiling instability in parallel microchannels, water in a D_h = 120-160 μm triangular pipe, Hetsroni et al. (2004).

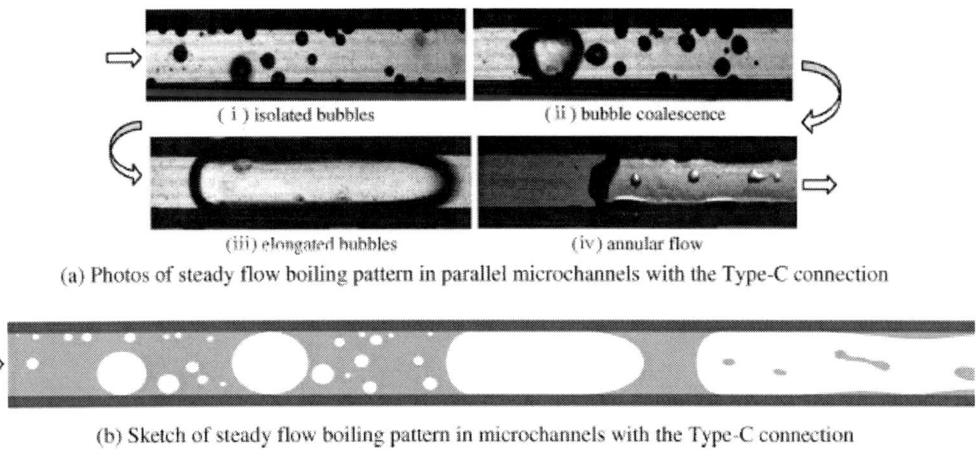

Figure 19. Photos and sketch of steady flow boiling patterns in parallel microchannels (D_h = 186 μm) with the Type-C connection (inlet orificing), Wang et al. (2008).

3.4. Flow Boiling Heat Transfer Perspectives

The complete picture of the "heat transfer map" for flow boiling heat transfer in microchannel has not yet been established. As outlined by Thome (2004, 2006), there are still a number of open issues which have to be addressed in order to have a clearer picture of the boiling heat transfer mechanisms in microchannel. Though not exhaustive, they are as follows:

- if nucleate boiling is the dominant regime as some studies suggest, then the surface finish of the microchannel has an important influence on the heat transfer and should be measured and reported;
- benchmark data with the same fluid, same tube diameter, same test conditions, has to be made to double-check the universality of the data;
- if the pressure drop is large in the test sections, as it is often the case, then the flashing effect on the enthalpy change has to be taken into account when determining and reporting the local vapour quality;
- the influence of the test section fluid inlet conditions on the measured data has not yet been thoroughly ascertained; two-phase flow structure inside the microchannel can be affected by subcooled boiling prior the test section entrance.

On the other hands, a new elongated bubble flow heat transfer model (called three-zone model) developed by Thome et al. (2004) has shown that transient evaporation of the thin liquid films surrounding elongated bubbles can be assumed as the dominant heat transfer mechanism. The three-zone flow boiling model describes the transient variation in the local heat transfer coefficient during sequential and cyclic passage of (i) a liquid slug, (ii) an evaporating elongated bubble, and (iii) a vapour slug when film dryout has occurred at the end of the elongated bubble. Figure 20 depicts a schematic of their model applicable to elongated bubble type flows. This phenomenological model contains five empirical constants: three to predict the bubble frequency, one to set the film dryout thickness δ_{min}, and one to correct the method they use to predict the initial film thickness δ_o. Their values were determined using a broad heat transfer database derived from the literature covering seven fluids. The model illustrates the strong dependency of heat transfer on the bubble frequency, the lengths of the bubbles and liquid slugs, and the initial liquid film thickness and its thickness at dryout. This model so far only covers heat transfer in the elongated bubble (slug) flow regime with and without intermittent dryout. Agostini and Thome (2005) have also made a preliminary extension to annular flows. On the other hand, there are numerous additional thermal phenomena one could imagine requiring a place within the model in the future.

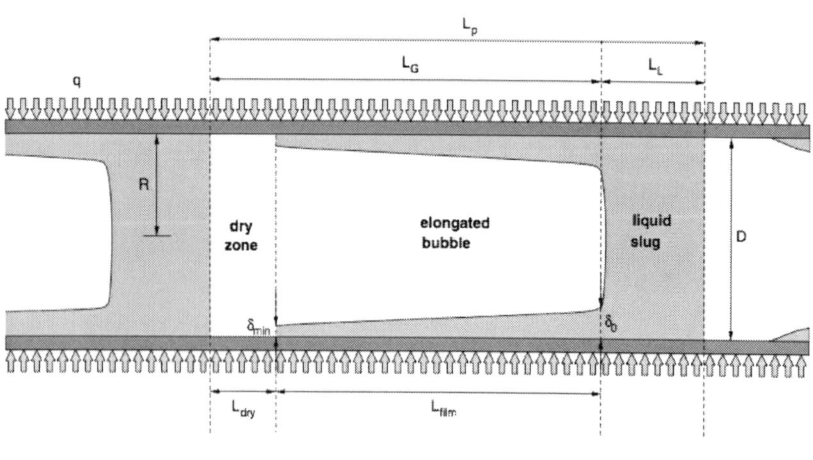

Figure 20. Schematic of the three-zone evaporation model for elongated bubble flow regime, Thome (2006).

This model is able to provide with a good prediction of some experimental data of heat transfer in microchannels, as reported, as an example, in figures 21 and 22, where the Bao et al. (2000) data and Shiferaw et al. (2006) data are well predicted, respectively.

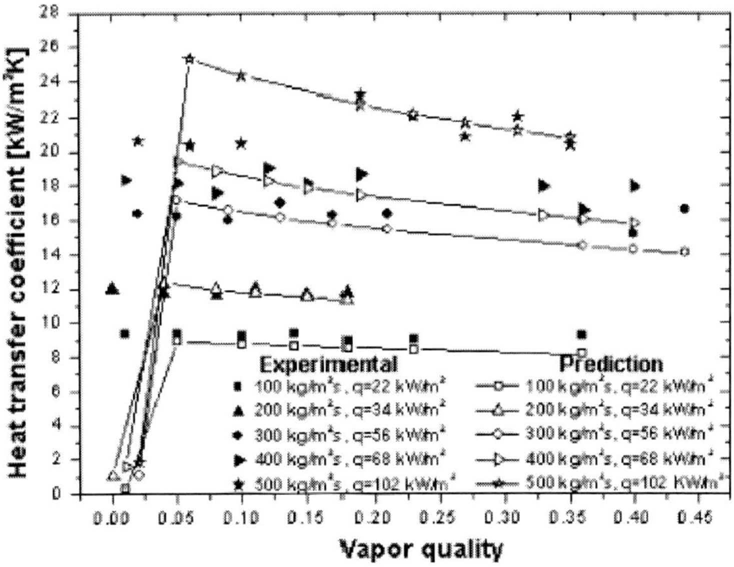

Figure 21. Thome et al. (2004) model predictions of the data of Bao et al. (2000) for R-11; model parameters set to $\Delta T_{nuc} = 28$ K ($r_{nuc} = 0.070$ μm) and $\delta_o = 12.5$ μm.

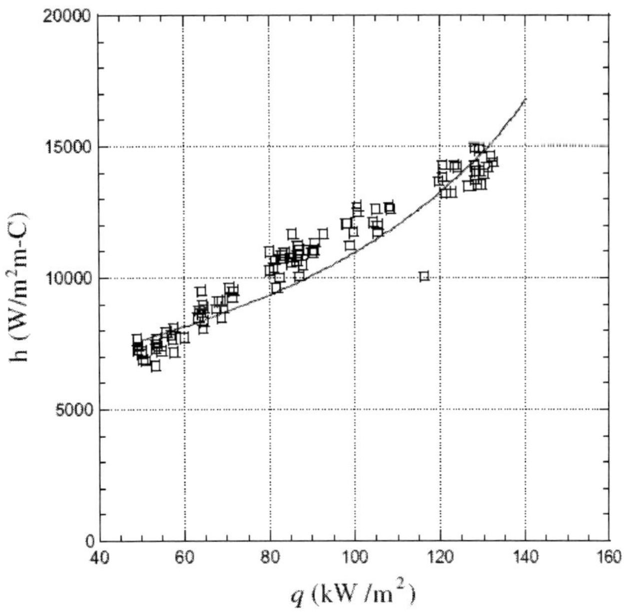

Figure 22. Comparison of local flow boiling heat transfer coefficients for R-134a in 2.01 mm stainless steel tube of Shiferaw et al. (2006) to the Thome et al. (2004) model.

In conclusion, the state-of-the-art of two-phase flow and heat transfer in microchannels is still in its infancy and much fundamental experimental and analytical work remains to be done. Main areas still worthy of a detailed research, in addition to what already reported here above, as also stated by Thome (2006) can be bulleted as follows:

- a more refined and universal method for identifying the threshold between microscale and macroscale is required;
- critical heat flux has to be investigated in detail for both single and multichannels elements, investigating the influence of non uniform heating and non-circular channel;
- the influence of geometry, physical properties, etc. on two-phase flow pattern transition has to be still investigated;
- flow instability and flow induced fluctuations need to be better understood, and a method to establishing stability criteria has to be developed;
- new and more sophisticated methods and models have to be developed for better predictions of local flow boiling heat transfer in microchannels, including the effects of flow pattern change, non-circular geometry, back flow, etc.;
- visualization of boiling heat transfer is encouraged for a better physical insight; usage of Indium-tin-Oxide coating, enabling Joule effect heating and simultaneous visualization of the boiling process (coating thickness is around 80-100 nm) may be of great help in this matter;
- there is a need for developing a growing database for two-phase frictional pressure drops, along with description of reliable methods for its prediction.

It is definitely possible to conclude here that flow boiling heat transfer in microchannels is quite far from having been thoroughly and exhaustively investigated. Different from single-phase flow heat transfer, the current knowledge of flow boiling heat transfer in macroscale cannot be extended tout court to microscale, where bubble confinement plays a relevant role reducing the channel geometry.

4. Concluding Remarks

The present manuscript presents a critical evaluation of existing works in the field of heat transfer of single (liquid) and two-phase flow (boiling) in microchannels. The literature survey provides with many works in this field, the accuracy of which is increasing with date, recent works being the most accurate.

As far as liquid single-phase flow heat transfer is concerned, there is more and more the general agreement on the validity of macroscale knowledge going down to microscale. No new phenomenon has been ascribed to the microscale, though some aspects typically neglected in macroscale must be accounted for in microscale for a proper evaluation of heat transfer data (this comment being valid also for single-phase fluid flow). The scale reduction gives relevance to some effects such as viscous dissipation (or viscous heating, due to the general large pressure drop associated with microchannel flow), thermal entrance (due to the generally short geometry of microchannels) and axial length (associated with the general large thickness of the microchannels), which becomes very important in microscale. Their neglectfulness is bounded to give poor analysis of heat transfer data misleading the possible correct conclusion of the experimental outcome. Once these so-called scaling effects are properly accounted for, microscale heat transfer in liquid single-phase flow looks well predictable using current knowledge of macroscale, such as Gnielinski (1978) correlation for turbulent flow and $Nu = 4.36$ equation for laminar flow. Occurrence of scaling effects can also be used to identify the threshold between micro- and macroscale, which depends not only on the geometry but also on fluid physical properties. More accurate experiments are still needed, though the knowledge looks now more assessed, especially for what concerns the axial conduction (assessment of numerical simulation), understanding of differences encountered in different heating type, transition from laminar-to-turbulent flow, non circular-geometry.

Boiling heat transfer in microscale is associated with bubble confinement in the reduced scale, which is definitely expected to alter the flow dynamics inside the small size channel. Identification of micro-to-macroscale threshold has been recently associated with bubble confinement, although a more refined correlation is called for a better and more universal definition of the transition. Because of that, divergence between macroscale knowledge (often based on empirical bases or experimental evidences) is clearly awaited and understandable. Also in this case some conclusions drawn from the experimental data in tubes, estimated from the findings of boiling heat transfer in macroscale (nucleate versus forced convective boiling), might have led to erroneous conclusions about the type of boiling regime in microscale. Additional care has to be taken about the stability of flow boiling both in parallel microchannels and single micropipes. Recent theoretical works, supported by first visualization works, have supposed a different boiling mechanisms for microchannels, based on elongated bubble existence and evaporation of thin liquid layer between the elongated bubble and the channel wall. Much work is still required to possibly validate this theory or produce new insights for a better mechanistic modelling of boiling heat transfer. Area worthy of new investigations have been identified and recommended, such as identification of micro-to-macroscale transition, critical heat flux, visualization, flow pattern transition, flow instability, two-phase pressure drop, etc.

The current status of the knowledge in single-phase and flow boiling heat transfer in pipes is also reflecting the knowledge we have in large pipes. Single-phase flow heat transfer is well established both in laminar and in turbulent flow, and the heat transfer coefficient can be evaluated from theoretically based equations (Polhausen solution in laminar flow and Reynolds-Colburn analogy in turbulent flow). As no new phenomenon is occurring in microscale, apart from the non-negligibility of scaling effects, we may reasonably expect that single-phase heat transfer in microscale can be well predicted from macroscale knowledge, as we have seen from recent accurate experiments. The knowledge of flow boiling heat transfer is far from being based on a theoretical approach and available correlations for the evaluation of the heat transfer coefficient are typically empirically based. Bubble confinement, representing the major features of flow boiling in microscale, brings available correlations out of range thus reducing inevitably their performances.

REFERENCES

Adams, T.M., Abdel-Khalik, S.I., Jeter, S.M. and Qureshi, Z.H., 1998, An Experimental Investigation of Single-Phase Forced Convection in Micro-Channels, *Int. J. Heat Mass Transfer*, 41, 851-857.

Agostini, B. and Thome, J.R., 2005, Comparison of an Extended Database for Flow Boiling Heat Transfer Coefficients in Multi-Microchannels Elements with the Three-Zone Model, ECI Heat Transfer and Fluid Flow in Microscale, *Castelvecchio Pascoli*, Italy, Sept. 25–30.

Balasubramanian P. and Kandlikar, S.G., 2005, Experimental Study of Flow Patterns, Pressure Drop and Flow Instabilities in Parallel Rectangular Minichannels, *Heat Transfer Engineering*, 26, 20-27.

Bao, Z.Y., Fletcher, D.F. and Haynes, B.S., 2000, Flow Boiling Heat Transfer of Freon R11 and HFCFC123 in Narrow Passages, *Int. J. Heat Mass Transfer*, 43, 3347–3358.

Bergles, A.E. and Kandlikar, S.G., 2005, On the Nature of Critical Heat Flux in Microchannels, *J. Heat Transfer*,127, 101-107.

Brutin, D., Topin, F. and Tadrist, L., 2003, Experimental Study of Unsteady Convective Boiling in Heated Minichannels, *Int. J. Heat Mass Transfer*, 46, 2957–2965.

Bucci A., Celata, G.P., Cumo, M., Serra, E. and Zummo, G., 2003, Water Single-Phase Fluid Flow and Heat Transfer in Capillary Tubes, *Thermal Science and Engineering*, 11, 81-89.

Celata, G.P., Cumo, M. and Zummo, G., 2004, Thermal-Hydraulic Characteristics of Single-Phase Flow in Capillary Pipes, *Experimental Thermal and Fluid Science*, 28, 87–95.

Celata, G.P., Cumo, M., Gervasi, M. and Zummo, G., 2007b, Flow Pattern Analysis of Flow Boiling in Microgravity, *Multiphase Science and Technology*, 19, 183-210.

Celata, G.P., Cumo, M., Guglielmi, M. and Zummo, G., 2002, Experimental Investigation of Hydraulic and Single Phase Heat Transfer in 0.130 mm Capillary Tube, *Microscale Thermophysical Engineering*, 6, 85-97.

Celata, G.P., Cumo, M., Marconi, V., McPhail, S.J. and Zummo, G., 2006b, Microtube Liquid Single-Phase Heat Transfer in Laminar Flow, *International Journal of Heat and Mass Transfer*, 49, 3538-3546.

Celata, G.P., Cumo, M., McPhail, S.J. and Zummo, G., 2007a, Single-Phase Laminar and Turbulent Heat Transfer in Smooth and Rough Microtubes, *Microfluidics and Nanofluididics,* 3, 697-707.

Celata, G.P., Morini, G.L., Marconi, V., McPhail, S.J. and Zummo, G., 2006a, Using Viscous Heating to Determine the Friction Factor in Microchannels – An Experimental Validation, *Experimental Thermal and Fluid Science*, 30, 725-731.

Chang K.H. and Pan, C., 2007, Two-Phase Flow Instability for Boiling in a Microchannel Heat Sink, *Int. J. Heat Mass Transfer*, 50, 2078-2088.

Chiou, J.P., 1980, The Advancement of Compact Heat Exchanger Theory Considering the Effects of Longitudinal Heat Conduction and Flow Non-Uniformity – Symposium on Compact Heat Exchangers, *ASME HTD* vol. 10, pp. 101-121.

Choi, S.B., Barron, R.F. and Warrington, R.Q., 1991, Fluid Flow and Heat Transfer in Micro-Tubes, in: D. Choi et al. (Eds.), *Micro-Mechanical Sensors, Actuators and Systems, ASME DSC*, 32, 121–128.

Consolini, L. and Thome, J.R., 2009, Micro-Channel Convective Boiling Heat Transfer with Flow Instabilities, ECI International Conference on Boiling Heat Transfer, *Florianopolis*, 3-7 May, Paper MC1.

Gnielinski, V., 1976, New Equations for Heat and Mass Transfer in Turbulent Pipe and Channel Flow, *Int. Chem. Eng.*, 16, 359-368.

Grohmann, S., 2005, Measurement and Modeling of Single-Phase and Flow Boiling Heat Transfer in Microtubes, *Int. J. Heat Mass Transfer*, 48, 4073-4089.

Hetsroni, G., Gurevich, M., Mosyak, A. and Rozenblit, R., 2004, Drag Reduction and Heat Transfer of Surfactants Flowing in a Capillary Tube, *Int. J. Heat Mass Transfer*, 47, 3797-3809.

Hetsroni, G., Klein, D., Mosyak, A., Segal, Z. and Progrebnyak, E., 2004, Convective Boiling in Parallel Microchannels, *Microscale Thermophysical Engineering*, 8, 403-421.

Hetsroni, G., Mosyak, A., Pogrebnyak, E. and Yarin, L.P., 2005, Heat Transfer in Micro-Channels: Comparison of Experiments with Theory and Numerical Results, *Int. J. Heat Mass Transfer*, 48, 5580-5601.

Holman, J.P., 1992, *Heat Transfer*, 7th ed., McGraw-Hill.

Huh, C. and Kim, M.H., 2006, Two-Phase Pressure Drop and Boiling Heat Transfer in a Single Horizontal Microchannel, *Proceedings of the 4th International Conference on Nanochannels, Microchannels and Minichannels*, ICNMM2006 2006 B, 1097-1104.

Huh, C., Kim J. and Kim, M.H., 2007, Flow Pattern Transition Instability During Flow Boiling in a Single Microchannel, *Int. J. Heat Mass Transfer*, 50, 1049-1060.

Huo, X., L. Chen, L., Tian, Y.S. and Karayiannis, T.G., 2004, Flow Boiling and Flow Regimes in Small Diameter Tubes, *Applied Thermal Engineering*, 24, 1225-1239.

Kandlikar, S.G., 2001, Two-Phase Flow Patterns, Pressure Drop and Heat Transfer During Boiling in Minichannel and Microchannel Flow Passages of Compact Heat Exchangers, In: *Compact Heat Exchangers and Enhancement Technology for the Process Industries–* –2001, Begell House, New York, pp. 319–334.

Kandlikar, S.G., 2002, Fundamental Issues Related to Flow Boiling in Minichannels and Microchannels, *Experimental Thermal Fluid Science*, 26, 389-407.

Kandlikar, S.G., Joshi S. and Tian S., 2003, Effect of Surface Roughness on Heat Transfer and Fluid Flow Characteristics at Low Reynolds Numbers in Small Diameter Tubes, *Heat Transfer Engineering*, 24, 4-16.

Kew, P.A. and Cornwell, K., 1997, Correlations for the Prediction of Boiling Heat Transfer in Small-Diameter Channels, *Applied Thermal Engineering*, 17, 705-715.

Lazarek, G.M. and Black, S.H., 1982, Evaporative Heat Transfer, Pressure Drop and Critical Heat Flux in a Small Vertical Tube with R-113, *Int. J. Heat Mass Transfer*, 25, 945-960.

Lelea, D., Nishio S. and Takano, K., 2004, The Experimental Research on Microtube Heat Transfer and Fluid Flow of Distilled Water, *Int. J. Heat Mass Transfer*, 47, 2817-2830.

Lin, S., Kew, P.A. and Cornwell, K., 2001, Two-Phase Heat Transfer to a Refrigerant in a 1 mm Diameter Tube, *Int. J. Refrigeration*, 24, 51–56.

Maranzana, G., Perry, I. and Maillet, D., 2004, Mini- and Micro-channels: Influence of Axial Conduction in the Walls, *Int. J. Heat and Mass Transfer*, 47, 3993-4004.

Mehendal, S.S., Jacobi, A.M. and Shah, R.K., 2000, Fluid Flow and Heat Transfer at Micro- and Meso-Scales with Application to Heat Exchanger Design, *Appl. Mech. Rev*. 53, 175–193.

Morini, G.L., 2004, Single-Phase Convective Heat Transfer in Microchannels: A Review of Experimental Results, *Int. J. Thermal Sciences*, 43, 631-651.

Morini, G.L., 2005, Viscous Heating in Liquid Flows in Microchannels, *Int. J. Heat and Mass Transfer*, 48, 3637-3647.

Morini, G.L., 2006, Scaling Effects for Liquid Flows in Microchannels, *Heat Transfer Engineering*, 27, 64-73.

Nguyen, N.T., Bochnia, D., Kiehnscherrf, R. and Dozel, W., 1996, Investigation of Forced Convection in Micro-Fluid Systems, *Sensors Actuators*, A, 55, 49–55.

Owhaib, W. and Palm, B., 2004, Experimental Investigation of Single-Phase Convective Heat Transfer in Circular Microchannels, *Experimental Thermal and Fluid Science*, 28, 105-110.

Owhaib, W., Martin-Callizo, C. and Palm, B., 2004, Evaporative Heat Transfer in Vertical Circular Microchannels, *Applied Thermal Engineering*, 24, 1241-1253.

Peles, Y.P., Yarin, L.P. and Hetsroni, G., 2001, Steady and Unsteady Flow in a Heated Capillary, *Int. J. Multiphase Flow*, 27, 577–598.

Peng, X.F. and Peterson G.P., 1996, Convective Heat Transfer and Flow Friction for Water Flow in Micro-Channel Structures, *Int. J. Heat Mass Transfer*, 39, 2599–2608.

Qi, S.L., Zhang, P., Wang R.Z. and Xu, L.X., 2007, Flow Boiling of Liquid Nitrogen in Micro-Tubes: Part I – The Onset of Nucleate Boiling, Two-Phase Flow Instability and Two- Phase Flow Pressure Drop, *Int. J. Heat Mass Transfer*, 50, 4999-5016.

Qu, W. and Mudawar, I., 2003, Measurement and Prediction of Pressure Drop in Two-Phase Micro-Channel Heat Sinks, *Int. J. Heat Mass Transfer*, 46, 2737-2753.

Revellin, R., Dupont, V., Ursenbacher, T., Thome, J.R. and Zun, I., 2006, Characterization of Diabatic Two-Phase Flows in Microchannels: Flows Parameter Results for R-134a in a 0.5 mm Channel, *Int. J. Multiphase Flow*, 32, 755–774.

Schlichting, H., 1979, *Boundary-Layer Theory*, McGraw-Hill, 7th Edition, pp. 615-617.

Shah, R.K. and London, A.L., 1978, Laminar flow forced convection in ducts, in T.F. Irvine, J.P. Hartnett (ed.), *Advances in Heat Transfer*, Academic Press, New York, pp. 51-52, 124-128.

Shiferaw, D., Karayiannis, T.G. and Kenning, D.B.R., 2006, A Comparison with the Three-Zone Model for Flow Boiling Heat Transfer in Small Diameter Tubes, *13th International Heat Transfer Conference*, Sydney, Australia, Aug. 13–18.

Thome, J.R., 2004, Boiling in Microchannels: A Review of Experiment and Theory, *Int. J. Heat Fluid Flow*, 25, 128-139.

Thome, J.R., 2006, State-of-the-Art Overview of Boiling and Two-Phase Flows in Microchannels, *Heat Transfer Engineering*, 27, 4-19.

Thome, J.R., Dupont, V. and Jacobi, A.M., 2004, Heat Transfer Model for Evaporation in Microchannels. Part I: Presentation of the Model, *Int. J. Heat Mass Transfer*, 47, 3375-3385.

Tiselj, I., Hetsroni, G., Mavko, B., Mosyak, A., Pogrebnyak, E. and Segal, Z., 2004, Effect of Axial Conduction on the Heat Transfer in Micro-Channels, *Int. J. Heat Mass Transfer*, 47, 2551–2565.

Tran, T.N., Wambsganss, M.W., France, D.M., 1996, Small Circular and Rectangular Channel Boiling with Two Refrigerants. *Int. J. Multiphase Flow*, 22, 485–498.

Ullmann, A. and Brauner, N., 2007, The Prediction of Flow Pattern Maps in Minichannels, *Multiphase Science and Technology*, 19, 49-73.

Xu, B., Ooi, K.T., Mavriplis, C. and Zaghloul, M.E., 2003, Evaluation of Viscous Dissipation in Liquid Flow in Microchannels, *Journal of Micromechanics and Microengineering*, 13, 53-57.

Yan, Y.Y. and Lin, T.F., 1998, Evaporation Heat Transfer and Pressure Drop of Refrigerant R134a in a Small Pipe, *Int. J. Heat Mass Transfer*, 30, 3072-3083.

Yu, D., Warrington, R., Barron, R. and Ameel, T., 1995, An Experimental and Theoretical Investigation of Fluid Flow and Heat Transfer in Microtubes, *ASME/JSME Thermal Engineering Conference*, vol. 1, ASME, New York.

Wang, G., Cheng, P. and Bergles, A.E., 2008, Effects of Inlet/Outlet Configurations on Flow Boiling Instability in Parallel Microchannels, *Int. J. Heat Mass Transfer*, 51, 2267–2281.

In: Frontier Research in Microscale and Nanoscale... ISBN 978-1-61324-665-8
Editor: Lixin Cheng

Chapter 3

AN EXPERIMENTAL STUDY ON FLOW BOILING HEAT TRANSFER OF R134A IN A 2.3 MM TUBE

Cristiano Bigonha Tibiriçá and Gherhardt Ribatski[*]
Departament of Mechanical Engineering, Escola de Engenharia de São Carlos (EESC)
University of São Paulo (USP), São Carlos, Brazil

ABSTRACT

This paper presents new experimental flow boiling heat transfer results in micro-scale tubes. The experimental data were obtained in a horizontal 2.32 mm I.D. stainless steel tube with heating length of 464 mm, R134a as working fluid, mass velocities ranging from 50 to 600 kg $m^{-2}s^{-1}$, heat flux from 5 to 55 kW m^{-2}, exit saturation temperatures of 22, 31 and 41 °C, and vapor qualities from 0.05 to 0.98. Flow pattern characterization was also performed from images obtained by high-speed filming. Heat transfer coefficient results from 2 to 14 kW m^{-2} K^{-1} were measured. It was found that the heat transfer coefficient is a strong function of the saturation pressure, heat flux, mass velocity and vapor quality. The experimental data were compared against the following flow boiling predictive methods from the literature: Liu and Winterton [3], Kandlikar and Balasubramanian [4], Zhang *et al.* [5], Saitoh *et al.* [6], Bertsch *et al.* [7] and Thome *et al.* [11]. Comparisons against these methods based on the data segregated according to flow patterns were also performed. Though not satisfactory, Saitoh *et al.* [6] worked the best and was able of capturing most of the experimental heat transfer trends.

Keywords: Microchannel; Heat Transfer Coefficient; Two-Phase Flow; R134a; Convective Boiling.

NOMENCLATURE

D	diameter, (m)

[*] E-mail: ribatski@sc.usp.br

E	heat rate, (W)
G	mass velocity, (kg m^{-2} s^{-1})
h	heat transfer coefficient, (W m^{-2} K^{-1})
i	enthalpy, (J kg^{-1})
L	length, (m)
m	mass flow rate, (kg s^{-1})
P	electrical power, (W)
p	pressure, (Pa)
q	heat flux, (W m^{-2})
Re	Reynolds number, (dimensionless), $\mathrm{Re} = \dfrac{G \cdot D}{\mu}$
T	temperature, (K)
x	vapor quality, (dimensionless)
z	axial position, (m)

Greek Symbols

μ	fluid viscosity, (Pa s)
Δp	pressure drop, (kPa)
ε	mean absolute error, (%)
λ	fraction of data predicted to within ±30%, (%)
ρ	fluid density, (kg m^{-3})

Subscripts

bottom	bottom of the tube
exit	exit of the test section
exp	experimental
in	inlet of the pre heater
local	at a specific position on the tube
middle	middle position on the cross section of the tube
nb	nucleated boiling
sat	saturation temperature
sp	single phase
top	top of the tube
tp	two-phase
w	inner wall of the tube

1. Introduction

Flow boiling heat transfer inside channels having hydraulic diameters of less than 3 mm, identified here as micro-scale channels, has been studied since the late 1950s [1], but only in recent years it has turned into a major research topic in the thermal fluid science field. Most of this interest is driven by the industrial demand for compact devices capable of dissipating extremely high heat fluxes, such as those expected for the next microprocessor generation. Despite the enormous benefits that can be achieved by using precise designing methods, they are still not available and flow-boiling based heat-spreaders are being developed based on prototype machining and testing. Over the last decade great effort has been put into identifying the main parameters governing micro-scale flow boiling heat transfer and also into quantifying their relevance.

In macro-scale flow boiling, different heat transfer mechanisms are dominant according to the vapor quality range, and heat flux and mass velocity levels. At low vapor qualities, nucleate boiling effects prevail while at high vapor qualities and prior to the liquid dryout, the heat transfer coefficient is mainly controlled by convective effects. The predominance of these mechanisms is commonly considered [2-7] when developing flow-boiling heat transfer coefficient predictive methods. Chen [2] was one of the pioneers of such an approach according to which the flow boiling heat transfer coefficient is calculated as follows:

$$h_{tp} = S \cdot h_{nb} + F \cdot h_{sp} \tag{1}$$

where h_{nb}, is given according to Foster and Zuber [8] nucleate pool boiling correlation, and h_{sp} is the liquid single-phase heat transfer contribution to the two-phase heat transfer coefficient calculated according to Dittus-Boelter [9] correlation. In Eq. (1) S is the nucleate boiling suppression factor that takes into account steeper temperature gradients near the wall due to the fluid motion which tends to suppress the number of bubble active sites. The convective enhancement factor, F, takes into account the increment of convective effects relative to that of the single-phase flow of the liquid, promoted by the flow acceleration due to the evaporation process. Since its proposal, several authors have adjusted their database to Chen's type correlations. This approach has been also adopted for micro-scale flow boiling.

Liu and Winterton [3] developed a method in order to predict saturated and subcooled flow boiling heat transfer based on Chen's basic hypothesis that both convective and nucleate boiling heat transfer mechanisms play a role in flow boiling. In their method, the convective and nucleate boiling contributions were added according to a power-type asymptotic approach with an exponent of 2. They adjusted their method according to a database that comprises more than 5000 experimental data points, nine fluids and tube internal diameters ranging from 2.92 mm to 32 mm. The micro-scale flow-boiling predictive method by Kandlikar and Balasubramanian [4] is only a modified version of the previous method for conventional channels developed by Kandlikar [10] in the early 1990's. The main modifications are related to the fact that laminar flows were taken into account and gravitational effects were neglected. The values for the empirical constant characteristic of the fluid/surface-material pair were kept the same as in the previous version.

Zhang *et al.* [5] modified the Chen method in order to predict micro-scale flow boiling heat transfer. In their approach, the correlation by Foster and Zuber [8] was maintained to predict the nucleate boiling heat transfer component. The boiling suppression factor proposed by Chen was also retained. Then, to determine the convective enhancement factor and the single-phase heat transfer coefficient, vapor and liquid flow conditions (laminar or turbulent) were taken into account. Their method was compared against 1203 independent experimental data points from the literature for water, R11, R12 and R113 and predicted the database with an average error of 18.3 %. By modifying the macro-scale flow boiling correlation by Chen and using a database comprising only R134a experimental results, Saitoh *et al.* [6] proposed a model which covers both macro- and micro-scale flow boiling heat transfer. In their method, emphasis was given to properly capture the tube diameter effect on the heat transfer coefficient through the Weber number. Empirical coefficients were adjusted based on a database consisting of tube diameters ranging from 0.51 mm to 11.0 mm.

Thome *et al.* [11] developed a micro-scale model that describes the heat transfer processes during the cyclic passage of elongated bubbles in a micro-scale channel. In this model, a time-averaged local heat transfer coefficient is obtained during the cyclic passage of (i) a liquid slug, (ii) an evaporating elongated bubble, and (iii) a vapor slug when present. The model includes five experimental parameters obtained by Dupont *et al.* [12] according to an experimental database with 1591 data taken from seven independent studies in the literature. Recently, Bertsch *et al.* [7] proposed a Chen-type predictive method based only on micro-scale flow boiling data which includes 3899 experimental data points covering 12 fluid and hydraulic diameters ranging from 0.16 to 2.92 mm. To evaluate the ability of the available methods to predict heat transfer coefficients in micro-scale channels, Tibiriçá and Ribatski [13] have compared 7 heat transfer predictive methods [3-7,11] against a database comprising 15 fluids and containing more than 2500 micro-scale flow boiling experimental data points from the literature. They found that the flow boiling heat transfer methods poorly predict the database; however, they also indicated that such a result was not surprising since large discrepancies between data from different authors had been observed.

Several aspects contribute to the huge discrepancies among experimental databases. Firstly, quite a large number of studies have been performed in the presence of thermal instabilities, whose effects may drastically affect the heat transfer trends, as shown by Consolini [14]. Experimental characteristics as the internal surface roughness of the channels are rarely mentioned, consequently, database comparisons are performed and heat transfer trends are suggested without an adequate knowledge of the experimental conditions. Moreover, erroneous data regression procedures are frequently adopted, as shown by Ribatski *et al.* [15]. Inherent difficulties verified in conventional flow boiling heat transfer measurements are incremented in the case of micro-scale conditions due to the reduced scales involved. Energy balances under single-phase flow conditions and single-phase flow heat transfer measurements to validate the experimental apparatus are not a common practice. Finally, even when using state-of-the-art instrumentation and calibration procedures, experimental heat transfer coefficient errors, most of the times higher than ±20% [14], still play an important role in the data analyses.

In this paper, new results of micro-scale flow boiling heat transfer are presented. Extreme care was taken in order to perform accurate heat transfer coefficient measurements. Flow boiling experiments were performed for R134a in a horizontal stainless tube with internal diameter of 2.32 mm and 464 mm long. The main heat transfer trends were identified and discussed taking into account two-phase flow visualizations. The experimental results were compared against the leading predictive methods of micro-scale flow boiling.

2. EXPERIMENTS

2.1. Experimental Apparatus

The experimental setup is comprised of refrigerant and ethylene-glycol circuits. The refrigerant circuit is shown schematically in Figure 1.

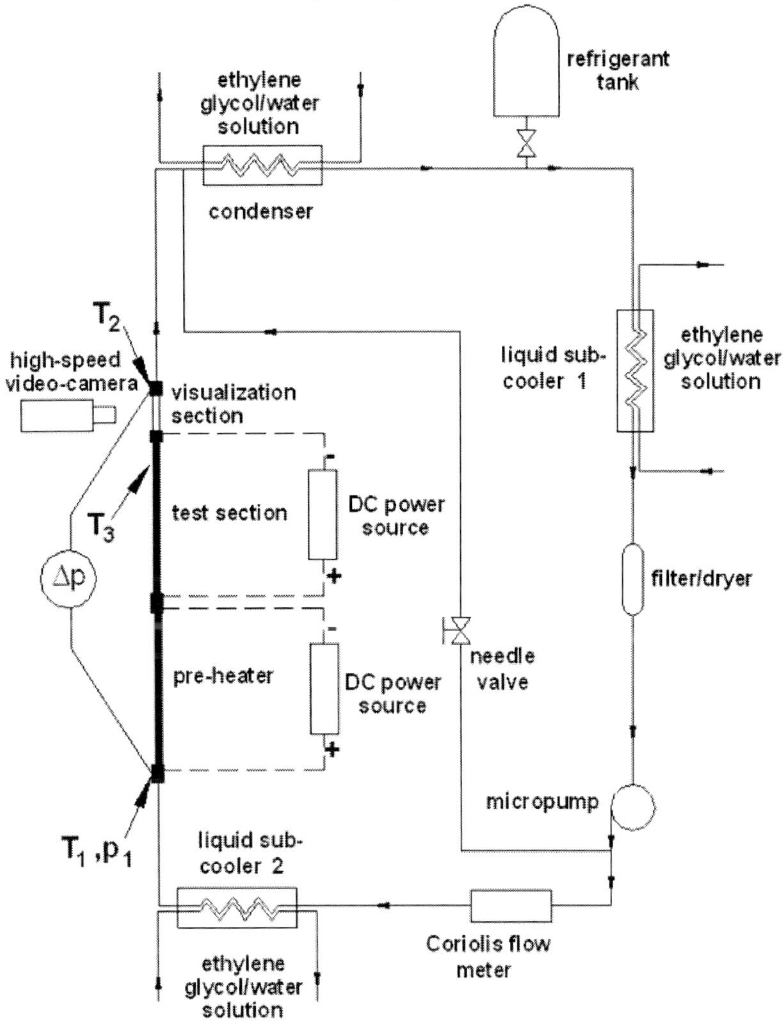

Figure 1. Schematic diagram of the refrigerant circuit.

It globally comprises a micropump to drive the working fluid through the circuit, a pre-heater to establish the experimental conditions at the inlet of the test section, a test section, a visualization section, a condenser to condense the vapor created in the heated sections, and a reservoir. The water-glycol circuit (not shown) is intended to condense and subcool the fluid in the refrigerant circuit. The cooling effect is obtained by a 60% solution of ethylene glycol/water, which operates as an intermediate fluid in a system that comprises electrical heaters actuated by PID controllers, 3 water/glycol tanks, heat exchangers and a refrigeration circuit. In the refrigerant circuit, starting from the subcooler 1, the test fluid flows through the filter to the micropump (selflubricating without oil). Downstream the micropump, a bypass piping line containing a needle-valve is installed, so that together with a frequency controller actuating on the micropump the desired liquid flow rate can be set. Then, there is the Coriolis mass flow meter and the subcooler 2 to assure that the fluid entering the pre-heater will be subcooled. Just upstream the pre-heater inlet, the enthalpy of the liquid is estimated from its temperature $T1$ by a 0.25 mm thermocouple within the pipe, and its pressure $p1$ by an absolute pressure transducer. At the pre-heater, the fluid is heated up to the desired condition at the test section inlet. The pre-heater and the test section are horizontal stainless steel tubes with an internal diameter of 2.32 mm and 464 mm of length. Both are thermally insulated and heated by applying a direct DC current to their surface. Their internal surface roughness was measured by a surface profiler using an open tube and a value of 0.33 μm was found given in terms of the CLA arithmetic average, Ra. The power is supplied to the pre-heater and the test section by two independent DC power sources controlled by the data acquisition system. The visualization section is a horizontal fused silica tube with inner diameter of 2.1 mm and length of 85 mm, and is located just downstream to the test section. The pre-heater, the test section and the flow visualization section are connected through junctions made of electrical insulation material and specially designed and machined to match up their ends and keep a smooth and continuous internal surface. Once the liquid leaves the visualization section, its temperature T_2 is determined from a 0.25 mm thermocouple within the pipe. The corresponding absolute pressure is estimated from a differential pressure transducer that provides the total pressure drop between the pre-heater inlet and the flow visualization section outlet, Δp. Then, the working fluid is directed to the tube-in-tube type heat exchanger where it is condensed and subcooled by exchanging heat with the anti-freezing ethylene glycol aqueous solution. The refrigerant tank containing a serpentine coil operates as a reservoir of the working fluid and is used to control the saturation pressure in the refrigerant circuit. The saturation pressure is set by acting on the temperature of the anti-freezing ethylene glycol aqueous solution, which flows through the serpentine coil within the tank.

Wall temperatures are measured through 0.25 mm type K thermocouples fixed along the test section. Rubber o-rings having an internal diameter smaller than the external diameter of the test section are used to fix the thermocouples on the surface by pressing them against the tube wall. At each measuring cross-section, the surface temperature is read at three locations, 90° spaced, from the bottom to the top of the tube, as indicated in Figure 2.

The experiments were conducted first by setting the temperature in the refrigerant tank. This temperature was kept almost constant by a thermal-controller system. Once the saturation pressure had been established in the refrigerant circuit by controlling the temperature in the refrigerant tank, the mass velocity was set through a frequency controller acting on the micropump.

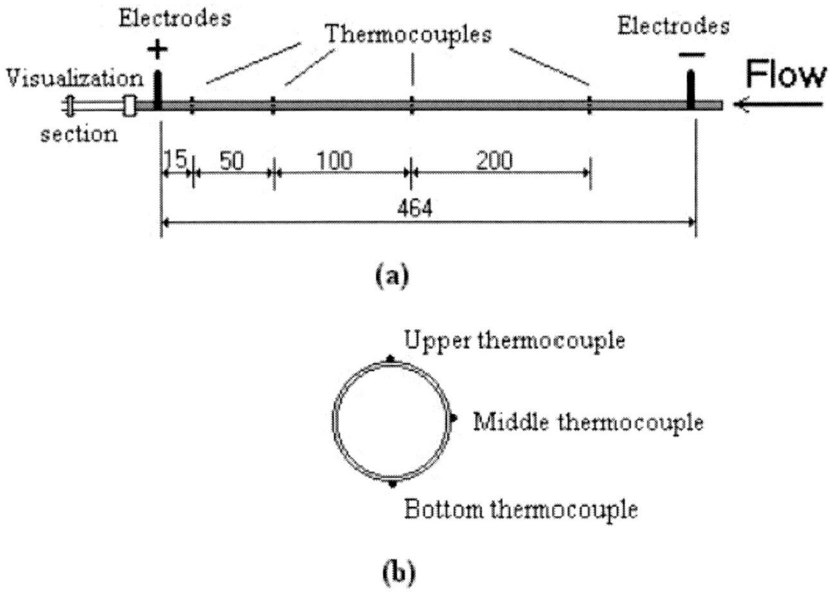

Figure 2. Details of the test section and thermocouples positioning along its surface.

Then, the desired experimental heat flux was imposed to the test section and by varying the power supplied to the pre-heater, experimental results for distinct vapor qualities were obtained keeping the remaining parameters fixed. Tests were conducted by gradually increasing the heat flux. The datum points were logged only under steady-state conditions, which were characterized by temperature measurement variations within ±0.15K of its average value.

During the thermocouple calibrations and the experimental campaign, the temperature of the room containing the test facility was maintained around 22°C. Such a procedure was adopted in order to minimize possible effects of the room temperature on the measurements accuracy.

2.2. Data Reduction

Heat transfer coefficient. Local heat transfer coefficients were calculated according to the Newton's cooling law as follows:

$$h_{local} = \frac{q}{T_w - T_{sat}(z)} \tag{2}$$

where T_w is the surface temperature of the inner tube wall estimated according to the Fourier's law based on the outer wall temperature measurements, and assuming both one-dimensional conduction and adiabatic external surface, and T_{sat} is the local saturation temperature of the refrigerant. To estimate this temperature, firstly, based on the heat flux and the measured temperature and pressure at the pre-heater inlet, the subcooled region length, the single-phase pressure drop over its length, and the saturation temperature at the beginning of the saturated region were calculated by solving simultaneously an equation of state relating p_{sat} and T_{sat} plus energy balance and single-phase pressure drop equations. The overall pressure drop over the saturated region was then determined by subtracting the single-phase pressure drop from the measured total pressure drop, Δp.

By addopting tthis procedure, boiling under subcooled conditions was assumed negligible. After that, a constant pressure drop gradient given by the ratio of the overall pressure drop over the saturated region and its length is assumed from the beginning of the saturated region until the end of the flow visualization section. Then, T_{sat} was calculated from the estimated local saturation pressure. The heat flux, q, was calculated as the ratio between the electrical power supplied to the test section and its internal area based on the heated length, where the electrical power was given by the product between the electrical current and voltage supplied by the DC power sources.

The perimeter-average heat transfer coefficient at each cross-section was calculated as follows:

$$\overline{h(z)} = \frac{h_{top}(z) + 2 \cdot h_{middle}(z) + h_{bottom}(z)}{4} \tag{3}$$

Vapor quality. The local vapor quality, $x(z)$ was obtained from energy balances over the pre-heater and the test section

2.3. Experimental Validation and Uncertainties

Compressible volume instabilities also termed in the literature by "explosive boiling" (as in [16]) are a common phenomenon in micro-scale flow boiling. These instabilities can promote severe pressure and temperature oscillations in the flow and seem to be related to some of the discrepancies observed when comparing experimental results from different authors (see Consolini *et al.* [17]). During the present experimental campaign, such instabilities were not observed and the fluctuations of the fluid temperature and pressure were within the uncertainty range of their measurements.

Single-phase flow experiments were performed in order to assure the accuracy of the estimated vapor quality and evaluate the effective rate of heat losses during single phase experiments, $(\Delta E/E)$, defined as follows:

$$\left(\Delta E/E\right) = \frac{\left(\left[\left(\pi D^2\right)/4\right] G\left(i_{out} - i_{in}\right)\right) - \left(P_1 + P_2\right)}{P_1 + P_2} \tag{4}$$

where i_{in} and i_{out} are the refrigerant enthalpies estimated at the pre-heater inlet and just downstream the visualization section, respectively.

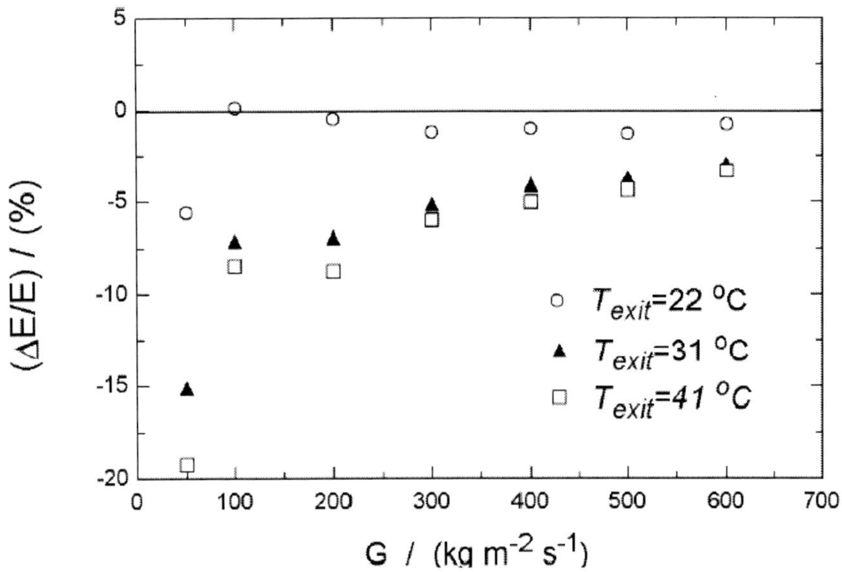

Figure 3. Evaluation of the effective rate of heat transferred to the single-phase flow refrigerant (T_{in} = 15 °C).

As one can seen in Figure 3, the heat losses decrease by increasing the mass velocity with an effective rate of heat transferred to the refrigerant lower than 10% for $G \geq 100$ kg m^{-2}s^{-1} and lower than 5% for $G \geq 400$ kg m^{-2}s^{-1}. Higher heat losses are observed for G=50 kg m^{-2}s^{-1}.

Table 1. Uncertainty of measured and calculated parameters

Parameter	Uncertainty	Parameter	Uncertainty
D	20 μm	P	4.5 kPa
L	1 mm	Δp	150 Pa
G	1.7 %	P_1, P_2	0.88 %
h	<30 %	T	0.15 °C
x	<5 %		

Temperature measurements were calibrated and the temperature uncertainty was evaluated according to the procedure suggested by Abernethy and Thompson [18]. Accounting for all instrument errors, uncertainties for the calculated parameter were estimated using the method of sequential perturbation according to Moffat [19]. The experimental uncertainties associated with the sensors and calculated parameters are listed in Table 1. The experimental heat transfer coefficient error was always less than 30% with an average absolute error of 8.5% calculated based on a total of 2508 experimental data points obtained in this study. In order to evaluate the accuracy of the measurements and validate the flow boiling results, single-phase heat transfer and pressure drop experiments were also performed and compared against the results provided by Gnielinski [20] and Petukhov [21] correlations, respectively. Figure 4 shows that the measured single-phase heat transfer coefficients agree reasonably well with those predicted by the correlation. Though not shown in this paper, good agreement was also obtained when comparing measured pressure drop against the predicted results.

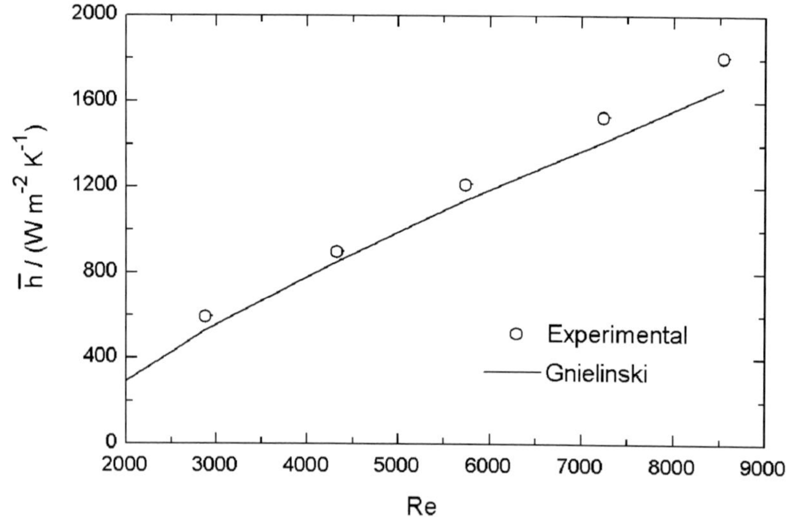

Figure 4. Comparison between the experimental single-phase heat transfer coefficient and the values predicted by Gnielinski [20] correlation (T_{exit}= 40°C).

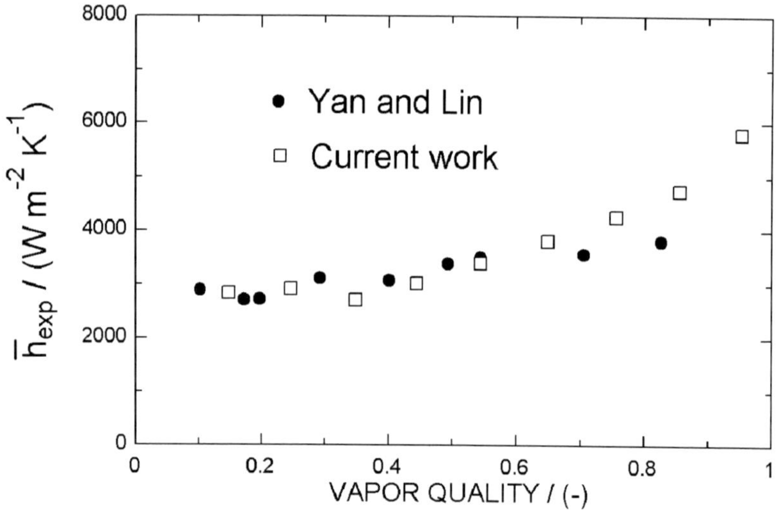

Figure 5. Comparison of the present flow boiling data against data from the literature [22] (G=200 kg m^2 s^{-1}, q=5 kW m^2 K^{-1}, T_{sat}=31°C)

To double-check the measurements performed in this study, a comparison against previous flow boiling data from the literature under conditions somewhat similar to those in the present work is displayed in Figure 5. In this figure, it is possible observe that the present data agree reasonably well with those by Yan and Lin [22]. These authors performed their experiments for R134a in a multi micro-channel test section composed of circular channels with internal diameters of 2.0 mm.

3. RESULTS

3.1. Heat Transfer Coefficient along the Tube Perimeter

Figure 6 shows the effect of the vapor quality on the heat transfer coefficient along the perimeter of the test section. Heat transfer coefficient measurements at the top, middle and bottom of the test tube for the thermocouples located 15 mm from the end of the heated region of the test section are displayed. Figure 6 also shows two-phase flow images corresponding to the heat transfer measurements. These images were obtained by high speed filming at the visualization section located just 50 mm downstream the heat transfer measurements. According to this figure, stratification effects are present and a thicker liquid film is observed in the tube's bottom region. This behavior affects the heat transfer coefficient as follows: (i) for vapor qualities up to 0.85, the heat transfer coefficient increases in the upward direction along the tube perimeter. This result is related to the lower thermal resistance in the upper part of the tube since the liquid film thickness decreases upwards as displayed in the two-phase flow images; (ii) for vapor qualities higher than 0.85, the heat transfer coefficient decreases in the upward direction. Such a behavior permits inferring the occurrence of a progressive surface dryout initiating in the upper part of the tube with increasing vapor quality. It is important to highlight that stratified flows were not detected during the current experimental campaign [23].

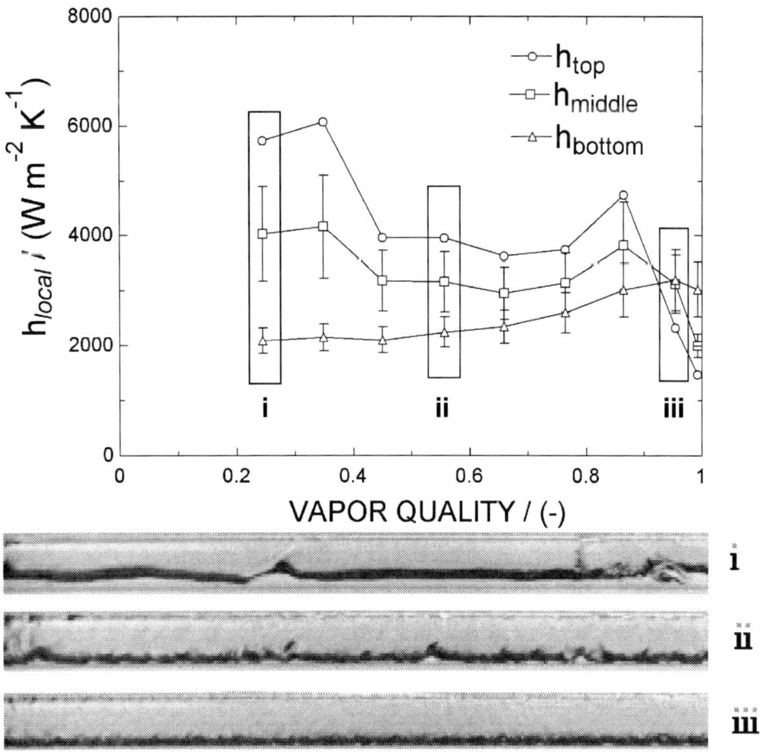

Figure 6. Heat transfer coefficient distribution along the tube perimeter and the corresponding two-phase flow images (G=100 kg m^{-2}s^{-1}, q=5kW m^{-2}, T_{sat}=22°C).

3.2. Parametric Analyses of the Experimental Results

In this item, a parametric investigation of the effect of the experimental variables on the heat transfer coefficient is presented. In this analysis, only perimeter-average heat transfer coefficients defined according to Eq. (3) are considered.

Mass velocity. Figure 7 illustrates the effect of the mass velocity on the heat transfer coefficient for different heat flux levels and a saturation temperature of 31°C. According to this figure, in general, the heat transfer coefficient increases with increasing mass velocity. This effect becomes more pronounced as the vapor quality increases. Distinct heat transfer behaviors with increasing vapor quality are observed according to a mass velocity threshold of 200 kg $m^{-2}s^{-1}$. For mass velocities higher than this threshold, the heat transfer coefficient increases with vapor quality until vapor quality values of up to 90%, whereas for mass velocities lower than 200 kg $m^{-2}s^{-1}$ the heat transfer coefficient presents a premature and smooth decrease with increasing vapor quality. The effects of mass velocity on the heat transfer behaviors are maintained independent of the heat flux level. A mass velocity threshold defining different heat transfer trends with vapor quality were also observed by Saitoh *et al.* [24] in their experimental data in a 3 mm ID tube. This threshold seems to be related to the addition of the following two mechanims: two-phase flow stratification and progressive dryout of the upper region of the tube. For lower mass velocities, according to Figure 6, the decrease of the heat transfer coefficient in the upper region of the tube, possibly due to the dryout effects, is much steeper than its increase in the lower region due to the decrease of the liquid film thickness.

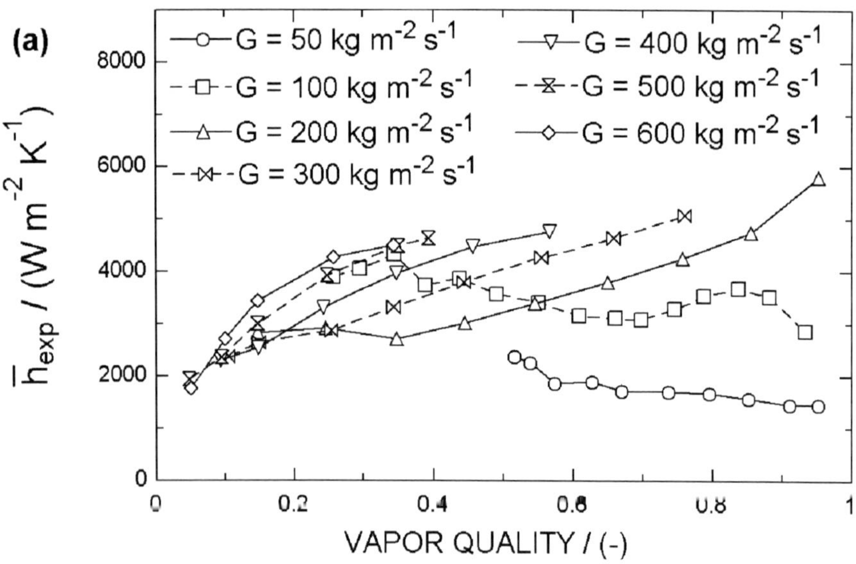

Figure 7a. Illustration of the effect of the mass velocity on the heat transfer coefficient for different heat fluxes and T_{sat}=31°C. (a) q=5 kW m^{-2}; (b) q=15 kW m^{-2}; (c) q=25kW m^{-2}.

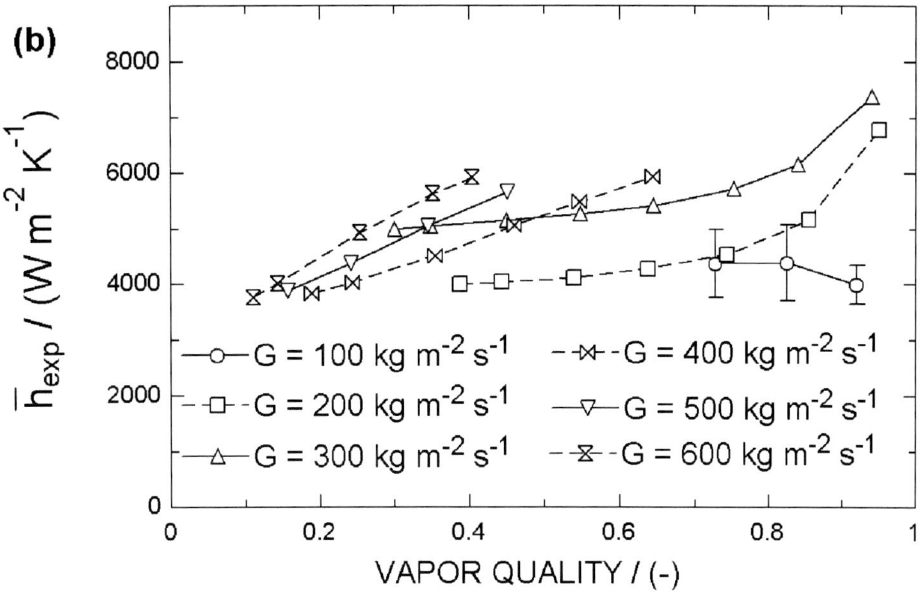

Figure 7b. Illustration of the effect of the mass velocity on the heat transfer coefficient for different hat fluxes and T_{sat}=31°C. (a) q=5 kW m^{-2}; (b) q=15 kW m^{-2}; (c) q=25kW m^{-2}.

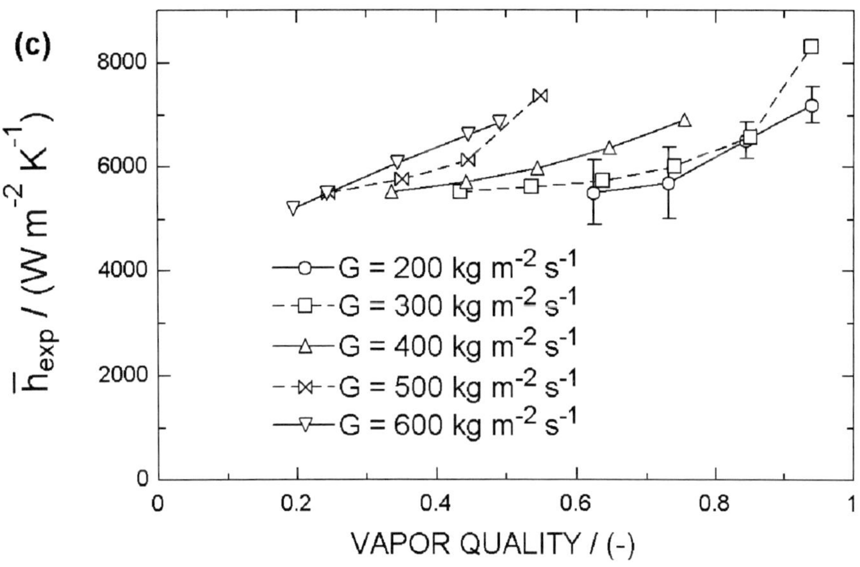

Figure 7c. Illustration of the effect of the mass velocity on the heat transfer coefficient for different hat fluxes and T_{sat}=31°C. (a) q=5 kW m^{-2}; (b) q=15 kW m^{-2}; (c) q=25kW m^{-2}.

These distinct heat transfer behaviors promote as final result the decrease of the perimeter average heat transfer coefficient with increasing vapor quality. The liquid film thickness uniformity along the tube perimeter increases with increasing the mass velocity and, so, the local heat transfer coefficient increases along all the tube perimeter with increasing vapor quality due to the reduction of the liquid film thickness.

Saturation temperature. Figure 8 displays the effects of the saturation temperature on the heat transfer coefficient for different experimental conditions.

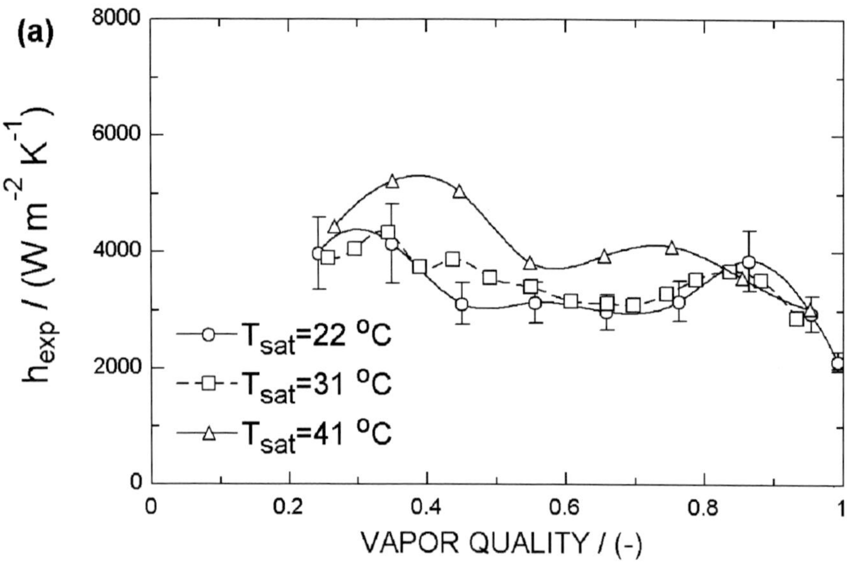

Figure 8a. Illustration of the effect of the saturation temperature on the heat transfer coefficient. (a) G=100 kg m^{-2}s^{-1}; q=5 kW m^{-2} (b) G=300 kg m^{-2}s^{-1}; q=15 kW m^{-2} (c) G=600 kg m^{-2}s^{-1}; q=15 kW m^{-2}.

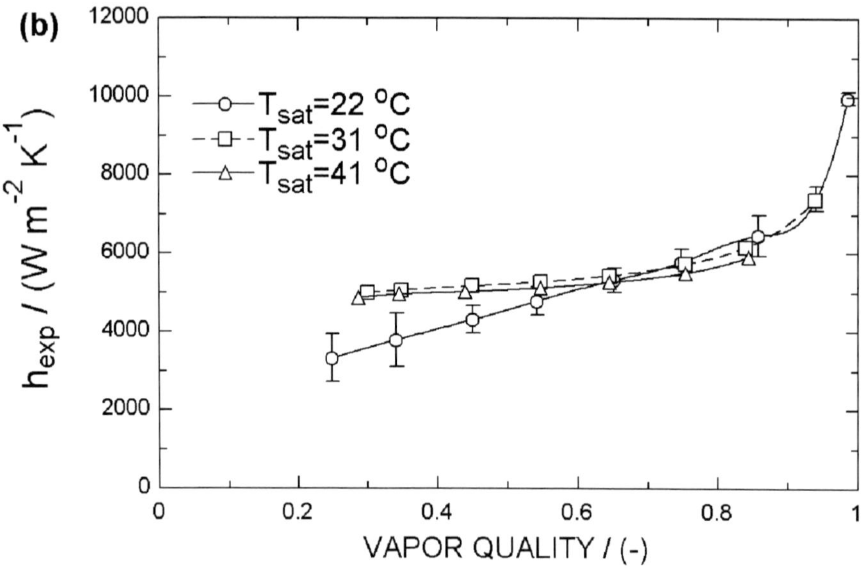

Figure 8b. Illustration of the effect of the saturation temperature on the heat transfer coefficient. (a) G=100 kg m^{-2}s^{-1}; q=5 kW m^{-2} (b) G=300 kg m^{-2}s^{-1}; q=15 kW m^{-2} (c) G=600 kg m^{-2}s^{-1}; q=15 kW m^{-2}.

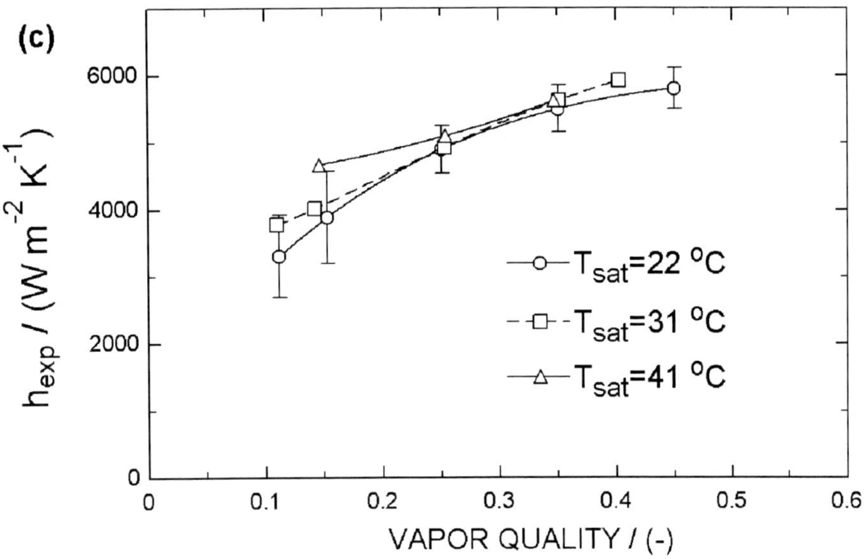

Figure 8c. Illustration of the effect of the saturation temperature on the heat transfer coefficient. (a) G=100 kg m^{-2}s^{-1}; q=5 kW m^{-2} (b) G=300 kg m^{-2}s^{-1}; q=15 kW m^{-2} (c) G=600 kg m^{-2}s^{-1}; q=15 kW m^{-2}.

It can be noted that, in general, the heat transfer coefficient increases with increasing saturation temperature. Such behavior is more pronounced at low vapor qualities and mass velocities and becomes almost negligible at high mass velocities and vapor qualities. These results are qualitatively in agreement with those data obtained for flow boiling in macro-scale channels. Thus, it is not surprising that macro-scale heat transfer prediction approaches, as the one proposed by Chen [2], are adopted for micro-scale flow boiling and that some of them seem to work satisfactorily under such conditions when compared against their own database.

Heat flux. According to Figure 9, except for some of the results at vapor qualities higher than 0.9, the heat transfer coefficient increases with heat flux independently of the mass velocity range. Such behavior differs from the one observed in macro-scale channels, according to which the heat transfer coefficient increases with increasing the heat flux, but only at low vapor qualities and mass velocities. Additionally, this behavior is related to the fact that nucleate boiling has been suggested as the dominant heat transfer mechanism in micro-scale channels.

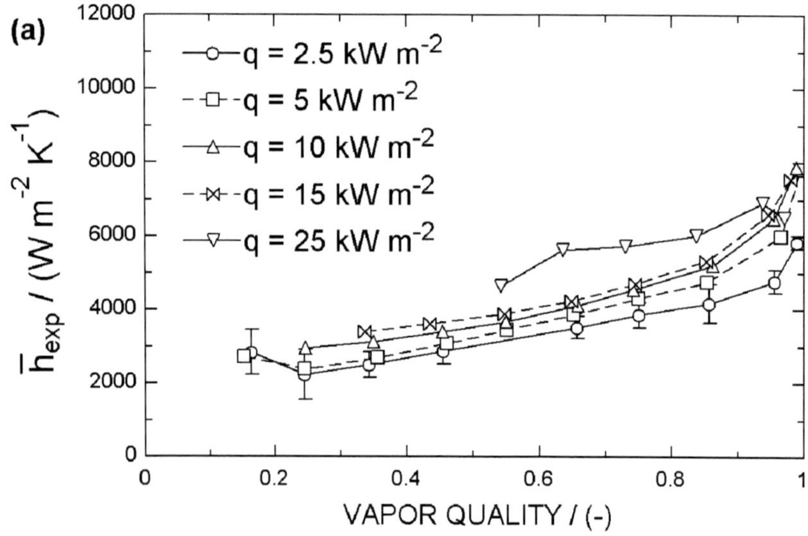

Figure 9a. Illustration of the effect of the heat flux on the heat transfer coefficient for T_{sat}=22°C. (a) G=200 kg m^{-2}s^{-1}; (b) G=400 kg m^{-2}s^{-1}; (c) G=600 kg m^{-2}s^{-1}.

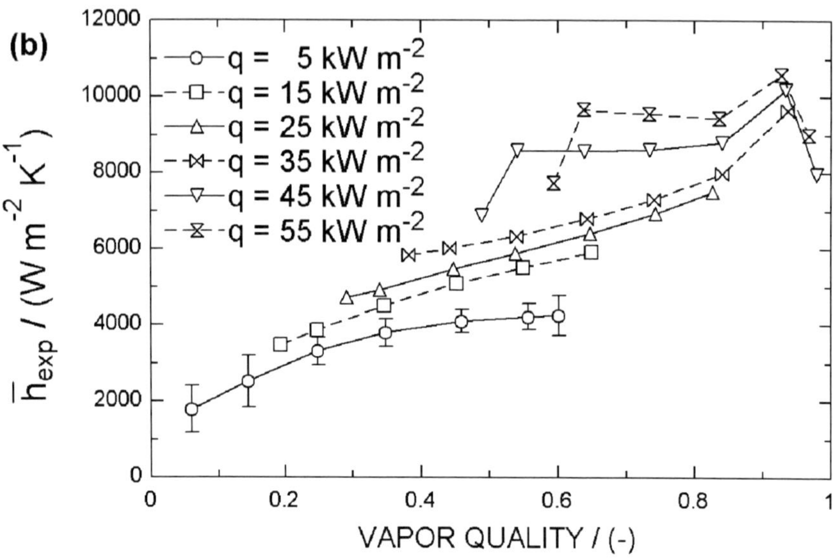

Figure 9b. Illustration of the effect of the heat flux on the heat transfer coefficient for T_{sat}=22°C. (a) G=200 kg m^{-2}s^{-1}; (b) G=400 kg m^{-2}s^{-1}; (c) G=600 kg m^{-2}s^{-1}.

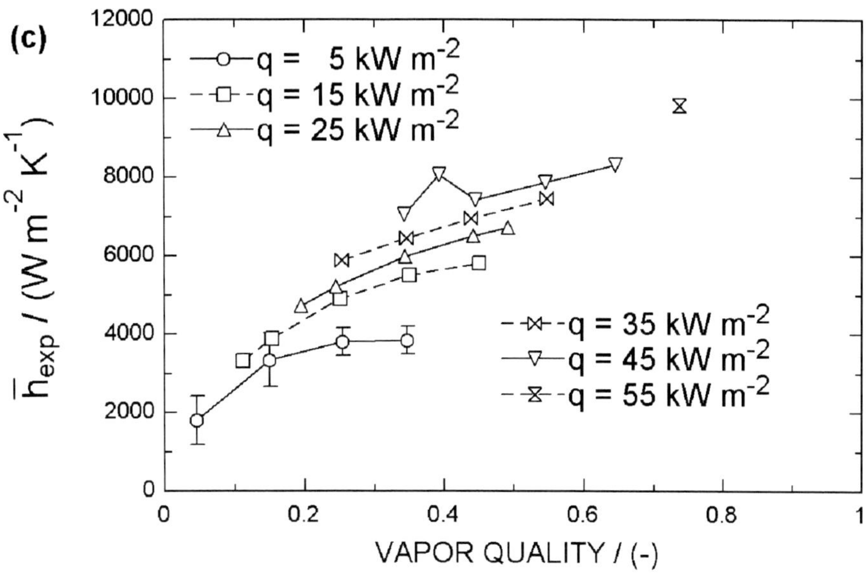

Figure 9c. Illustration of the effect of the heat flux on the heat transfer coefficient for T_{sat}=22°C. (a) G=200 kg m^{-2}s^{-1}; (b) G=400 kg m^{-2}s^{-1}; (c) G=600 kg m^{-2}s^{-1}.

This statement results from the misconception that an evaporation process dependent on the heat flux necessarily means that nucleate boiling is the controlling mechanism. Thome *et al.* [11], for elongated bubble flow, have shown that the evaporation of a thin liquid film can also result that the heat transfer coefficient increases with heat flux. According to Thome's model, the heat transfer coefficient increases with increasing heat flux is mainly related to the fact that the time average film thickness during the passage of an elongated bubble decreases with increasing heat flux.

3.3. Comparison of the Experimental Data and the Prediction Methods

In order to evaluate the capability of the current flow boiling prediction methods to predict the present database, a comparison between the experimental data and six flow boiling heat transfer prediction methods [3-7,11] was performed. This comparison includes both macro- and micro-scale methods and a total of 2508 experimental data points of which 627 have their flow pattern identified by high speed recording. The flow patterns were visually identified as elongated bubble, churn and annular flows. The predictive methods were evaluated according to two criteria: the fraction of data predicted to within ±30%, λ, and the mean absolute error, ε.

Table 2 depicts the statistical comparisons of the methods for the overall database and also comparisons for the data segregated according to each flow pattern. Bubbly flows were not considered since they were observed only at calculated thermodynamic equilibrium vapor quality just above zero and, under this condition, saturated flow boiling conditions were not assured due to the vapor quality uncertainties. According to Table 2, Saitoh *et al.* [6] method worked the best when compared against the overall database.

Table 2. Comparison between heat transfer prediction methods and the experimental database

Flow pattern (n° of data points)		PREDICTIVE METHOD					
		Zhang et al. (2004)	Thome et al. (2004)	Kandlikar and Balasubramanian (2004)	Liu and Winterton (1991)	Saitoh et al. (2007)	Bertscha et al. (2008)
Elongated bubbles (26)	ε	46.9	51.1	62.9	50.9	46.3	45.4
	λ	23.1	30.8	23.1	26.8	26.9	26.9
Churn (63)	ε	36.3	36.5	62.0	38.3	36.4	38.2
	λ	69.8	46.0	19.0	61.9	71.4	30.2
Annular (538)	ε	24.5	29.1	39.9	17.9	21.8	44.1
	λ	77.3	58.6	37.9	81.8	78.1	27.1
Overall database (2508)	ε	22.6	32.5	43.1	22.7	22.0	39.7
	λ	75.8	51.1	35.6	73.3	77.4	31.2

This result may be related to the fact that the empirical constants in their method were adjusted based solely on R134a data and for internal tube diameters ranging from 0.5 mm to 11 mm, coinciding with the experimental conditions of the present study. Saitoh *et al.* [6] method is also ranked as the best to predict flow boiling heat transfer under churn flow conditions and Liu and Winterton [3] the best for annular flow. Thome *et al.* [11] and Bertsch *et al.* [7] provided the best heat transfer predictions for churn and elongated bubble flows.

A good predictive method should be not only statistically accurate, but also able of capturing the main trends of the experimental results. Taking this into account, Figure 10 displays the evolution of the heat transfer coefficients vs. vapor quality in comparison to the heat transfer predictive methods for different experimental conditions.

According to the experimental results, the heat transfer peak at low vapor qualities indicated by the 3-Zone model [11] seems to occur only in Figure 10d displaying the data obtained for the highest heat flux and mass velocity. High experimental heat transfer coefficients at lower vapor qualities are also displayed in Figure 10b for G=50 kg m^{-2}s^{-1}, however, the heat transfer coefficient falloff with increasing mass velocity is much steeper than the one indicated by the 3-Zone model. Zhang *et al.* [5] methods predicts reasonably well the increase in the heat transfer coefficient with vapor quality at high vapor qualities displayed in Figures 10b and 10c.

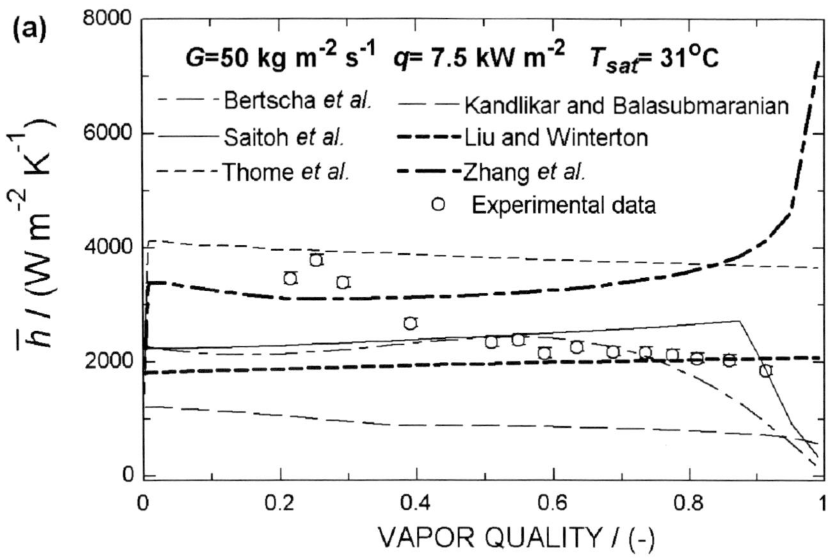

Figure 10a. Heat transfer coefficient evolution with vapor quality according to the experimental data and predictive methods. (a)G=50 kg m^{-2}s^{-1}; q=7.5 kW m^{-2}; T_{sat}=31 °C; (b)G=200 kg m^{-2}s^{-1}; q=15 kW m^{-2};T_{sat}=22 °C (c)G=400 kg m^{-2}s^{-1};q=35 kW m^{-2}; T_{sat}=22 °C; (d) G=600 kg m^{-2}s^{-1} q=45 kW m^{-2}; T_{sat}=41 °C.

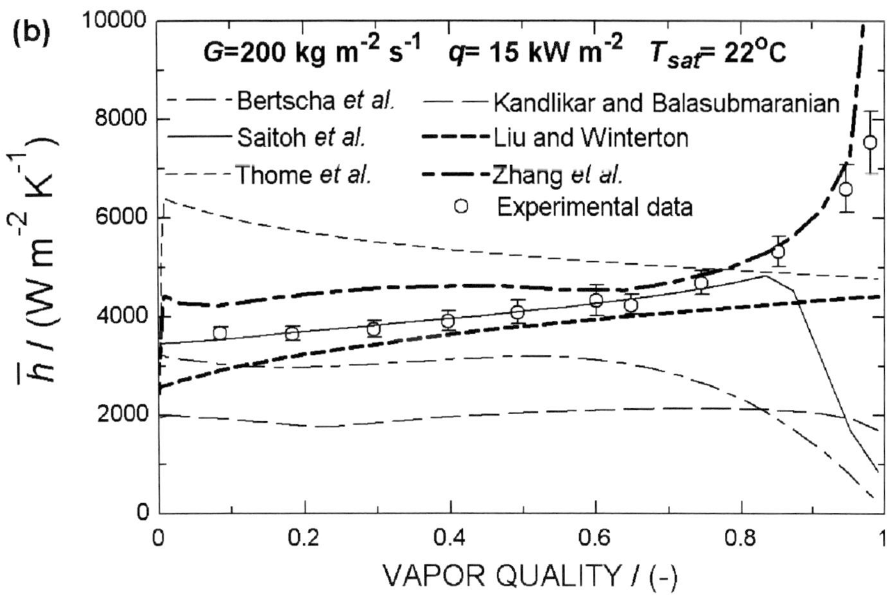

Figure 10b. Heat transfer coefficient evolution with vapor quality according to the experimental data and predictive methods. (a)G=50 kg m^{-2}s^{-1}; q=7.5 kW m^{-2}; T_{sat}=31 °C; (b)G=200 kg m^{-2}s^{-1}; q=15 kW m^{-2};T_{sat}=22 °C (c)G=400 kg m^{-2}s^{-1};q=35 kW m^{-2}; T_{sat}=22 °C; (d) G=600 kg m^{-2}s^{-1} q=45 kW m^{-2}; T_{sat}=41 °C.

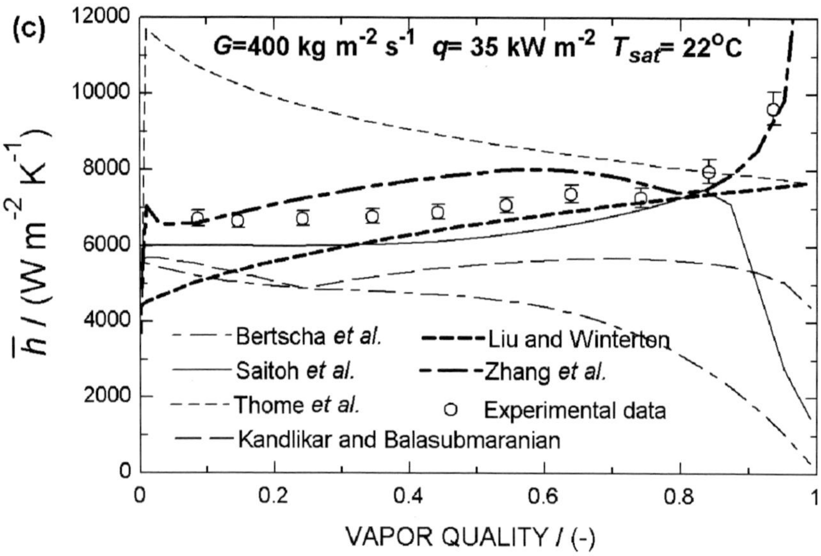

Figure 10c. Heat transfer coefficient evolution with vapor quality according to the experimental data and predictive methods. (a)G=50 kg m^{-2}s^{-1}; q=7.5 kW m^{-2}; T_{sat}=31 °C; (b)G=200 kg m^{-2}s^{-1}; q=15 kW m^{-2};T_{sat}=22 °C (c)G=400 kg m^{-2}s^{-1};q=35 kW m^{-2}; T_{sat}=22 °C; (d) G=600 kg m^{-2}s^{-1} q=45 kW m^{-2}; T_{sat}=41 °C.

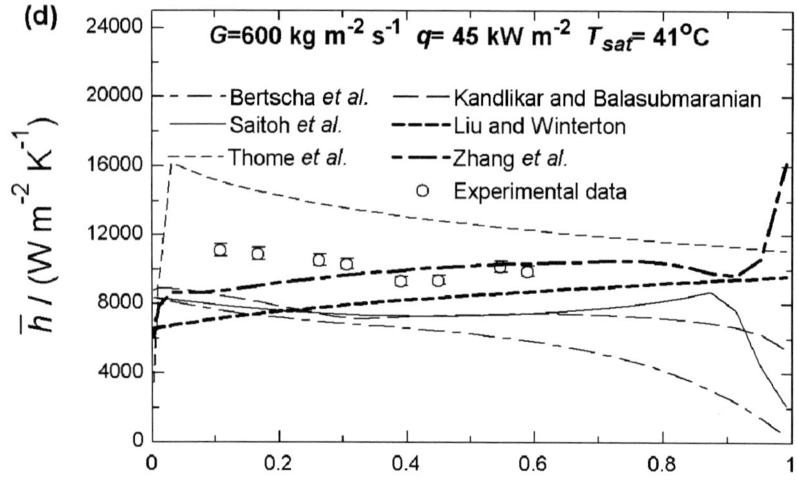

Figure 10d. Heat transfer coefficient evolution with vapor quality according to the experimental data and predictive methods. (a)G=50 kg m^{-2}s^{-1}; q=7.5 kW m^{-2}; T_{sat}=31 °C; (b)G=200 kg m^{-2}s^{-1}; q=15 kW m^{-2};T_{sat}=22 °C (c)G=400 kg m^{-2}s^{-1};q=35 kW m^{-2}, T_{sat}=22 °C; (d) G=600 kg m^{-2}s^{-1} q=45 kW m^{-2}; T_{sat}=41 °C.

CONCLUSIONS

New accurate R134a flow boiling heat transfer data in a 2.32 mm circular tube were obtained and presented. The experimental data were parametrically analyzed and compared against predictive methods. The conclusions can be summarized as:

1. Flow stratification was visualized and local wall temperature measurements revealed its effect on the circumferential variation of the heat transfer coefficient along the tube perimeter.
2. Generally speaking, the heat transfer coefficient increases with increasing heat flux, mass velocity and saturation temperature. Distinct heat transfer behaviors with increasing vapor quality are observed according to a mass velocity threshold of 200 kg $m^{-2}s^{-1}$.
3. The predictive method proposed by Saitoh *et al.* [6] worked the best predicting the overall database with an average error of 22.0 %. Zhang *et al.* [5] accurately captured the fact that under certain heat flux and mass velocity conditions the heat transfer coefficient increases with increasing vapor quality at high vapor qualities.

ACKNOWLEDGMENTS

The authors gratefully acknowledge the financial support under contract numbers 05/60031-0, 06/52089-1 and 07/53950-5 given by FAPESP (The State of São Paulo Research Foundation, Brazil). The technical support given to this investigation by Mr. José Roberto Bogni is also appreciated and deeply recognized.

REFERENCES

[1] W. Lowdermilk, C. Lanzo, and L, Siegel, Investigation of boiling burnout and flow stability for water flowing in tubes, *NACA*. TN 4382, 1958.

[2] J.C. Chen, Correlation for boiling heat transfer to saturated fluids in convective flow, I*&EC Process Des. Dev* 5 (1966) 322–329.

[3] Z. Liu and R. H. S. Winterton, A general correlation for saturated and subcooled flow boiling in tubes and annuli based on a nucleate pool boiling equation, *Int. J. Heat Mass Transfer* 34 (1991) 2759–2766.

[4] S.G. Kandlikar and P. Balasubramanian, An extension of the flow boiling correlation to transition, laminar, and deep laminar flows in minichannels and microchannels, *Heat Transfer Engineering* 25 (2004) 86–93.

[5] W. Zhang, T. Hibiki and K. Mishima, Correlation for flow boiling heat transfer in mini-channels, *Int. J. Heat and Mass Transfer* 47 (2004) 5749– 5763.

[6] S. Saitoh, H. Daiguji and H. Hihara, Correlation for boiling heat transfer of R-134a in horizontal tubes including effect of tube diameter, *Int. J. Heat Mass Transfer* 50 (2007) 5215–5225.

[7] S.S. Bertsch, E.A.Groll, and S.V. Garimella, A composite heat transfer correlation for saturated flow boiling in small channels, *Int. J. Heat Mass Transfer* 52 (2008) 2110-2118.

[8] H.K. Foster and N. Zuber, Bubble dynamics and boiling heat transfer, *AIChe Journal 1* (1955) 531-535.

[9] F.W. Dittus and L.M.K. Boelter, *Heat transfer in automobile radiators of the tubular type*, University of California Publications in engineering 2 (1930) 443-461.

[10] S.G. Kandlikar, A model for predicting the two-phase flow boiling heat transfer coefficient in augmented tube and compact heat exchanger geometries, *J. Heat Transfer* 113 (1991) 966–972.

[11] J.R. Thome, V. Dupont and A.M. Jacobi, Heat transfer model for evaporation in microchannels, Part I: Presentation of the model, *Int. J. Heat Mass Transfer* 47 (2004) 3375–3385.

[12] V. Dupont, J.R. Thome, A.M. Jacobi, Heat transfer model for evaporation in microchannels. Part II: Comparison with the database, *Int. J. Heat Mass Transfer* 47 (2004) 3387–3401.

[13] C.B. Tibiriçá and G. Ribatski, Análise de dados experimentais e métodos para previsão do coeficiente de transferência de calor em micro-canais, in: *Proceedings of the 1st Brazilian Meeting on Boiling, Condensation and Multiphase Flow*, Florianópolis, Brazil, paper MF-103, 2008.

[14] L. Consolini, Convective boiling heat transfer in a single micro-channel. PhD thesis, Swiss Federal Institute of Technology Lausanne (EPFL), Switzerland, 2008.

[15] G. Ribatski, W. Zhang, L. Consolini; J. Xu, J. R. Thome. On the prediction of heat transfer in micro-scale flow boiling, *Heat Transfer Engineering* 28 (2007) 842-851.

[16] G. Hetsroni, A. Mosyak, E. Pogrebnyak, and Z. Segal, Explosive boiling of water in parallel micro-channels, *Int. J. Multiphase Flow* 31 (2005) 371-392.

[17] L. Consolini, G. Ribatski, Z. Wei, J. Xu, and J.R. Thome, Heat transfer in confined forced flow-boiling, *Heat Transfer Engineering* 28 (2007) 826-833.

[18] R.B. Abernethy and J.W. Thompson, Handbook uncertainty in gas turbine measurements, Arnold Engineering Development Center, Arnold Air Force Station, Tennessee, 1973.

[19] R.J. Moffat, Describing the Uncertainties in Experimental Results, *Exp. Thermal Fluid Science* 1 (1988) 3-17.

[20] V. Gnielinski, New equations for heat and mass transfer in turbulent pipe and channel from, *Int. Chem. Engineering* 16 (1976) 359-368.

[21] B. S. Petukhov, in T. F. Irvine and J. P. Hartnett, Eds., *Advances in Heat Transfer*, vol. 6, Academic Press, New York, 1970.

[22] Y.-Y. Yan, T.-F. Lin, Evaporation heat transfer and pressure drop of refrigerant R-134a in a small pipe, *Int. J. Heat Mass Transfer* 41 (1998) 4183–4194.

[23] A.A. Arcanjo, J.O. Freitas, C.B. Tibiriça and G. Ribatski, Two-phase flow characteristics during flow boiling of halocarbon refrigerants in micro-scale channels, *ECI International Conference on Boiling Heat Transfer*, Florianópolis, Brazil, May 2009.

[24] S. Saitoh, H. Daiguji, E. Hihara, Effect of tube diameter on boiling heat transfer of R-134a in horizontal small-diameter tubes, *Int. J. Heat Mass Transfer* 48 (2005) 4973–4984.

In: Frontier Research in Microscale and Nanoscale... ISBN 978-1-61324-665-8
Editor: Lixin Cheng

Chapter 4

INTERFACIAL-AREA TRANSPORT OF VERTICAL UPWARD BUBBLY FLOW IN MINI PIPES

***T. Hazuku*[1,*]*, T. Takamasa*[1] *and T. Hibiki*[2]**
[1]Faculty of Marine Technology,
Tokyo University of Marine Science and Technology,
2-1-6 Etchujima, Koto, Tokyo 135-8533, Japan
[2]School of Nuclear Engineering, Purdue University, 400 Central Drive,
West Lafayette, IN 47907-2017, U. S.

ABSTRACT

In relation to the development of the interfacial-area transport equation, this study collected data sets on the axial development of area-averaged (one-dimensional) void fraction and interfacial-area concentration in vertical upward-developing bubbly flows using three mini pipes. The test pipe inner diameters were 0.55, 0.79 and 1.02 mm, and the lengths were 500 mm. Image processing was used to characterize the flow at five axial locations, z/D = 15.0, 75.0, 150, 250 and 450, where z and D are the axial distance from the inlet and inner pipe diameter, respectively. In the experiment, the superficial liquid velocity and the void fraction ranged from 0.395 m s^{-1} to 4.89 m s^{-1} and from 0.958 % to 28.6 %, respectively. The one-dimensional interfacial-area transport equation with a sink term due to wake entrainment was evaluated by using the newly obtained datasets. It was confirmed that the interfacial-area transport equation could reproduce the proper trend of the axial interfacial-area transport and predict the measured interfacial-area concentrations well. It was found that the present model shows promise for predicting the interfacial-area transport of mini-pipe bubbly flows in the test conditions of this study.

Keywords: interfacial-area transport, two-fluid model, wake-entrainment, gas-liquid bubbly flow, multiphase flow, microchannel, minichannel.

---

* Corresponding author: Tel.: +81 3 5245 7727, fax: +81 3 5245 7410. E-mail address: hazuku@kaiyodai.ac.jp

NOMENCLATURE

a_i	interfacial-area concentration, m^{-1}
C_D	drag coefficient
D	pipe diameter, m
f	friction factor
g	gravitational acceleration, m s^{-2}
j	mixture volumetric flux, m s^{-1}
j_f	superficial liquid velocity, m s^{-1}
j_g	superficial gas velocity, m s^{-1}
M_F	frictional pressure gradient in the multi-particle system, Pa m^{-1}
N_{Ref}	liquid Reynolds number
p	pressure, Pa
r_b	bubble radius, m
t	time, s
v	velocity, m s^{-1}
v_i	interfacial velocity, m s^{-1}
v_r	relative velocity between phases, m s^{-1}
z	axial position, m

Greek Symbols

α	void fraction
ε	energy dissipation rate per unit mass, m^2 s^{-3}
ϕ_B	rates of change of bubble number density due to bubble breakup, m^{-3} s^{-1}
ϕ_C	rates of change of bubble number density due to bubble coalescence, m^{-3} s^{-1}
ϕ_P	rates of change of bubble number density due to phase change, m^{-3} s^{-1}
Φ_B	rates of change of interfacial-area concentration due to bubble breakup, m^{-1} s^{-1}
Φ_C	rates of change of interfacial-area concentration due to bubble coalescence, m^{-1} s^{-1}
Φ_P	rates of change of interfacial-area concentration due to phase change, m^{-1} s^{-1}

Φ_V	rates of change of interfacial-area concentration due to void transport, $m^{-1} s^{-1}$
Φ_{WE}	rates of change of interfacial-area concentration due to wake entrainment, $m^{-1} s^{-1}$
Γ_{WE}	adjustable coefficient
ν	kinematic viscosity, $m^2 s^{-1}$
ρ	density, $kg\ m^{-3}$
ψ	factor depending on the bubble shape

Subscripts

f	liquid phase
g	gas phase
m	mixture

Mathematical Symbols

$\langle\ \rangle$	area-averaged quantity
$\langle\langle\ \rangle\rangle$	void fraction-weighted-averaged quantity
$\langle\langle\ \rangle\rangle_a$	interfacial-area concentration-weighted-averaged quantity

1. Introduction

The development of high-power electronic devices and compact heat exchangers has made it important to understand the characteristics of gas-liquid two-phase flow in a minichannel. In a minichannel, the effect of axial frictional pressure loss is significant and the force of surface tension often dominates the force of gravity. For example, the rising velocity of bubbles in stagnant water becomes zero in a capillary tube [1]. Thus, it is anticipated that the characteristics of gas-liquid two-phase flow differ from those in a conventional pipe with a larger inner diameter, and that this fact significantly affects boiling heat transfer. The investigation of minichannel flow has been driven both by its engineering importance and by scientific curiosity. Reliable and extensive databases and correlations of various flow parameters including void fraction, frictional pressure loss, heat transfer coefficient and critical heat flux are being developed [2-16].

One of the interesting flow characteristics in minichannels is a continuous developing flow. The flow structure of two-phase flow in minichannels changes continuously with flow development in the axial direction because the axial pressure changes are very large compared with those in conventional-size pipes. Significant axial pressure loss along the flow direction acts as flow-induced body acceleration and leads to rapid axial changes in the two-phase flow structure. Thus, we need to construct precise databases on local two-phase flow parameters in the axial direction and develop proper mechanistic models for the axial change of two-phase flow parameters in minichannels. However, most existing databases for minichannels have been obtained from flow measurements performed at a fixed axial location.

On the other hand, practical thermal-hydraulic phenomena are often dominated by interfacial transport. Lack of proper mechanistic models for the interfacial structure and interfacial transfer processes leads to inaccurate predictions of these phenomena, and thus becomes a major concern in current two-phase flow modeling practice. In view of great importance to two-fluid model formulation, local measurements of interfacial-area concentration have been performed in a bubbly flow in conventional-size pipes over the past 15 years. Some correlations have been also proposed to predict volume averaged interfacial-area concentration (for example, [17-20]). However, because of considerable difficulties in terms of measurements and modeling, reliable and accurate closure relations for the interfacial transfer terms are not fully developed.

In the present thermal-hydraulic system analysis codes, the interfacial-area concentration is given by the correlation. The following shortcomings are caused by the correlations based on traditional two-phase flow regimes and regime transition criteria [21]:

(1) The flow regime transition criteria are algebraic relations for steady-state, fully developed flows. They do not fully reflect the true dynamic nature of changes in the interfacial structure. Hence, the effects of the entrance and developing flow cannot be taken into account correctly, nor the gradual transition between regimes.
(2) The method based on the flow regime transition criteria is a two-step method, which requires flow configuration transition criteria and interfacial-area correlations for each flow configuration. The compound errors from the transition criteria and interfacial-area correlations can be very significant.
(3) The transition criteria and flow-regime-dependent interfacial correlations are valid in limited parameter ranges for certain specific operational conditions and geometries. Most of them are obtained from simple experiments and phenomenological models. Often the scale effects of geometry and fluid properties are not correctly taken into account. When applied to high-to-low-pressure steam-water transients, these models may cause significant discrepancies, artificial discontinuities and numerical instability.

In minichannels, the flow changes continuously with flow development in the axial direction due to significant axial pressure loss. Thus, the above shortcomings of the correlation would be more enhanced in the prediction of interfacial-area concentration in minichannels.

To solve such problems, the introduction of the interfacial-area transport equation has been recommended [22, 23]. The interfacial-area transport equation can be obtained by considering the fluid particle number density transport equation analogous to Boltzmann's transport equation. It can replace the traditional flow regime maps and regime transition criteria. Changes in the two-phase flow structure can be predicted mechanistically by introducing the interfacial-area transport equation. Thus, a successful development of the interfacial-area transport equation can provide a significant improvement in the two-fluid model formulation.

In the first stage of the development of the interfacial-area transport equation, flows in conventional-size pipes were the focus, and the equation was developed successfully for conventional-size pipes by modeling the sink and source terms of the interfacial-area concentration due to bubble coalescence and breakup [24, 25]. In order to extend this success to various pipe sizes, extensive efforts have been made to gather data for relatively large and small diameter pipes [26-29]. However, almost no data of the interfacial-area concentration in minichannels are available in spite of its significance in two-phase flow formulation. This is mainly due to the difficulty to measure the interfacial-area concentration in minichannels with a non-intrusive method.

In this context, the present study aims to measure the axial development of the void fraction and the interfacial-area concentration of adiabatic bubbly flows in mini pipes with pipe sizes of 0.55, 0.79 and 1.02 mm. Image processing was used to characterize the flow at five axial locations, z/D = 15.0, 75.0, 150, 250 and 450, where z and D are the axial distance from the inlet and inner pipe diameter, respectively. A detailed discussion of the axial development of the flow parameters is provided. Although considerable effort has been expended in the past to study the characteristics of two-phase flow in minichannels, very few detailed measurements have been made of the phenomena occurring in a developing flow in minichannels, despite the fact that these are considered important for industrial design and useful for the improvement of existing analytical models and empirical correlations to aid in predicting flows. As the first step in conducting an analytical study of interfacial-area transport in minichannels, the interfacial-area transport equation [30] will be examined in the light of the data obtained from mini pipes.

2. Interfacial-Area Transport Equation in Relatively Small Diameter Pipe

The interfacial-area transport equation can be deduced by considering the fluid particle number density transport equation, analogous to Boltzmann's transport equation [31]. By applying cross-sectional area-averaging, the one-dimensional interfacial-area transport equation becomes:

$$\frac{\partial\langle a_i\rangle}{\partial t}+\frac{d}{dz}\left(\langle a_i\rangle\langle\langle v_i\rangle\rangle_a\right)=\frac{1}{3\psi}\left(\frac{\langle\alpha\rangle}{\langle a_i\rangle}\right)^2\left(\phi_B-\phi_C+\phi_P\right)+\frac{2}{3}\left(\frac{\langle a_i\rangle}{\langle\alpha\rangle}\right)\left\{\frac{\partial\langle\alpha\rangle}{\partial t}+\frac{d}{dz}\left(\langle\alpha\rangle\langle\langle v_g\rangle\rangle\right)\right\}$$
$$=\Phi_B-\Phi_C+\Phi_P+\Phi_V$$

where

$$\Phi_B\equiv\frac{1}{3\psi}\left(\frac{\langle\alpha\rangle}{\langle a_i\rangle}\right)^2\phi_B,\quad \Phi_C\equiv\frac{1}{3\psi}\left(\frac{\langle\alpha\rangle}{\langle a_i\rangle}\right)^2\phi_C,\quad \Phi_P\equiv\frac{1}{3\psi}\left(\frac{\langle\alpha\rangle}{\langle a_i\rangle}\right)^2\phi_P$$

and

$$\Phi_{\mathrm{V}} \equiv \frac{2}{3}\left(\frac{\langle a_{\mathrm{i}}\rangle}{\langle\alpha\rangle}\right)\left\{\frac{\partial\langle\alpha\rangle}{\partial t}+\frac{d}{dz}\left(\langle\alpha\rangle\langle\langle v_{\mathrm{g}}\rangle\rangle\right)\right\} \tag{1}$$

The symbols ai, t, vi, vg, ψ, and □ denote the interfacial-area concentration, time, interfacial velocity, gas velocity, a factor related to the bubble shape (ψ = 1/(36□) for spherical bubbles), and void fraction, respectively. ϕB, ϕC and ϕP are the rates of change of bubble number density due to bubble breakup, bubble coalescence, and phase change, respectively. □B, □C, □P and □V are the rates of change of interfacial-area concentration due to bubble breakup, bubble coalescence, phase change, and void transport, respectively. $\langle\ \rangle$, $\langle\langle\ \rangle\rangle_{a}$, and $\langle\langle\ \rangle\rangle$ are, respectively, area-averaged, interfacial-area concentration-weighted-averaged, and void-fraction-weighted-averaged quantities. When there is no phase change, ϕP and □P become zero. The sink and source terms of the interfacial-area concentration, □B and □C should be modeled mechanistically based on possible bubble coalescence and breakup mechanisms.

In the previous study [25], the major mechanism for the interfacial-area transport of adiabatic bubbly flow in medium-size pipes with inner diameters of 25.4 and 50.8 mm was modeled successfully by taking into account the bubble coalescence due to random bubble collisions driven by liquid turbulence and the bubble breakup due to the impact of turbulent eddies. However, the dominant mechanism of bubble coalescence in a small-diameter pipe with an inner diameter of 9.0 mm was found by visual observation to be wake entrainment [32]. The difference in the bubble coalescence mechanisms in medium and small pipes was attributed to the relatively high bubble size-to-pipe diameter ratio and the restricted radial movement of bubbles in a small-diameter pipe, which might not allow random bubble collisions. On the other hand, visual observation suggested that for low liquid velocities, bubble breakup could be considered negligible because of weak turbulence [32]. Thus, the mechanism of interfacial-area transport in a small-diameter pipes with low liquid velocities could be modeled successfully by taking into account the bubble coalescence due to wake entrainment [32].

When bubbles enter the wake region of a leading bubble, they will accelerate and may collide with the leading bubble (see Figure 1). Based on this mechanism, the one-dimensional form of the bubble wake-entrainment term Φ_{WE} is formulated by [32, 33]

$$\left\langle\Phi_{\mathrm{WE}}\right\rangle = \Gamma_{\mathrm{WE}} C_{\mathrm{D}}^{1/3}\left\langle a_{\mathrm{i}}\right\rangle^{2}\left\langle v_{\mathrm{r}}\right\rangle \exp\left(-\frac{K_{\mathrm{C}}\rho_{\mathrm{f}}^{1/2}\left\langle D_{\mathrm{b}}\right\rangle^{5/6}\left\langle\varepsilon\right\rangle^{1/3}}{\sigma^{1/2}}\right) \tag{2}$$

where Γ_{WE}, C_{D}, and v_{r} are, respectively, the adjustable coefficient, drag coefficient, and relative velocity between phases.

In the bubbly-flow regime, it can be approximated that $\langle\langle v_g\rangle\rangle \approx \langle\langle v_i\rangle\rangle_a$ [25, 34]. For $D_b / D \geq 1/3$, the steady one-dimensional interfacial-area transport equation at low liquid velocity for mini pipes is expressed as

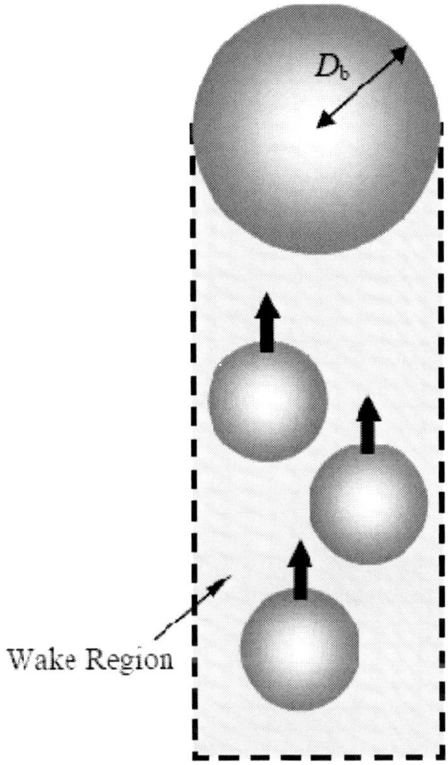

Figure 1. Schematic diagram of wake-entrainment model [32].

$$\frac{d}{dz}\left(\langle a_i\rangle\langle\langle v_g\rangle\rangle\right)=\frac{2}{3}\left(\frac{\langle a_i\rangle}{\langle\alpha\rangle}\right)\frac{d}{dz}\left(\langle\alpha\rangle\langle\langle v_g\rangle\rangle\right)-\langle\Phi_{WE}\rangle \tag{3}$$

The wake-entrainment term can be evaluated solely by data obtained under conditions such as low liquid velocity and $D_b/D \geq 1/3$.

In the previous study, the interfacial-area transport equation was developed, taking the effect of gravity into account, and the constitutive equation for the sink term of the interfacial-area concentration due to wake entrainment was freshly formulated by considering the body acceleration due to frictional pressure loss [30]. In what follows, the results from the earlier paper are summarized.

Equation (2) is a general form of the wake-entrainment term. In our previous work [35], we formulated the drag coefficient in a confined channel by taking into account the effect of a frictional pressure gradient due to liquid flow:

$$C_D = \frac{8}{3}\frac{\langle r_b \rangle}{\rho_f \langle v_r \rangle^2}\left\{\Delta\rho g(1-\langle\alpha\rangle) + M_F\right\} = \frac{8\left\{\Delta\rho g(1-\langle\alpha\rangle) + M_F\right\}\langle\alpha\rangle}{\rho_f \langle v_r \rangle^2 \langle a_i \rangle} \tag{4}$$

where $\langle r_b \rangle$ and g are, respectively, the area-averaged bubble radius expressed by $3\langle\alpha\rangle/\langle a_i \rangle$ and gravitational acceleration. The two-phase frictional pressure gradient M_F is given by

$$M_F = \left(-\frac{dp}{dz}\right)_F \tag{5}$$

where p is pressure. The validity of Eq. (4) has been verified by comparing the drift-flux model incorporated in Eq. (4) with experimental data obtained in microgravity conditions [36].

Substituting Eq. (4) into Eq. (2) yields

$$\langle \Phi_{WE} \rangle = \Gamma_{WE}\left[\frac{8\left\{\Delta\rho g(1-\langle\alpha\rangle) + M_F\right\}}{\rho_f}\right]^{1/3} \langle\alpha\rangle^{1/3}\langle a_i \rangle^{5/3}\langle v_r \rangle^{1/3} \times \exp\left(-\frac{K_C \rho_f^{1/2}\langle D_b \rangle^{5/6}\langle\varepsilon\rangle^{1/3}}{\sigma^{1/2}}\right) \tag{6}$$

An expression for the energy dissipation rate per unit mass can be obtained from the mechanical energy balance on the assumption that the dissipation of turbulent energy in the flow is equal to its production [37]. The expression can be extended by considering the effect of a frictional pressure gradient due to a liquid flow to be

$$\langle\varepsilon\rangle = \left\{g + \frac{M_F}{\Delta\rho(1-\langle\alpha\rangle)}\right\}\left|\langle j_g \rangle\right| \exp\left(-0.0005839 N_{Re_f}\right) + \frac{\left|\langle j \rangle\right|}{\rho_m} M_F \left\{1 - \exp\left(-0.0005839 N_{Re_f}\right)\right\} \tag{7}$$

where j_g and j are, respectively, the superficial gas velocity and mixture volumetric flux. Mixture density ρ_m and liquid Reynolds number N_{Ref} are defined by Eqs. (8) and (9), respectively:

$$\rho_m = \rho_g\langle\alpha\rangle + \rho_f(1-\langle\alpha\rangle) \tag{8}$$

$$N_{Re_f} \equiv \frac{\langle j_f \rangle D}{\nu_f} \tag{9}$$

where ρ_g, j_f, and ν_f are, respectively, the gas density, superficial liquid velocity, and kinematic viscosity of the liquid phase.

3. EXPERIMENTAL

Figure 2 is a schematic diagram of a flow loop. In this experiment, non-intrusive image processing was used to measure the axial development of flow parameters. The test section was a round pipe made of fluorinated ethylene propylene (FEP) with an index of refraction of 1.34, similar to that of water (1.33) to avoid image distortion. Because the tolerance of inner diameter of FEP pipe is relatively large such as ±0.1 mm based on manufacturer specifications, there is uncertainty of pipe diameter variation in axial direction. Thus the inner diameter of test pipes should be evaluated as the averaged value in pipe axial direction. It is, however, quite difficult to measure the local pipe diameter in axial direction directly because cutting plane of the FEP pipe is deformed due to its soft material. Therefore, the inner diameter was determined by using the analytical solution of the friction factor for laminar flow as follows [38]. The friction factor for laminar flow in a round pipe is given by the following well-known equation for Hagen-Poiseuille flow as

$$f = \frac{64}{N_{Re_f}}. \tag{10}$$

The inner diameter was determined from the above equation by using the measured friction factor, measured flow velocity and the fluid properties. The error in the diameter so obtained was estimated to be within ±1 %. The test pipe inner diameters were 0.55 mm, 0.79 mm and 1.02 mm. The length of all test pipes was 500 mm.

Nitrogen gas was supplied from a nitrogen bottle and was introduced into a mixing chamber through a gas injector. The gas injector consisted of a bubble injection nozzle with inner and outer diameters of 0.1 and 0.3 mm, and a tapered acrylic cylinder. No significant swirl flow near the test section inlet was observed in video images of the trajectories of dispersed bubbles. The nitrogen gas and purified water were mixed in the mixing chamber and the mixture then flowed through the test section.

Gas and liquid flow meters were installed at the upstream of the test section inlet and used as the indicator to set the predetermined gas and liquid flow rates. After flowing through the test section, the nitrogen gas and water were collected to measure their volumes. As shown in Figure 2, the nitrogen gas was collected by a measuring cylinder placed in the gas-liquid separator, and the volume collected per unit time was measured. The overflow water from the gas-liquid separator was also collected by a measuring cylinder, and the volume collected per unit time was measured. Thus the gas and liquid flow rates were determined by the collected gas and liquid volumes. The gas and liquid flow rates measured by the measuring cylinder agreed with those measured by gas and liquid flow meters within ±10 %. It should be noted that the maximum solubility of nitrogen in water is negligible in these experimental conditions. The loop temperature was kept at 20 °C within ±0.5 °C. The difference between the inlet and outlet temperatures was within ±0.5 °C, and thus the temperature drop due to the gas expansion along the test section could be considered negligible. The pressure and differential pressure measurements were made with a pressure sensor and differential pressure cell, respectively. The pressure devices detected the system pressure and differential pressure through a tiny hole in the test section wall (diameter 0.2 mm). Because the locations of the pressure tap were away from the test section inlet and outlet, the pressure disturbance due to the inlet and outlet was not considered. Measurement accuracy was conservatively estimated to be ±1 %. An electrical conductivity meter was installed at the gas-liquid separator to monitor water quality. The experiment was performed at an electrical conductivity less than 1 μS cm^{-1}.

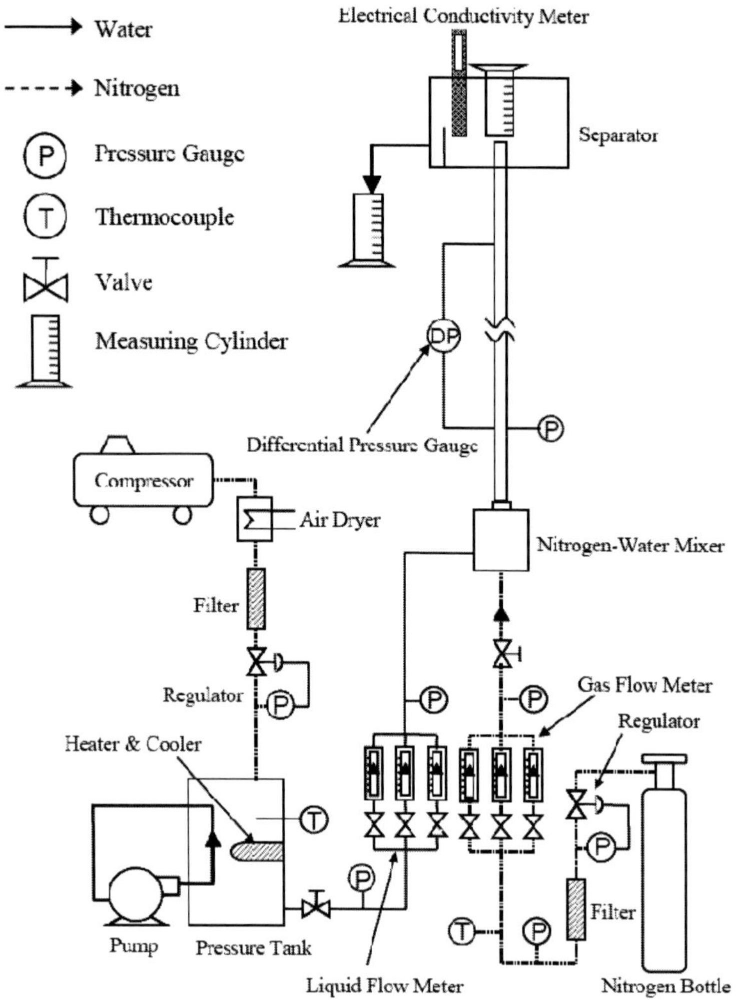

Figure 2. Schematic diagram of flow loop.

Flow measurements were performed by image processing using a digital video camera and a stroboscope with a flash duration of 0.3 μs at z/D = 15.0, 75.0, 150, 250, and 450. The water boxes were placed at the measuring stations to minimize the image distortion due to refraction. Bubble images taken by the digital video camera were transmitted to a personal computer equipped with an image processing board. The image processing to obtain dimensions of bubbles was performed by the following procedure.

1. Reading and bordering the images,

2. Image enhancement and data smoothing for noise reduction,
3. Median-filtering the images at eight-neighbor pixels of a specific pixel,
4. Binarization of the pixels to extract the outer boundaries of the bubbles,
5. Labeling the bubbles,
6. Measuring bubble dimensions such as major and minor axes.

The void fraction, interfacial-area concentration, and bubble Sauter mean diameter were calculated from the obtained images with an assumption of an axisymmetric bubble, and the bubble number density defined as the total number of bubbles per unit volume was simply determined by counting the number of bubbles.

Typical bubble images are shown in Figure 3.

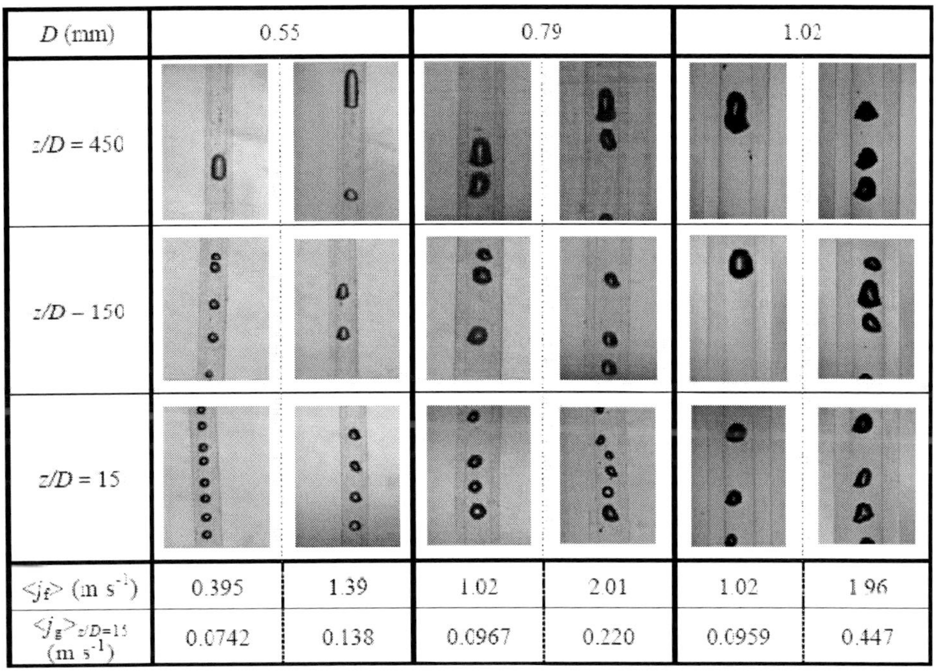

Figure 3. Typical bubble images.

As can be seen in Figure 3, the bubble shapes appeared to deviate from being axisymmetric to some degree. However, as will be described later, measured flow parameters agreed with those measured by other reference methods within ±15 %, by conservative estimate. Thus the assumption of an axisymmetric bubble was acceptable within the measurement error of ±15 %. The methodology of the image processing method was detailed in our previous papers [26].

About 3000-4000 bubbles were sampled to maintain similar statistics between the different combinations of experimental conditions [16]. The void fractions measured by the image processing method agreed with those obtained in a 1.09-mm round pipe by neutron radiography [38] within the averaged relative deviation of 12.3 %. The image processing method for the interfacial-area concentration measurement was also validated by a double-sensor conductivity probe method. A separate test was performed in a 25.4 mm round pipe, yielding good agreement for the interfacial-area concentration measurement, within the averaged relative deviation of 6.95 % [39]. Since the measurement accuracy of the double sensor probe method is reported to be ±7 % [40], the measurement accuracy of the image processing method is within ±15 % by conservative estimate. The experimental conditions in each test pipe are tabulated in Tables 1-3.

Table 1. Experimental conditions in 0.55 mm pipe

Flow parameters		Run 1	Run 2	Run 3	Run 4	Run 5	Run 6
$\langle j_f \rangle$	m s^{-1}	0.395	0.395	0.395	0.864	0.864	0.864
$\langle j_g \rangle_{z/D=15}$	m s^{-1}	0.0742	1.39	0.204	0.0621	0.116	0.165
C_{A0}	-	0.106	0.213	0.253	0.0380	0.0785	0.109
C_{A1}	-	6.02×10^{-5}	7.17×10^{-5}	-1.96×10^{-5}	2.40×10^{-5}	4.56×10^{-5}	-1.23×10^{-5}
C_{A2}	-	-4.52×10^{-8}	-2.80×10^{-8}	2.09×10^{-7}	5.72×10^{-8}	5.54×10^{-8}	1.25×10^{-7}
C_{J0}	m s^{-1}	0.0751	0.141	0.215	0.0627	0.114	0.170
C_{J1}	m s^{-1}	2.25×10^{-5}	3.98×10^{-5}	1.07×10^{-5}	-2.44×10^{-6}	7.83×10^{-5}	-5.74×10^{-6}
C_{J2}	m s^{-1}	1.60×10^{-8}	4.35×10^{-8}	1.47×10^{-7}	7.96×10^{-8}	-1.59×10^{-8}	1.90×10^{-7}
$\langle a_i \rangle_{z/D=75}$	m^{-1}	1870	3170	3540	838	1150	1400
M_F	kPa m^{-1}	98.8	121	135	193	210	259
$\langle j_f \rangle$	m s^{-1}	1.39	1.39	1.39	2.13	2.13	2.13
$\langle j_g \rangle_{z/D=15}$	m s^{-1}	0.0462	0.0921	0.138	0.0471	0.0842	0.123
C_{A0}	-	0.0175	0.0375	0.0559	0.0117	0.0234	0.0318
C_{A1}	-	2.98×10^{-5}	1.17×10^{-5}	3.42×10^{-5}	6.59×10^{-6}	1.28×10^{-5}	7.90×10^{-6}
C_{A2}	-	-1.86×10^{-8}	4.01×10^{-8}	3.77×10^{-8}	1.16×10^{-8}	7.39×10^{-9}	3.24×10^{-8}
C_{J0}	m s^{-1}	0.0467	0.0918	0.137	0.0481	0.0834	0.128
C_{J1}	m s^{-1}	1.50×10^{-5}	4.21×10^{-5}	7.72×10^{-5}	4.81×10^{-6}	4.91×10^{-5}	-2.57×10^{-5}
C_{J2}	m s^{-1}	3.19×10^{-8}	3.65×10^{-8}	5.06×10^{-8}	5.94×10^{-8}	5.29×10^{-8}	1.96×10^{-7}
$\langle a_i \rangle_{z/D=75}$	m^{-1}	394	653	974	220	430	610
M_F	kPa m^{-1}	331	306	341	444	474	539

Figures 4-6 show the axial development of the void fraction measured in 0.55, 0.79 and 1.02 mm inner diameter pipes, respectively. The experimental conditions in each test pipe are tabulated in Tables 1-3. As shown in Figures 4-6, a significant increase in the void fraction along the flow direction is observed in the tested experimental conditions. This is mainly attributable to the pressure reduction along the flow direction.

Table 2. Experimental conditions in 0.79 mm pipe

Flow parameters		Run 1	Run 2	Run 3	Run 4	Run 5	Run 6
$\langle j_f \rangle$	m s^{-1}	1.02	1.02	1.02	2.01	2.01	2.01
$\langle j_g \rangle_{z/D=15}$	m s^{-1}	0.0967	0.180	0.251	0.0901	0.162	0.220
C_{A0}	-	0.0583	0.0649	0.136	0.0185	0.0365	0.0523
C_{A1}	-	-3.68×10^{-5}	4.97×10^{-4}	1.07×10^{-4}	4.84×10^{-5}	5.64×10^{-5}	1.19×10^{-4}
C_{A2}	-	1.66×10^{-7}	-7.18×10^{-7}	-5.86×10^{-9}	-3.96×10^{-8}	2.33×10^{-8}	-7.33×10^{-8}
C_{J0}	m s^{-1}	0.103	0.166	0.265	0.0889	0.158	0.218
C_{J1}	m s^{-1}	-6.64×10^{-5}	2.21×10^{-4}	1.25×10^{-4}	1.28×10^{-5}	1.45×10^{-4}	1.46×10^{-4}
C_{J2}	m s^{-1}	2.16×10^{-7}	-2.42×10^{-7}	-7.48×10^{-10}	1.35×10^{-7}	1.75×10^{-8}	1.99×10^{-7}
$\langle a_i \rangle_{z/D=75}$	m^{-1}	665	1050	1350	358	497	766
M_F	kPa m^{-1}	70.8	83.9	95.1	139	160	179
$\langle j_f \rangle$	m s^{-1}	3.08	3.08	3.08	3.98	3.98	3.98
$\langle j_g \rangle_{z/D=15}$	m s^{-1}	0.0937	0.162	0.290	0.0801	0.155	0.302
C_{A0}	-	0.0149	0.0139	0.0270	0.0137	0.00646	0.0284
C_{A1}	-	2.13×10^{-5}	1.36×10^{-4}	1.85×10^{-4}	-3.66×10^{-5}	1.07×10^{-4}	1.55×10^{-4}
C_{A2}	-	5.42×10^{-9}	-1.19×10^{-7}	-1.03×10^{-7}	9.92×10^{-8}	-9.93×10^{-8}	-1.58×10^{-7}
C_{J0}	m s^{-1}	0.0968	0.151	0.270	0.0769	0.149	0.296
C_{J1}	m s^{-1}	-5.86×10^{-6}	1.71×10^{-4}	2.86×10^{-4}	8.00×10^{-5}	1.54×10^{-4}	2.30×10^{-4}
C_{J2}	m s^{-1}	1.80×10^{-7}	1.69×10^{-7}	3.64×10^{-7}	6.66×10^{-8}	2.70×10^{-7}	1.10×10^{-7}
$\langle a_i \rangle_{z/D=75}$	m^{-1}	322	389	636	300	321	665
M_F	kPa m^{-1}	273	267	316	398	446	447

4. Results and Discussion

4.1. Database and Flow Parameters Used in Interfacial-Area Transport Calculation

Figure 3 shows typical flow images along the test pipes. These images clearly indicate that the flow regime and the interfacial-area concentration in mini pipes changes continuously with flow development in the axial direction and the fully developed flow assumption is not suitable for flow in mini pipes. It can be also said that the prediction of interfacial-area concentration based on the interfacial-area transport equation could become very important for solving two-phase flow problems for mini pipe flow applications.

Table 3. Experimental conditions in 1.02 mm pipe

Flow parameters		Run 1	Run 2	Run 3	Run 4	Run 5	Run 6
$\langle j_f \rangle$	m s^{-1}	1.02	1.02	1.02	1.96	1.96	1.96
$\langle j_g \rangle_{z/D=15}$	m s^{-1}	0.0959	0.292	0.472	0.0877	0.283	0.447
C_{A0}	-	0.0535	0.142	0.180	0.0276	0.0675	0.109
C_{A1}	-	4.36×10^{-5}	-8.74×10^{-5}	3.26×10^{-4}	2.40×10^{-5}	1.87×10^{-4}	2.06×10^{-4}
C_{A2}	-	-4.94×10^{-8}	3.37×10^{-7}	-3.99×10^{-7}	-6.55×10^{-9}	-2.45×10^{-7}	-2.37×10^{-7}
C_{J0}	m s^{-1}	0.0961	0.292	0.473	0.0880	0.284	0.449
C_{J1}	m s^{-1}	3.17×10^{-5}	1.09×10^{-4}	1.89×10^{-4}	3.67×10^{-5}	1.28×10^{-4}	2.08×10^{-4}
C_{J2}	m s^{-1}	2.85×10^{-8}	1.15×10^{-7}	2.24×10^{-7}	4.70×10^{-8}	1.90×10^{-7}	3.36×10^{-7}
$\langle a_i \rangle_{z/D=75}$	m^{-1}	367	724	998	232	553	828
M_F	kPa m^{-1}	57.4	69.0	76.4	100	114	122
$\langle j_f \rangle$	m s^{-1}	2.88	2.88	2.88	3.92	3.92	3.92
$\langle j_g \rangle_{z/D=15}$	m s^{-1}	0.0817	0.229	0.428	0.0741	0.236	0.398
C_{A0}	-	0.0192	0.0501	0.0787	0.0119	0.0356	0.0580
C_{A1}	-	1.12×10^{-5}	5.11×10^{-5}	2.20×10^{-4}	8.93×10^{-6}	1.28×10^{-4}	5.17×10^{-5}
C_{A2}	-	2.88×10^{-8}	-4.92×10^{-9}	-2.20×10^{-7}	3.48×10^{-8}	-1.29×10^{-7}	1.85×10^{-7}
C_{J0}	m s^{-1}	0.0826	0.232	0.434	0.0764	0.244	0.412
C_{J1}	m s^{-1}	4.26×10^{-5}	1.20×10^{-4}	2.24×10^{-4}	3.22×10^{-5}	9.67×10^{-5}	1.53×10^{-4}
C_{J2}	m s^{-1}	1.14×10^{-7}	3.65×10^{-7}	7.42×10^{-7}	2.24×10^{-7}	7.65×10^{-7}	1.37×10^{-6}
$\langle a_i \rangle_{z/D=75}$	m^{-1}	189	377	623	156	368	493
M_F	kPa m^{-1}	197	213	224	318	337	350
$\langle j_f \rangle$	m s^{-1}	4.89	4.89	4.89			
$\langle j_g \rangle_{z/D=15}$	m s^{-1}	0.0754	0.232	0.392			
C_{A0}	-	0.00966	0.0228	0.0515			
C_{A1}	-	1.53×10^{-5}	9.41×10^{-5}	3.63×10^{-5}			
C_{A2}	-	2.90×10^{-8}	2.04×10^{-8}	2.03×10^{-7}			
C_{J0}	m s^{-1}	0.0796	0.249	0.412			
C_{J1}	m s^{-1}	1.18×10^{-5}	-6.82×10^{-6}	2.99×10^{-5}			
C_{J2}	m s^{-1}	3.75×10^{-7}	1.40×10^{-6}	2.13×10^{-6}			
$\langle a_i \rangle_{z/D=75}$	m^{-1}	151	291	468			
M_F	kPa m^{-1}	466	488	502			

Figures 7-9 show the axial development of interfacial-area concentration. The experimental conditions in each test pipe are tabulated in Tables 1-3. The error bar indicates the measurement error, ±15%. As shown in the figures, the axial interfacial-area concentration transport appears to be dependent on the flow conditions. As can be seen in Eq.(3), the interfacial-area concentration change is governed by two competing mechanisms: void fraction change due to axial pressure reduction (source term) and bubble coalescence due to wake entrainment (sink term). As shown in Figure 7(a), the wake entrainment effect prevails over the void fraction effect at low liquid velocities, $\langle j_f \rangle$=0.395 m s^{-1}. As the superficial liquid velocity increases, the void fraction change becomes dominant over wake entrainment. Some reasons for this may be insufficient time for approach of a trailing bubble in the wake region to the preceding bubble, and a significant void fraction change due to the large frictional pressure drop for high superficial liquid velocities.

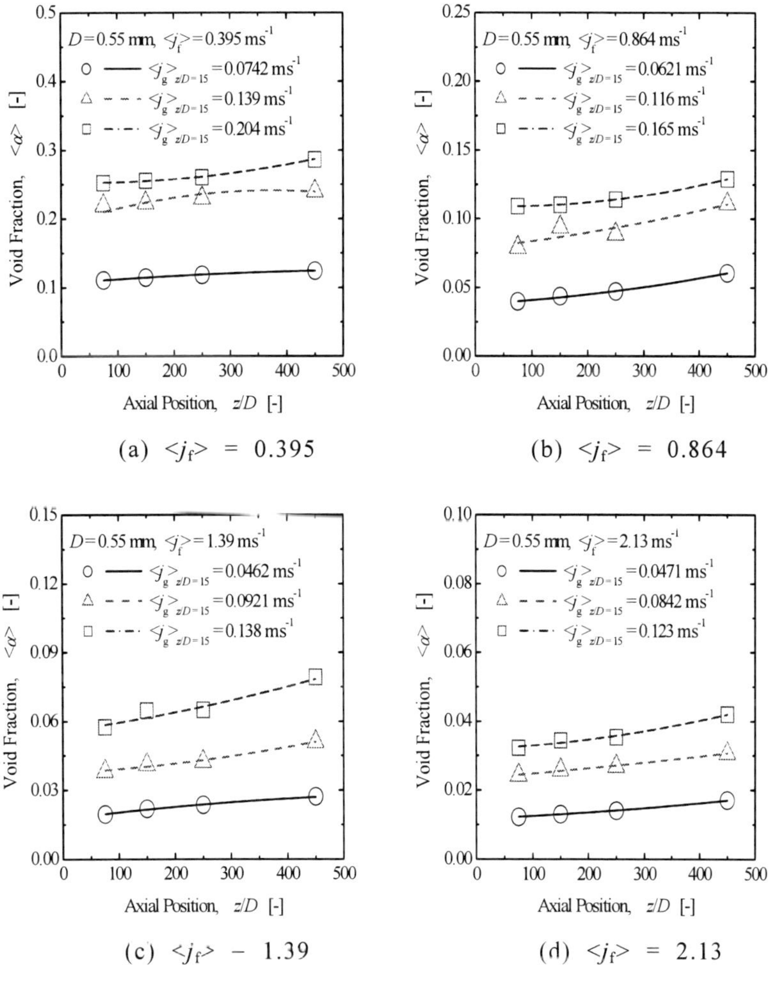

Figure 4. Axial development of void fraction in 0.55 mm-diameter pipe.

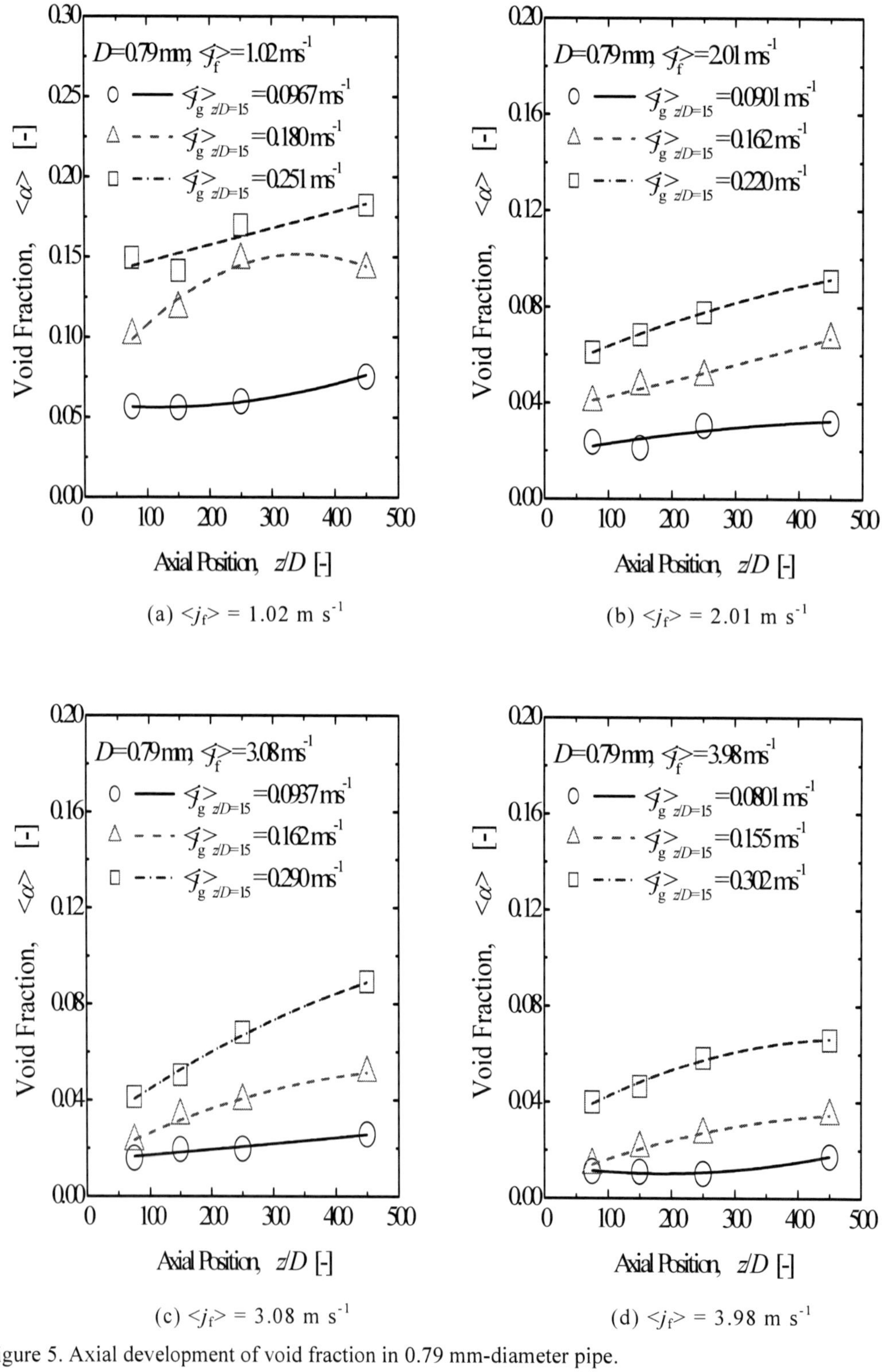

(a) $\langle j_f \rangle$ = 1.02 m s^{-1}

(b) $\langle j_f \rangle$ = 2.01 m s^{-1}

(c) $\langle j_f \rangle$ = 3.08 m s^{-1}

(d) $\langle j_f \rangle$ = 3.98 m s^{-1}

Figure 5. Axial development of void fraction in 0.79 mm-diameter pipe.

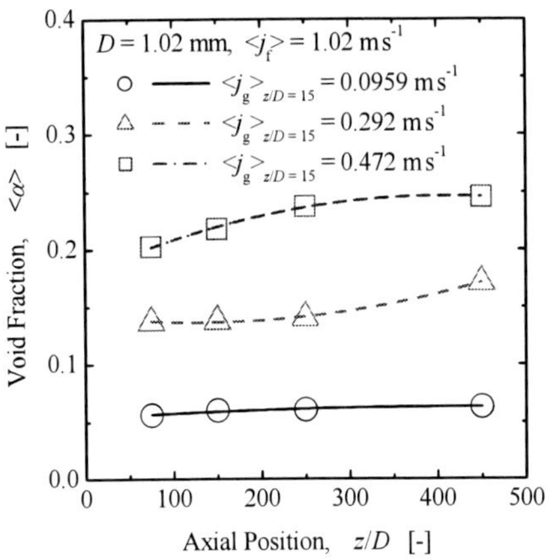

(a) $<j_f>$ = 1.02 m s^{-1}

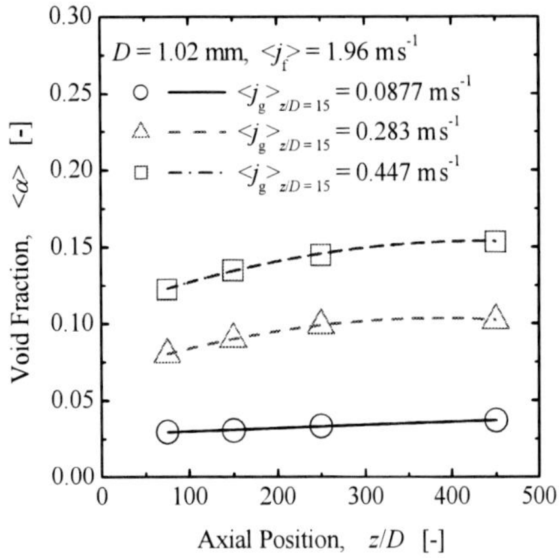

(b) $<j_f>$ = 1.96 m s^{-1}

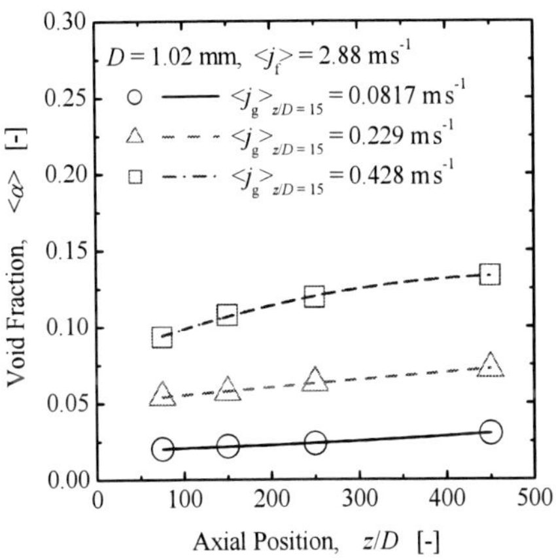

(c) $<j_f>$ = 2.88 m s^{-1}

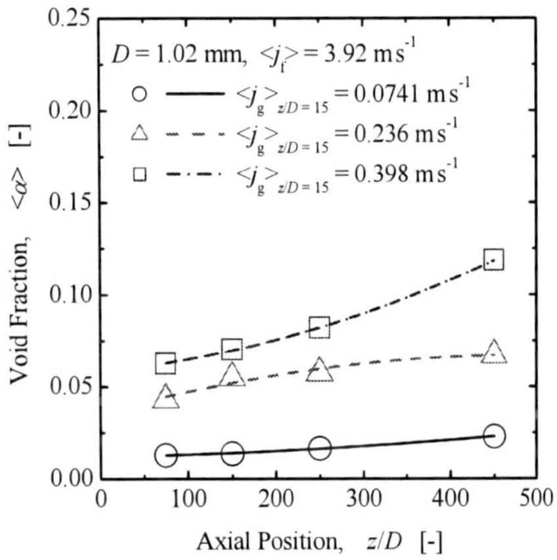

(d) $<j_f>$ = 3.92 m s^{-1}

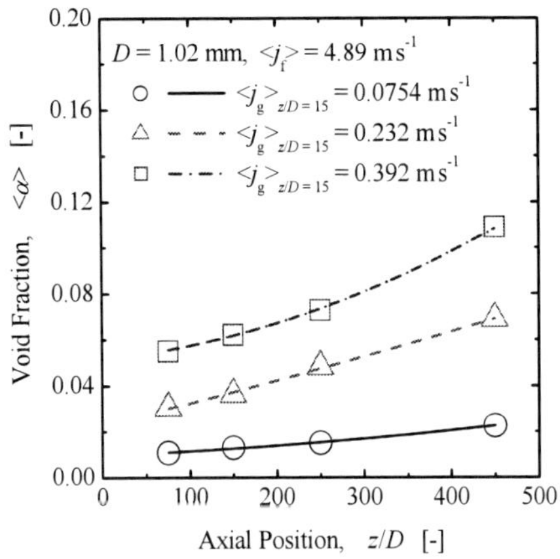

(e) $<j_f>$ = 4.89 m s^{-1}

Figure 6. Axial development of void fraction in 1.02 mm-diameter pipe.

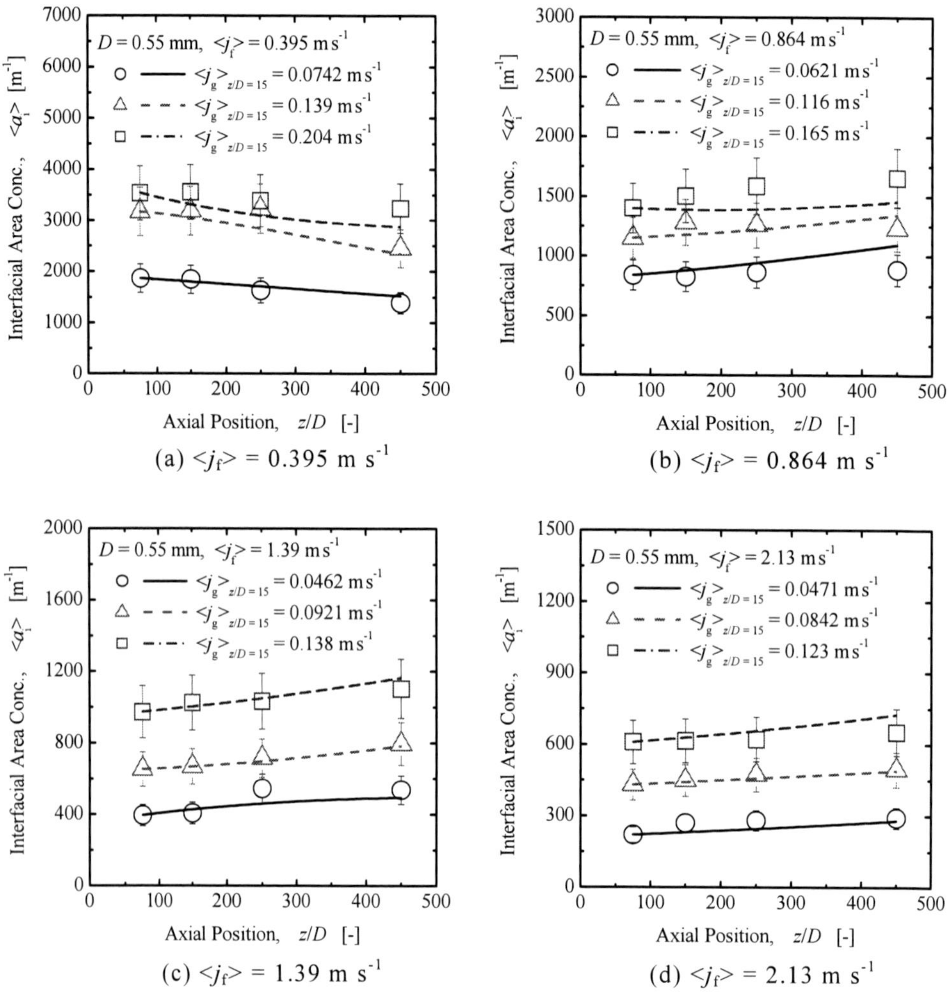

Figure 7. Comparison of interfacial-area transport equation with measured interfacial-area concentrations in 0.55 mm-diameter pipe.

To perform the interfacial-area transport calculation, the axial values of void fraction $\langle\alpha\rangle$, gas velocity $\langle\langle v_g \rangle\rangle$, superficial gas velocity $\langle j_g \rangle$, initial interfacial-area concentration $\langle a_{i,0} \rangle$, and two-phase frictional pressure gradient M_F should be given, either from constitutive relations or measured values. To evaluate the sink term in the wake entrainment model, the flow parameters in the model should be as accurate as possible. Therefore the void fraction and superficial gas velocity are given here by the following fitting functions obtained from the measured flow parameters:

$$\langle\alpha\rangle = C_{A0} + C_{A1}\left(z/D\right) + C_{A2}\left(z/D\right)^2, \tag{11}$$

$$\langle j_g \rangle = C_{J0} + C_{J1}\left(z/D\right) + C_{J2}\left(z/D\right)^2. \tag{12}$$

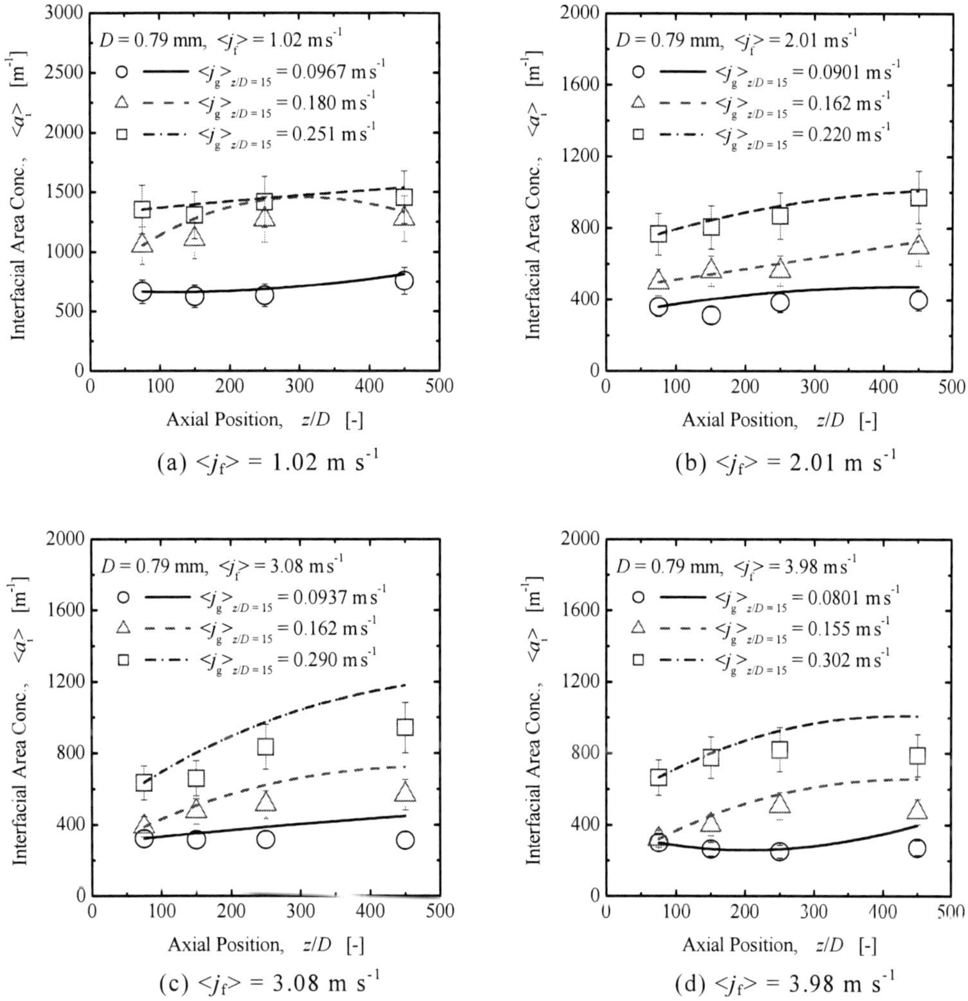

Figure 8. Comparison of interfacial-area transport equation with measured interfacial-area concentrations in 0.79 mm-diameter pipe.

Coefficients in Eqs. (11) and (12) for each test pipe are listed in Tables 1-3. The lines in Fiures 4-6 indicate the axial changes in void fraction calculated by Eq. (11). The axial changes of gas velocity $\langle\langle v_g \rangle\rangle$ can be calculated by

$$\langle\langle v_g \rangle\rangle = \langle j_g \rangle / \langle \alpha \rangle. \tag{13}$$

Here, the superficial gas velocity and void fraction in Eq. (13) are, respectively, given by Eq. (11) and Eq. (12). Because the data obtained around the inlets of test pipes such as the first measuring station (z/D=15) were scattered due to some entrance effects, the interfacial-area concentration measured at the second measuring station (z/D=75) is used as the initial value of interfacial-area concentration, $\langle a_{i,0} \rangle$. In the interfacial-area transport calculation, the

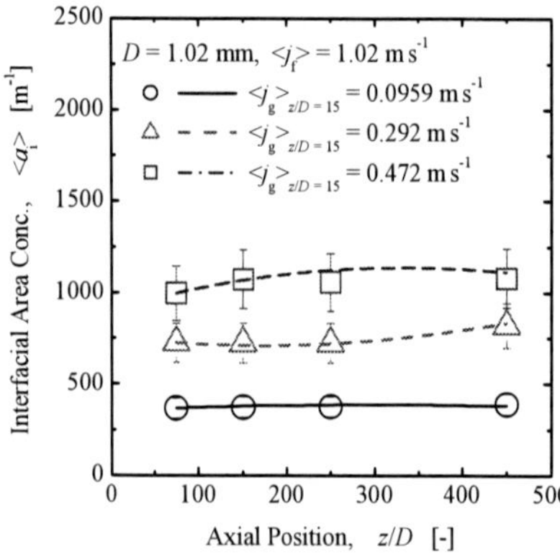

(a) $\langle j_f \rangle$ = 1.02 m

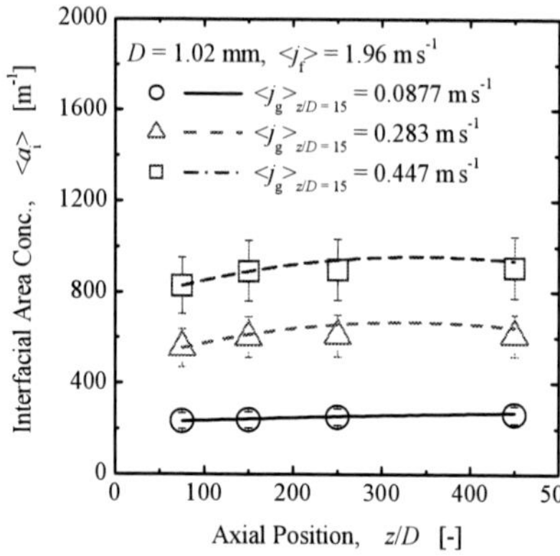

(b) $\langle j_f \rangle$ = 1.96 m

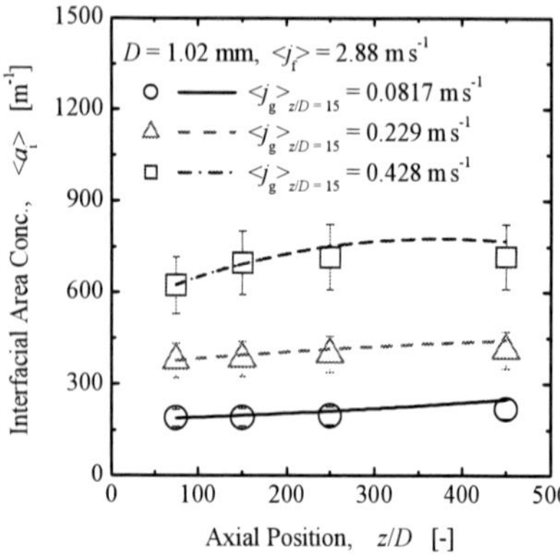

(c) $\langle j_f \rangle$ = 2.88 m

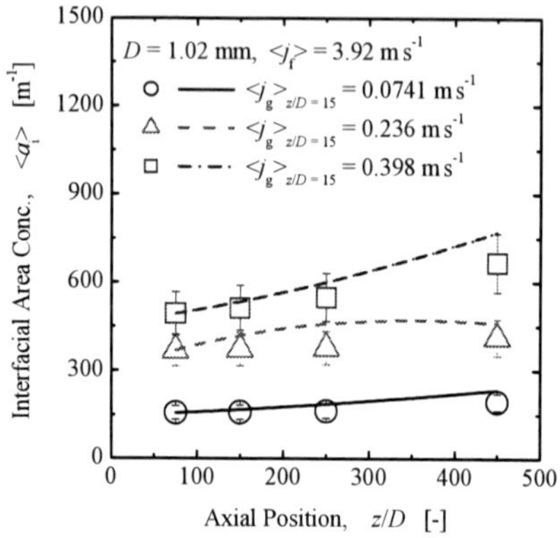

(d) $\langle j_f \rangle$ = 3.92 m

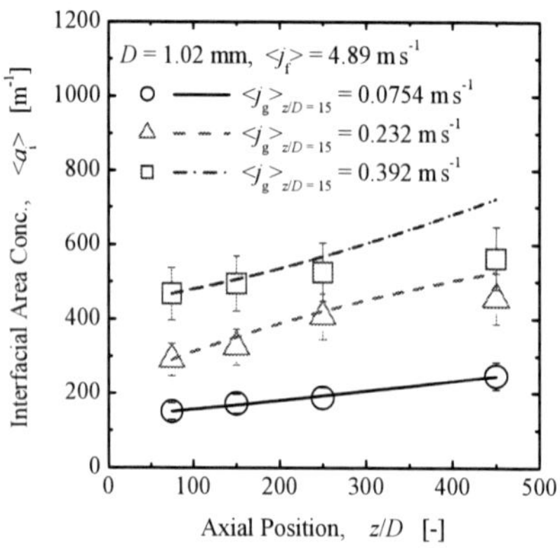

(e) $\langle j_f \rangle$ = 4.89 m

Figure 9. Comparison of interfacial-area transport equation with measured interfacial-area concentrations in 1.02 mm-diameter pipe.

two-phase frictional pressure gradient M_F in Eq. (6) is given by the measured frictional pressure gradient. The measured values of M_F for all flow conditions in each test pipe are listed in Tables 1-3.

4.2. Applicability of the Interfacial-Area Transport Equation to Flow in Mini Pipes

The one-dimensional interfacial-area transport equation with a sink term due to wake entrainment is evaluated below by using the experimentally obtained database. The sink and source terms of the interfacial-area concentration due to random bubble collisions and turbulent impacts are neglected in the interfacial-area transport calculation. These approximations are supported by a preliminary calculation of these terms and by visual observation. The axial change in interfacial-area concentration is computed from Eqs. (3) and (6). As can be seen from these equations, there is only one adjustable coefficient, Γ_{WE}, in the interfacial-area transport equation. The value of Γ_{WE} is found to be 0.0042 using 39 data sets from the present experiments. The computed axial changes of the interfacial-area concentrations in each test pipe are indicated by lines in Figures 7-9. As shown in the figures, the interfacial-area concentrations computed with Eqs. (3) and (6) agree with the measured interfacial-area concentrations very well. However, the present equation tends to slightly overestimate the measured values in downstream region where some Taylor bubbles are formed. The present equation is formulated based on the assumption of spherical bubble shape and the effect of change in bubble shape on the interfacial-area change at bubbly-to-slug flow transition is not considered in the model. This may be the main reason of the discrepancy between the calculated and measured values in downstream region. For such flow, the two-group interfacial-area transport equation [41] which was formulated by considering two-groups of bubbles such as spherical/distorted bubbles and cap/slug bubbles should be utilized. Further discussion on the applicability of two-group interfacial-area transport equation to mini pipe flows will be required.

In order to evaluate the contribution of bubble coalescence and void transport to interfacial-area transport, typical changes of the interfacial-area concentration due to each mechanism along the axial direction in 0.55, 0.79 and 1.02 mm-diameter pipes are shown in Figures 10, 11 and 12. Open symbols and solid lines indicate measured and computed interfacial-area concentration changes from the interfacial-area concentration at $z/D=75$, $\langle a_i \rangle - \langle a_i \rangle_{z/D=75}$, respectively. Dashed-dotted and broken lines represent, respectively, the interfacial-area concentration changes due to void-fraction change and wake entrainment. There are essentially two mechanisms that contribute to axial interfacial-area concentration change due to void fraction change along the flow direction. These mechanisms are bubble expansion due to pressure reduction along the flow direction and void distribution change along the flow direction. The bubble expansion has a positive impact, increasing the void fraction along the flow direction. Because the pressure reduction in mini pipes is relatively large compared to that in conventional size pipes, the axial change of interfacial-area concentration due to bubble expansion is often dominant in mini pipes. The void fraction is also changed by the way the void distribution changes along the flow direction. In an adiabatic bubbly flow, the maximum void-fraction change due to void distribution change is generally 30% at bubbly-to-slug flow transition along a test section [42]. In mini pipes, this effect would be negligibly small because radial bubble movement would be restricted by the presence of a pipe wall, so radial movement causes insignificant void distribution change along the flow direction. Thus, the bubble expansion may provide the largest contribution to the axial interfacial-area concentration change due to void fraction change.

As shown in Figures 10-12, the axial interfacial-area concentration change due to void-fraction change along the flow direction is significant for these cases. For low liquid flow rates (Figure 10a), the wake entrainment effect dominates in the axial change of interfacial-area concentration and enhances the axial decrease of the interfacial-area concentration.

In contrast, for high liquid flow rates (Figures 10b, 11a,b and 12a,b), the axial changes of the interfacial-area concentration due to wake entrainment are insignificant and the void-fraction change along the pipe is the dominant contributor to the axial changes in the interfacial-area concentration. This is because the residence time of two-phase flow in the length of the pipes tested, and bubble-coalescence efficiency, are reduced at higher liquid flow rates, resulting in suppressed bubble coalescence due to wake entrainment.

The validity of the interfacial-area transport equation with a wake entrainment term has been verified by data sets obtained in mini pipes as shown in Figures 7-12, although the equation tends to slightly overestimate the measured values in downstream region where some Taylor bubbles are formed. The averaged prediction errors of the interfacial-area transport equation for data points obtained at 3 axial locations, z/D=150, 250 and 450 in 0.55, 0.79 and 1.02 mm-diameter pipe are ±7.43 %, ±12.1 % and ±6.37 %, respectively.

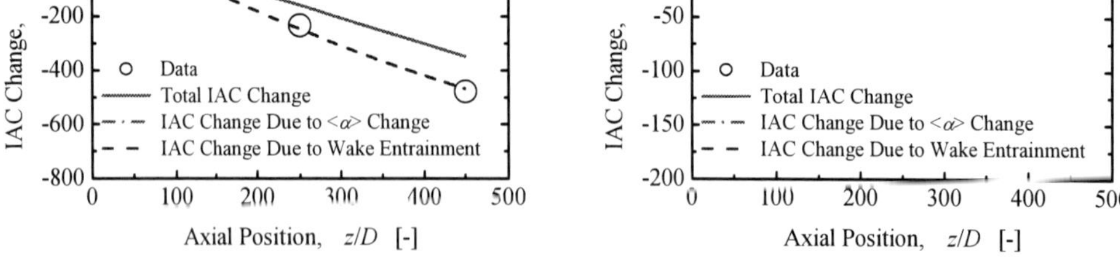

(a) $\langle j_f \rangle = 0.395$ m s^{-1}, $\langle j_g \rangle_{z/D=15} = 0.0742$ m s^{-1}

(b) $\langle j_f \rangle = 2.13$ m s^{-1}, $\langle j_g \rangle_{z/D=15} = 0.0842$ m s^{-1}

Figure 10. Contributions of bubble coalescence and void transport to interfacial-area transport in 0.55 mm-diameter pipe.

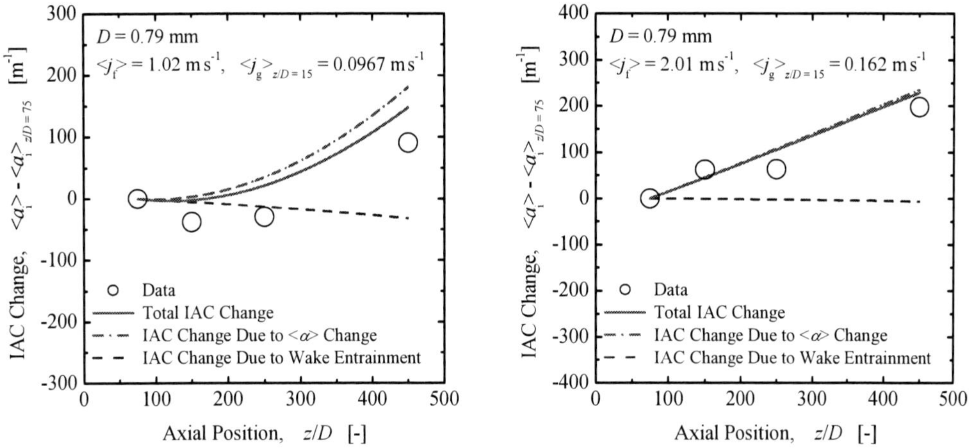

(a) $\langle j_f \rangle = 1.02\ \mathrm{m\ s^{-1}}$, $\langle j_g \rangle_{z/D=15} = 0.0967\ \mathrm{m\ s^{-1}}$

(b) $\langle j_f \rangle = 2.01\ \mathrm{m\ s^{-1}}$, $\langle j_g \rangle_{z/D=15} = 0.162\ \mathrm{m\ s^{-1}}$

Figure 11. Contributions of bubble coalescence and void transport to interfacial-area transport in 0.79 mm-diameter pipe.

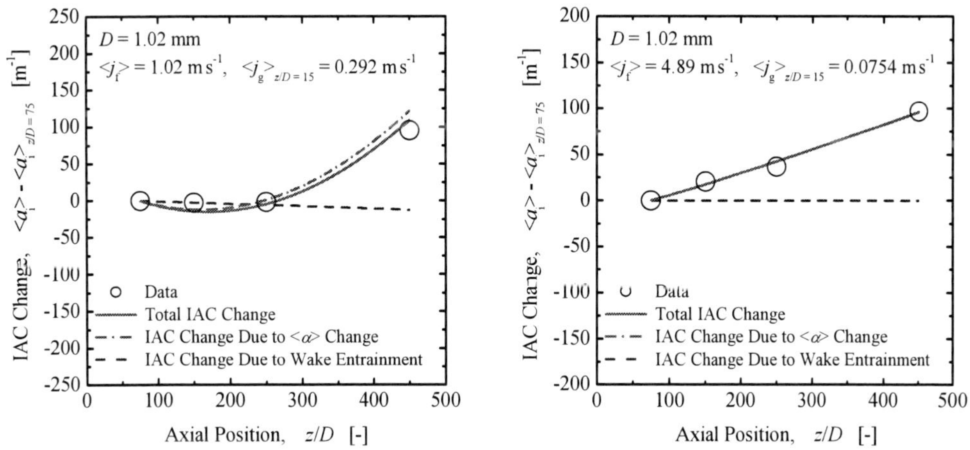

(a) $\langle j_f \rangle = 1.02\ \mathrm{m\ s^{-1}}$, $\langle j_g \rangle_{z/D=15} = 0.292\ \mathrm{m\ s^{-1}}$

(b) $\langle j_f \rangle = 4.89\ \mathrm{m\ s^{-1}}$, $\langle j_g \rangle_{z/D=15} = 0.0754\ \mathrm{m\ s^{-1}}$

Figure 12. Contributions of bubble coalescence and void transport to interfacial-area transport in 1.02 mm-diameter pipe.

CONCLUSION

Accurate prediction of the flow parameters is essential to successful development of the interfacial transfer terms in the two-phase flow formulation. Mechanistic modeling of the interfacial transfer terms relies entirely on accurate local flow measurements over an extensive range of flow conditions and channel geometries. To advance this process, axial measurements of flow parameters were performed by image processing, and databases were constructed with information on the axial development of void fraction and interfacial-area concentration for vertical upward bubbly flows in 0.55, 0.79 and 1.02 mm-diameter pipes. The one-dimensional interfacial-area transport equation with a sink term due to wake entrainment was evaluated by using the obtained datasets. Comparing the existing interfacial-area transport equation with experimental data in mini pipes shows satisfactory agreement. The averaged prediction errors of the interfacial-area transport equation for data points obtained at 3 axial locations in 0.55, 0.79 and 1.02 mm-diameter pipe are ±7.43 %, ±12.1 % and ±6.37 %, respectively.

ACKNOWLEDGMENTS

The authors are thankful to Messrs. Y. Fukuhara, A. Kamura and Mrs. Y. Ohkubo of Tokyo University of Marine Science and Technology for their assistance in the experiment.

REFERENCES

[1] A.H. Gibson, On the Motion of Long Air-bubbles in a Vertical Tube, *Phil. Mag.* 26 (1913) 952–965.

[2] H.J. Lee, S.Y. Lee, Pressure Drop Correlations for Two-phase Flow within Horizontal Rectangular Channels with Small Heights, *Int. J. Multiphase Flow* 27 (2001) 783–796.

[3] H.J. Lee, S.Y. Lee, Heat Transfer Correlation for Boiling Flows in Small Rectangular Horizontal Channels with Low Aspect Ratios, *Int. J. Multiphase Flow* 27 (2001) 2043–2062.

[4] S.G. Kandlikar, Two-phase Flow Patterns, Pressure Drop, and Heat Transfer during Boiling in Minichannel Flow Passages of Compact Evaporators, *Heat Transfer Eng.* 23 (2002) 5–23.

[5] S.G. Kandlikar, Fundamental Issues Related to Flow Boiling in Minichannels and Microchannels, *Exp. Thermal Fluid Sci.* 26 (2002) 389–407.

[6] A. Serizawa, Z.P. Feng, Z. Kawahara, Two-phase Flow in Microchannels, *Exp. Thermal Fluid Sci.* 26 (2002) 703–714.

[7] A. Kawahara, P. M. -Y. Chung, M. Kawaji, Investigation of Two-phase Flow Pattern, Void Fraction and Pressure Drop in a Microchannel, *Int. J. Multiphase Flow* 28 (2002) 1411-1435.

[8] W.L. Qu, I. Mudawar, Measurement and Prediction of Pressure Drop in Two-phase Micro-channel Heat Sinks, *Int. J. Heat Mass Transfer* 46 (2003) 2737–2753.

[9] S.M. Ghiaasiaan, Gas-liquid Two-phase Flow and Boiling in Mini and Microchannels, *Multiphase Sci. Tech.* 15 (2003) 323–336.

[10] M. Kawaji, P.M. -Y. Chung, Adiabatic Gas-liquid Flow in Microchannles, Microscale *Thermophysical Eng.* 8 (2004) 239–257.

[11] W. Zhang, T. Hibiki, K. Mishima, Correlation for Flow Boiling Heat Transfer in Mini-channels, *Int. J. Heat Mass Transfer* 47 (2004) 5749–5763.

[12] W. Zhang, T. Hibiki, K. Mishima, Correlation for Flow Boiling Heat Transfer at Low Liquid Reynolds Number in Small Diameter Channels, *J. Heat Transfer* 127 (2005) 1214–1221.

[13] W. Zhang, T. Hibiki, K. Mishima, Correlation of Critical Heat Flux for Flow Boiling of Water in Mini-channels, *Int. J. Heat Mass Transfer* 49 (2006) 1058–1072.

[14] H. Ohta, Boiling and Two-phase Flow in Channels with Extremely Small Diameters: *A Review of Japanese Research, Microfluidics and Nanofluidics* 1 (2005) 94–107.

[15] [15] H. J. Lee, D. Liu, S. C. Yao, Y. Alyousef, Generalized Two-phase Pressure Drop and Heat Transfer Correlations in Evaporative Micro/Mini-Channels, *Proc. 2007 ASME-JSME Thermal Engineering Summer Heat Transfer Conf.*, HT2007-32157 (2007).

[16] T. Hibiki, T. Hazuku, T. Takamasa, M. Ishii, Some Characteristics of Developing Bubbly Flow in a Vertical Mini Pipe, *Int. J. Heat and Fluid Fl*ow 28 (2007) 1034-1048.

[17] G. Kocamustafaogullari, W. D. Huang, J. Razi, Measurement of Modeling of Average Void Fraction, Bubble Size and Interfadal Area, *Nucl. Eng. Des*. 148 (1994) 437-453.

[18] J. M. Delhaye, P. Bricard, Interfacial Area in Bubbly Flow: Experimental Data and Correlations, *Nucl. Eng. Des*. 151 (1994) 65-77.

[19] M. Millies, D. A. Drew, R.T. Lahey Jr., A First Order Relaxation Model for the Prediction of the Local Interfacial Area Density in Two-Phase Flows, *Int. J. Multiphase Flow* 22 (1996) 1073-1104.

[20] T. Hibiki, M. Ishii, Interfacial Area Concentration in Steady Fully-Developed Bubbly Flow, *Int. J. Heat Mass Transfer* 44 (2001) 3443-3461

[21] T. Hibiki, M. Ishii, Interfacial Area Transport Equations for Gas-Liquid Flow, *J. Computational Multiphase Flows* 1-1 (2009) 1-22.

[22] M. Ishii, Thermo-Fluid Dynamic Theory of Two-Phase Flow, *Eyrolles* (1975).

[23] M. Ishii, T. Hibiki, Thermo-Fluid Dynamics of Two-Phase Flow, *Springer* (2005).

[24] Q. Wu, S. Kim, M. Ishii, S.G. Beus, One-group Interfacial-area Transport in Vertical Bubbly Flow, *Int. J. Heat Mass Transfer* 41 (1998) 1103–1112.

[25] T. Hibiki, M. Ishii, One-group Interfacial-area Transport of Bubbly Flows in Vertical Round Tubes, *Int. J. Heat Mass Transfer* 43 (2000) 2711–2726.

[26] T. Takamasa, T. Goto, T. Hibiki, M. Ishii, Experimental Study of Interfacial-area Transport of Bubbly Flow in Small-diameter Tube, *Int. J. Multiphase Flow* 29 (2003) 291–304.

[27] T. Hibiki, M. Ishii, Experimental Study on Interfacial Area Transport in Bubbly Two-phase Flows, *Int. J. Heat Mass Transfer* 42 (1999) 3019-3035.

[28] T. Hibiki, M. Ishii, Z. Xiao, Axial Interfacial Area Transport of Vertical Bubbly Flows, *Int. J. Heat Mass Transfer* 44 (2001) 1869-1888.

[29] X. Sun, T.R. Smith, S. Kim, M. Ishii, J. Uhle, Interfacial Structure of Air-water Two-phase Flow in a Relatively Large Pipe, *Exp. Fluids* 34 (2003) 206–219.

[30] T. Hibiki, T. Hazuku, T. Takamasa, M. Ishii, Interfacial-area Transport Equation at Reduced-gravity Conditions, *AIAA Journal* 47 (2009) 1123–1131.

[31] G. Kocamustafaogullari, M. Ishii, Foundation of the Interfacial-area Transport Equation and its Closure Relations, *Int. J. Heat Mass Transfer* 38 (1995) 481–493.

[32] T. Hibiki, T. Takamasa, M. Ishii, Interfacial-area Transport of Bubbly Flow in a Small Diameter Pipe, *J. Nuclear Science and Technology* 38 (2001) 614–620.

[33] M. Ishii, T. Hibiki, Thermo-Fluid Dynamics of Two-Phase Flow, *Springer*, New York, 2006.

[34] T. Hibiki, M. Ishii, Development of One-Group Interfacial-area Transport Equation in Bubbly Flow Systems, *Int. J. Heat Mass Transfer* 45 (2002) 2351–2372.

[35] T. Hibiki, M. Ishii, One-dimensional Drift-flux Model and Constitutive Equations for Relative Motion Between Phases in Various Two-Phase Flow Regimes, *Int. J. Heat Mass Transfer* 46 (2003) 4935–4948.

[36] T. Hibiki, T. Takamasa, M. Ishii, K.S. Gabriel, One-dimensional Drift-flux Model at Reduced Gravity Conditions, *AIAA Journal* 44 (2006) 1635–1642.

[37] T. Hibiki, M. Ishii, Interfacial-area Concentration of Bubbly Flow Systems, *Chemical Engineering Science* 57 (2002) 3967–3977.

[38] K. Mishima, T. Hibiki, Some Characteristics of Air-water Two-phase Flow in Small Diameter Vertical Tubes, *Int. J. Multiphase Flow* 22 (1996) 703–712.

[39] T. Hibiki, S. Hogsett, M. Ishii, Local Measurement of Interfacial-area, Interfacial Velocity and Liquid Turbulence in Two-phase Flow, *Nucl. Eng. Des.* 184 (1998) 287–304.

[40] Q. Wu, M. Ishii, Sensitivity Study on Double-sensor Conductivity Probe for the Measurement of Interfacial-area Concentration in Bubbly Flow, *Int. J. Multiphase Flow* 25 (1999) 155–173.

[41] T. Hibiki, M. Ishii, Two-Group Interfacial Area Transport Equations at Bubbly-to-Slug Flow Transition, *Nucl. Eng. Des.* 202 (2000) 39-76.

[42] T. Hibiki, M. Ishii, Distribution Parameter and Drift Velocity of Drift-flux Model in Bubbly Flow, *Int. J. Heat Mass Transfer* 45 (2002) 707-721.

In: Frontier Research in Microscale and Nanoscale... ISBN 978-1-61324-665-8
Editor: Lixin Cheng

Chapter 5

THE FEASIBILITY STUDY OF A MICRO-SCALE VIBRATION-INDUCED DROPLET ATOMIZATION AND DROPWISE CONDENSATION MEMBRANE DEVICE

Adam K. Robinson and J. N. Chung*
Department of Mechanical and Aerospace Engineering
University of Florida,
Gainesville, FL, U. S.

ABSTRACT

This study demonstrated the feasibility of simultaneously using a vibrating piezoelectric driver as an atomizer and condensing surface. The vibrations were of appropriate frequency and intensity to produce Vibration Induced Droplet Atomization (VIDA) in the condensate on the condenser surface.

The effect of VIDA was to disperse the vast majority of the condensate as a chaotic spray. The driver was a brass shim 21 mm in diameter and 0.1 mm thick. Driving signal was sinusoidal at 35 V_{p-p} and the frequency ranged from 2.0 to 4.0 kHz. The working fluid was water. Saturation temperature ranged from 23 to 32°C (0.028 to 0.032 bar). Condenser surface subcooling ranged from 0 to 17°C. The cold side of the driver was considered insulated. The piezoelectric driver/condenser surface was oriented in a vertical plane and was non-isothermal.

Surface temperature measurements revealed slight resistive heating in the piezoelectric element for the VIDA cases, which raised the temperature of over half the condenser surface above the saturation temperature of the vapor. This surface temperature change invalidated the theoretical model of the heat transfer process.

Despite this resistive heating, and resulting reduction in active condenser surface area, the VIDA cases still showed roughly equivalent overall heat transfer rate to the non-VIDA cases, indicating a more effective heat transfer process in the former.

---

* Corresponding Author, e-mail : jnchung@ufl.edu, Fax : 352-392-1071

Keywords: Dropwise Condensation, Droplet Atomization, Microscale, Piezoelectric Membrane, Vibration-Induced, Feasibility Study.

NOMENCLATURE

c_P	Constant-pressure specific heat, kg^{-1} K^{-1}
q	Total heat rate, Watt
Tatm	Atmospheric temperature just above the test surface, assumed to be equal to saturation temperature, °C
To	Average edge temperature of the driver, °C
T	sat Saturation temperature of test vapor, °C
ΔT	Maximum subcooling of the driver, Tatm – To, °C
$\dot{V}$	Volumetric flow-rate, m^3 s^{-1}
r_{eff}	Effective maximum departing droplet radius, m

ABBREVIATION

DWC	Dropwise condensation
FWC	Filmwise condensation
HF	Heat flux
HT	Heat transfer
HTC	Heat transfer coefficient
THR	Total heat rate
VIDA	Vibration-induced droplet atomization

I. INTRODUCTION/LITERATURE SURVEY

Progressively smaller and faster processing electronic devices require the dissipation of higher heat flux. In 2001, the Semiconductor Industry Association reported that the high-performance electronics category required near 40W/cm^2 heat flux dissipation. The required heat flux dissipation predicted for the same category in 2016 more than doubled to 96W/cm^2 [1].

The challenge in this growth of required heat flux lies in the fixed temperature drop over which it must occur. Since the maximum junction temperature – the highest temperature in the device – of these semiconductor-based devices is near 85°C, the source-to-sink temperature drop must be less than about 60°C. According to Mudawar [2], in his assessment of high-heat-flux thermal management techniques, the above constraints demand a two-phase cooling scheme. However, reliance on a two-phase cooling solution brings new problems of thermal shock associated with incipient temperature drop and the added complexity of fluid handling in a sensitive electronic environment.

François, Shyy, and Chung [3] proposed a theoretical model of a MEMS device, completing a vapor compression cooling cycle with a fluorinert chemical as the working fluid. The proposed system was to be driven by a piezoelectric driver that would also act as the condensing surface. This scheme allowed for spatial control of the condensate drops through surface tension. This step was taken so that the device would be gravity in-sensitive. This analysis presumed dropwise condensation (DWC) on the driver surface, which has been shown to offer heat transfer coefficient (HTC) an order of magnitude higher than filmwise condensation (FWC) [4].

François [5], in a numerical simulation, looked in detail at a droplet impacting a heated wall and the heat transfer aspects of the evaporative side of the proposed device. However, analysis of the condensation side of the cell was left as an assumption that DWC would offer a certain order of magnitude heat transfer. Aside from a simple energy analysis, detailed consideration of the condensation and droplet ejection processes was left for future study.

Preliminary effort at droplet ejection on the meso length-scale showed surface tension to be a major barrier to ejecting the majority of the primary drop as one mass from a vibrating piezoelectric driver [6]. Thus this study sought an alternative method for droplet ejection in Vibration Induced Droplet Atomization (VIDA), as described by Vukasinovic [7]. The process of VIDA occurs when a stationary sessile droplet is subjected to periodic axial accelerations of a certain frequency and amplitude. This frequency corresponds to the resonant frequency of the drop-driver system, while the VIDA process exhibits a threshold behavior with respect to the peak amplitude of accelerations experienced by the drop. Below this threshold of acceleration amplitude, the drop surface shows standing axial and longitudinal waves, which grow in amplitude with increasing acceleration. Above the threshold acceleration, the amplitude of the surface waves increases until the momentum of the liquid at the wave tips overcomes the surface tension and the much smaller satellite drops pinch off, or are ejected. This ejection process has a time scale of milliseconds, essentially atomizing the primary drop instantly. The VIDA method for condensate dispersal was attractive in that it also provided for control of the liquid condensate through surface tension.

The first major question to come from this course of design was how a vibrating surface would function as a condenser. Some work has been done on the effects of low frequency (0.1 to 100 Hz), high amplitude (1.5 to 13+ mm) axial and perpendicular oscillations with external FWC on a cylindrical surface, as in [8–13]. Some work had also been done on mechanical vibrations in the ultrasonic frequency range, but only on the liquid-to-gas [14] and solid-to-liquid [15] phase change regimes. One study was found to specifically examine DWC of steam on a vertically oriented, perpendicularly vibrating condenser plate [16]. However, the vibrations here were similar to previously reviewed FWC studies in both frequency ($\leq$15 Hz) and amplitude (5 $mm_{p\text{-}p}$). All of the condensation with vibration studies above made a common observation, that if the mechanical vibrations caused a significant deformation of the condensate film or otherwise reduced the total amount of condensate on the condenser surface, then overall heat transfer was increased, due to lower conduction resistance in the liquid phase.

Therefore if VIDA-range accelerations were applied to a condenser surface, one would expect that much of the condensate would be ejected. This would then leave the remaining condensate held in droplets of much smaller size on the condenser surface. According to the majority of experimental results [17–23], smaller maximum droplet size for a condenser surface undergoing DWC leads to a higher HTC. Tanasawa et al. [24] applied a periodic wiper blade to clear away condensate from their DWC surface. The effect of the wiper blade was to artificially reduce the maximum droplet size on the condenser surface. They noted just after the wiper blade passed over the test surface that the HTC peaked to 100 W/cm^2 – between 2.5 and 3.5 times the time averaged value. In a numerical simulation, Burnside and Hadi [22] noted a peak HTC of 200 W/cm^2, for a reduced maximum departing droplet size. However, no existing work was found concerning VIDA with multiple droplets on the driver surface or VIDA with condensation heat transfer. Heffington et al. [25] attempted a heat dissipation device based on pure VIDA, but modified that arrangement with a porous cover plate, yielding an entirely different ejection process.

Typically some method of surface preparation is required to achieve DWC. Surface preparation is necessary to reduce the wettability of the condenser surface – in other words to decrease the contact angle of the vapor-liquid-solid interface or contact line. Three main modes of surface preparation were used in the literature:

1. Direct application of organic or other compounds to the condenser surface
2. Addition of these same promoters to the bulk liquid reservoir supplying the steam to the condenser surface
3. Physical or electro-chemical manipulation of the surface makeup, including the following:
   - Noble metal plating or condenser surface
   - Teflon coating of condenser surface with and without ion implantation
   - Copper condenser surface with ion implantation
   - Self-assembled organic monolayers and L-B films

Direct application of organic and other promoters to the surface was generally most effective in a laboratory setting, in that it produced the highest HTC. However, the DWC promotion was generally short lived and extremely sensitive to impurities in the test vapor or surface contamination [24, 26-32]. Addition of promoter to the liquid reservoir as the promotion method generally found similar drawbacks to the direct application cases, although the promotion effect could often be regenerated for a few cycles by addition of promoter to the reservoir [20, 27, 33, 34]. Noble metal or chromium plating of the condenser surface found the same drawbacks of relatively short lifetimes and high susceptibility to contamination as the organic promoters [21, 31, 35, 36]. Self-assembled monolayers and Langmuir-Blodgett films (several monolayers on top of one another) of organic promoters such as *n*-octadecyl mercaptan and stearic acid have also shown promise for promotion methods [37-40]. Substrate bonding problems were experienced by those that studied Teflon coated copper tubes with and without ion-implantation [41-44]. Some promise was shown with ion implantation of a copper surface, as this was the only DWC promoter surface preparation to prove durable in a test under industrial conditions [45, 46].

II. Apparatus

The piezoelectric driver used as a resonator in this study was an American Piezoelectric Corporation APC# FT-20T-3.6A1. The driver was a circular brass shim 0.004 inch (0.1 mm) thick and 0.75 inch (19 mm) in diameter. Figure 1 shows the piezoelectric driver located in the center of a circular Delrin plate, 0.5 inch thick. The visible side of the shim faced the hot side of the pressure vessel and acted as the condenser surface. The other side of the shim faced the cool side of the pressure vessel and held the piezoelectric ceramic patch. The area behind the cool side of the driver was hollwed out and vented to allow free driver movement. About 1 mm in the radial dirction was allowed for clamping the driver edge.

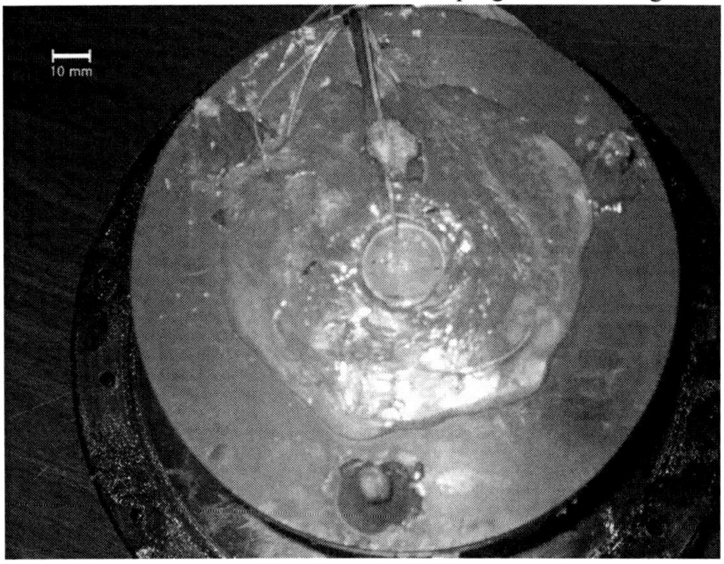

Figure 1. Photograph of piezoelectric driver mounting and insulation.

The driving signal was generated by a standard B+K, model 4017A, 10 MHz function generator. Only the sinusoidal wave-form was used. This signal was amplified by a Krohn-Hite DCA-10. This amplifier worked with frequencies from DC to 10 MHz with continuously variable gain from 1 to 10. The applied voltage for each applicable case was 35.0±0.3 V_{p-p}. Temperature measurements were made with type-T, Teflon-insulated thermocouples from Omega Engineering, Inc. Thirty-six gauge thermocouple wires were used for the surface temperature measurements of the driver, while 30 gauge wires were used for all other measurements. The accompanying data acquisition system was a Measurement Computing PCI-DAS-TC, which included an isothermal terminal block. Cold-junction compensation was internal and automatic.

Three small-gauge thermocouples were attached to the hot-side driver surface along a radial line and equally spaced. RTV silicone sealant was the adhesive used. Care was also taken to insulate the wire leads of the thermocouple bead, so that they would not act as fins to influence the bead temperature [47]. One thermocouple was placed about ½ inch above the condenser surface to measure the atmospheric temperature.

Four thermocouples were equally spaced around the circumference of the driver. The beads of these thermocouples were held in contact with the top surface of the driver edge by the clamping ring, of polycarbonate. This same piece also served to insulate the thermocouple beads from their surroundings.

One thermocouple each was inserted in the inlet and outlet coolant tubes, for a total of ten temperature measurements. The coolant flow-rate was determined from a float-type flow-meter. The schematic of the experimental system is given in Figure 2. The pressure vessel was an aluminum cylinder with four circular observation windows, evenly placed around the circumference. There was also a circular observation window parallel and directly adjacent to the test surface. This arrangement allowed for both overhead and oblique views of the condenser surface. However, condensation on the interior window surface obscured the view of the test surface when the saturation temperature rose sufficiently above the ambient room temperature.

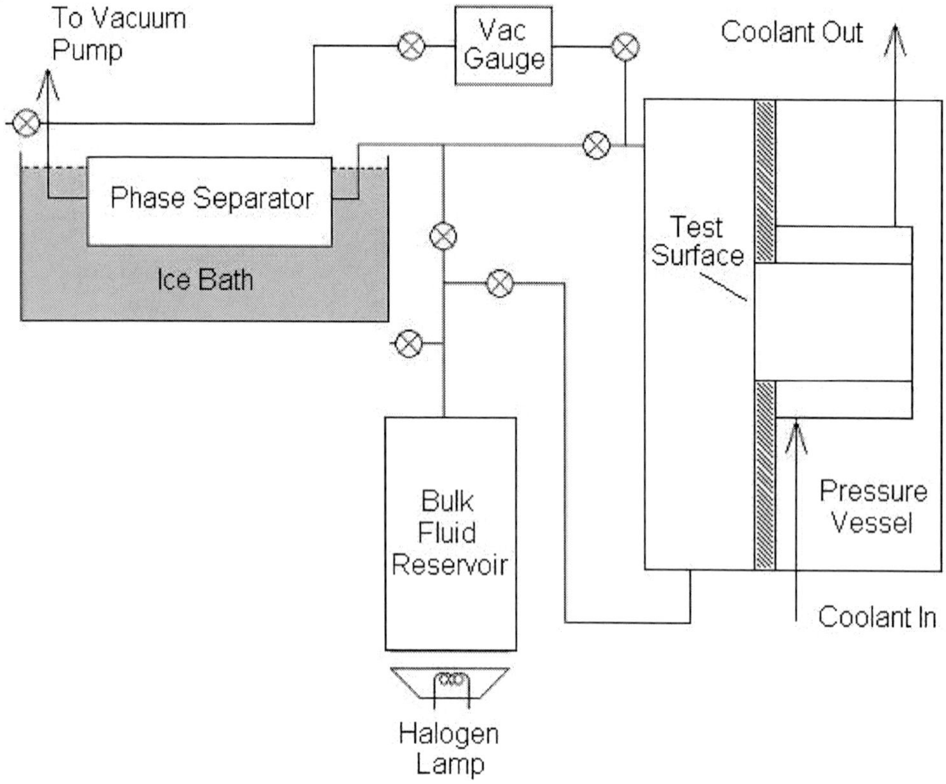

Figure 2. Schematic of apparatus.

The bulk fluid reservoir was a galvanized steel container connected to the rest of the apparatus by vacuum tubing and various shutoff valves. The reservoir was thus thermally isolated from the main vessel. The heat source to the reservoir was a 250W halogen lamp, manually controlled by a precision rheostat. A phase separator minimized the amount of liquid condensing in the vacuum pump oil. Vacuum seals were made with Teflon tape, pipe threads, buna-n rubber o-rings, JB Weld, RTV silicon sealant, Dupont high vacuum grease, and Copper-Bond epoxy. The coolant was water from an ice bath at atmospheric pressure.

III. Theoretical Model

Most studies of condensation heat transfer either focus on a constant-temperature, flat, vertical condenser surface or on the overall performance of a single tube or an array of tubes. This study was unique in that it used a flat, vertical condenser surface that was not isothermal. The theory used to predict the DWC HF was the Rose theory [48-50]. The Rose theory assumed an isothermal condenser surface and provided an estimate of average surface HF as a function of saturation temperature, surface temperature, and vapor properties. Because of the theory's isothermal condenser surface assumption, a radial temperature profile had to be assumed in order to make an estimate of the theoretical total heat transfer rate (THR) of the driver at the given experimental conditions. A two dimensional model (radial and axial position) of the conduction in the thin brass shim showed that the maximum temperature differences in the axial direction would be much less than 1°C, thereby justifying the 1-dimensional heat conduction model used.

This radial temperature profile was assumed to take the nonlinear form, $T(r) = C/r^n + B$, where *B*, *C*, and *n* were constants (Figure 3). Three boundary conditions were needed to define a radial temperature profile based on experimental conditions. The temperature of the saturated vapor represented the highest temperature to which the shim was exposed. The edge temperature of the shim, at the inside diameter of the cooler barrel, was the lowest temperature to which it was exposed.

The cold side of the driver was assumed insulated. Since all the heat conducted through the shim would then have to pass through the small edge area, this area was estimated to have a constant HF. This constant HF was the THR for the driver, divided by the small cross-sectional area at the edge (circumference x thickness) of the exposed portion of the driver surface. This constant HF defined the temperature gradient at the driver edge, a boundary condition for the assumed temperature profile. An initial guess for the THR was required to begin the iterative process of determining the theoretical THR for a given set of experimental vapor and edge temperatures.

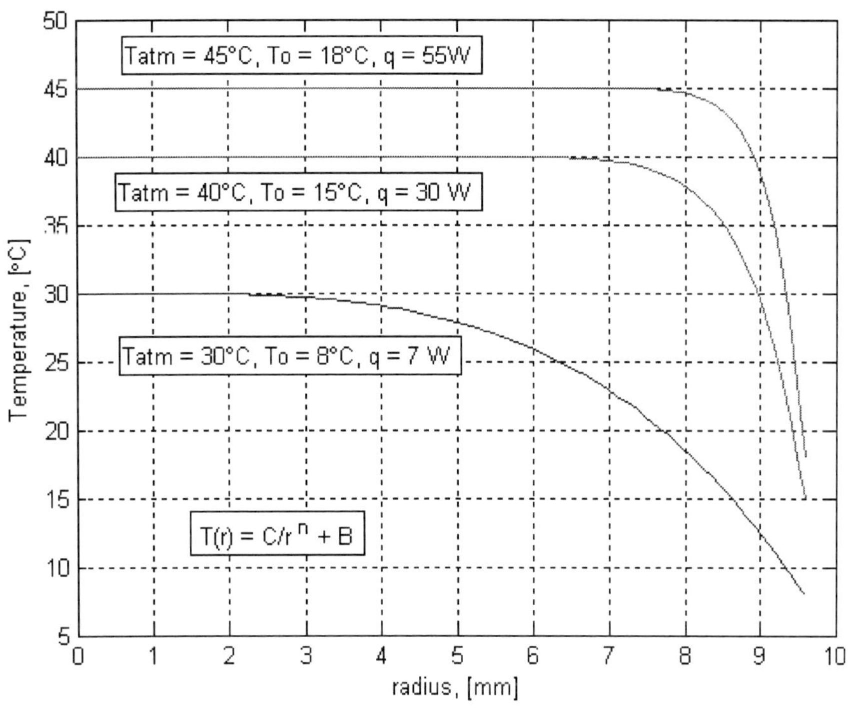

Figure 3. Assumed radial temperature profiles for three experimental data sets.

The resulting temperature profile was then input to the Rose theory – a function of vapor-to-surface temperature drop and fluid properties – to obtain a radially-distributed estimate of the HF coming into the driver (Figure 4). The radial HF distribution was then integrated over the surface of the shim to get the theoretical estimate of the THR. The THR value for this iteration was then averaged with the originally assumed value, used to create the temperature profile for this iteration.

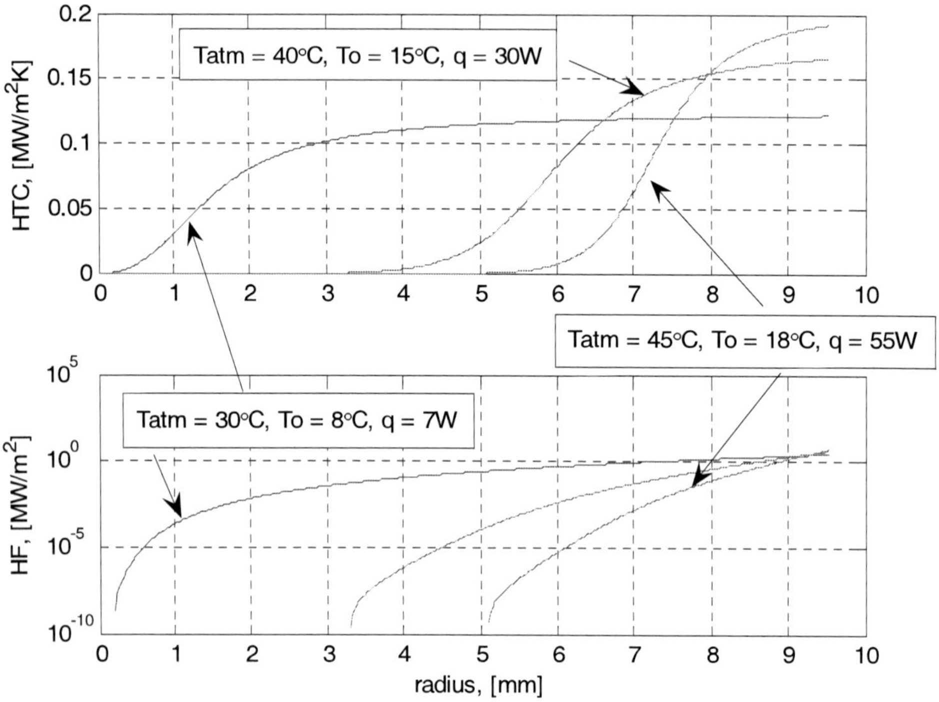

Figure 4. Theoretical radial heat transfer coefficient and heat flux distribution for three sets of experimental data.

The change between the resulting, averaged value and the original assumption for the iteration was noted. If the change was too large, a new temperature profile was assumed with the current THR estimate in place of the old and the process of constructing the theoretical estimate repeated. If the change in estimated theoretical THR was sufficiently small for that step, then the value was accepted and iterations terminated.

It should be noted that the assumed exponential form of the radial temperature distribution does not satisfy the cylindrical differential Heat Conduction Equation. However, it does match well with the available surface temperature data for the non-vibrational case. Also, the data points presented in this study showed that the theoretical THR values obtained in this manner matched the experimental THR values, to well within experimental uncertainty.

Seven different tables were made, each assuming a different value for the maximum departing drop size and covering applicable ranges of saturation and subcooling temperatures for the study. Then the experimental atmospheric and condenser edge temperatures were used to interpolate the theoretical THR which would result if the given maximum departing droplet size for each table were assumed. These seven THR values for a given pair of edge and saturation temperatures were then used to construct an exponential curve fit as a function of maximum departing droplet radius. The maximum departing droplet radius from the curve fit which yielded the measured THR was then recorded as the effective maximum departing droplet radius. If this effective maximum departing droplet radius was smaller than the theoretical one, given by the Rose theory, then the DWC HT was more efficient than the theoretical model predicted.

There was no way to account for the effects of vibration in the heat conduction model without more detailed information about the driver surface temperature distribution.

Measured THR was determined by,

$$THR = \rho \dot{V} c_p \Delta T \tag{1}$$

where ρ was the density of the coolant (water), $\dot{V}$ was the volumetric flow-rate read from the flow meter, c_p was the constant-pressure specific heat of the coolant, and ΔT was the temperature drop between the coolant inlet and outlet to the pressure vessel.

IV. RESULTS AND DISCUSSION

Before each test run, the driver surface was cleaned with an abrasive cleaner (Brasso) and rinsed with distilled water to remove contaminants. Other than this, no special effort was made to promote DWC. Nevertheless, the observed mode of condensation was distinctly dropwise. The apparatus was never able to produce filmwise condensation.

This result was likely due to a residue of the RTV silicone sealant or the Dupont high vacuum grease that was used on the vessel seals. After the completion of each set of tests, an oily film was apparent on all interior surfaces of the vessel. The specific agent of promotion was not identified.

A quiescent driver undergoing DWC was completely covered by hemi-spherical drops of varying size (Figure 5-a), i.e., no condensate film was apparent. The largest of these drops were near 0.5 to 1.0 mm in radius, determined by graphical analysis of video images. The observed condensation process on the quiescent driver was typical of DWC. In a continual process, in various temporal stages across the condenser surface, drop embryos nucleated, grew from direct contact condensation to the point where they touched, and then merged with neighboring drops. This process progressed until the drop mass exceeded the surface tension holding it to the vertical surface, and it swept down the condenser surface. This size was referred to as the maximum departing droplet radius. Once a large droplet departed (Figure 5-b) it absorbed all droplets in its path as it ran down the condenser surface (Figure 5-c), leaving behind exposed condenser surface area (Figure 5-d). On this exposed condenser surface, new droplets nucleated and grew (Figure 5-e) until the condenser surface appeared as before (Figure 5-f). When the driver signal was applied in an appropriate voltage and frequency range to incite VIDA, the vast majority of the condensate on the condenser surface was cleared off within a few milliseconds, leaving the maximum droplet size on the surface less than one tenth of the previous value. Video frames, taken with a high-speed camera, offered a detailed look at the onset of VIDA for a condenser surface undergoing DWC (Figure 6). Before the driver was switched on, the droplet size and spatial distribution was typical for quiescent DWC (Figure 6-a). Condensation on the viewing window partially obscured the camera view of the condenser surface. Ejected droplets also contributed to obscuring the camera view. Just at the onset of vibration (4ms elapsed from the first visual evidence of movement), the outlines of the individual droplets became less distinct, due to their induced oscillations (Figure 6-b).

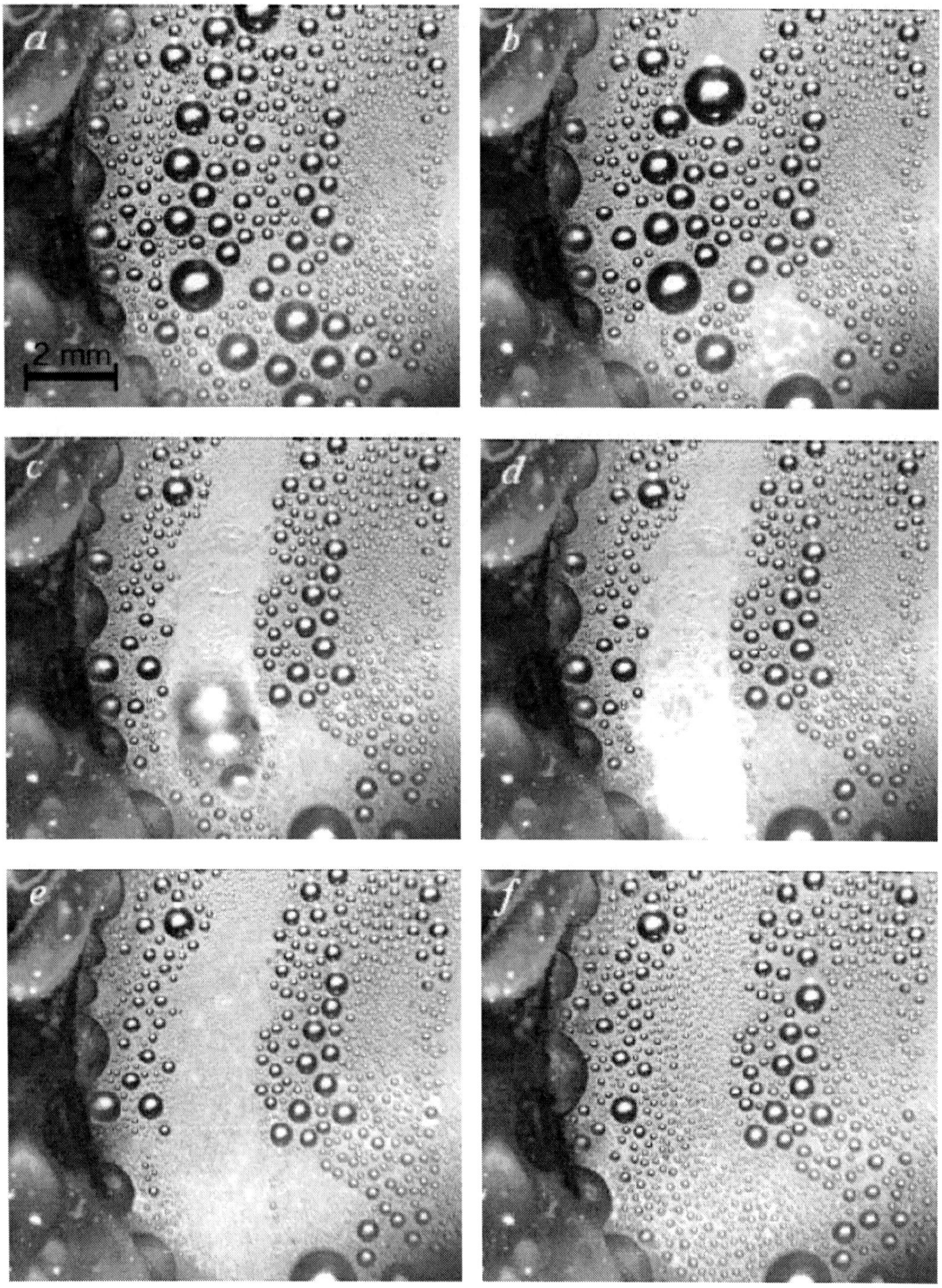

Figure 5. Droplet departure and regeneration in DWC without vibration. Series of frames from a video of condensation without VIDA taken on March 15-16, 2005. Conditions were THR of 27 W, Tatm of 35°C and To of 16°C.

Next, the oscillating droplets began to merge, 6 ms elapsed (Figure 6-c). Centripetal acceleration from the driver oscillations caused droplets above a certain size (mass) to be propelled toward the center of motion, 16 ms elapsed (Figure 6-d).

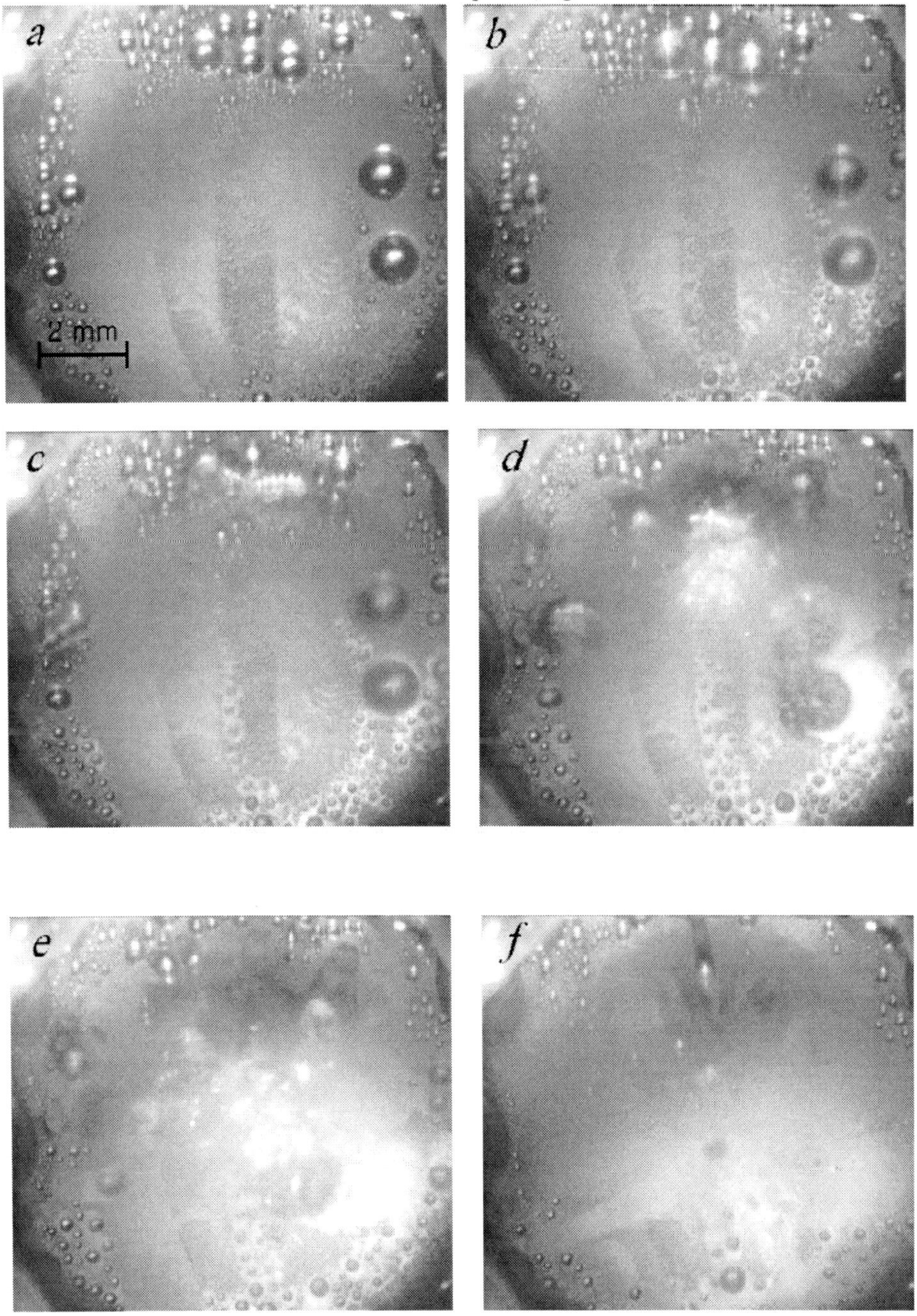

Figure 6. Onset of VIDA, front view. Conditions were THR of 27 W, Tatm of 35°C and To of 16°C. Video taken at 500 fps on March 18, 2005.

This acceleration was more intense closer to the center of motion, therefore the largest stable drop sizes grew smaller toward the center of the driver. The centripetal acceleration was strong enough to compel drops, against the force of gravity, directly upward on the vertically oriented driver. The propelling of droplets toward the center of motion aided the effectiveness of VIDA of clearing the condensate from the driver surface.

As the merged droplets moved toward the center of the resonating driver, their surface was constantly oscillating and ejecting droplets. At the center of the driver, this ejection was most intense. The bulk of fluid was quickly dissipated in this manner, leaving a surface with significantly less condensate on it, 24 ms elapsed (Figure 6-e). The observed steady state droplet distribution, 120 ms elapsed, with applied VIDA-range vibrations showed at least an order-of-magnitude decrease in the maximum departing droplet size (Figure 6-f). However, there was also a noted absence of even small droplets on the center area of the condenser surface, a condition which remained during steady state operation.

The onset behavior of the DWC with VIDA was similar to steady state operation (Figure 7). In a typical experimental image set, a large droplet was propelled from the outer edge of the driver toward the center, and a small residual drop was present in the center (Figure 7-a). Streaks issued from the drop as it progressed toward the center, 3 ms elapsed (Figure 7-b).

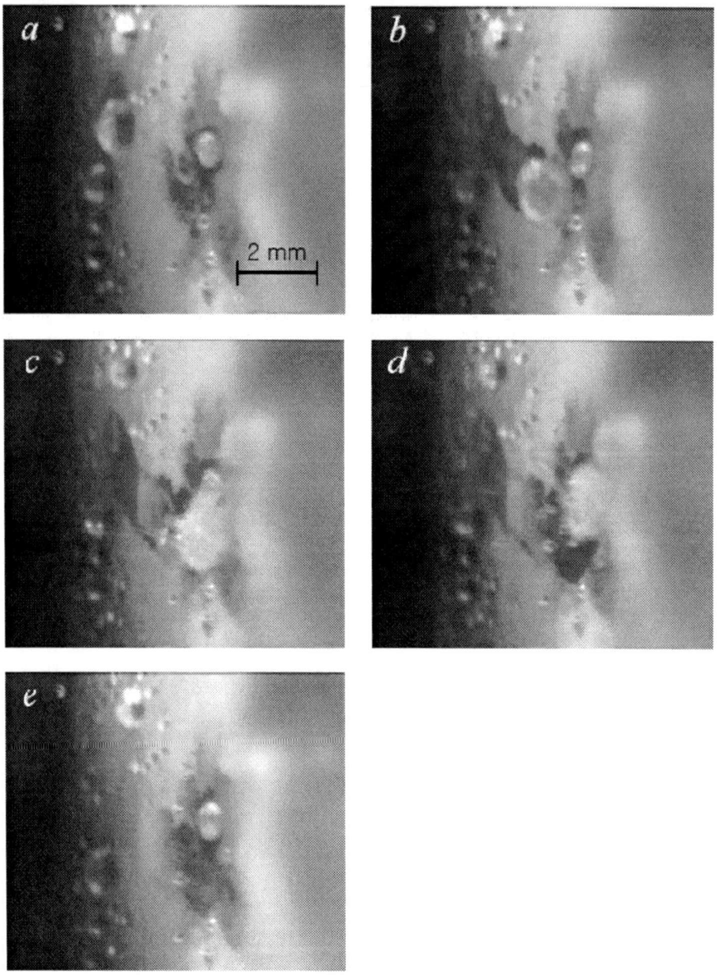

Figure 7. Droplet departure and dissipation by VIDA during steady state DWC. Conditions were THR of 27 W, Tatm of 35°C, and To of 16°C. Video taken at 1000 fps on March 23, 2005.

These streaks were satellite droplets being ejected from the primary drop. During the position change of the drop, the liquid-vapor interface was undulating and chaotic, due to the induced wave structure from the driver oscillations, rather than hemispherical, as it would have been on a quiescent condenser surface. As it reached the center of the driver, the primary drop quickly merged with the residual drop, present in the first frame, 6 ms elapsed (Figure 7-c). The merged drop quickly dissipated, 9 ms elapsed (Figure 7-d) by ejection of satellite drops until only a mass similar in size to the original residual drop remained, 26 ms elapsed (Figure 7-e). The size of this residual drop decreased for a given increase in vibration intensity. Intensity of vibration was highest at the exact resonant frequency of the driver-condensate system. However, many variables likely affected this quantity, such as temperature of the piezoelectric element, the mass of condensate, the spatial distribution of the condensate, age of the driver, as well as the frequency of the applied signal. Therefore frequent adjustment of the applied signal frequency was required to maintain good vibration intensity, as judged by the intensity of the sound produced.

For DWC on a quiescent driver, the experimental surface temperature measurements matched the theoretical model well (Figure 8). The surface temperature data points for the quiescent case (circles) showed a slight downward trend with increasing radius compared to the assumed profile. However the difference between theory and experiment was well within experimental uncertainty. Surface temperature measurements for DWC in the presence of VIDA showed that the assumed temperature profile under–predicted for this case. Though the experimental parameters, and therefore the assumed temperature profiles, were nearly indistinguishable in the two comparison cases, the data points for the VIDA case surface temperatures (diamonds) (Figure 8) clearly were higher for the first two points than the assumed temperature profile.

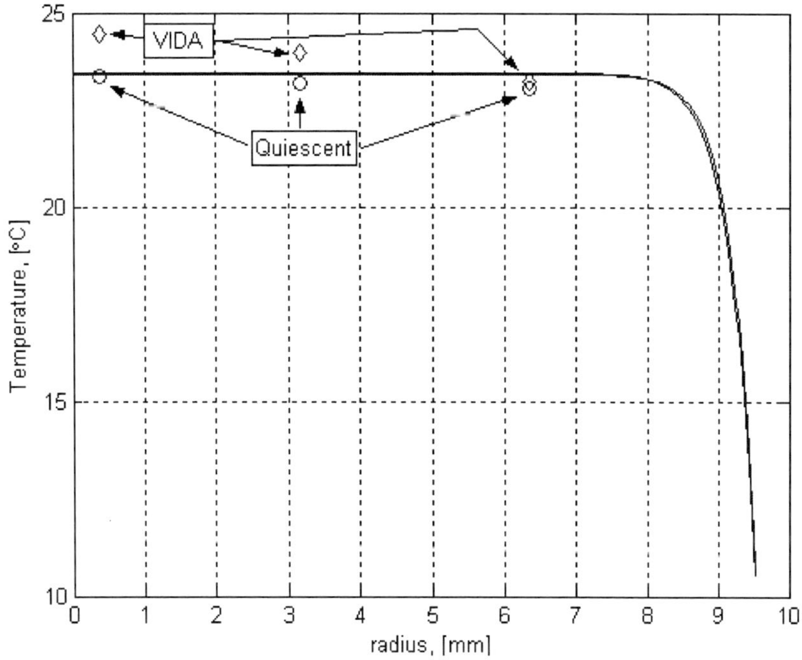

Figure 8. Comparison of experimental surface temperature measurements and assumed temperature profile for both the quiescent and VIDA cases.

The two surface temperature measurements closest to the center of the driver increased above the saturation temperature when vibration was applied, indicating that condensation was not possible in that region. This condition was likely due to resistive heating of the piezoelectric element. This result agreed with the observation that the center of the driver under steady VIDA conditions appeared free from even very small drops. With the information gathered, it was not possible to estimate the resistive heating of the piezoelectric element.

The elevated surface temperature removed at least the center 2/3 of the driver from the condensation process and therefore reduced the active heat transfer area by about 40 to 50%. So if there was no beneficial effect of vibration, one would expect the overall heat transfer also would be reduced. However, this change would not be proportional to the area since the bulk of the heat transfer occurred near the outer edge of the driver. In future applications a special cooling mechanism could be designed such that more cooling is available in the central area.

Despite this expected decrease in active heat transfer area, the experimental THR for VIDA conditions and DWC showed an increase over the quiescence DWC condition in about half of the cases (Figure 9). The other half of the vibrational cases were intermingled with the spread of data for the quiescent case. This under-performance for half of the VIDA cases may have been due to trace presence of air in the test vapor. The process of DWC has been shown to be extremely sensitive to and retarded by the presence of even tiny amounts of non-condensable gas species, especially at reduced pressures, as in this study [26, 27]. Concentrations as low as 0.1% NCG in the test vapor, between 0.03 and 0.1 bar, reduced HTC by 50%, [27].

No definite effect of vibration was apparent from an examination of experimental THR as a function of maximum subcooling of the driver surface (Figure 10). Maximum subcooling was defined as the measured saturation temperature less than the average edge temperature of the driver. However, an increase in observed overall heat transfer rate was observed for a given increase in maximum subcooling, as expected from the theory.

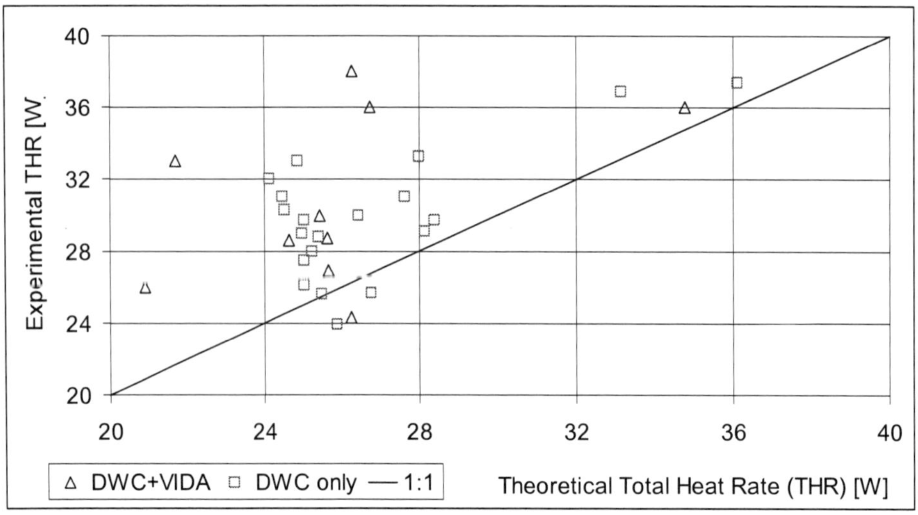

Figure 9. Experimental THR vs. Theoretical THR. Both axes in W.

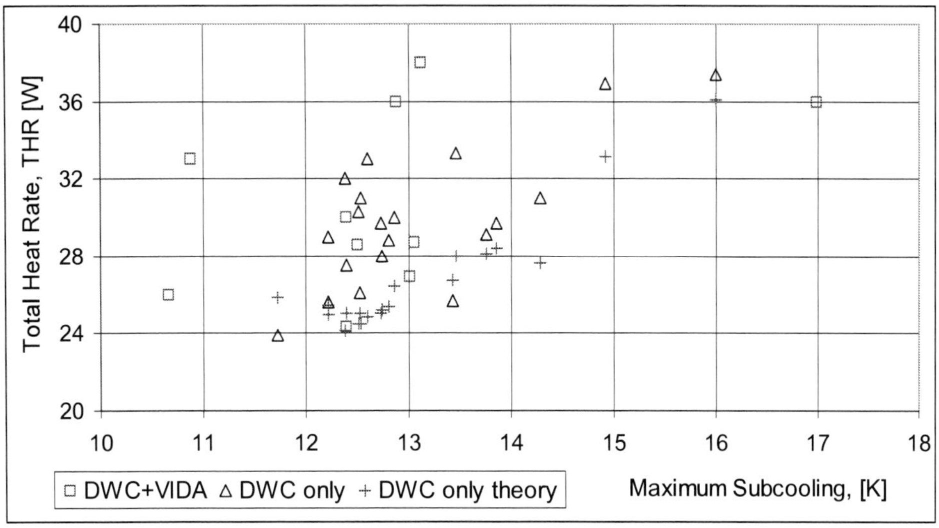

Figure 10. THR vs. Maximum Subcooling, ΔTmax = Tsat – To, in W and K.

As stated earlier, the theoretical model of the DWC HT process was proven invalid by surface temperature measurements for the VIDA case. However, experimentally measured THR for the VIDA cases actually was higher than the theoretical estimate (Table 1). For this reason, the theoretical estimate of THR was included in the data summary, only as a case by case comparator. The non-VIDA cases showed effective maximum departing droplet radii (r_{eff}) 20 to 140% of the theoretical value, indicating the model was basically effective. The VIDA cases showed effective maximum departing droplet radii 12 to 80% of the theoretical value of 1mm. (Smaller effective maximum departing droplet radius indicates higher HTC.) So, despite the reduced HT area of the VIDA cases, the DWC HT was as good as or better than the non-VIDA case.

Table 1. Data summary

VIDA	Atmospheric Temperature	Maximum Subcooling (T_{atm} – To)	Total Heat Rate, Measured	Total Heat Rate, Theoretical	r_{eff}
	°C	°C	W	W	mm
N	30.7	11.7	24	26	1.7
N	25.6	12.2	29	25	0.5
N	27.4	12.2	26	26	1.1
N	25.2	12.4	28	25	0.6
N	24.7	12.5	26	25	0.9
N	26.1	12.9	30	26	0.5
N	27.2	13.5	33	28	0.4
N	26.1	13.8	29	28	0.9
N	26.2	13.9	30	28	0.9
N	27.3	14.3	31	28	0.6
N	30.6	14.9	37	33	0.6
N	31.8	16.0	37	36	0.9

Table 1. (Continued)

VIDA	Atmospheric Temperature	Maximum Subcooling (T_{atm} – To)	Total Heat Rate, Measured	Total Heat Rate, Theoretical	r_{eff}
N	23.1	12.4	32	24	0.2
N	23.0	12.5	31	25	0.3
N	23.3	12.5	30	25	0.3
N	23.5	12.7	30	25	0.4
N	24.1	12.8	29	25	0.5
N	23.9	12.6	33	25	0.2
N	23.7	12.7	28	25	0.6
N	24.4	13.4	26	27	1.4
Y	24.2	10.7	26	21’	0.3’
Y	24.7	10.9	33	22’	0.08’
Y	28.3	12.4	24	26’	1.7’
Y	26.2	12.4	30	25’	0.4’
Y	26.8	12.9	36	27’	0.18’
Y	25.3	17.0	36	35’	0.9’
Y	23.6	13.1	29	26’	0.6’
Y	23.7	13.0	27	26’	0.8’
Y	24.7	13.1	38	26’	0.12’
Y	23.8	12.5	27	25’	0.5’

‘ – Theoretical model not valid for VIDA cases. These values included only for comparison.

V. UNCERTAINTY

The full set of 10 type-T thermocouples was calibrated simultaneously against two precision, lab-grade mercury thermometers (± 0.1 and ± 0.05°C) from near 0 to 47°C at 37 points. Individual calibration curves were constructed for each thermocouple. The maximum root mean squared of the error for each calibrated thermocouple over the calibration data set was less than 0.2°C, relative to the thermometer readings. The maximum discretization error of the A/D converter at the settings used was less than 0.02°C. The maximum absolute error reported by the manufacturer for these thermocouples was 1.0°C. Therefore, the absolute uncertainty of the thermocouple readings was 1.0°C. The calibration data set confirmed this, in that the maximum deviance from the reference temperature for any single reading was ± 1.0°C over 380 data points from -1 to 47°C. The maximum deviation for the calibrated thermocouples was 0.1°C over the calibration set. This value was then used as the uncertainty in calculations.

The coolant flow rate was measured from an analog, float-type flowmeter, with an uncertainty of ±0.1 gph. This combined with the temperature uncertainty to yield an uncertainty in measured (experimental) THR of ±18%. The uncertainty in the theoretical THR, resulting from the uncertainty in input temperatures was ±5%, while that of the r_{eff} was quite high at 200%. The extremely high uncertainty in r_{eff} was due to both the high uncertainty in the experimental THR and the exponential character of the curve fit. Therefore, the effective maximum departing droplet radius was useful only as a first approximation.

VI. Conclusions

A vibrating piezoelectric driver was successfully used as a condenser surface. Though the basic temperature profile of the driver was different in the case with vibration – active condenser surface area was smaller due to resistive heating of the piezoelectric element – the overall heat transfer was roughly equivalent to or better than the non-vibrational cases.

Vibration Induced Droplet Atomization was demonstrated to be an effective method for removing condensate from a piezoelectric driver, which is also used as a condenser surface. It is believed that the VIDA decreased the transport resistance and increased the effectiveness of the heat transfer mechanism.

Acknowledgments

The first author acknowledged the support from the University of Florida in the form of a University Presidential Fellowship. This project was partially supported by the Andrew H. Hines, Jr./Progress Energy Endowment funds.

References

[1] SIA, Semiconductor Industry Association, International Technology Road Map, Assembly and Packaging Section, Technology Requirements (2001) 4–8.

[2] Mudawar, Assessment of high-heat-flux thermal management schemes, *IEEE Transactions on Components and Packaging Technologies* 24 (2001) 122–142.

[3] M.M. François, W. Shyy, J.N. Chung, A Micro-Scale Membrane-Actuated Condenser/Evaporator for Enhanced Thermal Management, *ASME Heat Transfer Division, International Mechanical Congress and Exposition*, HDT (2000) 366-2.

[4] J.W. Rose, Surface tension effects and enhancement of condensation heat transfer, *Chem. Eng. Research and Design* 82 A4 (2004) 419–429.

[5] M.M. François, W. Shyy, *Computations of Drop Dynamics with Heat Transfer*, Doctoral Thesis, University of Florida, Department of Mechanical Engineering (2002).

[6] M. Friedman, J.N. Chung, *Feasibility Study on a Piezoelectric Droplet Ejector*, unpublished paper, University of Florida Department of Mechanical and Aerospace Engineering, November (2001).

[7] B. Vukasinovic, A. Glezer, *Vibration-Induced Droplet Atomization*, Doctoral Thesis, Georgia Technical University, Department of Mechanical Engineering, March (2002).

[8] J.C. Dent, Effect of vibration on condensation heat transfer to a horizontal tube, Proc. Inst. Mech. Eng. 184 Part 1 (1969) 99–105.

[9] J.C. Dent, The calculation of heat transfer coefficient for condensation of steam on a vibrating vertical tube, *International Journal of Heat and Mass Transfer* 12 (1969) 991–996.

[10] Yu.M. Brodov, R.Z. Savel'yev, V.A. Permyakov, V.K. Kuptsov, A.G. Gal'perin, The effect of vibration on heat transfer and flow of condensing steam on a single tube, *Heat Transfer – Soviet Research* 9 (1977) 152–156.

[11] R.Z. Savel'yev, Yu.M. Brodov, L.G. Gal'perin, M.A. Nirenshteyn, A.Y. Ryabchikov, Heat transfer and flow in condensation on vibrating tubes, *Heat Transfer - Soviet Research* 12 (1980) 1–5.

[12] K. Nishiyama, A. Murata, T. Kajikawa, The study of condensation heat transfer performance due to periodic motion of a vertical tube, *Heat Transfer – Japanese Research* 15 (1986) 32–43.

[13] K. Nishiyama, Study of heat transfer performance of a periodic moving condenser, *Heat Transfer – Japanese Research* 17 (1988) 32–42.

[14] H.Y. Kim, Y.G. Kim, B.H. Kang, Enhancement of natural convection and pool boiling heat transfer via ultrasonic vibration, *International Journal of Heat and Mass Transfer* 47 (2004) 2831–2840.

[15] Y.K. Oh, S.H. Park, Y.I. Cho, A study of the effect of ultrasonic vibrations on phase-change heat transfer, *International Journal of Heat and Mass Transfer* 45 (2002) 4631–4641.

[16] H. Shiraiwa, T. Hosokawa, T. Nakata, K. Itoh, Heat Transfer Characteristics on a Horizontal(ly) Vibrating Plate Under Dropwise Condensation, *5th Int. Conf. Multiphase Flow, Yokohama*, Japan, May 30 – June 4 (2004) paper No. 283.

[17] L.R. Glicksman, A.W. Hunt, Jr., Numerical simulation of dropwise condensation, *International Journal of Heat and Mass Transfer* 15 (1972) 2251–2269.

[18] J.W. Rose, Further aspects of dropwise condensation theory, *International Journal of Heat and Mass Transfer* 19 (1976) 1363–1370.

[19] Tanasawa, Y. Utaka, Measurement of condensation curves for dropwise condensation of steam at atmospheric pressure, *Transactions of the ASME Journal of Heat Transfer* 105 (1983) 633–638.

[20] A.S. Gavrish, V.G. Rifert, A.I. Sardak, V.L. Podbereznyy, A new dropwise condensation promoter for desalination and power plants, *Heat Transfer – Soviet Research* 25 (1993) 82–86.

[21] C. Yamali, H. Merte Jr., Influence of sweeping on dropwise condensation with varying body force and surface subcooling, *International Journal of Heat and Mass Transfer* 42 (1999) 2943–2953.

[22] B.M. Burnside, H.A. Hadi, Digital computer simulation of dropwise condensation from equilibrium droplet to detectable size, *International Journal of Heat and Mass Transfer* 42 (1999) 3137–3146.

[23] Y.T. Wu, Yang, C.X., and Yuan, X.G., A theoretical study of the effect of surface thermal conductivity on heat transfer coefficient in dropwise condensation, *Numerical Heat Transfer*, Part A (Applications) 40 (2001) 169–179.

[24] Tanasawa, Y. Funawatashi, J. Ochiai, *Experimental study on dropwise condensation-effect of maximum drop size upon the heat transfer coefficient*, Sixth International Heat Transfer Conference (1978) 477–482.

[25] S.N. Heffington, W.Z. Black, A. Glezer, *Vibration-Induced Droplet Atomization Heat Transfer Cell for High-Heat Flux Dissipation*, Doctoral Thesis, Georgia Institute of Technology, Department of Mechanical Engineering, December (2000).

[26] E.J. Le Fevre, J.W. Rose, An experimental study of heat transfer by dropwise condensation, *International Journal of Heat and Mass Transfer* 8 (1965) 1117–1133.

[27] D.W. Tanner, D. Pope, C.J. Potter, D. West, Heat transfer in dropwise condensation at low steam pressures in the absence and presence of non-condensable gas, *International Journal of Heat and Mass Transfer* 11 (1968) 181–182.

[28] E. Citakoglu, J.W. Rose, Dropwise condensation – some factors influencing the validity of heat-transfer measurements, *International Journal of Heat and Mass Transfer* 11 (1968) 523–537.

[29] E. Citakoglu, J.W. Rose, Dropwise condensation – the effect of surface inclination, *International Journal of Heat and Mass Transfer* 12 (1969) 645–650.

[30] S.A. Stylianou, J.W. Rose, Dropwise condensation on surfaces having different thermal conductivities, *Transactions of the ASME Journal of Heat Transfer* 102 (1980) 477–482.

[31] S.S. Finnicum, J.W. Westwater, Dropwise vs filmwise condensation of steam on chromium, *International Journal of Heat and Mass Transfer* 32 (1989) 1541–1549.

[32] Y.L. Lee, The wettability of solid surfaces modified by vacuum deposition of stearic acid, *Colloids and Surfaces A: Physicochemical and Engineering Aspects* 155 (1999) 221–229.

[33] R.J. Hannemann, B.B. Mikic, An experimental investigation into the effect of surface thermal conductivity on the rate of heat transfer in dropwise condensation, *International Journal of Heat and Mass Transfer* 19 (1976) 1308–1317.

[34] K.J. Kim, A.M. Lefsaker, A. Razani, A. Stone, The effective use of heat transfer additives for steam condensation, *Applied Thermal Engineering* 21 (2001) 1863–1874.

[35] S. Hatamiya, H. Tanaka, Dropwise condensation of steam at low pressures, *International Journal of Heat and Mass Transfer* 30 (1987) 497–507.

[36] G.A. O'neill, J.W. Westwater, Dropwise condensation of steam on electroplated silver surfaces, *International Journal of Heat and Mass Transfer* 27 (1984) 1539–1549.

[37] Q. Zhao, G.M. Wang, D.C. Zhang, J.F. Lin, Dropwise condensation on L-B film surface, *Chemical Engineering Processing* 35 (1996) 473–477.

[38] G. Pang, J.D. Dale, D.Y. Kwok, An integrated study of dropwise condensation heat transfer on self-assembled organic surtaces through Fourier transform infra-red spectroscopy and ellipsometry, *International Journal of Heat and Mass Transfer* 48 (2005) 307–316.

[39] S. Vemuri, K.J. Kim, An experimental study on the concept of dropwise condensation, *International Journal of Heat and Mass Transfer*, Article in Press (2005) xx–xx.

[40] S. Vemuri, K.J. Kim, B.D. Wood, S. Govindaraju, T.W. Bell, Long-term testing for dropwise condensation using self-assembled monolayer coatings of *n*-octadecyl mercaptan, *Applied Thermal Engineering* 26 (2006) 421–429.

[41] V.N. Bochagov, A.R. Dorokhov, Dropwise condensation on a polymer surface, *Heat Transfer - Soviet Research* 15 (1983) 73–80.

[42] Y. Utaka, A. Saito, T. Tani, H. Shibuya, K. Katayama, Study on dropwise condensation curves (measurement of propylene glycol vapor on PTFE coated surfaces), *Bulletin Japan Society Mechanical Engineers* 28 (1985) 1150 1157.

[43] X.H. Ma, B.X. Wang, D.Q. Xu, J.F. Lin, Lifetime test of dropwise condensation on polymer-coated surfaces, *Heat Transfer – Asian Research* 28 (1999) 551–558.

[44] C. Ma, J. Chen, D. Xu, J. Lin, C. Ren, Z. Long, Influence of processing conditions of polymer film on dropwise condensation heat transfer, *International Journal of Heat and Mass Transfer* 45 (2002) 3405-3411.

[45] Q. Zhao, D. Zhang, J.F. Lin, Surface materials with dropwise condensation made by ion implantation technology, *International Journal of Heat and Mass Transfer* 34 (1991) 2833–2835.

[46] Q. Zhao, B.M. Burnside, Dropwise condensation of steam on ion implanted condenser surfaces, *Heat Recovery Systems and CHP* 14 (1994) 525–534.

[47] J.W. Rose, Heat-transfer coefficients, Wilson plots and accuracy of thermal measurements, *Experimental Thermal and Fluid Science* 28 (2004) 77–86.

[48] E.J. Le Fevre, J.W. Rose, A Theory of Heat Transfer by Dropwise Condensation, *Proceedings of the 3rd International Heat Transfer Conference*, American Institute of Chemical Engineers, Chicago, IL 2 (1966) 362–375.

[49] J.W. Rose, Some aspects of condensation heat transfer theory, *International Communications in Heat and Mass Transfer* 15 (1988) 449–473.

[50] J.W. Rose, Dropwise condensation theory and experiment: a review, *Proceedings of the Institution of Mechanical Engineers, Part A (Journal of Power and Energy)* 216 A2 (2002) 115–128.

In: Frontier Research in Microscale and Nanoscale... ISBN 978-1-61324-665-8
Editor: Lixin Cheng

Chapter 6

LTCM LABORATORY DESCRIPTION

John R. Thome*
Laboratory of Heat and Mass Transfer
Ecole Polytechnique Fédérale de Lausanne
Lausanne, Switzerland

MICROSCALE TWO-PHASE FLOW AND HEAT TRANSFER RESEARCH AT THE LTCM LAB AT EPFL

The main fields of research at LTCM "Heat and Mass Transfer Laboratory" are: Two-phase flow and heat transfer (boiling and condensation) in microchannels, convective boiling of pure fluids and mixtures in plain and enhanced tubes, condensation of pure fluids and mixtures, falling film evaporation and condensation on enhanced tubes and tube arrays, boiling on tube bundles of plain and enhanced tubes, two-phase flows and modeling of flow regime transitions, flow visualization of two-phase phenomena and void fraction measurements with use of laser sheets, turbulence modeling of annular two-phase flows, and numerical modeling of adiabatic and condensing microscale two-phase flows. The LTCM laboratory was founded in 1998 by Prof. Thome upon his appointment at the EPFL. Currently, the laboratory has 12 Ph.D. students (plus one visiting) and 7 post-doctoral researchers.

The LTCM laboratory is well-known for its experimental work and prediction models for numerous types of two-phase heat transfer and two-phase flow processes, typically publishing more than 15 journal papers per year. LTCM hosts (Prof. Thome is the director and Dr. N. Borhani is the secretary) the European Coordination Centre of ERCOFTAC (the *E*uropean *R*esearch *C*ommunity *O*f *F*low, *T*urbulence *A*nd *C*ombustion) with over 120 universities and institutional research organizations as members and publishes a quarterly ERCOFTAC bulletin (in both printed and online versions). LTCM hosts an international week long course on Microscale Heat Transfer each year, having welcomed more than 150 participants from

* Address correspondence to John Thome, Laboratory of Heat and Mass Transfer, EPFL-STI-IGM-LTCM, Mail 9, CH-1015 Lausanne, Switzerland. E-mail: john.thome@epfl.ch

over 20 different countries in the past five years. Professor Thome is also the author of four books, with his 3rd book notable as a free widely used e-book with 20 chapters on two-phase flow and two-phase heat transfer, including chapters on microscale two-phase flow and heat transfer and with over 200 embedded videos (many illustrating two-phase flow phenomena in microchannels, see www.wlv.com for Wolverine Databook III). Prof. Thome received the ASME Heat Transfer Division's Best Paper Award in 1998 for a 3-part paper on flow boiling heat transfer of refrigerants in horizontal tubes, published in the *ASME Journal of Heat Transfer*. Recently, he has received the JandE Hall Gold Medal from the U.K. Institute of Refrigeration in February, 2008 for his contributions to the development of microscale refrigeration heat transfer. His Ph.D. student (R. Revellin) was awarded the Eurotherm Award for the best European thermal science thesis in the period 2005-2008 for his experimental work on characterization of two-phase flows in microchannels.

Microchannel Test Facilities. LTCM possesses six microscale heat transfer test facilities: a single-channel two-phase flow and boiling test facility built as part of a European Madame Curie project (see Figure 1), two multi-microchannel flow boiling test facilities (one located at the IBM Zürich Research Laboratory), one multi-microchannel condensation test facility, one microchannel two-phase cooling test prototype facility for blade servers, and one air-water micro-PIV test unit for flows in small channels. Current projects include three Swiss National Science Foundation grants, including a $4.3 million Nano-Tera project entitled *CMOSIAC* of which LTCM is the project coordinator on the topic of cooling and thermal modeling of future 3D computer chips. Figure 2 shows simplified diagrams of such a 3D-IC processor. Figure 3 shows two diagrams of boiling of R134a in parallel microchannels with poor and excellent flow distribution for the cooling of 2D computer chips, increasing critical heat flux from 115 W/cm^2 to 314 W/cm^2.

Microchannel flow boiling heat transfer, void fraction, critical heat flux and flow pattern map models: This family of LTCM models includes the three-zone flow boiling heat transfer model for evaporation in the elongated bubble regime in microchannels, a physically based model to predict void fractions in the elongated bubble regime as a function of bubble length, two methods for prediction of the critical heat flux in microchannels (one empirical and the other numerical, including hot spot simulation), and two diabatic flow pattern maps for microchannels (one empirical and one mechanistically based). A new 1-d annular flow turbulence model provides a unified method for predicting the liquid film velocity profile, the film thickness, the liquid entrainment fraction and pressure drop for microscale *and* macroscale channels…this work is the stepping stone to the future development of turbulence based annular flow models for CHF, boiling heat transfer and convective condensation in microchannels.

Other LTCM Test Facilities. The LTCM *Falling Film Test Facility* is utilized to obtain local heat transfer coefficients and to visualize inter-tube flow regimes during evaporation or condensation on vertical arrays of horizontal tubes (up to 10 rows deep and 3 rows wide). Numerous types of enhanced tubes with a myriad of re-entrant microchannels have been tested and prediction methods developed. This work is supported by the LTCM Falling Film Research Club (a consortium of refrigeration companies and enhanced tube manufacturers). The LTCM *Flow Boiling Test Facility* is designed for in-tube macrochannel two-phase flow and boiling experiments, including tests on internally microfinned tubes with fins on the order of 0.1-0.4 mm in height. The facility is used to measure two-phase pressure drops, void fractions, heat transfer coefficients, flow regimes and critical heat flux. Currently it is being used to investigate two-phase pressure drops and flow patterns in and after U-bends, sponsored by ASHRAE. The LTCM *Bundle Boiling Test Stand* is designed to conduct two-phase flow boiling experiments on tube bundles, including enhanced tubes with a microscale surface enhancements that increase heat transfer by up to a factor of 15 times compared to plain surfaces. Flow visualization techniques allow for flow visualization within the tube bundle and characterization of flow patterns (using tubes with local glass sections). The LTCM *Blade Server Cooling System* is a prototype for cooling of an actual IBM blade server with all the associated instrumentation and control to operate as a standalone system.

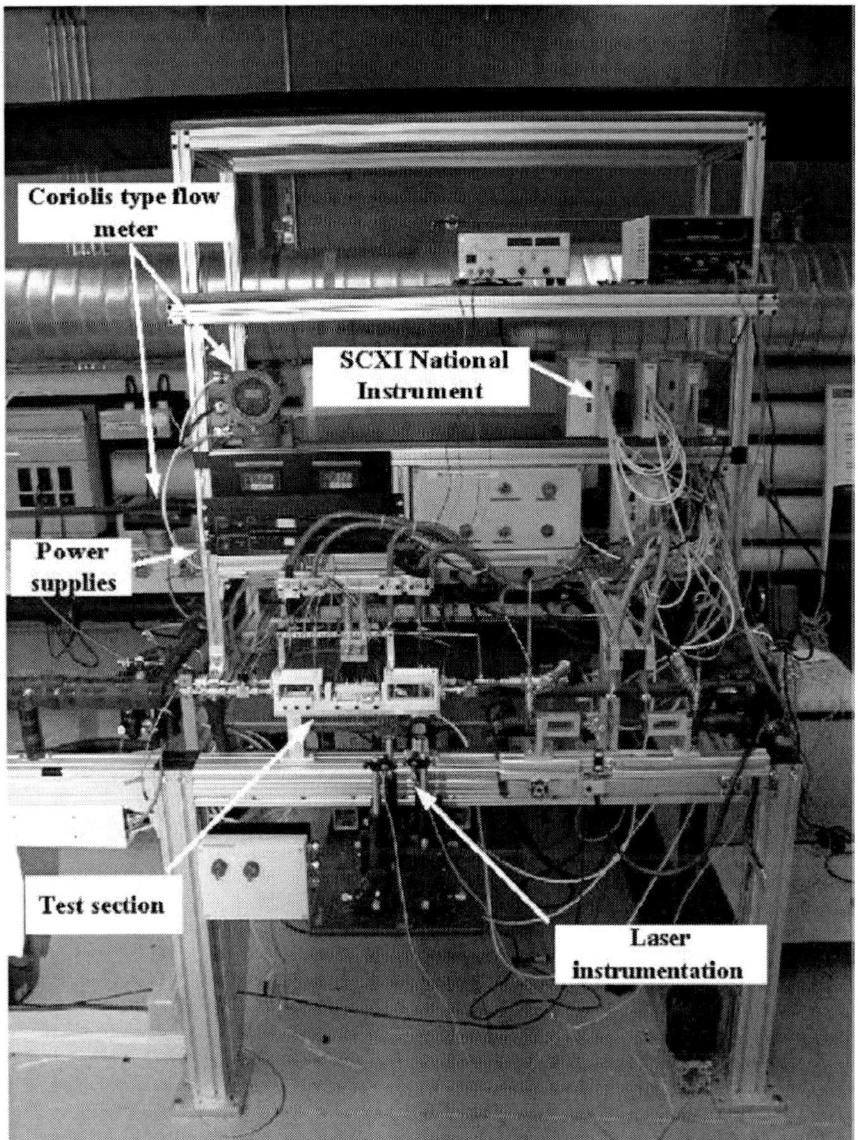

Figure 1. Photograph of LTCM microchannel test facility used to obtain two-phase flow visualizations, characterize flow phenomena (bubble frequencies, lengths, velocities, etc.), local flow boiling heat transfer coefficients, two-phase pressure drops and critical heat fluxes.

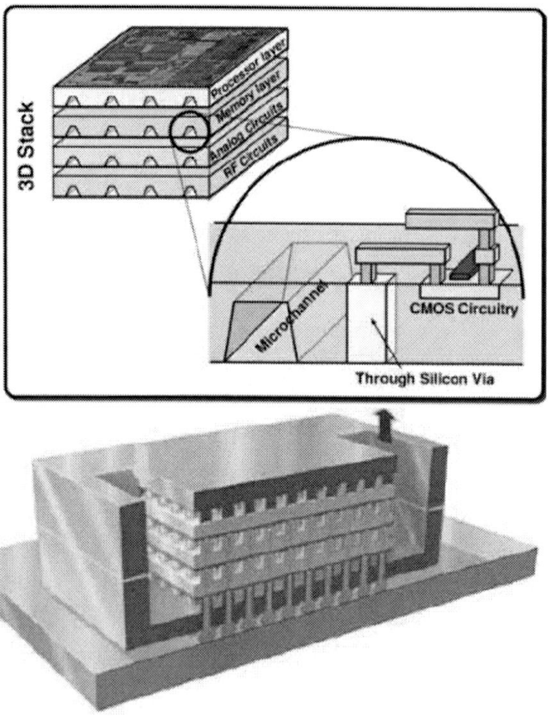

Figure 2. Top: through silicon via connection; bottom: 3D computer chip with micro cooling channels.

Figure 3. Left: flow maldistribution without inlet flow control with large dry zone at upper right; Right: uniform flow distribution using inlet orifices for each channel. (red zone is the heated zone; channels 0.45mm wide and 4.0 mm deep).

Some representative LTCM microscale publications are listed below:

[1] Jacobi, A.M. and Thome, J.R. (2002). Heat Transfer Model for Evaporation of Elongated Bubble Flows in Microchannels, *J. Heat Transfer*, Vol. 124, pp. 1131-1136.

[2] Thome, J.R., Dupont, V. and Jacobi, A. (2004). Heat Transfer Model for Evaporation in Microchannels, Part I: Presentation of the Model, *Int. J. Heat Mass Transfer*, Vol. 47, pp. 3375-3385; Part II: Comparison with the Database, *Int. J. Heat Mass Transfer*, Vol. 47, pp. 3387-3401.

[3] Revellin, R., Dupont, V., Ursenbacher, T., Thome, J.R. and Zun, I. (2006). Characterization of Two-Phase Flows in Microchannels: Optical Measurement Technique and Flow Parameter Results for R-134a in a 0.5 mm Channel, *Int. J. Multiphase Flow*, Vol. 32, pp. 755-774.

[4] Revellin, R. and Thome, J.R. (2007). A New Type of Diabatic Flow Pattern Map for Boiling Heat Transfer in Microchannels, *J. of Micromechanics and Microengineering*, Vol. 17, pp. 788-796.

[5] Revellin, R. and Thome, J.R. (2008). A Theoretical Model for the Prediction of the Critical Heat Flux in Heated Microchannels, *Int. J. Heat Mass Transfer*, Vol. 51, pp. 1216-1225.

[6] Agostini, B., Revellin, B. and J.R. Thome (2008). Elongated Bubbles in Microchannels. Part I: Experimental Study and Modeling of Elongated Bubble Velocity, *Int. J. Multiphase Flow*, Vol. 34, pp. 590-601; Part II: Experimental Study and Modeling of Bubble Collisions, *Int. J. Multiphase Flow*, Vol. 34, pp.602-613.

[7] Cioncolini, A. and Thome, J.R. (2009). Algebraic Turbulence Modeling in Adiabatic Gas-Liquid Annular Two-Phase Flow, *Int. J. Multiphase Flow*, Vol. 35, pp. 580-596.

Received 17 November 2009

Professor John R. Thome, Director of the LTCM Lab, EPFL, Switzerland and an editorial member of IJMNTFTP

In: Frontier Research in Microscale and Nanoscale... ISBN 978-1-61324-665-8
Editor: Lixin Cheng

Chapter 7

REVIEW ON FLOW BOILING IN MICROCHANNELS

***Sujoy Kumar Saha*[1]*, Giuseppe Zummo and Gian Piero Celata* [2*]**

[1]Mechanical Engineering Department, Bengal Engineering and Science University, Shibpur, Howrah, West Bengal, India

[2]ENEA Casaccia Research Centre, Institute of Thermal Fluid Dynamics S. M. Galeria, Roma, Italy

ABSTRACT

An extensive review of flow boiling in microchannels has been made. Channel classification has been shown. Onset of nucleate boiling, liquid film thickness, flow pattern and bubble growth, heat transfer, pressure drop, critical heat flux and boiling instability with particular emphasis on different types of instabilities have been discussed. Direction of further research has been delineated in the conclusion.

Keywords: Flow boiling, Microchannel, Onset of nucleate boiling, Flow pattern, Bubble growth, Instability, Critical heat flux.

NOMENCLATURE

A	area, m^2
Bo	boiling number, dimensionless
Ca	Capillary number, dimensionless
Co	Confinement number, dimensionless
D	tube inner diameter, m
Ec	Eckert number, dimensionless
Eo	Eotvos number, dimensionless

* Corresponding author: ENEA Casaccia Research Centre, Institute of Thermal Fluid Dynamics, Office Building F-20, Via Anguillarese 301, 00123 S. M. Galeria, Roma, Italy. Tel: +39 06 30483905, Fax: +39 06 30483026, Email: celata@enea.it

F	forced convection heat transfer enhancement factor, dimensionless
Fr	Froude number
G	mass velocity, mass flux, kg m^{-2} s^{-1}
g	gravity acceleration, m s^{-2}
h	heat transfer coefficient, W m^{-2} K^{-1}, specific enthalpy, J kg^{-1}
h_{fg}	latent heat of vaporization, J kg^{-1}
Kn	Knudsen number, dimensionless
k	thermal conductivity, W m^{-1} K^{-1}
L	test section length, m
M	Morton number, Mach number, dimensionless; molecular weight
Ma	Marangoni number, dimensionless
N	number of microchannels
Nu	Nusselt number, dimensionless
P	pressure, N m^{-2}
Pr	Prandtl number, dimensionless
p_r	reduced pressure (absolute pressure/critical pressure), dimensionless
Q	heat transfer rate to the channel, W
q	heat flux W m^{-2}
Re	Reynolds number, dimensionless
S	suppression factor, dimensionless
T	temperature, K
t	time, s
v	specific volume, m^3 kg^{-1}
W	total mass flow to the channel, kg s^{-1}
We	Weber number, dimensionless
x	local vapor quality, dimensionless
Z	axial distance, m

Greek Symbols

α	void fraction, dimensionless
δ	liquid film thickness, μm
Δ	difference
φ	heat flux, W m^{-2}
μ	dynamic viscosity, kg m^{-1} s^{-1}
ρ	density, kg m^{-3}
σ	surface tension, kg s^{-2}

Subscript

b	bulk fluid
c	critical, cross-section

e	exit
f	saturated
fg	from liquid to vapor
i, in	inlet
l	liquid
lo	liquid only
lv	difference between liquid and vapor
L	all the mass flow rate taken as liquid
pool	pool boiling
SAT	saturation
sub	difference between saturation and inlet
tp, TP	two phase
v	vapor
w	wall

1. INTRODUCTION

Microchannel heat sinks are extensively used in high heat flux systems, e.g. microelectronic, optical and microfluidic devices, high performance supercomputers, electric vehicles, advanced military avionics, power and laser devices, etc. The heat dissipation rate is extremely high in these devices. Phase-change processes like flow boiling are very important in such devices. In addition, two-phase microchannel heat sinks are one of the strongest candidates for heat removal devices in high heat flux environments by virtue of their large surface area to volume ratio, compact dimensions and low flow rate requirements. Flow boiling in microchannels has drawn considerable attention from the researchers worldwide. Microchannels are used in MEMS and high-end microprocessors cooling applications.

There is a scale effect which becomes evident when the hydraulic diameter is less than the capillary length $\left[\sigma / g(\rho_l - \rho_v)\right]^{0.5}$.

In microchannels, the hydraulic diameter of the channels is smaller than the capillary length. Identifying the threshold, or the transition band, beyond which a two phase flow may be considered 'micro' is an open issue. The channel classification has been given by Kandlikar and Grande [1]. This classification should be used as a mere guide indicating the size range, rather than rigid demarcations based on specific criteria. Relevant literature in this direction would be Kawaji and Chung [2]. Confinement number (Co) as the distinguishing parameter, gives only a rough idea:

$$Co = \frac{1}{D}\sqrt{\frac{\sigma}{(\rho_l - \rho_v)g}} \quad (1)$$

Suo and Griffith [3] set Co_{th} =3.3, Kew and Cornwell [4] suggest Co_{th} =0.5 whereas that suggested by Brauner and Moalem Maron [5] is 0.16.

Bubble dynamics in microchannel is quite different from that in an ordinary sized channel. Bubble growth in a microchannel is restrained by the channel wall and in the transverse direction. The two-phase flow pattern, observed simultaneously in the channels may be different, depending on heat and mass fluxes.

Microchannels invariably involve cooling channels in blocks, as opposed to mini and larger diameter normal size channels that have individual confining walls and are usually thermally well controlled. Commonly accepted dimensions of microchannels are 10 mm long with 10-200 μm hydraulic diameter. The channels are usually cut in a block. For silicon, microelectronic fabrication techniques are used. For copper or other metals, an end mill or a lamination and bonding process is used. Typically, there will be 100 or so parallel channels. Heat is usually supplied to one side of the block. If the channels are cut from one side, a cover plate is provided on that side and inlet and outlet headers couple the microchannel heat exchanger to the flow system.

Boiling is desirable in microchannel heat sinks for more probable uniform wall temperature since then wall temperature is constrained to the fluid saturation temperature. The flow rate in microchannels is low to contain the pressure drop. Therefore, large enthalpy rise of the fluid for high heat flux is inevitable.

Small microchannel flow boiling systems are sensitive to oscillatory instabilities. The oscillations may be

(1) parallel channel instabilities in the case of parallel channel evaporators with common inlet headers, or
(2) the compressible volume instabilities which may develop even in single channel systems.

Parallel channel instabilities are primarily due to the interaction of the two-phase flow among adjacent channels. Compressible volume instabilities may propagate in the presence of a degree of compressibility (i.e. a body of vapor, flexible hosing, etc.) upstream or within the heated length.

Thermally induced two-phase flow instability must be avoided by all means to prevent any disastrous effect in the thermal systems. Two-phase flow instabilities in minichannels and microchannels are more intense than that in conventional channels due to the low flow velocities and confined space available for bubble growth. These instabilities result in a lower critical heat flux (CHF) than would be obtained with stable flow in the microchannels based heat sink, Bergles and Kandlikar [6].

Boiling in mini and microchannels seems to occur as a combination of both nucleate and convective boiling mechanisms. Some overviews of the progress made in the field of boiling in macro and small channels can be found in [7-30].

Agostini and Bontemps [31] worked with 11 parallel rectangular minichannels (3.28mm × 1.47 mm, hydraulic diameter: 2.01 mm) and they observed greater heat transfer than that for conventional normal sized channels. They also observed dryout at lower vapor qualities.

The use of mini channel heat exchangers (hydraulic diameter about 1 mm) in compact heat exchangers improves heat transfer coefficients and thermal efficiency while requiring a lower fluid mass. They are widely used in condensers for automobile air conditioning and are now being used in evaporators and other applications like domestic air conditioning systems.

Models for prediction of incipient boiling heat flux and wall superheat have been discussed in [32-42]. Thermodynamic and mechanical equilibrium on a curved vapor-liquid interface requires a certain degree of superheat in order to maintain a given curvature. Characteristics of homogeneous and heterogeneous nucleation can be estimated by the classical kinetics of nucleation theory. The vapor temperature in the bubble can be computed from equations, Cole [35]. Theoretical models and experiments have been compared in Bergles and Rohsenow [43], Davis and Anderson [36], Hsu [39], Hapke et al. [44], Hetsroni et al. [45], Hino and Ueda [46], Kennedy et al. [47], Li and Cheng [40] and Liu et al. [41]. Wall superheat effect on pressure has been studied. Wall superheat predicted by classical theory of kinetics of nucleation has been compared to experimental results. Effect of inlet velocity on wall superheat has been studied in Celata et al. [34], Hino and Ueda [46], Liu et al. [41] and Kennedy et al. [47]. Effect of inlet velocity and inlet temperature on incipient boiling heat flux has been studied in Liu et al. [41] and Qu and Mudawar [42]. Under some conditions, boiling incipience in surfactant solutions may be quite different from that in Newtonian fluids, Hetsroni et al. [48, 49]. The onset of nucleate boiling (ONB) in parallel microchannels, physical model of the explosive boiling, effect of dissolved gases on ONB during flow boiling of water and surfactant solutions and effect of roughness in microchannels have been discussed in great detail in [42, 50-60]. State of the art of the dynamics of vapor bubble in single and parallel microchannels can be obtained from [17, 18, 32, 37, 45, 57, 61-74]. Pressure drop, heat transfer and CHF in two phase flow boiling in microchannels database may be obtained from [45, 59, 60, 69, 75-110]. Information on explosive boiling of water in parallel microchannels, quasi-periodic boiling in a certain single microchannel of a heat sink, the initial thickness of the liquid film, system that contains a number of parallel microchannels and average heat transfer coefficient may be drawn from [59, 69, 102, 103, 105, 111-113].

Celata et al. [114] studied burnout in highly subcooled water flow boiling in small diameter tubes. Kandlikar [115] experimentally determined nucleation characteristics and analyzed stability during flow boiling in microchannels. Liu and Winterton [116] developed a general correlation for saturated and subcooled flow boiling in tubes and annuli based on a nucleate pool boiling equation.

Among the various aspects encountered in small channels, two phase flow heat transfer, flow pattern and pressure drop have been extensively investigated [17, 18, 47, 50-52, 112, 117-121]. Hetsroni et al. [50] evidenced quasi-periodical rewetting and refilling of the microchannels at high heat fluxes. Kandlikar [17-18] noted large amplitude fluctuations in multi-channel evaporators.

It is evident from the literature in the field that the thermophysical phenomenon is very complicated and it is far from being completely understood. In this paper, an attempt has been made to take a glance at the state-of-the-art of the flow boiling heat transfer in microchannels, focusing particularly in the flow reversal and instability aspect. At some points, minichannels and conventional size normal channels have also been discussed and such works have been referred. The structure of this paper is as follows:

Section 2 deals with the onset of nucleate boiling. Flow pattern and bubble growth is discussed in section 3 and liquid film thickness is dealt in section 4. Heat transfer, pressure drop, CHF and instability are discussed in sections 5, 6, 7 and 8, respectively. Conclusions are drawn in section 9.

In the following, some of the more important literature will be discussed in detail.

2. Onset of Nucleate Boiling (ONB)

Hetsroni [75] has given a predictive analytical model of ONB in microchannels taking into consideration of the functional dependence of wall superheat on fluid inlet velocity, subcooling and microchannel dimensions.

The approach developed by investigators [34, 36, 38-42, 122] is widely used to determine the onset of nucleate boiling (ONB) in conventional size normal channels and in microchannels. These models consider the behavior of a single bubble by solving the one-dimensional heat conduction equation.

The ONB for microchannels is significantly different from that in normal size conventional channels. Flow distribution among parallel channels in case of heat exchangers with microchannels is very important. Also, in case of microchannels, it is difficult to estimate the actual heat flux and temperature distributions at the channel wall due to the conjugate effects and circumferential and axial heat conduction in the material forming the channel. Microchannels, typically used by the investigators, have one unheated side wall covered by a transparent plate and heat is applied from the bottom resulting in circumferentially non-uniform heat flux and wall temperature distribution.

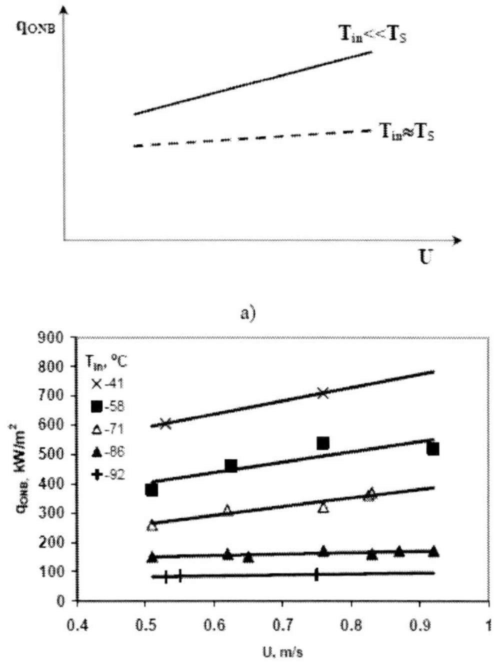

Figure 1. Variation of incipient boiling heat flux with inlet flow velocity. (a) analytical predictions, (b) experimental data by Liu et al. [41], (from Hetsroni [75]).

Qu and Mudawar [42] have shown that, in microchannels, $q_{ONB} \propto \Delta T_{S,ONB}$, whereas in case of conventional channels, $q_{ONB} \propto \Delta T^2_{S,ONB}$ and is independent of mass flux as shown by Bergles and Rohsenow [43]. The integral characteristics [75] for the circular channels, the rectangular channels with four and three heated walls are:

$$\frac{q_{ONB}}{\rho U_m C_p (T_S - T_{in})} \frac{2L_{ONB}}{r_o} = I_{c,ONB} \tag{2}$$

$$\frac{q_{ONB}}{\rho U_m C_p (T_S - T_{in})} \frac{L_{ONB}(a+b)}{ab} = I_{4,ONB} \tag{3}$$

$$\frac{q_{ONB}}{\rho U_m C_p (T_S - T_{in})} \frac{L_{ONB}(2a+b)}{2ab} = I_{3,ONB} \tag{4}$$

$I_{c,ONB}$, $I_{3,ONB}$ and $I_{4,ONB}$ are representative of a specific temperature field in heated microchannels and these are the criteria for the relative microchannel length. Since $I_{c,ONB}$ as found by Hetsroni [75] is very small, the ONB in the experiments by Celata et al. [34], although not mentioned explicitly in their paper, occurred at high liquid subcooling.

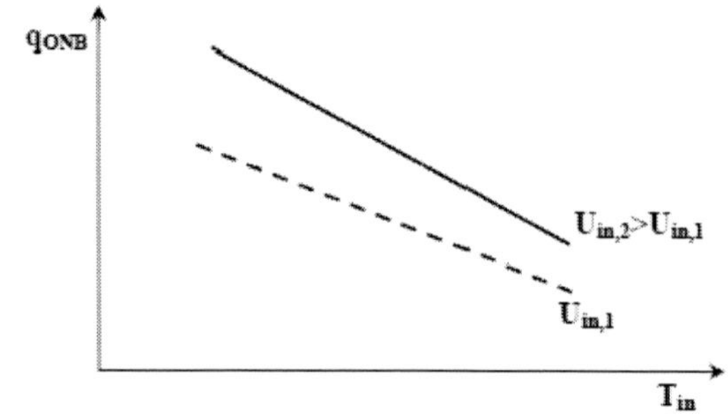

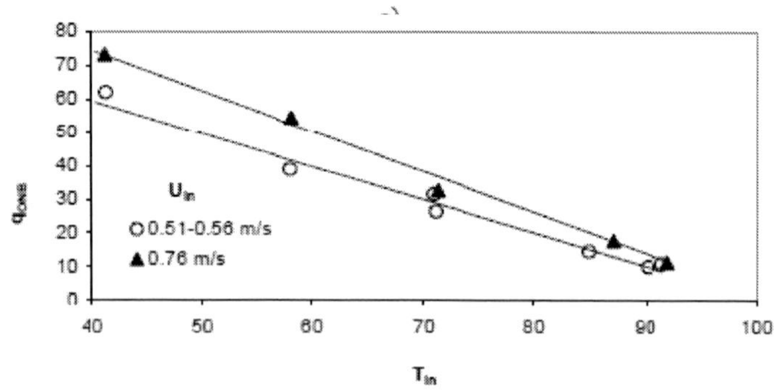

b)

Figure 2. Variation of incipient boiling heat flux with inlet flow temperature. analytical predictions, (b) experimental data by Liu et al. [41], (from Hetsroni [75]).

Figure 1 and Figure 2 show the variation of incipient boiling heat flux with inlet flow velocity and temperature, respectively as reported in analytical results of Hetsroni [75] vis a vis the experimental data given by Liu et al. [41].

The ONB of Hetsroni et al. [45] is established when the first bubbles are observed and the channel wall temperature exceeds the saturation temperature. As expected, the ONB location moves upstream as the heat flux increases. First a single bubble grows along the channel. Then the bubble grows preferentially in the axial streamwise direction, finally the elongated bubble is vented out and rewetting occurs. The life time of the elongated bubble is very short (typically 0.01 s) and at a given mass flow rate, the life time sharply decreases with increasing heat flux. The periodic flow reversal in microchannels has been termed explosive boiling process by Hetsroni et al. [45]. The system departs from stable operating conditions and hydrodynamic instability occurs. Liquid-vapor alternating flow in the vicinity of ONB occurs.

Peng and Wang [123] carried out experiments on the flow boiling of water in stainless steel microchannels with the rectangular cross-section of 600 μm × 700 μm. Velocity and liquid subcooling had no obvious effects on the fully developed nucleate boiling and no bubbles were observed at a heat flux much greater than 10 W/cm^2 due to evaporating space and fictitious boiling. Peng and Wang [123] have defined two hypothetical concepts, 'evaporating space' and 'fictitious boiling' to describe the physical processes and fundamental phenomena in boiling in microchannels. The 'evaporating space' is the space necessary for evaporation. 'Fictitious boiling' refers to the attainment of conventional nucleate boiling conditions yet internal evaporation and bubble growth have not started and the microbubbles cannot be visualized by ordinary means. Liquids with greater liquid/vapor density differences, higher latent heats and larger thermal diffusion coefficients need larger heat fluxes to initiate nucleation. The liquid is in a highly non-equilibrium state with an exceptional capability to absorb, transfer and transport thermal energy. If the microchannel size is smaller than the 'evaporating space', then 'fictitious boiling' occurs before nucleation is initiated. Otherwise, normal nucleate boiling occurs.

Qu and Mudawar [42] measured the incipient boiling heat flux in a copper heat sink containing 21 rectangular microchannels of 231 μm × 713 μm using deionized water. They identified boiling incipience using a microscope when bubbles were first detected near the outlet of the microchannels. They found that the bubbles grow and depart into the liquid flow instead of collapsing locally as in normal size channels. Both the incipient boiling heat flux and the associated wall superheat increase with increasing inlet velocity and decreasing inlet temperature. They compared their experimental data with their numerical prediction of incipient boiling heat flux.

Liquid nitrogen is used in micro-cryogenic surgery apparatus, the cooling of high temperature superconductivity and cryogenic cables. Qi et al. [124] studied the flow boiling of liquid nitrogen in the microtubes with hydraulic diameter of 531 μm, 834 μm, 1042 μm and 1931 μm. They studied ONB, two phase flow instability, pressure drop and heat transfer characteristics and CHF. At ONB, mass flux dropped suddenly, pressure drop increased and apparent wall temperature hysteresis in the range of 1.0-5.0 K occurred. Figure 3 shows boiling onset oscillations of 1042 μm microchannel of Qi et al. [124].

Hsu [39] postulated the minimum superheat criterion for the ONB in pool boiling. The criterion states that the bubble nucleus can grow only if the temperature surrounding the bubble is higher than the saturated temperature corresponding to the pressure inside the bubble. Sato and Matasumura [122] and Bergles and Rohsenow [43] extended the criterion of Hsu [39] to the flow boiling in conventional size channels. Hapke et al. [44] studied the ONB in a minichannel with i.d. of 1500 μm using a thermographic measuring method and they observed that the relatively high wall superheat is necessary to initiate the bubble nucleation. Ghiaasiaan and Chedester [38] found that the macroscale models underestimated the heat flux at ONB, and they suggested that the thermocapillary force may suppress the bubble nucleation from the wall cavities. Liu et al. [41] developed an analytical model to predict the heat flux for ONB of water in the rectangular microchannels of size 275 μm width and 636 μm depth.

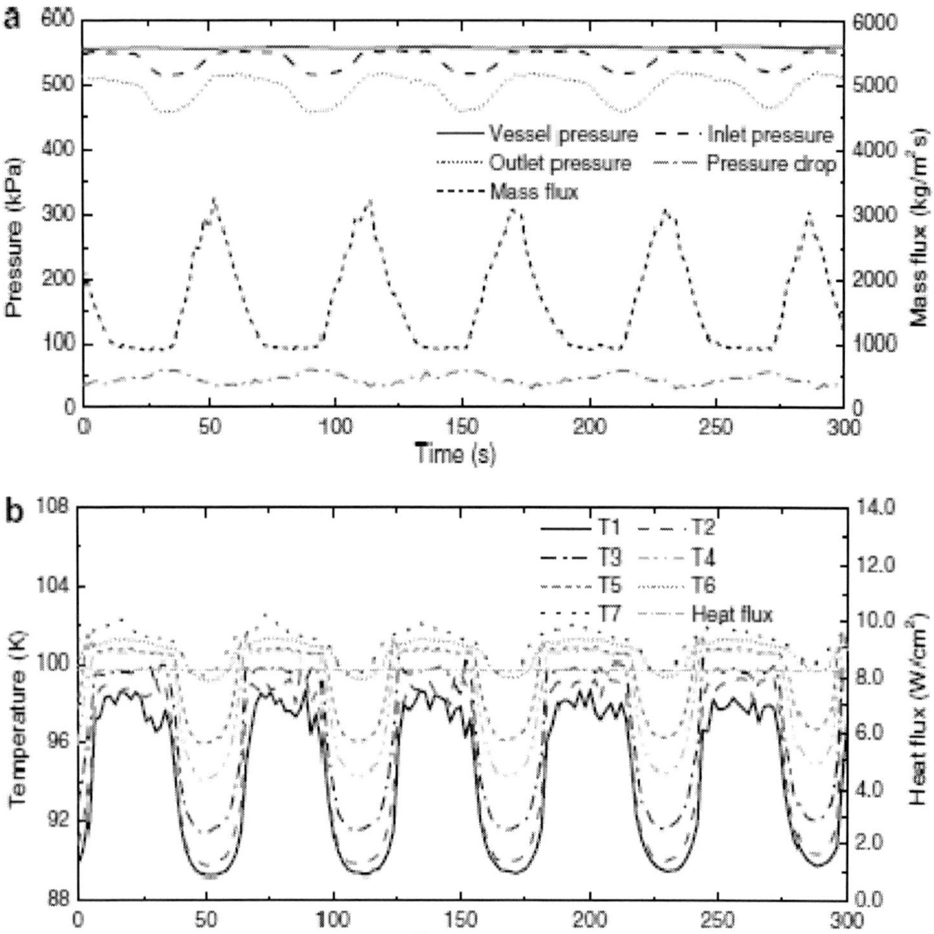

Figure 3. Boiling onset oscillations of 1042 μm microchannel of Qi et al. [124].

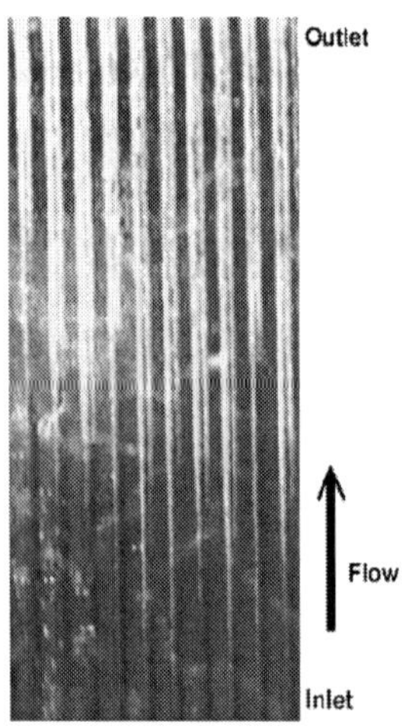

Figure 4. Flow pattern observed by Hetsroni et al. [45].

3. Flow Pattern and Bubble Growth

Quasi-diabatic two-flow pattern visualizations and measurements of elongated bubble velocity, frequency and length were performed by Arcanjo et al. [125]. The tests were run for R134a evaporating in a stainless steel tube with diameter of 2.32 mm, mass velocities from 50 to 600 kg/m^{2}s and saturation temperatures of 22°C, 31°C and 41°C. Images from a high-speed video-camera (8000 frames/s) obtained through a transparent tube just downstream of the heated section were used to identify the following flow patterns: bubbly, elongated bubbles, churn and annular. Dryout conditions were also characterized. Local heat transfer results were considered when investigating the presence of stratified flows. The visualized flow patterns were compared against the predictions provided by Revellin and Thome [126]. Additionally, elongated bubble velocities, frequencies and lengths were determined based on an analysis of high speed videos. Results suggested that the elongated bubble velocity depends on mass velocity, vapor quality and saturation temperature, and is independent of bubble length. The bubble velocity increases with increasing mass velocity and vapor quality and decreases with increasing saturation temperature. Additionally, bubble velocity was correlated as a linear function of the two-phase superficial velocity.

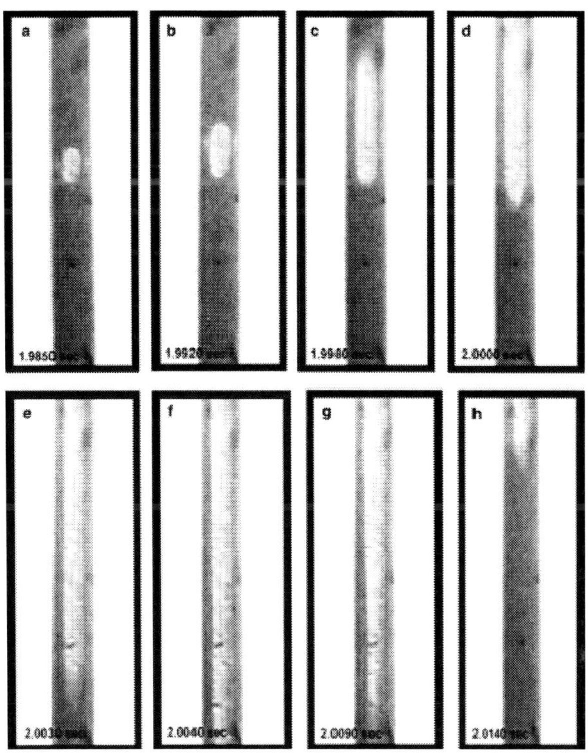

Figure 5. Bubble growth near outlet manifold of Hetsroni et al. [45].

Ory et al. [57] numerically predicted the growth and venting of a vapor bubble in 62.5μm diameter 2 mm long circular tube. The tube was open at both ends and connected two liquid reservoirs at constant pressure. Their model simulated the effect of an intense, localized brief heating of the liquid causing the nucleation and growth of a bubble. Hetsroni et al. [45, 50, 51, 53, 77] performed visualization and they made measurements of flow boiling in silicon triangular parallel microchannels. They observed periodic wetting and rewetting phenomena.

Generally two-phase microchannel heat sinks have straight constant cross-section. Subcooled liquid at the inlet is converted into liquid-vapor mixture at the outlet. The vapor slug with elongated bubble has rapid bubble growth and reverse flow occurs when the liquid-vapor interface is pushed at both ends of the channel. Figure 4 shows the flow pattern observed by Hetsroni et al. [45] and flow patterns were periodic. This explosive boiling was triggered by venting of elongated bubble due to very rapid expansion. They also observed dryout in the channel. Figure 5 and Figure 6 show the typical bubble growth near outlet manifold and flow pattern near inlet manifold, respectively [45]. Figure 5, [45], shows a sequence of bubble formation in one of the parallel triangular channels with hydraulic diameter 129 μm. Figure 6, [45], shows alternative two-phase flow pattern at a distance of 1000-1500 μm downstream from the inlet of the test section. Serizawa et al. [127] observed the similar results in steam-water flow in a 50 μm silicon tube. They reported that at low liquid flow rates, partially continuous liquid film flow changed to the rivulets or even to discrete liquid lumps or large liquid droplets.

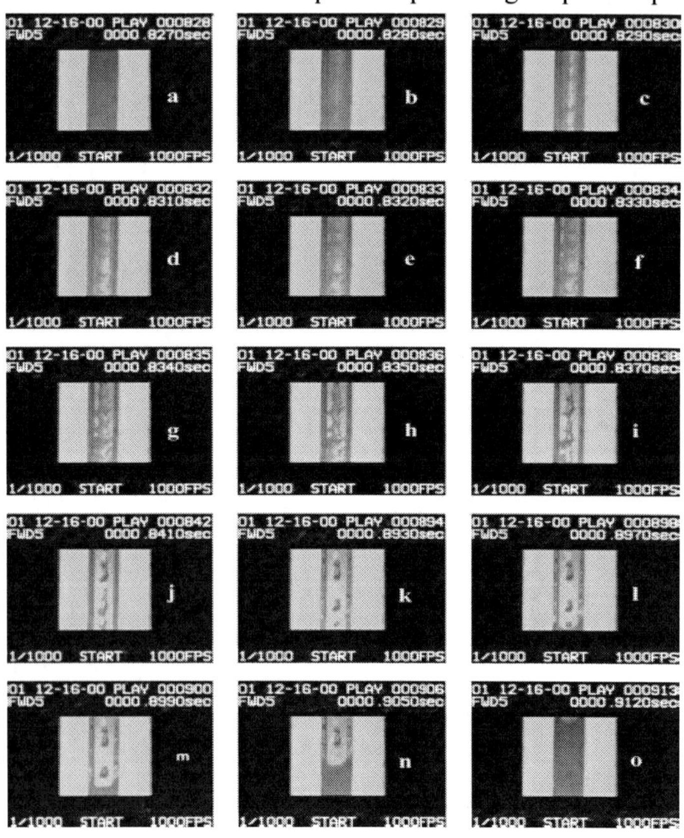

Figure 6. Flow pattern near inlet manifold, from Hetsroni et al. [45].

Flow boiling patterns at various heat flux and mass flux conditions were observed by Wang et al. [128] using a microscope and a high speed video recording system. Figure7 shows a map, in terms of heat flux versus mass flux, showing stable and unstable flow boiling regimes in parallel microchannels.

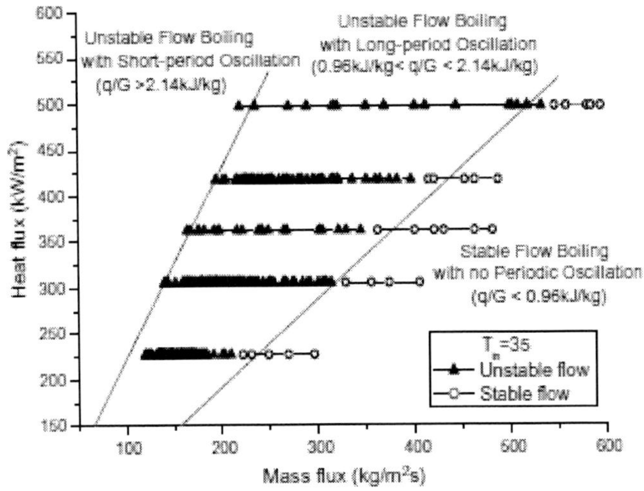

Figure 7. Flow pattern map, in terms of heat flux versus mass flux, showing stable and unstable flow boiling regimes in parallel microchannels, from Wang et al. [128].

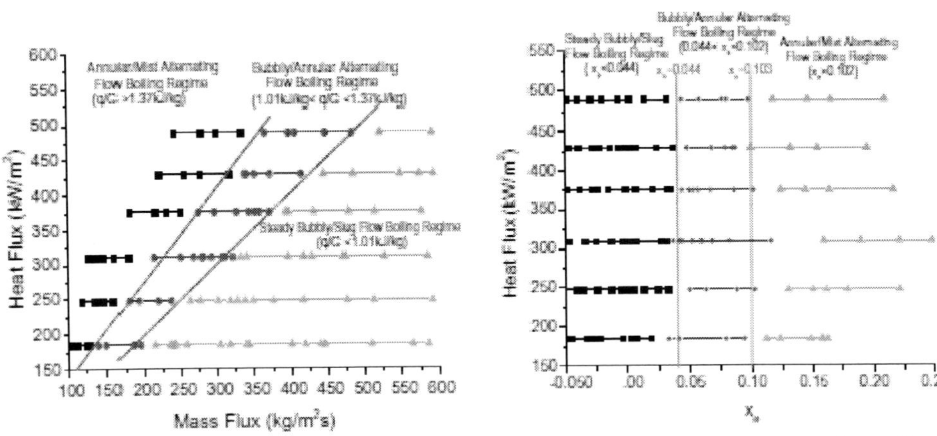

(a) Heat flux versus mass flux at T_{in} = 35 °C (b) Heat flux versus vapor quality at T_{in} = 35 °C

Figure 8. Different diabatic flow boiling patterns in parallel microchannels, Wang et al. [74].

Figure 8 shows different diabatic flow boiling patterns in parallel microchannels as observed by Wang et al. [74] with type B connection shown in Figure 9. The inclined straight lines in Figure 8 (a) become vertical straight lines at x_e = 0.044 and x_e = 0.103 in Figure 8 (b) because x_e depends on both the value of q/G and T_{inlet} given by the Eq.(5) below:

$$x_e = \frac{h_e - h_f}{h_{fg}} = \frac{h_{in} + \left(A_w / A_c \right) q / G - h_f}{h_{fg}} = \frac{h_{in} - h_f}{h_{fg}} + \frac{A_w Bo}{A_c} \tag{5}$$

where $Bo = \dfrac{q}{Gh_{fg}}$ is the boiling number.

$$h_e = h_{in} + \frac{q.N.A_w}{G.N.A_c} \tag{6}$$

Figure 10 and Figure 11 show measurements of inlet/outlet water and wall temperatures and inlet/outlet pressures in parallel microchannels bubbly/annular and annular/mist alternating flow boiling regimes, respectively as observed by Wang et al. [74].

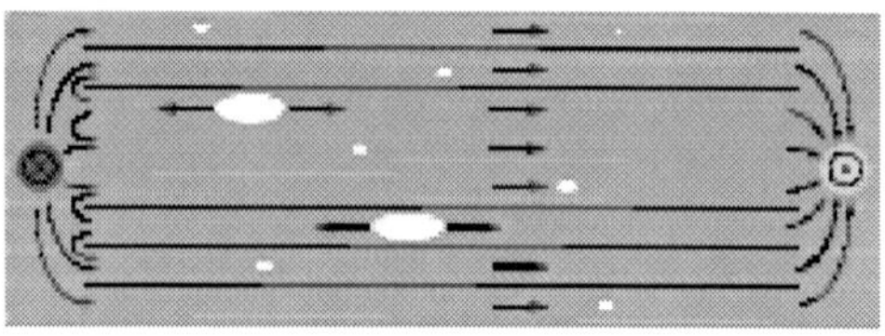

(a) Type-A connections: flow entering and exiting from parall microchannels with restrictions because inlet/outlet conduits perpendicular to microchannels

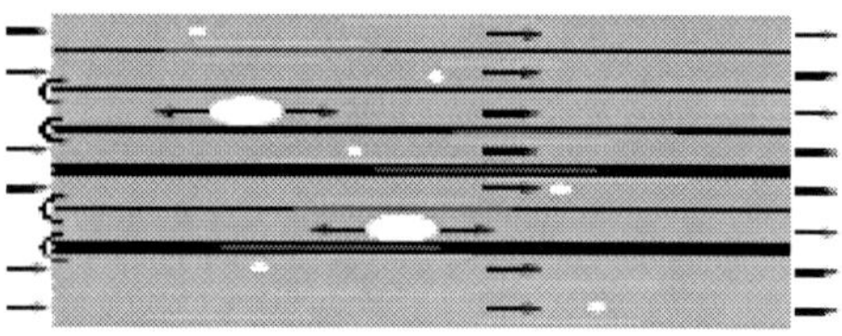

(b) Type-B connection: flow entering to and exiting from microchannels freely without restriction

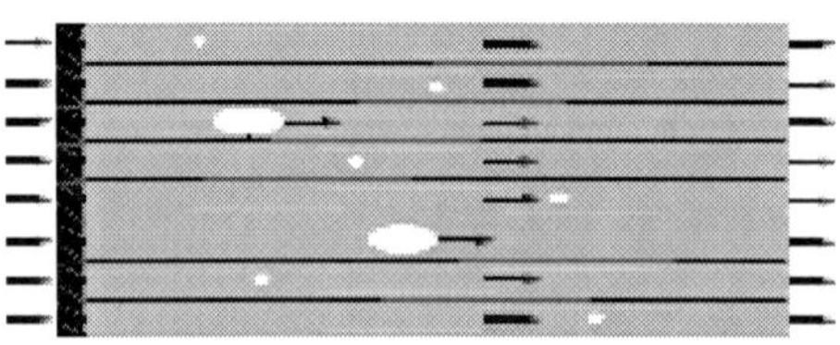

(i) Arrangement of parallel microchannels with restrictive inlet

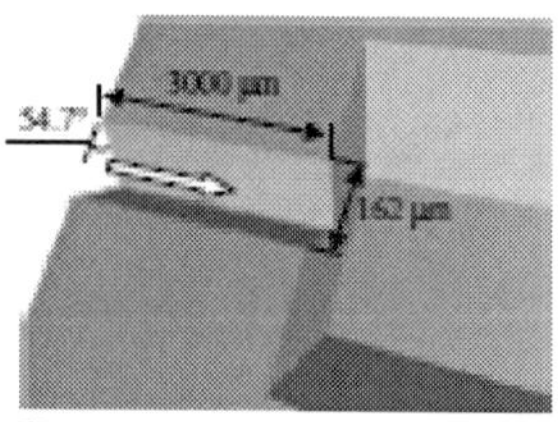

(ii) Sketch of the microchannel with a restrictive inlet

(c) Type-C connection: flow entering with restriction and exiting without restriction in microchannels

Figure 9. Types of connections, Wang et al. [74].

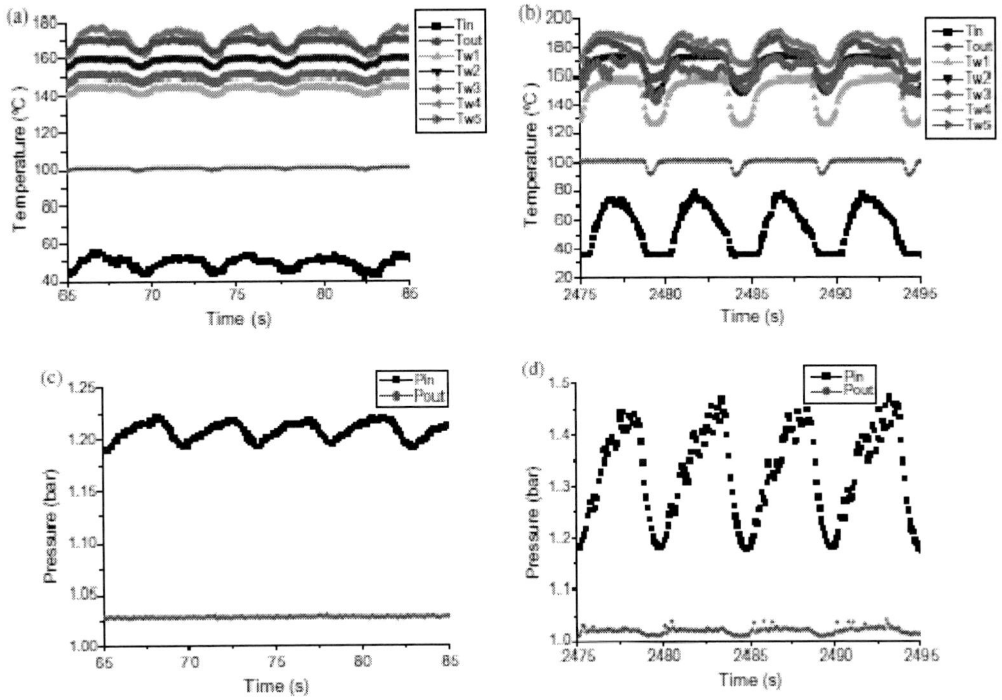

(a) Type-B with x_c= 0.096 and T_{in}= 35 °C (b) Type-A with x_c= 0.099 and T_{in}= 35 °C

Figure 10. Inlet/outlet water and wall temperatures and inlet/outlet pressures in parallel microchannels bubbly/annular alternating flow boiling regime, Wang et al. [74].

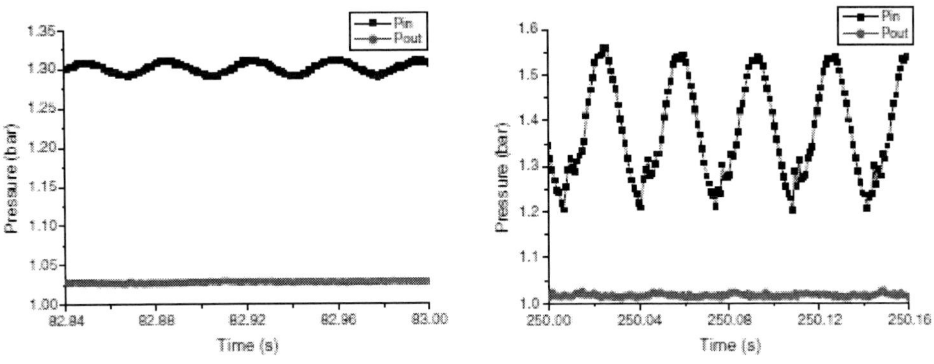

(a)Type-B connection with x_e = 0.745 and T_{in} = 35 °C (b)Type-A connection with x_e = 0.762 and T_{in} = 35 °C

Figure 11. Inlet/outlet pressures in parallel microchannels annular/mist alternating flow boiling regime, Wang et al. [74]

Tabatabai and Faghri [129] suggest that only patterns dominated by surface tension (bubble and slug) occur in tubes of 100 μm diameter or less. However, when the bubble size approaches the channel size (which is the case in microchannels), the result is the annular flow.

Wu and Cheng [103] investigated flow boiling of water in parallel silicon microchannels of trapezoidal cross-section having hydraulic diameter of 158.8 μm and

82.8 μm. They observed alternating in time two phase flow and single phase liquid flow after the boiling heat transfer was established. Wu and Cheng [103] observed bubbly flow, slug flow, churn flow and other peculiar flow patterns with the aid of a microscope and high speed video recording system. The flow patterns observed by Wu and Cheng [105] are similar to the earlier findings of Wu and Cheng [103].

Slug flow, churn flow and dryout condition have been observed by Hetsroni et al. [76] and Zhang et al. [130]. The long vapor bubble occurring in a microchannel is similar to annular flow with intermittent slugs of liquid between two long vapor trains.

Jiang et al. [131] dealt with flow boiling in silicon microchannels of triangular cross-section with hydraulic diameter 26 μm and 53 μm. They observed that local nucleation boiling occurred in the microchannels at a lower input power and stable annular flow occurred in higher input power. They did not observe any bubbly flow in their microchannels.

Zhang et al. [130] also did not observe any bubbly and slug flow typically observed in the microchannels. They carried out flow boiling experiment in rectangular silicon microchannels having hydraulic diameter 25 μm – 60 μm. Boiling occurred in their experiment with less than 5°C superheat and they observed mostly annular flow with a very thin layer of liquid.

Hetsroni et al. [77] investigated flow boiling of water in the silicon triangular microchannels having hydraulic diameter 103 μm and 129 μm. They observed periodic annular flow and the periodic dry steam flow. However, they did not present any temperature and pressure fluctuation data.

In the unstable flow boiling mode, Wang and Cheng [132] observed elongated bubbly/slug flow pattern. The flow pattern also behaved like an annular or semi-annular flow, in which the thin liquid film evaporated between the vapor core and the heating wall. When long bubbles occupied the channel, they blocked the two-phase flow and caused higher pressure drop. Later, it was observed that upstream vapor plug broke through the liquid front, reaching the long bubble downstream. Local dryout occurred due to the depletion of liquid film between vapor core and heating wall. Subsequent to this, annular/mist flow (alternating dryout and rewetting phenomena) with imminent burnout was observed. Average wall temperature in bubbly/elongated bubbly/slug flow was lower than that in the semi-annular flow and annular/mist flow.

Balasubramanian and Kandlikar [133] have also calculated the velocity of the liquid-vapor interface associated with bubble growth from the image sequences. The velocity was very high, on the order of 3.5 m/s. As the bubble nucleated and remained confined to the wall, it experienced a sudden reduction in its growth rate. After the bubble became confined to the channel wall, it started to grow toward the sides at a very high velocity. An increase in velocity at this stage was due to an increase in the evaporation rate of the bubble as it got closer to the channel walls.

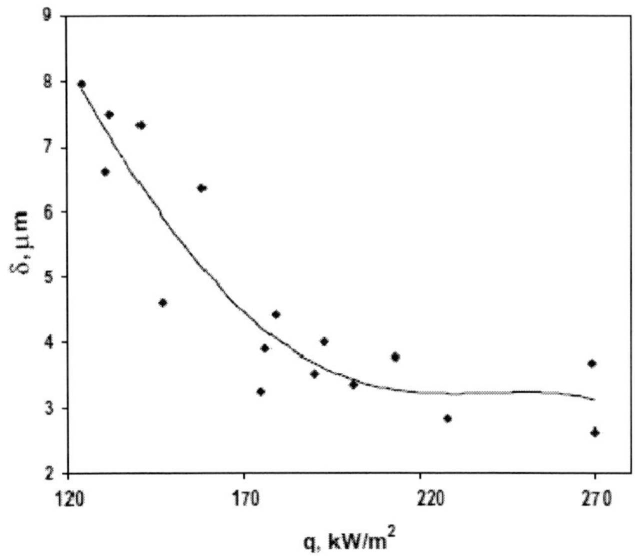

Figure 12. Dependence of average film thickness on heat flux, Hetsroni et al. [45].

Zivi [134], Agostini et al. [135] and Sobierska et al. [136] made experimental study and modeling of elongated bubble velocity and observed different flow patterns. Hapke et al. [137] studied flow patterns during flow boiling of water and N-Heptane in microchannels. Map of flow boiling of different fluids in parallel and single minichannels was developed in [138-143]. Taitel and Dukler [144] reported a model for predicting flow regime transitions in horizontal and near horizontal gas liquid flow.

4. Liquid Film Thickness

Moriyama and Inoue [111] performed experiments to measure the thickness of the liquid film formed by a growing flattened bubble in a narrow gap, whose width ranged between 100 μm--- 400 μm. A narrow gap caused faster growth of the bubble. The film thickness was about 2-7 μm depending on the gap size and initial superheat.

The typical nature [45] of the dependence of average liquid film thickness on heat flux is shown in Figure 12. This is comparable to those obtained by Moriyama and Inoue [111] and Qu and Mudawar [95]. Decreasing liquid film thickness with increasing heat flux is a distinct feature of dryout during flow boiling in microchannels. The time, during which the heated surface located upstream of the ONB is covered by liquid, decreases with an increase in the heat flux, the average film thickness decreases and the heat transfer coefficient decreases as well.

A recent literature review on film thickness measurements and on the available techniques for its measurements was performed by Tibirica et al. [145]. They focused on two-phase flows in micro- and minichannels.

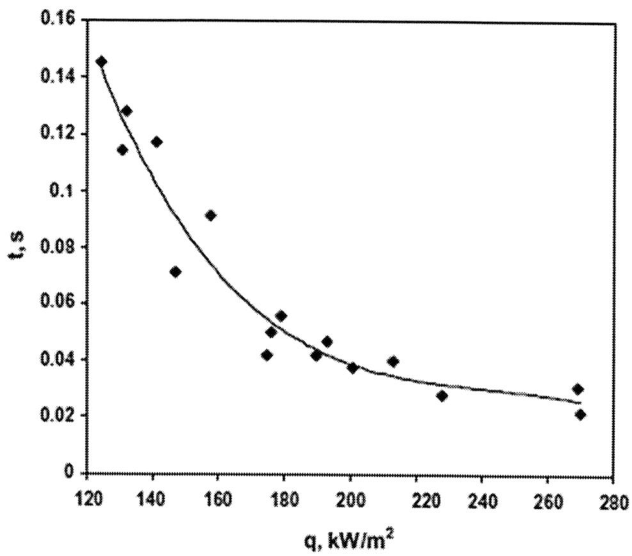

Figure 13. Dependence of time interval between cycles on heat flux, Hetsroni et al. [45].

5. Heat Transfer

Hetsroni et al. [45] visualized the flow pattern and measured heat transfer coefficient during explosive boiling of water in parallel triangular microchannels. The behavior of long vapor bubbles at low Reynolds number was not similar to annular flow with interposed intermittent slugs of liquid between two long vapor trains. They termed this process as explosive boiling with periodic wetting and dryout. Dryout takes place during the whole period upstream of ONB (i.e. in the saturated boiling region). The time interval between cycles is described in statistical terms due to the unsteady character of the boiling process. For each regime, the time period is obtained using histograms of statistical distribution. Figure 13 shows the dependence of time interval between cycles on heat flux obtained from the histograms of statistical distribution as observed by Hetsroni et al. [45]; increasing the heat flux about two fold leads to increasing the cycles of dryout about sevenfold.

Figure 14 shows the typical nature of the dependence of the saturated flow boiling heat transfer coefficient on the thermodynamic equilibrium vapor quality at the outlet of the test section. As observed by Hetsroni et al. [45] and other investigators [94, 95, 146-148], h decreases appreciably with increasing x. However, Hetsroni et al. [45] observed that heat transfer coefficient does not depend on mass flux. This observation of Hetsroni et al. [45] contradicts the observations of Qu and Mudawar [94, 95] who pointed to annular flow as the dominant two-phase flow pattern in microchannels at Re=60—300 and concluded that the saturated flow boiling heat transfer coefficient is a strong function of the mass flux and only a weak function of the heat flux. Hetsroni et al. [45] concluded that explosive boiling in microchannels could not be clearly attributed to any single non-dimensional number like Reynolds number, Weber number or Ohnesorge number. A decrease in heat transfer coefficient with increasing heat flux does not support the nucleate boiling mechanism observed in normal size conventional channels with hydraulic diameter > 3 mm, which is normally associated with a significant increase in the heat transfer coefficient with increasing heat flux.

The data of Wang et al. [74] is well predicted by the correlation of Mishima and Hibiki [92] for macrochannel with, of course, some underprediction by the correlation for high vapor quality ($x_e > 0.1$). The reason for underprediction is that the flow behavior is different in microchannels. The coalesced bubbles in microchannels are confined, elongated and finally they form the annular flow downstream, while in a macrochannel, many bubbles grow and flow along the channel without restriction. Also local dryout occurs in microchannels at high vapor quality causing vapor blocking in the channels and increased pressure drop from additional friction related to the elongated bubbles. However, the investigators do not have consensus that the local dryout occurs in microchannel at high vapor qualities. Several investigators have pointed out that the dryout vapor quality decreases with decreasing tube diameter and this also depends on the operational mass velocity range.

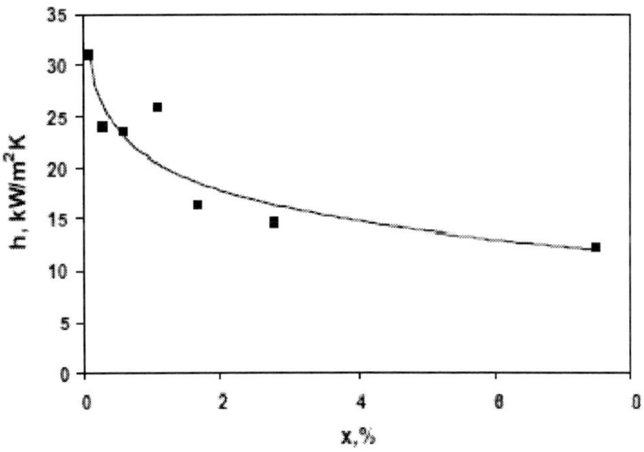

Figure 14. Dependence of the saturated flow boiling heat transfer coefficient on thermodynamic equilibrium quality at the outlet of the test section, Hetsroni et al. [45].

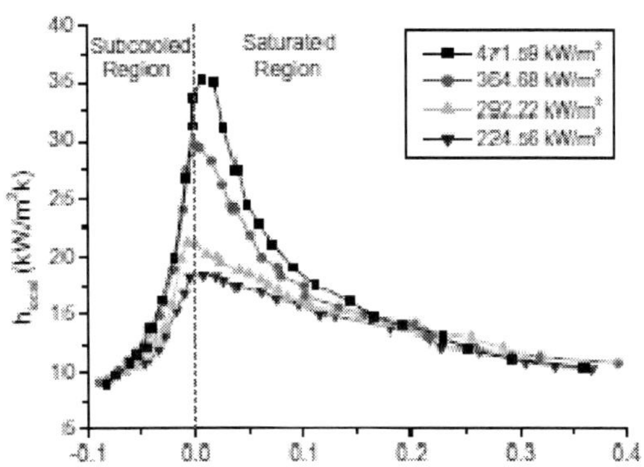

Figure 15. Heat transfer coefficient versus exit quality, Wang et al. [74].

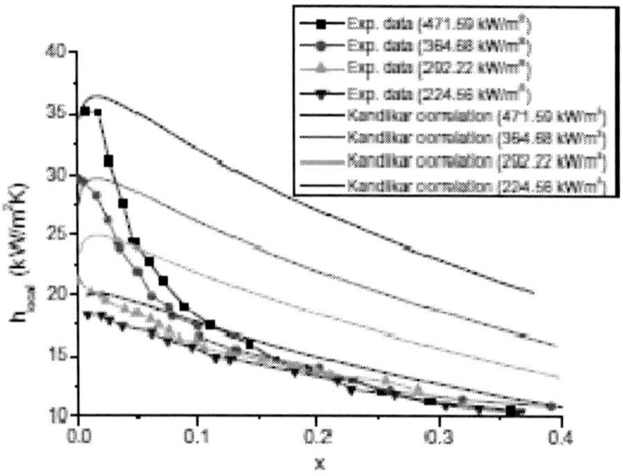

Figure 16. Comparison of boiling heat transfer coefficient with Kandlikar [150] correlation, Wang et al. [74].

The heat transfer coefficient increases, as shown in Figure 15 [74], with the increase in vapor quality in the subcooled boiling region ($x_e < 0$) due to a transition from partial to fully developed nucleate boiling. For saturated boiling region, ($x_e > 0$), h decreases with increasing x. This finding corroborates the experience of Hetsroni et al. [45] and Qu and Mudawar [94] but it is contradictory to the observations of Chen and Garimella [149]. Kandlikar [150] correlation overpredicts heat transfer coefficient data of Wang et al. [74] as shown in Figure 16.

Kandlikar [146] discussed comprehensively heat transfer mechanisms during flow boiling in microchannels. The forces due to surface tension and momentum change during evaporation, in conjunction with the forces due to viscous shear and inertia, govern the two phase flow patterns and heat transfer characteristics during flow boiling in microchannels. In low Reynolds number flows occurring in microchannels, nucleate boiling systemically emerges as the dominant mode of heat transfer. Kandlikar [146] observes that the high degree of wall superheat required to initiate nucleation in microchannels leads to rapid evaporation and flow instabilities, often resulting in flow reversal [18] in multiple parallel channel configuration. The usual nucleate boiling heat transfer mechanisms including liquid film evaporation and transient heat conduction in the liquid adjacent to the contact line region play important role. The liquid film under the large vapor slug evaporates completely at downstream locations and a dryout condition arises periodically with the passage of each large vapor slug. Kandlikar [146] has observed non-dimensional groups relevant to two phase flow studies in microchannels. These are Martinelli parameter, Convection number, Boiling number, Bond number, Eotvos number, Capillary number, Ohnesorge number, Weber number and Jakob number. While studying the effects of these numbers individually, Kandlikar [146] has found that following two non-dimensional groups K_1 and K_2 [151] are capable to represent some of the key flow boiling characteristics including CHF in microchannels:

$$K_1 = \left(\frac{q}{Gh_{fg}} \right)^2 \frac{\rho_L}{\rho_G} \tag{7}$$

$$K_2 = \left(\frac{q}{h_{fg}} \right)^2 \frac{D}{\rho_G \sigma} \tag{8}$$

Kandlikar [146] has made an attempt to see the range of K_1 and K_2 in the data available [152-155] for minichannels and microchannels.

Kandlikar et al. [156] have observed that actual nucleation wall superheat may decrease due to the presence of dissolved gases in the liquid. The periodic wetting, dryout and rewetting phenomena in microchannels have been observed by Hetsroni et al. [76] for parallel microchannels of hydraulic diameter 103 μm -129 μm.

Kandlikar and Steinke [157] observed with their water data that at low Reynolds number, the convective boiling does not become the dominant mechanism in the entire vapor quality range. The flow boiling mechanism is quite complex in microchannels as seen from the experimental data and flow visualization studies. Although simple models given by Jacobi and Thome [158] and others do a fairly good job under certain idealized conditions, phenomena such as rapid bubble growth, flow reversal and periodic dryout and also flow distribution through parallel microchannels introduce significantly greater complexities.

Diaz and Schmidt [159] observed that the heat transfer coefficient decreases with increasing quality of water. This trend is consistent with the results of others and indicates that nucleate boiling dominates the low quality region. However, for ethanol, different trends are observed after the low quality region. The heat transfer coefficient increases with increasing quality at low heat fluxes. With increasing heat flux, the heat transfer coefficient decreases with quality.

Xu et al. [160] used acetone as the working fluid and they proposed the boiling number as the key for establishing the relative dominance of boiling mechanisms, since it gives information on the evaporation momentum force relative to the inertia force.

Kandlikar and Balasubramanian [147, 161] experimented with three gravitational orientations under the same operating conditions. Bubbly, thin film nucleation, slug flow and churn flow have been observed in all three orientations. Time dependent flow patterns have been found. They have observed decreasing trends in heat transfer coefficients with increasing quality. These trends are in consonance with observations where nucleating bubbles are at the wall during liquid flow and also in the liquid film near the wall in plug or annular flow.

Dupont and Thome [162] and Tiselj et al. [163] studied influence of the channel diameter and axial conduction on heat transfer during flow boiling in microchannels. Harirchian and Garimella [164] studied microchannel size effects on local flow boiling heat transfer to a dielectric fluid. Investigators [165-168] dealt with scale effects on flow boiling in microchannels, the applicability of the flow boiling correlation to low

Reynolds number flows in microchannels and effects of gravitational orientation and channel geometry on flow boiling in parallel minichannels. Wang and Cheng [169], Xu et al. [160, 170] studied microbubble emission, transient flow pattern and heat transfer mechanism in microchannels.

Consolini [22] proposed a one-dimensional heat transfer model, based on the effect of interfacial shear on the temporal evolution of film thickness, for slug flow which is dominant flow mode at low and intermediate vapor qualities in microchannel evaporators.

Lazarek and Black [86] showed that the heat transfer coefficient remains unaffected by vapor quality, but it is a function of heat flux, leading them to conclude that nucleate boiling is dominant heat transfer mechanism as in macro-scale flow boiling. This trend as well as contradictory findings has been discussed at length in Agostini and Thome [171]. The relative importance of nucleate boiling and convection in the individual flow patterns that are characteristic to a microchannel flow is thus still unclear, [6, 112, 113, 118, 127, 146, 158, 172, 173]. Vapor formation and flow development in microchannels have been discussed at length in [16, 31, 41, 61, 89, 90, 96, 97, 115, 152, 154, 155, 160, 170, 174-183]. Heat transfer prediction methods in microchannels are discussed in [28, 147, 161, 184, 185].

Berstsch et al. [186] studied flow boiling heat transfer in two copper microchannel cold plate evaporators with hydraulic diameter 1090 μm and 540 μm with R-134a and R-245fa refrigerants. Heat transfer coefficient varied significantly with heat flux and vapor quality. But h was a weak function of saturation pressure and mass flux. However, other investigators have observed that, under nucleate boiling dominant conditions, the heat transfer coefficient is also a strong function of the saturation temperature. Berstsch et al. [186] observed that nucleate boiling dominated the heat transfer.

The heat transfer coefficients were often seen [187] to increase with heat flux, although not in all cases. For sufficiently high heat fluxes, the heat transfer coefficients became less sensitive to heat flux, and may at times decrease with higher heating levels. Consolini and Thome [187] concluded that one has to group data separately for stable and unstable categories for the successful predication of results.

Balasubramanian and Kandlikar [133] observed the periodic thin film nucleate boiling phenomenon giving an important clue to the flow boiling heat transfer mechanism. This clearly established the presence of nucleate boiling under annular flow conditions. This is to be incorporated in developing a mechanistic model for flow boiling in microchannels.

Aritomi et al. [188] studied thermohydraulics at low velocities of flow boiling two phase flow in high conversion light water reactors. Effects of heat flux, mass flux, vapor quality and saturation temperature on flow boiling heat transfer in microchannels were studied in [189-193]. Bertsch et al. [194] studied refrigerant flow boiling heat transfer in parallel microchannels as a function of local vapor quality.

Saturated boiling correlation given by Lazarek and Black [86] is as follows:

$$Nu = 30\,\mathrm{Re}^{0.857}\,Bo^{0.714} \tag{9}$$

where

$$Re = \frac{GD}{\mu_l} \tag{10}$$

$$Bo = \frac{\varphi}{Gh_{fg}} \tag{11}$$

$$Nu = \frac{hD}{k_l} \tag{12}$$

Eqn. (9) is applicable for the following range of parameters:

φ:	$1.4 \times 10^4 - 3.8 \times 10^5$ W/m^2
G:	$125 - 750$ kg/m^{2}s
Re:	$860 - 5500$
Bo:	$2.3 \times 10^{-4} - 76 \times 10^{-4}$
static pressure:	$1.3 \times 10^5 - 4.1 \times 10^5$ N/m^2

A modified Lazarek-Black [86] equation form has been suggested by Kew and Cornwell [4] to allow for an observed increase in the heat transfer coefficient with the vapor quality in larger tubes:

(13)

Another Lazarek-Black [86] based correlation was developed by Sun and Mishima [195] using regression method taking the effect of Weber number into account:

$$Nu = \frac{6\,Re_{lo}^{1.05}\,Bo^{0.54}}{We_l^{0.191}\left(\frac{\rho_l}{\rho_g}\right)^{0.142}} \tag{14}$$

where the Weber number for liquid phase is defined as:

$$We_l = \frac{G^2 D_h}{\sigma \rho_l} \tag{15}$$

6. Pressure Drop

Figure 17, [45], shows the dependence of (RMS ΔP)/ΔP on vapor quality. Hetsroni et al. [45] gave a physical model as shown in Figure 18 for the explosive boiling and dryout based on the high-speed video capture of the flow. A single bubble cyclically growing and collapsing away from the outlet manifold is capable of inducing a mean unidirectional fluid flow. This observation of Hetsroni et al. [45] is also similar to the trend observed by Ory et al. [57].

Qu and Mudawar [93] observed severe pressure drop oscillation threatening premature CHF. They have developed a correlation for the prediction of pressure drop considering the effects of both channel size and coolant mass velocity. A theoretical annular two phase flow model was developed describing various transport processes occurring in the microchannel.

A large number of correlations [87, 92, 93, 148, 196] have been developed under the framework of the homogeneous equilibrium flow model and the separated flow model to evaluate the pressure drop in the microchannel two phase region. Macrochannel correlations usually do a poor job in case of microchannels and give large mean absolute error (MAE) because the macrochannel correlations have been developed for turbulent flow prevalent in macrochannels, while the low coolant flow rate and small channel size of microchannels yield mostly laminar flow for liquid phase and laminar or turbulent flow for the vapor phase. However, microchannels predictive methods work accurately only when these are developed based on the investigators' own database, [28, 197]. Also, some investigators have obtained good predictions by comparing macrochannels correlations against microchannel experimental results. Qu and Mudawar [93] have developed their correlation by the modification of Mishima and Hibiki [92] correlation incorporating a mass velocity term and assuming the annular two phase flow dominant at moderate to high heat fluxes in microchannels. The correlation of Qu and Mudawar [93] does a fairly good job for microchannel hydrodynamics.

Triplett et al. [118] measured pressure drop and void fraction in microchannels with air water adiabatic flows and showed that the homogeneous model best predicted their results in the slug, churn and slug-annular flow patterns.

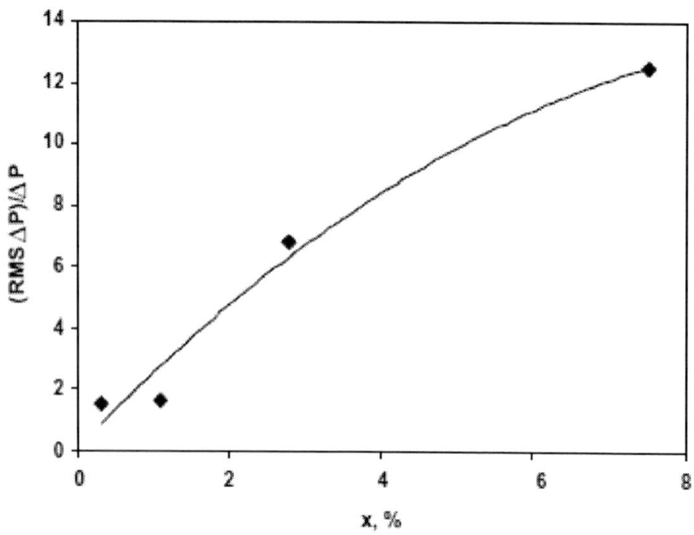

Figure 17. Dependence of (RMSΔP)/ΔP on vapor quality, Hetsroni et al. [45].

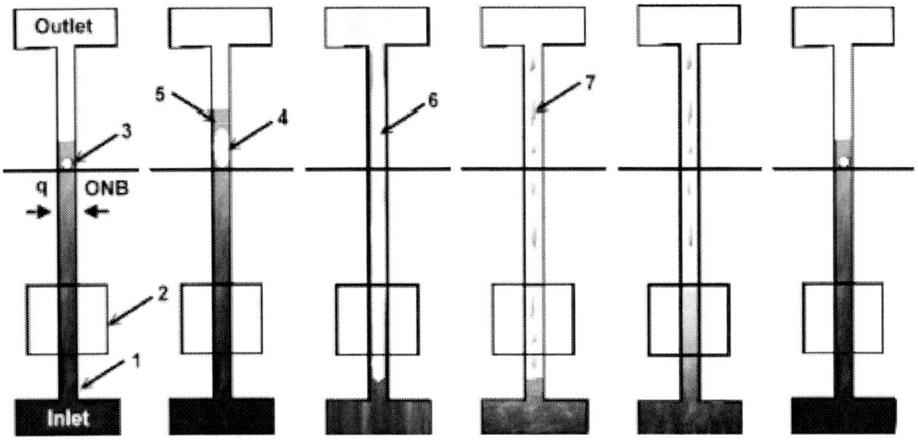

Figure 18. Physical model for the explosive boiling and dryout, Hetsroni et al. [45].

Two phase pressure drop characteristics were obtained for several heat fluxes, [81]. They have found unstable regime in the first zone of the N-shape curve and higher amplitude fluctuations as the inlet Reynolds number became smaller.

Most researchers, [89, 126, 148, 176, 198] agree that pressure drop correlations for the conventional channels are not applicable to the microchannels because the channel diameters affect the flow pattern transition. Wen and Kenning [198] observed that the parameter C in L-M model needs modification. Yu et al. [148] used a multiplier for pressure drop calculation in microchannels. Lee and Mudawar [89] experimental data did not match with the existing models. Zhang and Webb [199] developed a new model based on Friedel correlation for two phase friction for refrigerants in small diameter tubes. Revellin and Thome [126] proposed a modified homogeneous model.

Balasubramanian and Kandlikar [133] also observed that the magnitude of the pressure drop increases with an increase in surface temperature for a given mass flux. The peak-to-peak variation of pressure drop remained almost constant for small changes in the surface temperature.

Agostini et al. [200] and Warrier et al. [201] studied friction and heat transfer characteristics of flow boiling in minichannels. Martinelli and Nelson [202], Cheol and Moo [203] and Kandlikar and Steinke [204] measured pressure drop and studied boiling heat transfer characteristics and flow patterns during flow boiling in a single microchannel and parallel microchannels. Chisholm [205] and Yun et al. [206] determined pressure gradients and measured pressure drop due to friction during the flow of two phase mixtures in smooth tubes and channels.

7. CHF

Hetsroni et al. [45] observed that after venting of the elongated bubble, the region of liquid droplets begins. The vapor phase occupies most of the channel core and periodic dryout and wetting occurs. The two-phase period of vapor mixed with micro-droplet clusters on the heated walls affects the wall temperature and heat transfer in microchannels. With the increase in heat flux, the two-phase period decreases and critical heat flux (CHF) becomes imminent. The periodic phenomenon reveals that the entire channel acts like the area beneath a growing bubble, going through periodic drying and rewetting. The cycle is repetitive with venting of the elongated bubble.

Bergles and Kandlikar [6] dealt at length with the CHF aspect in microchannels. They have stressed on the need of modeling of saturated flow boiling CHF in microchannels under stable conditions. Accurate experimental data under stable operating conditions are essential for validating the models or even making any new models. Parallel microchannels (commonly used) must not be designed based on the heat transfer and pressure drop data for single channels which fail to capture the excursive or Ledinegg [223] instability resulting from the unique pressure drop characteristics of a boiling channel. Bergles and Kandlikar [6] have also emphasized that the heat transfer coefficient beyond the CHF condition (post dryout behavior), whatever be the cause, has to be studied to assess the consequences of exceeding the CHF. However, there are the technology limitations for the instrumentation in this regard. Several researchers [207-211] observed CHF and flow pattern for low pressure water and liquid nitrogen flowing in tubes.

Shah [212-213] presented a general correlation of CHF for subcooled and saturated boiling in vertical tubes. The upstream conditions correlation (UCC) covers the situation where CHF at a location depends on the upstream conditions, e.g., inlet subcooling and distance from tube inlet. The local conditions correlation (LCC) relates CHF to local quality, except for very short tubes. Hall and Mudawar [83] designed a statistical correlation based on almost all available subcooled CHF databases in the literature. Zhang et al. [108] developed correlation of critical heat flux for flow boiling of water in mini channels. Hetsroni et al. [69] also developed a correlation to predict heat transfer coefficient in terms of Eotvos number and boiling number. However, the correlations are very complex and [69, 83, 108, 212, 213] may be referred to for the correlations.

8. Instability

Two-phase flow instabilities in a microchannel can be dynamic and static instabilities. The dynamic instability generally reflects the feed back control process of a number of parameters such as pressure, mass flux and temperature. The unsteady behaviors caused by three types of commonly recognized oscillations: pressure drop, density wave and thermal oscillations can trigger a boiling crisis that causes mechanical damages.

A flow is stable if any perturbations in the systems eventually die down and the system asymptotically returns to the original operating conditions. If the system, when perturbed, returns to a new equilibrium condition different from the original equilibrium condition, then the system is subject to static instability. A system having delayed feedback and sufficient interaction between the inertia of flow and compressibility is said to be subject to a dynamic instability. Dynamic instability may also result from multiple feedbacks between flow rate, pressure drop and the change in density.

Many mechanisms can trigger boiling instabilities in macrosystems and microsystems. Depending on the mechanism, various types of static and dynamic instabilities, e.g. nucleation instabilities, flow pattern instabilities, Ledinegg instabilities, thermal oscillations, density-wave type oscillations and pressure drop type oscillations may arise, [214-219]. Early works [42, 72, 78-80, 119, 131, 220] on boiling in microchannels and minichannels focused on stable boiling. Studies [49, 53, 77, 130] of flow boiling instability in microchannels are relatively more recent.

Dynamic instabilities can be characterized, Boure et al. [24] as

(a) Density-wave type oscillations
(b) Pressure-Drop type oscillations
(c) Acoustic oscillations
(d) Thermal oscillations

Pressure drop type oscillations occur in boiling two phase flow systems where the amount of upstream compressible volume is significant. The period of this type of oscillation is usually long. The oscillations of pressure and mass flux are out of phase.

Density wave type oscillations are related to kinematic wave-propagation phenomena and, therefore, the alternative name for this oscillation is flow-void feedback instability. Its period is on the order of the time required for a density wave to travel through the system. Alternate fluid waves of higher and lower density travel in the system. The oscillation amplitudes and periods of pressure, mass flux and wall temperature are generally small.

Thermal oscillations are compound dynamic instabilities related to the instability of the liquid film next to the tube walls. Temperature fluctuations occur. The amplitudes and periods of pressure and mass flux oscillations are very small. The occurrence of different oscillations is function of G, q and T_{inlet}.

Figures 19-21 show pressure drop type, density wave type and thermal oscillations, respectively observed by Ding et al. [215] in a normal size (hydraulic diameter 10.9 mm) 1060 mm long horizontal channel for flow boiling of R-11. The oscillations are in a way similar to those observed in microchannels. These oscillations are strong functions of q, G and T_{inlet}.

Pressure drop type oscillations are associated with large amplitudes of fluctuation of flow parameters such as inlet pressure, wall temperature and inlet mass flux. Under a constant heat flux, the amplitudes and oscillation periods decrease as mass flux decreases. They may even remain unchanged within certain ranges. Higher heat input increases the oscillation amplitudes and reduces their oscillation periods since higher heat input supplies more energy to the fluid pushing the upstream compressible volume with larger magnitude and faster speed. Lower T_{inlet}, i.e. higher degree of inlet subcooling, generally increases the oscillation amplitudes and periods as explained by Liu and Kakac [221]. Peaks for different T_{inlet} are very close to each other, whereas valleys are far apart. The pressure difference is, therefore, larger with lower T_{inlet} and smaller with higher T_{inlet}. Thus larger amplitudes are found with lower T_{inlet} and consequently take a longer time.

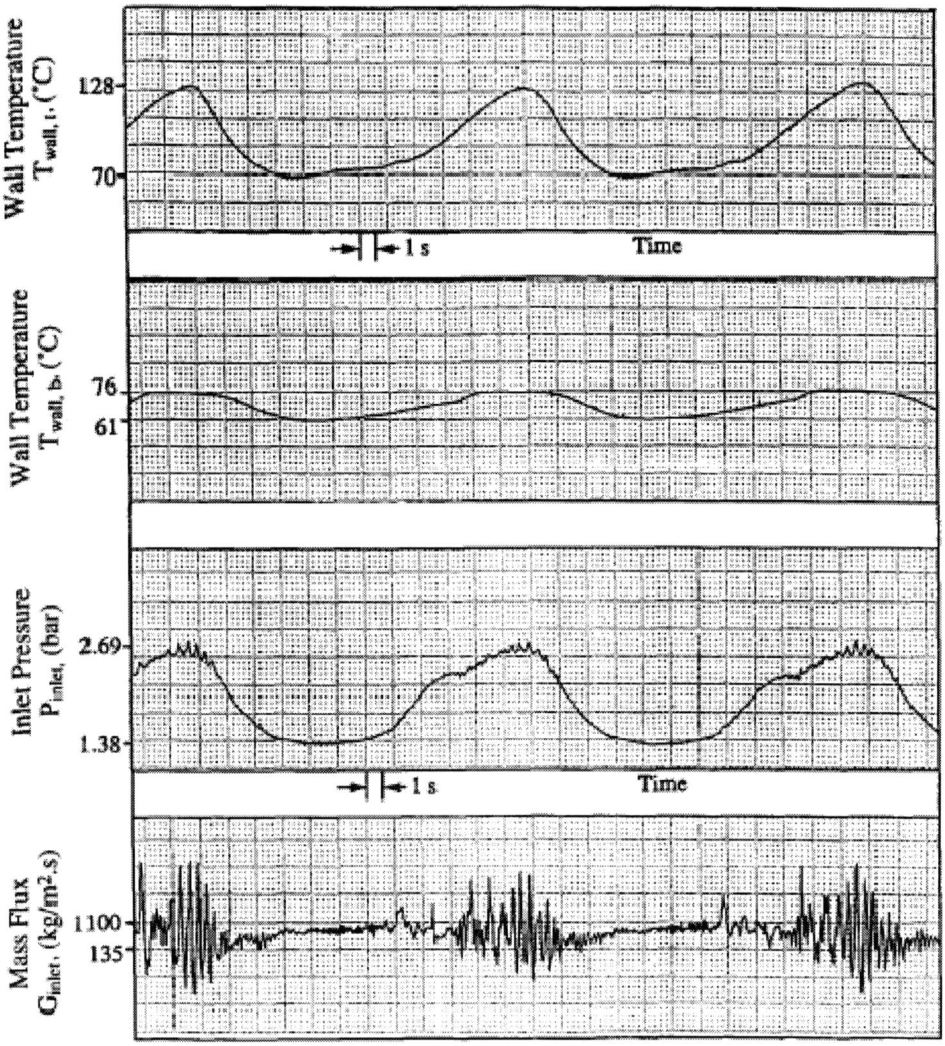

Figure 19. Pressure drop type oscillations, Ding et al. [215].

In a vertical upflow system [58], sustained density wave type oscillations occur on the steady state characteristic curve where the slope is positive. Pressure and flow rate oscillate in phase. However, in a horizontal flow system of Ding [222] and Ding et al. [215], this has been observed in the negative slope region and the variation of inlet pressure and mass flux is also in phase as shown in Figure 20. Density wave type of oscillations is enthalpy waves of high and low densities passing alternatively through the system. For high density wave and with large mass flux, the liquid content is substantially high owing to the large supply of liquid. The oscillation amplitudes are thus quite large. However, with lower mass flux, the amount of liquid is limited resulting in smaller oscillation amplitude. The oscillation periods of density wave type oscillations has been explained by Ding et al. [215] by means of flow stratification in horizontal channel. Depending up on the heat flux, periods of oscillation may increase (large heat flux case) or remain constant (low heat flux case) with the increase of mass flux.

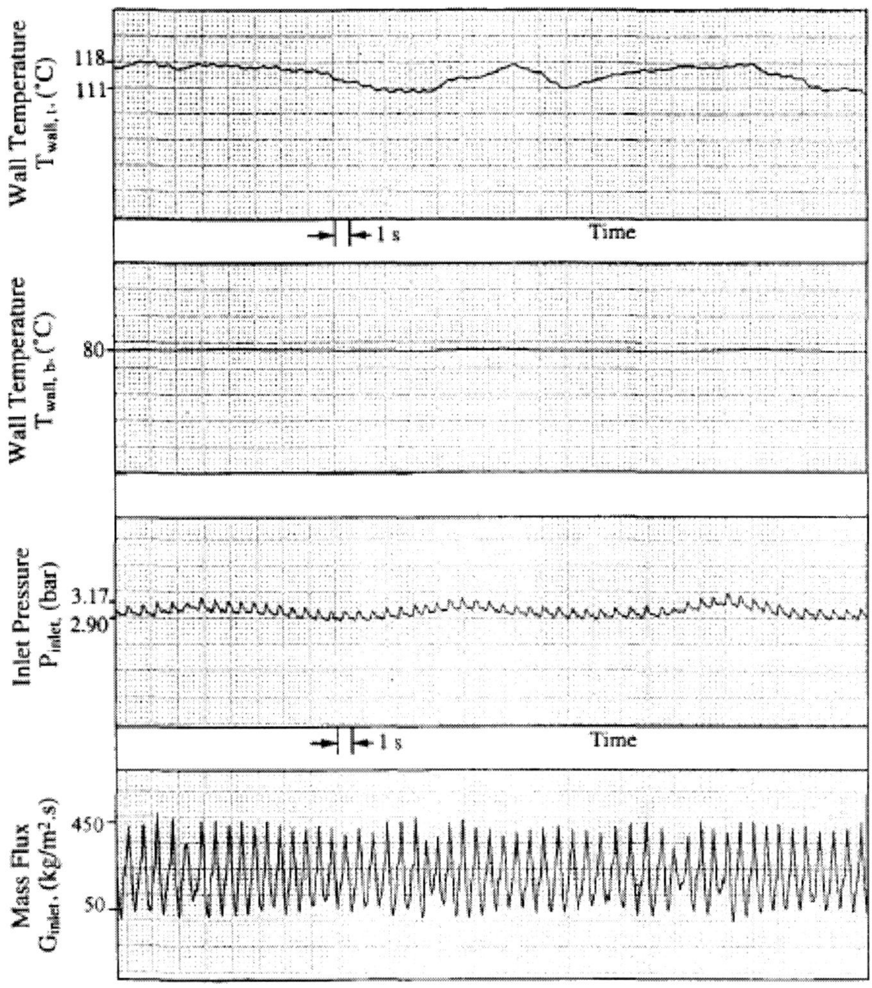

Figure 20. Density wave type oscillations, Ding et al. [215].

Higher heat flux in the system leads to a smaller density difference and hence a smaller amplitude. The periods of oscillations decrease with increase of heat flux since the heating effect to accelerate the fluid particles outperforms the effect caused by the reduction of flow rate. Higher inlet temperatures reduce the density difference between the high and low density waves resulting in smaller amplitude. Higher inlet temperature also accelerates the fluid particles more in the longer two phase region resulting in a shorter time to travel through the entire system. Figure 22 shows the oscillation boundaries observed by Ding et al. [215].

Ledinegg instability [223] involves a sudden change in the flow rate to a lower value. It occurs when the internal channel demand pressure drop versus flow rate characteristic curve is with negative slope and steeper than the external loop supply pressure drop versus flow rate characteristic curve and multiple intersections of the internal and external characteristic curves occur.

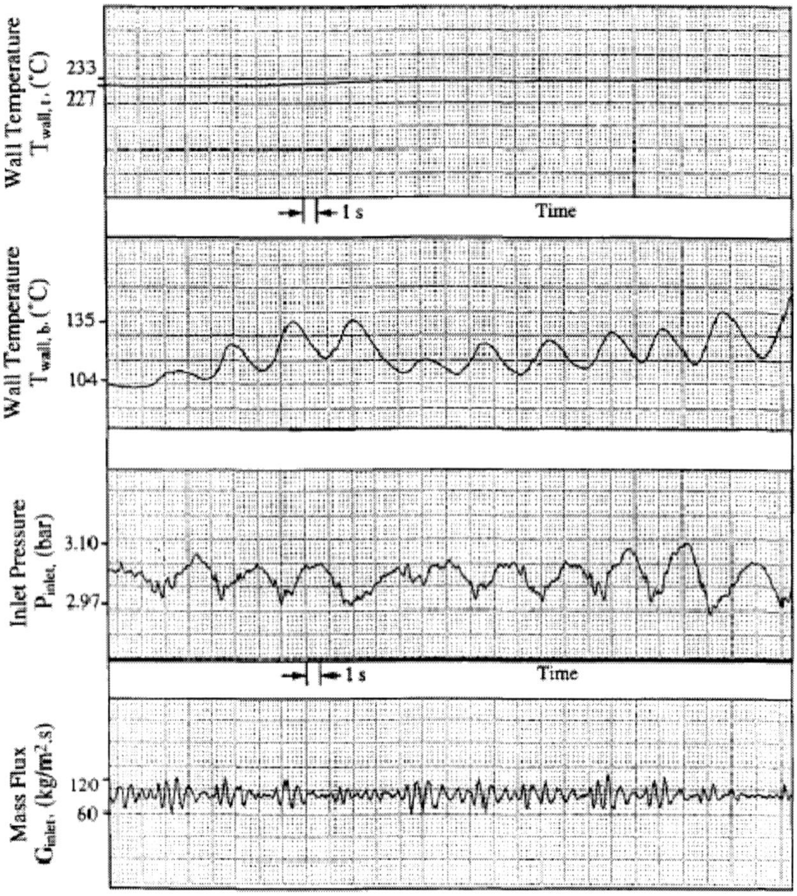

Figure 21. Thermal oscillations, Ding et al. [215].

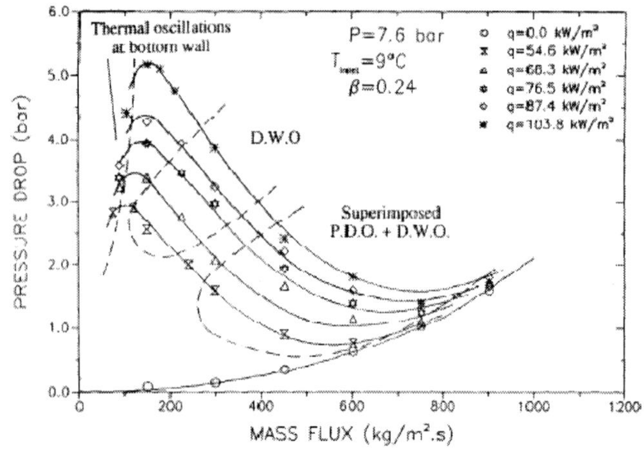

Figure 22. Oscillations boundaries, Ding et al. [215].

Mechanisms of different types of instability have been explained in Stenning [224]. The advances in stability analysis of any physical system have always been due to the milestone achievements in mathematics. Stability analysis can be made by the direct numerical analysis and frequency domain analysis approaches, Liu [225]; either by direct methods of finite difference or Laplace transform as applied in the control theory. Some pioneering works [58, 226-228] have been made in two phase flow dynamics using tools in non-linear dynamical systems. There have been several extensive reviews of instabilities in two phase flow systems, [24, 26, 29, 229-233]. Some works on two phase flow instabilities are [27, 218, 229, 234-268].

Hetsroni et al. [45] observed that alternative single-phase liquid flow and two-phase liquid-vapor flow in the microchannel leads to pressure drop fluctuations. However, they did not observe the long period (up to 78 s) pressure and temperature fluctuations reported by Wu and Cheng [102-105]. Oscillation amplitudes increase with increasing vapor quality. Vapor quality increases due to an increase in the heat flux and the time of dryout resulting in decrease in the heat transfer coefficient.

Wang et al. [128] have performed simultaneous visualization and measurement study to investigate dynamic instabilities in flow boiling of water in parallel 30 mm long microchannels as well as in a single microchannel of trapezoidal cross-section with hydraulic diameter 186 μm. They observed that after incipient boiling, both stable and unstable flow boiling occurred in parallel microchannels and in a single microchannel depending on the heat to mass flux ratio q/G. Different flow patterns caused unstable flow boiling with long (> 1s) as well as short (< 0.1s) period oscillations and Boure et al. [24] term these instabilities due to transition of flow patterns as dynamic instabilities.

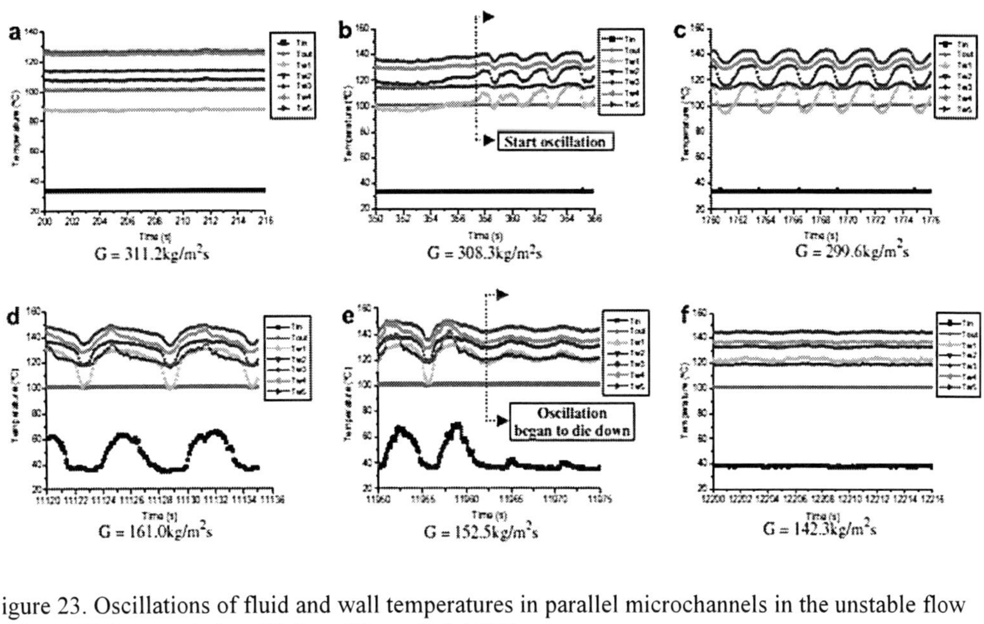

Figure 23. Oscillations of fluid and wall temperatures in parallel microchannels in the unstable flow regime with long period oscillation, Wang et al. [128].

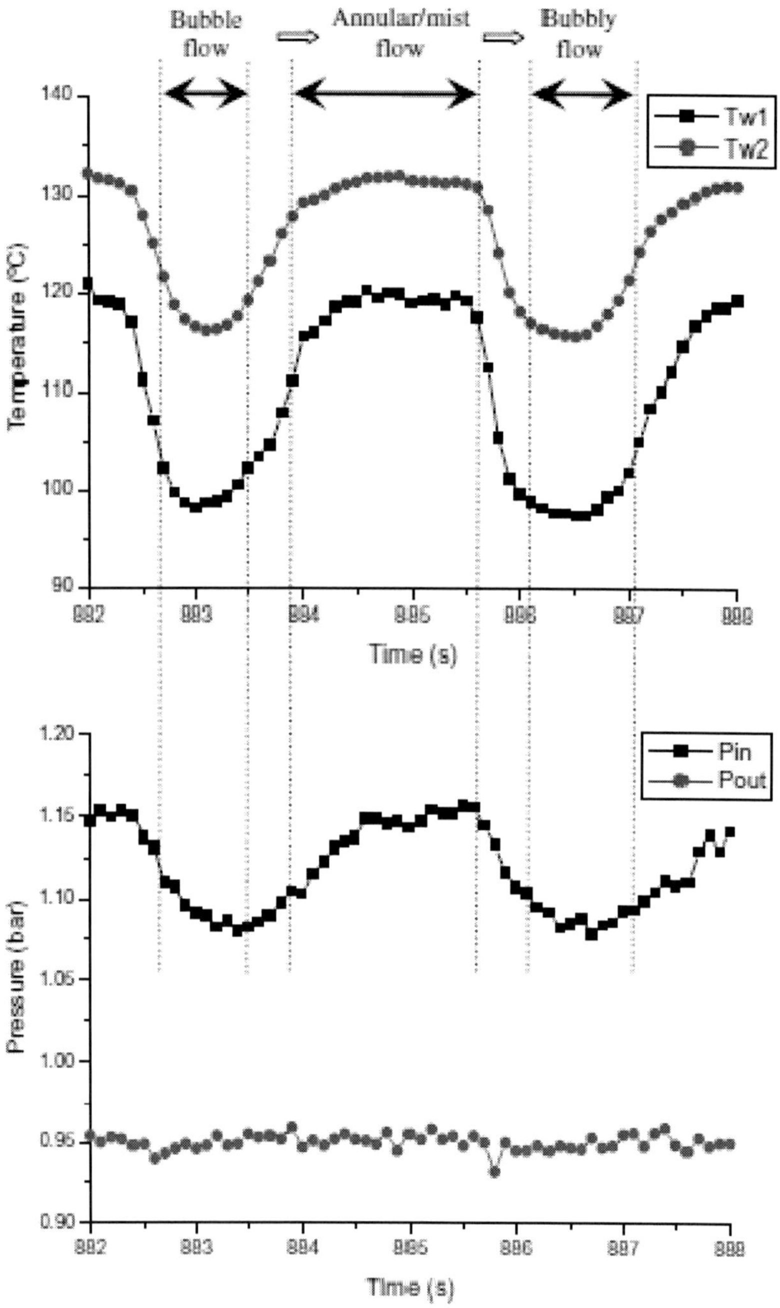

Figure 24. Oscillations of wall temperatures and inlet/outlet pressure in parallel microchannels in the unstable flow boiling regime with long period oscillation, Wang et al. [128].

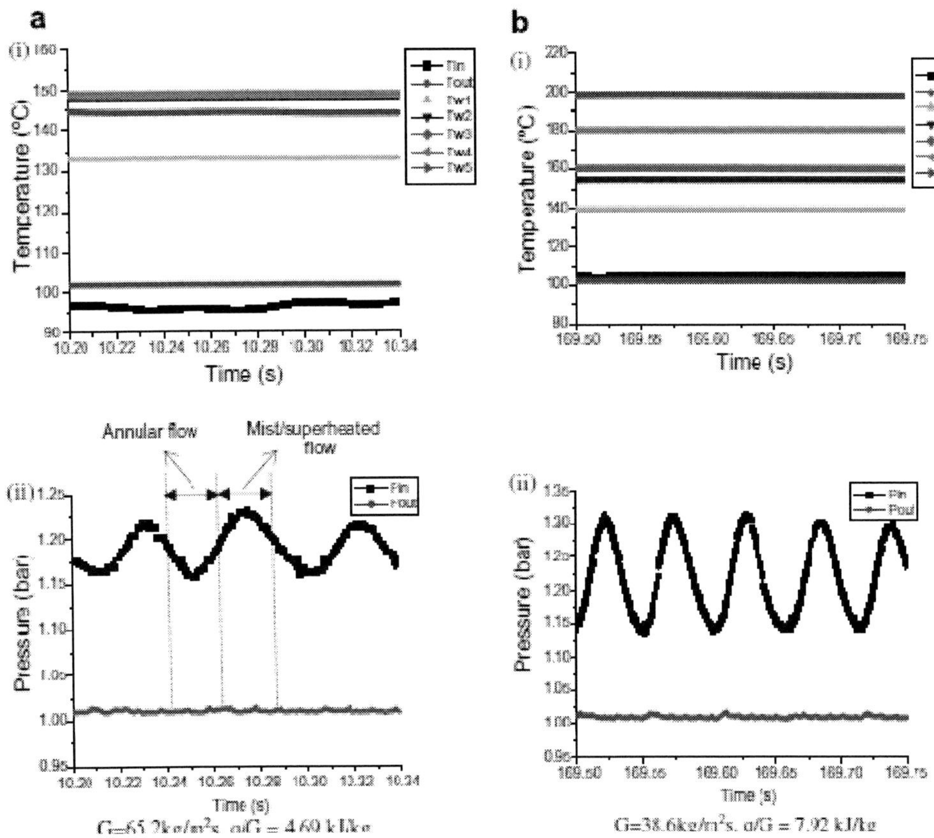

Figure 25. Inlet/outlet water and wall temperatures and inlet/outlet pressure in parallel microchannels in the unstable flow boiling regime with short period oscillation at lower heat flux, Wang et al. [128].

Wang et al. [128] measured periods and amplitudes of temperatures, pressures and mass flux fluctuations in these dynamic instabilities. Figures 23-26 show oscillations of fluid and wall temperatures and inlet/outlet pressure in parallel microchannels in the unstable flow regime with long period and short period oscillations observed by Wang et al. [128]. Figure 27 and Figure 28 show the observation of Wang et al. [128] for oscillation periods of temperature versus heat-to-mass flux ratio at different heat flux in parallel microchannels and in a single microchannel, respectively in the unstable flow boiling regime with long-period oscillation. Figures 29-31 show stable and unstable flow boiling regimes and oscillations in a single microchannel observed by Wang et al. [128]. They have observed isolated bubbles in the stable flow boiling regime. These bubbles are pushed away by the incoming subcooled liquid. Long-period oscillation is due to the expansion of vapor bubble from downstream while the short-term oscillation is owing to the flow pattern transition from annular to mist flow. Flow interaction effects in parallel microchannels from neighboring channels at the headers are significant.

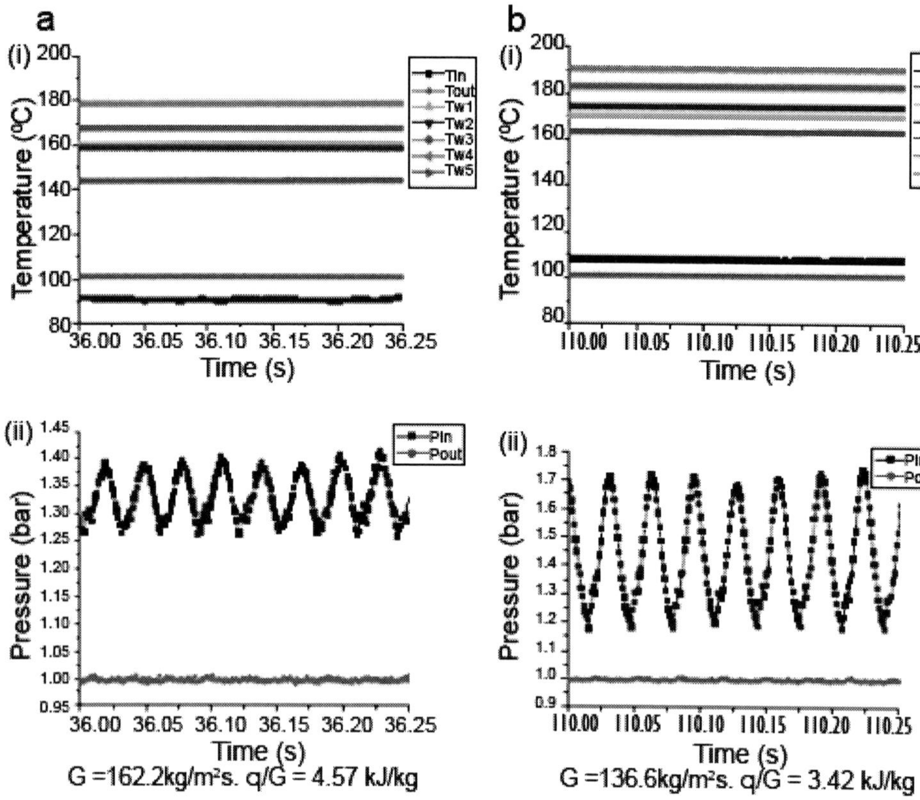

Figure 26. Inlet/outlet water and wall temperatures and inlet/outlet pressure in parallel microchannels in the unstable flow boiling regime with short period oscillation at higher heat flux, Wang et al. [128].

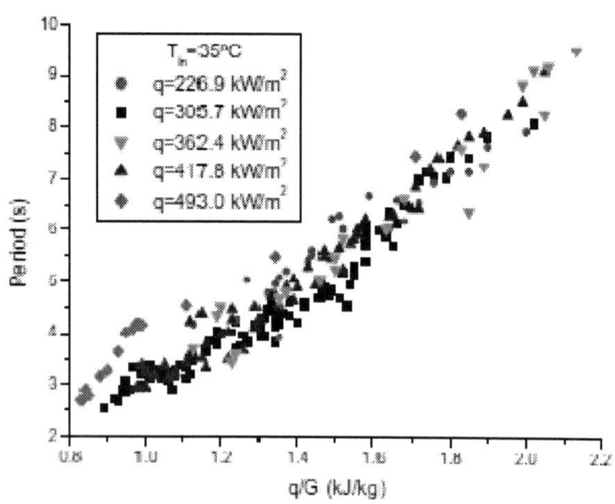

Figure 27. Oscillation periods of temperature vs. heat to mass flux ratio at different heat flux in parallel microchannels in the unstable flow boiling regime with long period oscillation, Wang et al. [128].

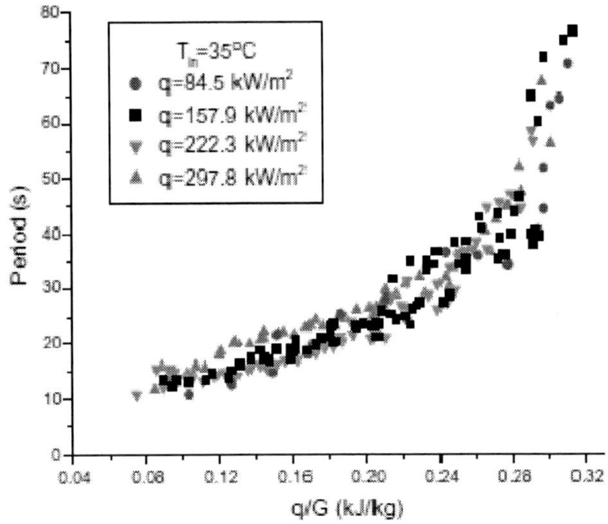

Figure 28. Oscillation periods of temperature vs. heat to mass flux ratio at different heat fluxes in a single microchannel in the unstable flow boiling regime with long period oscillation, Wang et al. [128].

Flow boiling instabilities in parallel microchannels with three types of connections are shown in Figure 9, Wang et al. [132]. In type A connection, inlet and outlet conduits are perpendicular to the parallel microchannels in the test section. In type B connection, fluid can flow into and exit from the parallel microchannels without any restriction and in type C connection, fluid enters each of the microchannels through an inlet restriction.

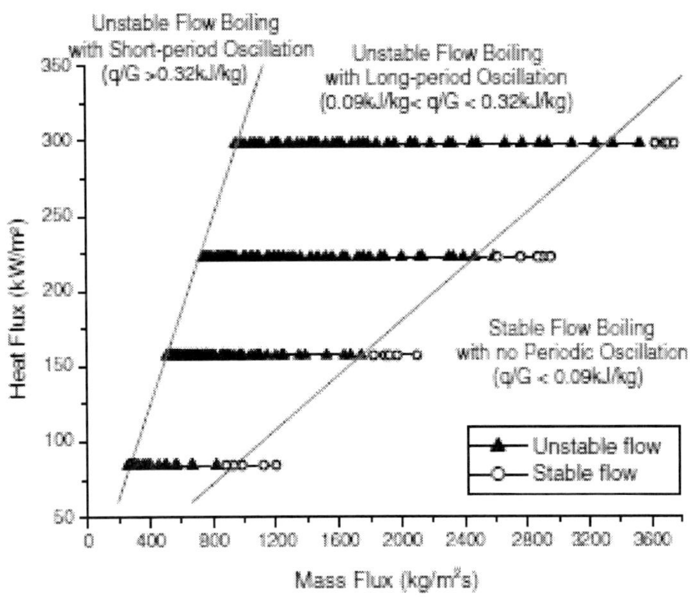

Figure 29. Stable and unstable flow boiling regimes in a single microchannel, Wang et al. [128].

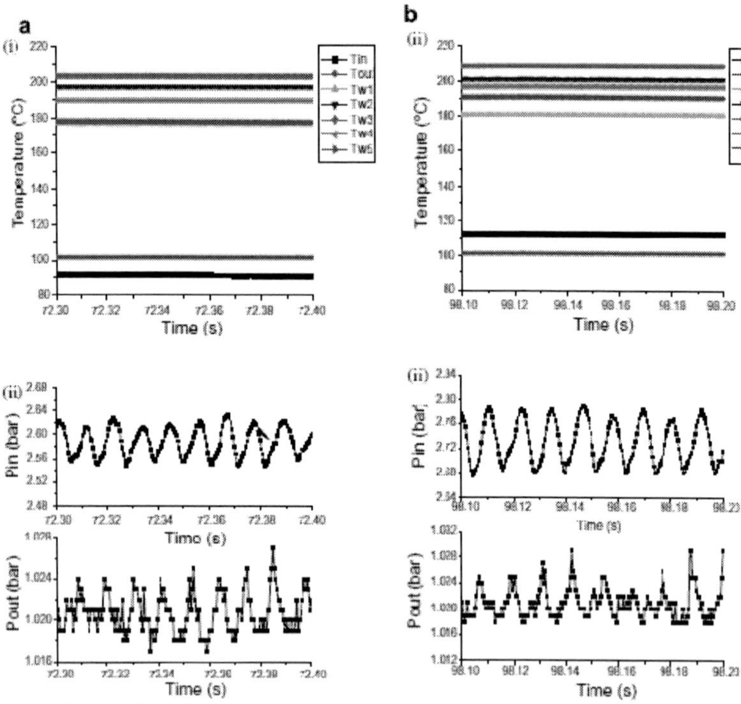

G = 711.8kg/ms², q/G = 0.43 kJ/kg G = 551.8kg/m²s, q/G = 0.56 kJ/kg

Figure 30. Temperatures and pressure in a single channel in the unstable flow boiling regime with short period oscillation at lower heat flux, Wang et al. [128].

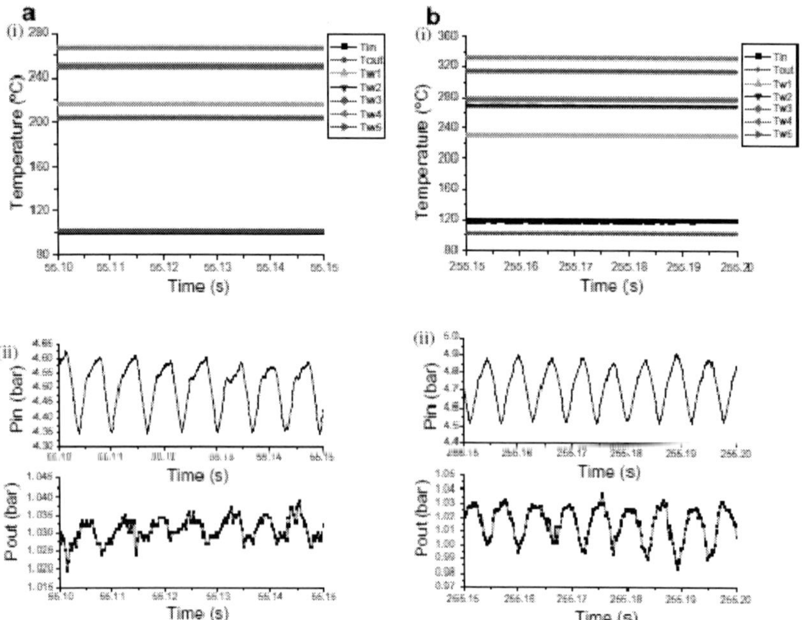

G = 895.2kg/m²s, q/G = 0.55 kJ/kg G = 679.5kg/m²s, q/G = 0.73 kJ/kg

Figure 31. Temperatures and pressure in a single channel in the unstable flow boiling regime with short period oscillation at higher heat flux, Wang et al. [128].

Temperature and pressure oscillations in microchannels with the Type-B connection, Wang et al. [132] are considerably less than those for the Type-A connection. Therefore, the microchannels with Type-B connection are more stable than those with the Type-A connection because the exit is more resistive in the latter case.

Bergles et al. [269] and Bergles and Kandlikar [6] recommend installation of inlet orifices in order to suppress both upstream compressible volume instability and excursive instability. Inlet orifices suppress the parallel channel interactions observed with both Type-A and Type-B connections, Figure 9, Wang et al. [132].

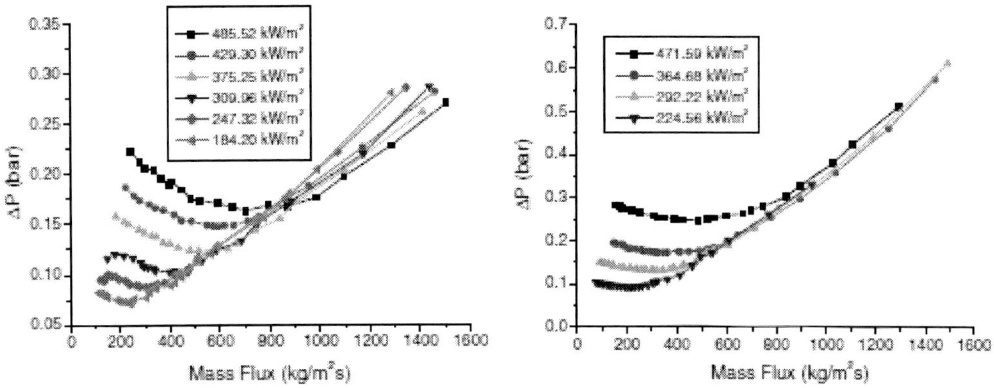

(a) Type-B connection (b) Type-C connection

Figure 32. Inlet pressure versus mass flux in parallel microchannels, Wang et al. [74].

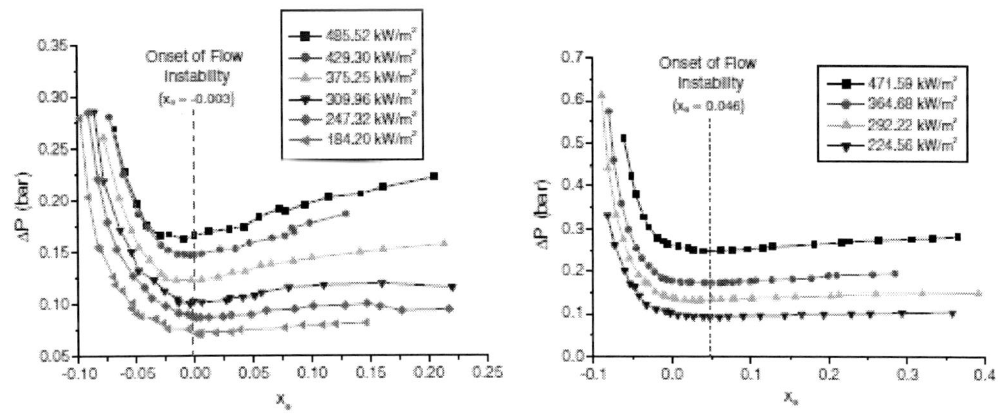

(a) Type-B connection (b) Type-C connection

Figure 33. Inlet pressure versus local vapor quality in parallel microchannels, Wang et al. [74].

Figure 32 and Figure 33 show measurements of inlet pressure versus mass flux in parallel microchannels for Type-B and Type-C connections of Wang et al. [132]. Steady flow boiling with no reversed flow of vapor bubble is achieved by fabrication of inlet restriction (Type-C connection); this increase in flow stability comes at the expense of higher pressure drop as is seen from the onset of flow instability (OFI) corresponding to the minimum point in Figure 33.

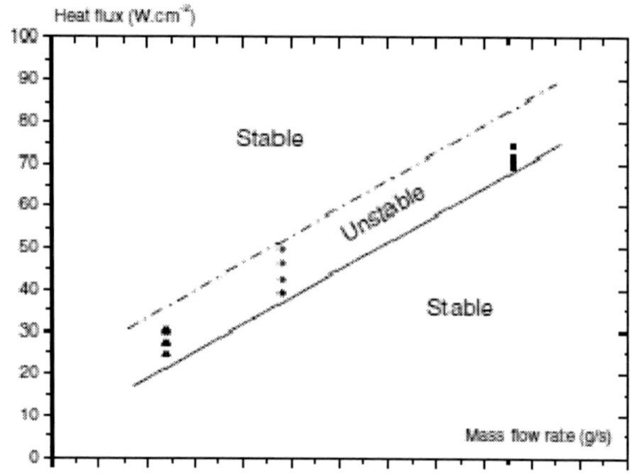

Figure 34. Stability diagram of heat flux supplied as a function of the mass flow rate, Brutin et al. [270].

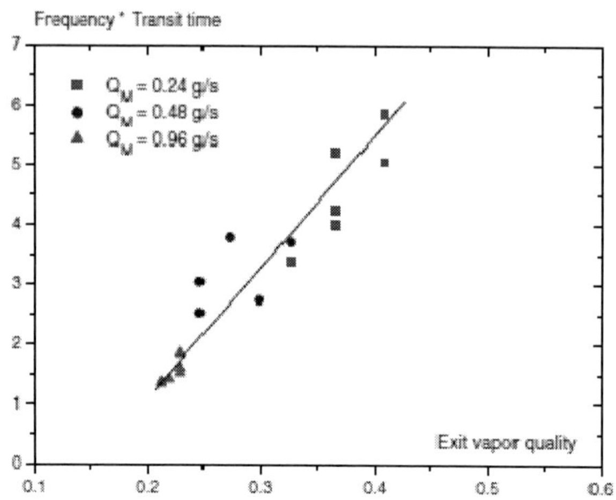

Figure 35. Oscillation frequency by transit time of a liquid particle as a function of exit vapor quality, Brutin et al. [270]

Figure 34, [270] shows the unsteady operating points characterized by the two parameters, heat flux and mass flux. The oscillation frequency of operating conditions by transit time of a liquid particle in the channel as a function of exit-vapor quality is shown in Figure 35 from the experiment of [270]. They defined a stability criterion. Low fluctuation amplitude (< 1 kPa) and no characteristic oscillation frequency identified by the spectrum analysis have been defined as steady state. The onset of flow instability (OFI) is characterized by a definite frequency and high fluctuation amplitude (> 1 kPa). Figure 36 and Figure 37 show typical steady boiling inlet and outlet pressure signals and boiling outlet and inlet temperature difference. The typical unsteady boiling inlet and outlet pressure signals observed by Brutin et al. [270] are shown in Figure 38.

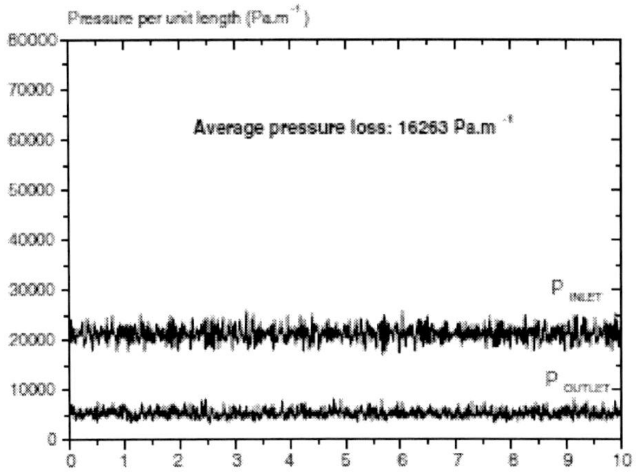

Figure 36. Typical steady boiling inlet and outlet pressure signals, Brutin et al. [270].

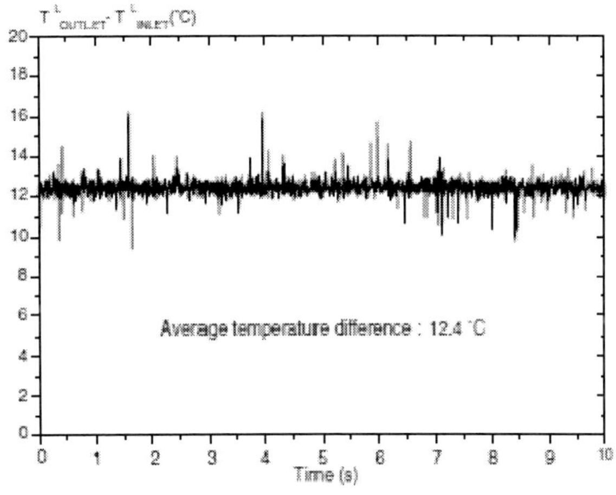

Figure 37. Typical steady boiling temperature difference, Brutin et al. [270].

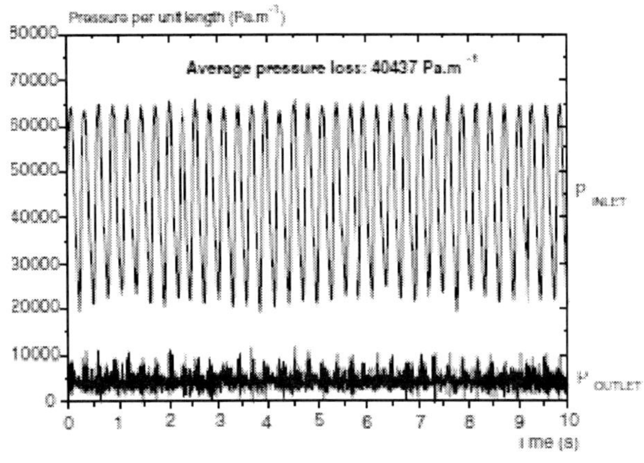

Figure 38. Typical unsteady boiling inlet and outlet pressure signals, Brutin et al. [270].

An important aspect of bulk boiling is the fluctuations in flow and pressure, because these fluctuations can initiate instabilities. Two phase flow noise is always present in flow bulk boiling due to bubble formation or the passage of liquid and vapor. A comprehensive discussion of fluctuations in narrow channels is available, Kandlikar [18].

Qu and Mudawar [93] investigated hydrodynamic instability and pressure drop in a water cooled two phase flow boiling microchannel heat sink containing 21 parallel 231 × 713 μm microchannels. The instability was attributed to interaction between vapor generation in channels and compressible volume in the flow loop upstream of the heat sink. They found that a throttling valve installed immediately upstream of the heat sink significantly reduces the instability.

Figures 39-41 show temperature, pressure and mass flux fluctuations, respectively observed by Wu and Cheng [103]. Large amplitude/long period fluctuations with time in wall and fluid temperatures, fluid pressures and fluid mass flux were measured. The fluctuation periods were found to be dependent on channel size, heat flux and mass flux. The fluctuations sustained when the fluctuations of pressure drop and mass flux had phase differences.

Wu and Cheng [105] investigated boiling instability modes of water flowing in microchannels at various heat flux and mass flux with the outlet of the channels at atmospheric pressure. Eight parallel silicon microchannels 30 mm long with hydraulic diameter 186 μm with an identical trapezoidal cross-section have been used. Three types of boiling instability were observed by them. These are:

- The liquid/two-phase alternating flow at low heat flux and high mass flux
- The continuous two phase flow at medium heat flux and medium mass flux
- The liquid/two phase/vapor alternating flow at high heat flux and low mass flux

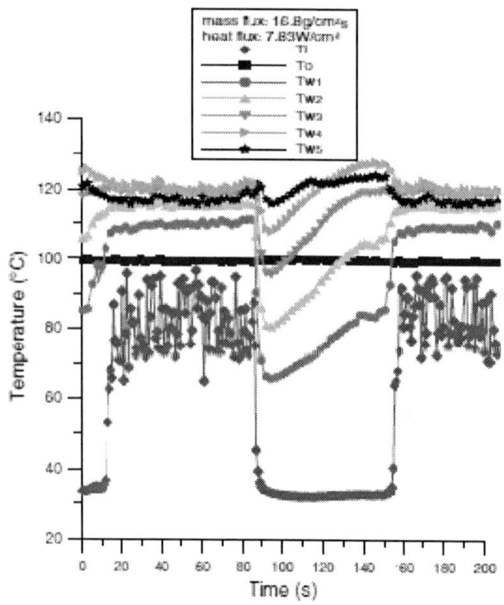

Figure 39. Temperature fluctuations during periodic boiling, Wu and Cheng [103].

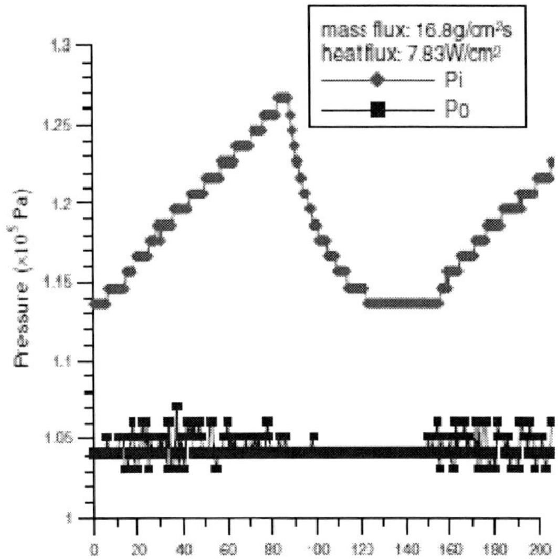

Figure 40. Pressure fluctuations during periodic boiling, Wu and Cheng [103].

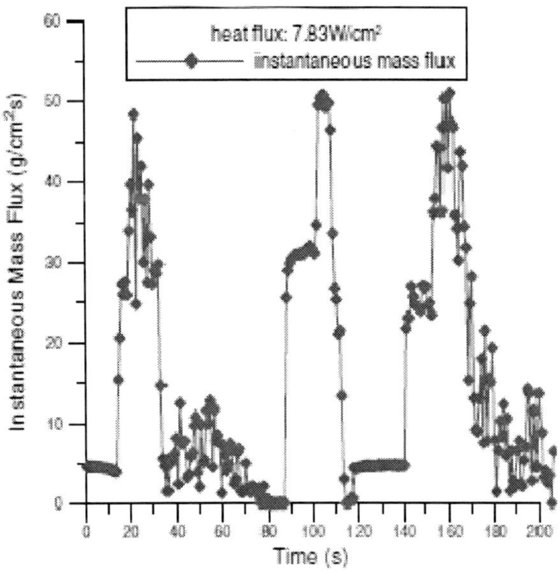

Figure 41. Mass flux fluctuations during periodic boiling, Wu and Cheng [103].

Periodic oscillations of wall temperature, inlet and outlet water temperatures and pressures and instantaneous mass flux were measured.

In the 3rd case, oscillation amplitudes were the largest with nearly out of phase pressure and mass flux. In the 2nd case, the oscillation amplitudes were the smallest with nearly in-phase oscillations of pressure and mass flux. In the 1st case, it was in between the 2nd and 3rd with oscillations of pressure and mass flux nearly out of phase. Oscillation periods are dependent greatly on amounts of heat flux and mass flux. Figure 42 shows typical oscillations observed by Wu and Cheng [105].

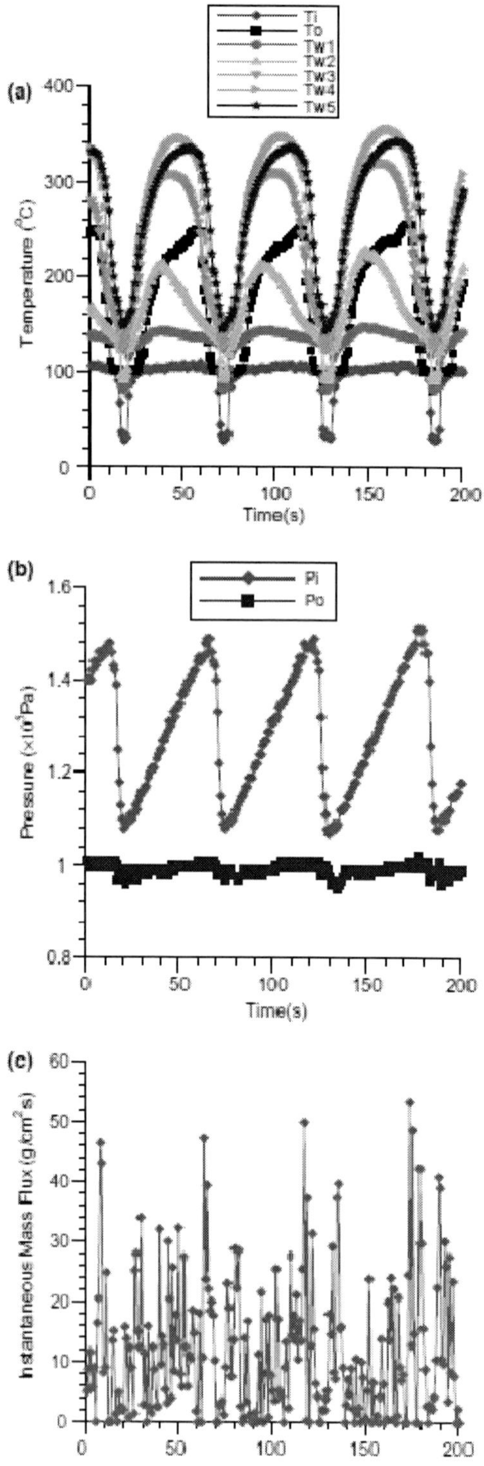

Figure 42. Oscillation of various measurements, Wu and Cheng [105].

Hetsroni et al. [53] reported short period (1-2 s) and low amplitude oscillation data in pressure drop and outlet temperature because of growth and collapse of vapor bubble. They carried out experiments on flow boiling with Vertrel XF fluid in silicon microchannels of triangular cross-section with hydraulic 130.14 μm. Some reports on boiling instability in the small channels with hydraulic diameter 889 μm – 5 mm are Wen et al. [271], Brutin et al. [272] and Suzuki et al. [273]. Jain [274] and Wedekind [275] studied self sustained hydrodynamic oscillations in two phase flow boiling loop.

Chang and Pan [276] investigated experimentally two phase flow instability for flow boiling in silicon based fifteen parallel rectangular microchannels with hydraulic diameter 86.3 μm for each channel. The width and depth of each channel were 99.4 μm and 76.3 μm, respectively. The fabrication was done by silicon bulk micromachining and anodic bonding process. Flow patterns were visualized using a high speed digital CCD camera. Chang and Pan [276] observed that for the stable cases, bubble nucleation, slug flow and slug or annular flow appears sequentially in the flow direction. Under unstable condition, forward or reversed slug/annular flows appear alternatively in every channel. They concluded that the length of bubble slug may oscillate for unstable cases with reversed flow demonstrating the suppressing effect of pressure field for bubble growth. The magnitude of pressure drop oscillations may be used as an idea for the appearance of reverse flow. A stability map on the plane of inlet subcooling number versus phase change number has been plotted by Chang and Pan [276], Figure 43. A very narrow region for stable two phase flow or mild two phase flow oscillations is present near the line of zero exit quality; this is significantly different from that for normal size boiling channel [236], which is an unstable region in the upper right part and otherwise stable on the plane of N_{sub} versus N_{pch},
where,

Phase change number, $$N_{pch} = \frac{Q_c v_{lv}}{W h_{lv} v_l} \tag{16}$$

Subcooling number, $$N_{sub} = \frac{h_{sub} v_{lv}}{h_{lv} v_l} \tag{17}$$

Figure 44 shows the pressure drop oscillations observed by Chang and Pan [276]. Figure45 demonstrates that bubble length grows exponentially, though not very smoothly, possibly due to the variation of pressure field around the bubble. The numerical value for the coefficient of exponent is in the same order of magnitude as those reported by Li et al. [277]. Li et al. [277] have demonstrated that the growth of the bubble slug in the microchannel is resulted from the balance between the bubble expansion due to the evaporation of thin liquid film next to the heating wall and the suppressing effect due to the pressure field around the bubble. The bubble length oscillates under unstable situations with significant pressure oscillations under unstable situations with significant pressure oscillations supporting the thin film evaporation model proposed by Li et al.

[277]. Stelling et al. [278] studied the onset of flow instability for downward flow in vertical channels.

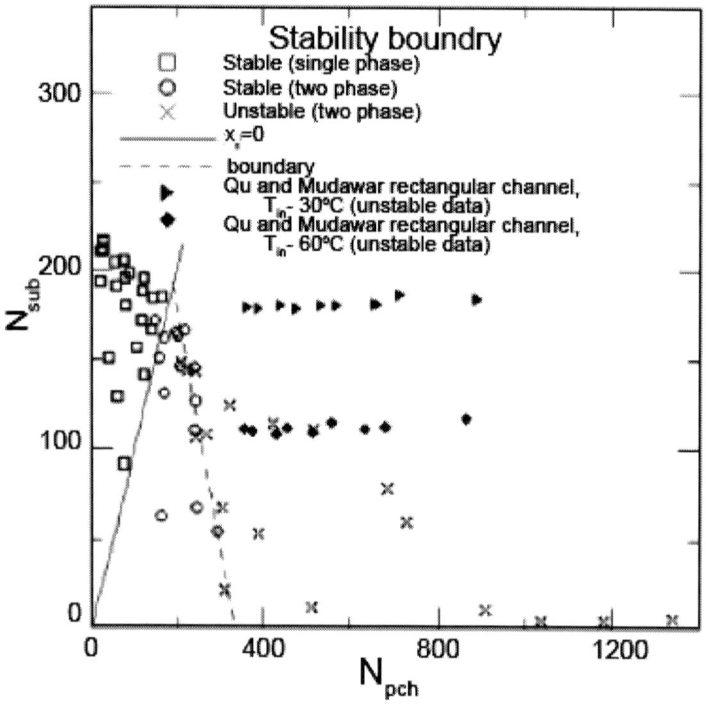

Figure 43. Stability boundary on the plane of subcooling number versus phase change number, Chang and Pan [276].

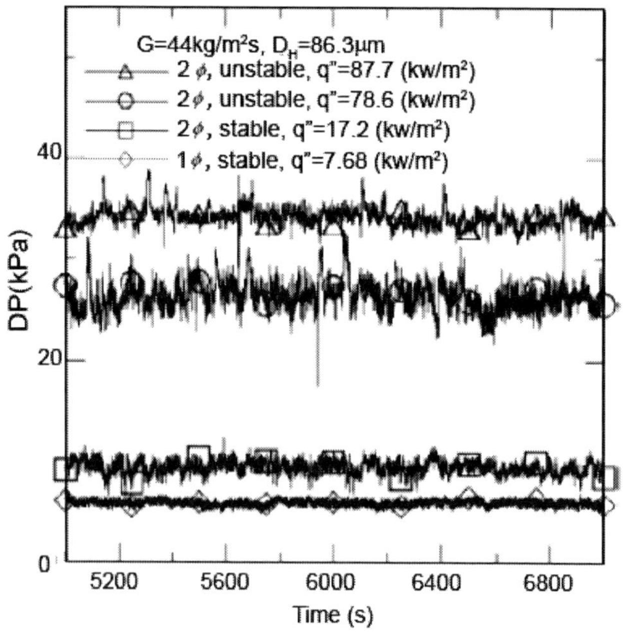

Figure 44. Significant pressure drop oscillations under unstable conditions, Chang and Pan [276].

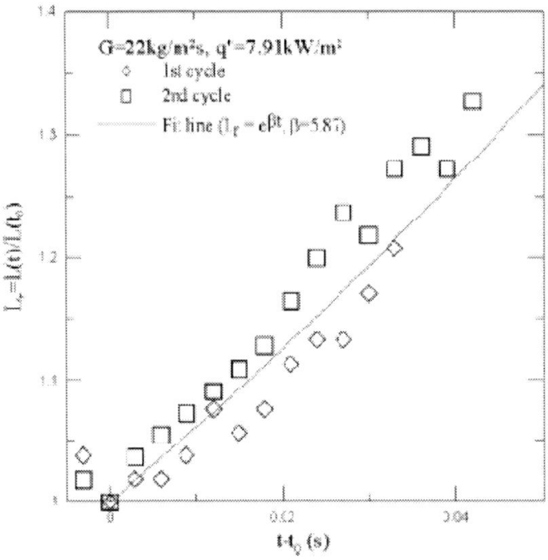

Figure 45. The exponential growth of bubble length, Chang and Pan [276].

The data of Qu and Mudawar [96] are also displayed in Figure 43. They attributed the occurrence of CHF in the microchannel to the appearance of significant two phase flow instability with reverse flow.

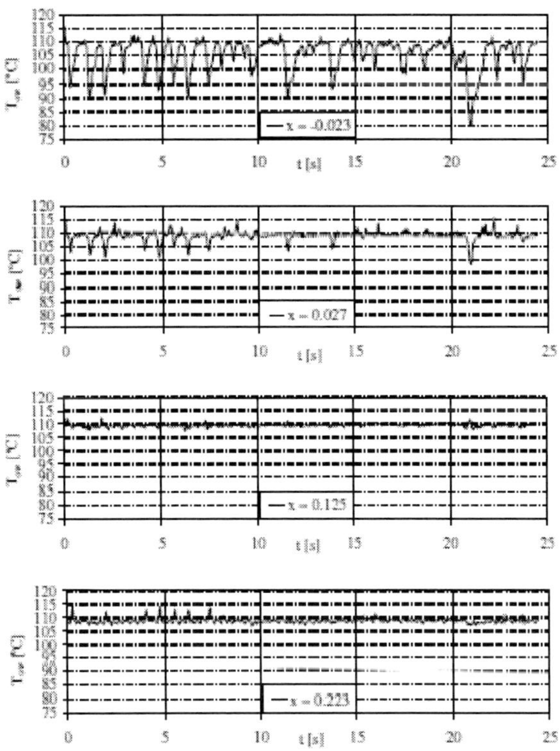

Figure 46. Outer wall temperature over time at lower heat and mass flux, Diaz and Schmidt [159].

Diaz and Schmidt [159] obtained microchannel flow boiling wall temperature distribution by infrared thermography. They have observed high amplitude oscillations for water in the region of subcooled boiling at qualities slightly below zero as shown in Figure 46. However, the amplitude decreased as quality increased and no characteristic frequency was observed. Flow boiling of ethanol yielded low amplitude oscillations of wall temperature, even within the subcooled boiling region as shown in Figure 47. Increase in quality resulted in increasing oscillation amplitude.

Kandlikar et al. [267] have found that an inlet pressure restrictor and fabricated nucleation sites completely eliminate two phase flow instabilities associated with the reverse flow in the flow boiling in microchannels. Kosar et al. [279] have shown that for sufficiently high inlet pressure restriction imposed by the inlet orifices, parallel microchannels and upstream compressible volume instabilities are eliminated. Kuo and Peles [280] found that high system pressure in microchannels moderates instabilities by reducing the void fraction, the superheat needed to activate bubble nucleation and bubble departure diameter. Local transient temperature measurements show lower magnitudes and higher frequencies of oscillations at high system pressure. Bogojevic et al. [23] made an experimental study on two phase flow instabilities inside a uniformly heated silicon heat sink consisting of 40 parallel rectangular microchannels with hydraulic diameter 194 μm, 15 mm long using water as cooling liquid. They observed high amplitude/low frequency and low amplitude/high frequency instabilities with appreciable pressure and temperature fluctuations.

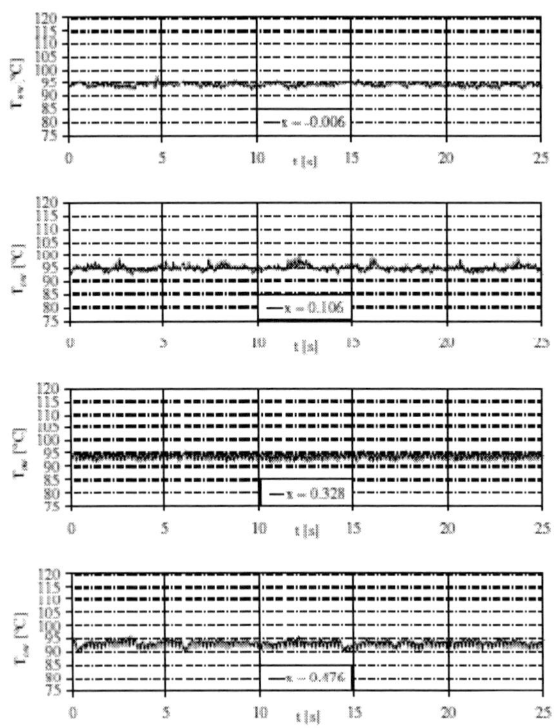

Figure 47. Outer wall temperature over time at higher heat and mass flux, Diaz and Schmidt [159].

Kandlikar [281], Bergles et al. [269] and Tadrist [26] gave some of the first interpretations of the two phase flow instabilities. When boiling occurs, similar phenomena like nucleation, coalescence, fragmentation and interfacial instabilities may be found whatever the channel hydraulic diameter size is. In a narrow channel, however, the vapor growth phase is limited in the radial direction because of the small thickness. Only the large direction allows vapor growth when boiling occurs. Small hydraulic diameter reduces the 3D effects of flow structures. Thus only the main processes on the development of the two phase flow instabilities will remain.

Two phase flow instabilities are also reported in single channels, Kenning and Yan [282], Brutin et al. [270]. Kennedy et al. [47] determined the onset of instabilities by using the pressure drop signal.

Flow reversal was observed during narrow channel flow boiling by Jones and Judd [283]. They concluded that boiling crisis and instability were due to the onset of CHF and these were the result of dryout and rewetting of the heated surface.

Brutin et al. [270] proposed a criterion for which the instability occurs based on the competition between the pressure difference induced by the vapor expansion and the pressure difference inducing the two phase flow in the channel.

If the oscillations in microchannels are duc to CHF mechanism, the oscillations characteristic frequency may be evaluated from Hamamura and Katto [284] model. This model allows the prediction of the time required for the rewetted heater surface to reach the CHF. The rewetting time must also be used for evaluating the period of oscillations.

Two phase oscillatory instabilities in microchannel evaporators are dealt with in great detail in [45, 159, 270, 282].

Huh et al. [285] focused on the boiling instability of deionized water in a 103.5 μm single microchannel at very low mass flux and heat flux conditions.

Yadigaroglu and Bergles [235] noted that two phase flow instabilities generated by unstable flow patterns, particularly slug flow, could amplify flow disturbances and lead to dryout. Therefore, pressure and flow rate oscillations in microchannels should be identified and prevented to ensure safe operations. Huh et al. [285] studied boiling instabilities and heat transfer mechanisms in a rectangular microchannel of hydraulic diameter 103.5 μm, 40 mm long made of polydimethylsiloxane (PDMS). They visualized flow patterns and measured momentum and thermal transport parameter. Flow boiling instabilities, pressure drop, mass flux, wall temperature and heat flux oscillations observed by Huh et al. [285] are shown in Figure 48 and Figure 49.

Wang and Cheng [132] observed that unstable flow boiling occurred in the single microchannel if the exit vapor quality was greater than 0.013. This critical value of the exit vapor quality is independent of inlet fluid temperature, heat and mass fluxes, the geometries of the microchannel and the microheater, as well as the physical properties of the fluid. For qualities below 0.013, stable flow boiling occurred in a single microchannel.

Wang and Cheng [132] observed that lower mass flux resulted in higher wall temperature and pressure drop, larger amplitude of in-phase temperature and pressure drop oscillations and longer oscillation period. It was pressure-drop type instability since long period of pressure drop and temperature oscillations were recorded. The oscillation periods were larger than 100 s.

Consolini and Thome [187] carried out an experimental investigation to study the characteristics of a circular micro-channel (510 μm) flow boiling of R-134a, R-236fa and R-245fa, affected by compressible volume instability. The flow boiling in small systems like microchannels is sensitive to oscillatory instabilities. The oscillations refer to either parallel channel instabilities or to compressible volume instabilities occurring in both parallel channels and single-channel system. Parallel channel instabilities are due to the interaction of the two-phase flow among adjacent channels. Compressible volume instabilities may be due to a body of vapor, flexible hosing, etc. upstream or within the heated length.

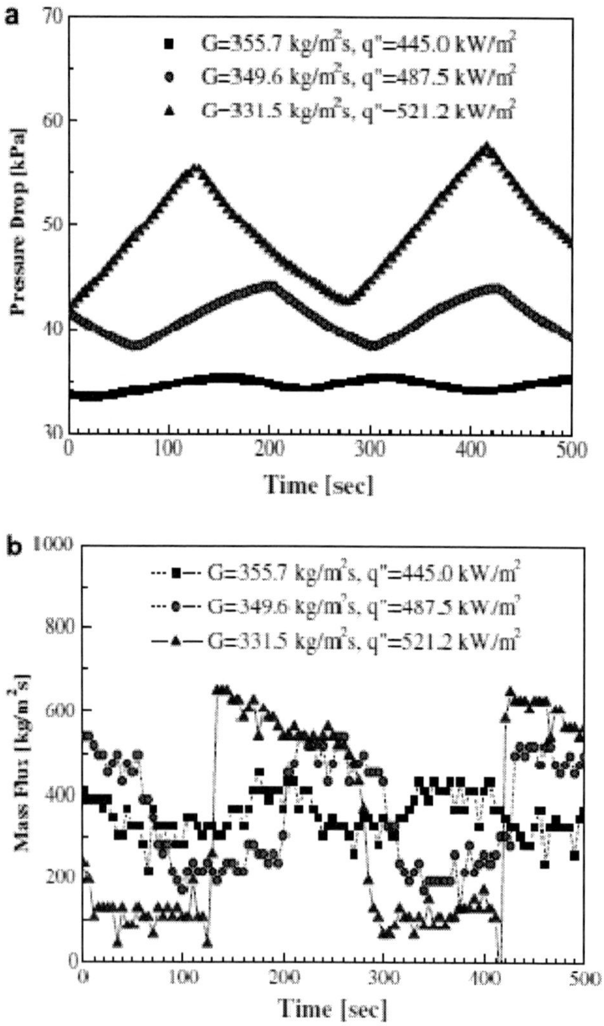

Figure 48. Comparison of (a) pressure drop and (b) mass flux fluctuations, Huh et al. [285].

Balasubramanian and Kandlikar [133] also observed change in the contact angles between the slug and the walls of the channels. When the vapor slug was flowing in the reverse direction, the head of the slug had a receding contact angle with the walls of the channel. However, when the vapor slug was flowing in the direction of the bulk flow, the head of the slug had an advancing contact angle with the walls of the channel. This was due to contact angle hysteresis. They experienced that pressure fluctuations are chaotic and do not follow a uniform pattern in parallel channels. It is extremely complex to map the flow patterns with the time-dependent pressure drop fluctuations because the effect of pressure drop caused by each channel gets superimposed on the other. They have analyzed the pressure fluctuations by Fourier transform method and observed that there is dominant frequency of oscillations which increases with an increase in the surface temperature. The increased activity of slug flow over nucleating bubbles at very high surface temperatures, however, may cause decrease in the dominant frequency signature in the pressure drop fluctuations. The bubbles nucleating at low surface temperatures have a higher frequency but lower amplitude, and the slug flow occurring at a higher surface temperature has a lower frequency but higher amplitude. Peak-to-peak variation of channel differential pressure remains almost invariant with an increase in surface temperature. Depending on the surface temperature, the slugs formed either at the inlet or toward the exit of the channels.

CONCLUSION

Following conclusions are drawn from the above discussion:

1. The bond number, which is a measure of the relative importance of the surface tension to gravitational effect, is a means to classify a channel as microchannel, minichannel or macrochannel.
2. Works on flow boiling in minichannels can be classified as stable flow or unstable flow. Effects of inlet/outlet configurations on flow boiling instability, flow patterns, pressure drop and heat transfer coefficients along with wall and fluid temperatures, model of evaporation of elongated bubble flow, CHF, subcooled or saturated flow boiling, microbubble emission, etc. have been studied.
3. Instability in parallel microchannels with horizontal inlet/outlet connections is partially reduced compared to that in vertical inlet/outlet connections since, in vertical constructions, flow cannot exit freely and the vapor bubbles tend to expand upstream because of the confinement of the wall. Temperature and pressure oscillations with vertical connections are more than those with horizontal connections. Thus systems with horizontal connections with rigid inlet connections rather than flexible inlet connections are more stable because, in case of horizontal rigid connections, there is less upstream compressible volume. The limit of exit quality for which stable flow boiling exists depends on the experimental set up, e.g. the inlet/outlet configurations, materials used, whether

the fluid is driven by a pump or a compressed gas tank. The oscillations depend on the amount of upstream compressible volume and specific time of measurements during the unstable flow boiling conditions.

4. Stable flow boiling with bubbly/slug flow exist at $x_e < 0.036$ and to extend the stability regimes, upstream volume stiffness increase by fabricating inlet restrictions or use of throttling/flow regulating valves upstream of the test section may be made. This suppresses the reverse flow of bubble.
5. Nucleation mechanism usually persists near the onset of boiling at the upstream with convective film vaporization in the Taylor bubble creating annular flow at downstream.
6. If the flow oscillations are delayed, isolated bubbly flow, bubbly-slug, slug, slug-annular, semi annular, annular and dryout regimes are observed.
7. Average pressure drop, calculated from the correlations like that of Mishima and Hibiki [92] predictions give the general trend but underpredicts the pressure drop for $x_e > 0.1$. This is due to different flow patterns caused by predominant surface tension force, tube wall surface material, the secondary flow occurring in sharp corners of the cross-section. Although, many bubbles grow and flow freely in mini and macrochannels, this is not the case in microchannels where confined bubbles block the free flow of bubbles.
8. Peak heat transfer coefficient occurs at $0 \leq x_e \leq 0.2$ in microchannels compared to $0.7 \leq x_e \leq 0.9$ in large channels [287] due to lengthwise but not along-the-diameter growth of confined bubbles forming elongated slug flow in microchannels. The elongated slug flow suppresses bubble nucleation at low vapor exit quality and causes local dryout at high vapor exit quality; both decrease heat transfer coefficient in microchannel. However, this observation may not be universally true. Contradictions have been observed; several investigators have observed a heat transfer coefficient peak at high vapor qualities under microscale conditions.
9. The models based on globally optimum parameters show large differences from experimental data, however, these models predict the trend of the data more accurately.
10. The CHF is higher for larger hydraulic diameter. For a given hydraulic diameter, CHF increases with the increase in mass velocity but decreases with increased heated length. More experimental data on subcooled flow boiling and microbubble emission boiling are required.
11. Incipient boiling, conjugate effects and character of bulk boiling need to be studied.
12. Contradictory results experienced by the researchers are dragging them to the finer aspects of the flow boiling and heat transfer in the microchannels.
13. Lastly, it may be commented that more intensive research, both experimental and theoretical, is necessary before the thermal-flow physics in flow boiling through microchannels is clearly understood, more insight into this micro-scale thermophysical phenomenon is developed and better services can be offered to

the practical engineers confronting with real life situations occurring in high heat flux systems.

ACKNOWLEDGMENTS

The first author gratefully acknowledges the generous financial support for the present research from the Abdus Salam International Centre for Theoretical Physics, Trieste, Italy and ENEA, Rome, Italy.

REFERENCES

Kandlikar, S. G. and Grande, W. J., 2003, Evolution of Microchannel Flow Passages---Thermohydraulic Performance and Fabrication Technology, *Heat Transfer Engineering*, 24(1), 3-17.

Kawaji, M. and Chung, P. M., 2003, Unique Characteristics of Adiabatic Gas Liquid Flows in Microchannels: Diameter and Shape Effects on Flow Patterns, Void Fraction and Pressure Drop, ICMM2003-1013, Proc. 1st Int. Conf. on *Microchannels and Minichannels*, N. Y., 115-127.

Suo, M. and Griffith, P., 1964, Two phase flow in capillary tubes, *ASME J. Basic Engg.*, 86, 576-582.

Kew, P. A. and Cornwell, K., 1997, Correlations for the prediction of boiling heat transfer in small diameter channels, *Appl. Thermal. Engg.*, 17, 705-715.

Brauner, N. and Moalem-Maron, D., 1992, Identification of the range of small diameter conduits regarding two phase flow transition, Int. Comm. *Heat Mass Transfer,* 19, 29-39.

Bergles, A. E. and Kandlikar, S. G., 2005, On the nature of critical heat flux in microchannels, *ASME J Heat Transfer,* 127, 101-107.

Groll, M. and Mertz, R., 2003, Minichannel heat transfer: an overview on activities in Europe, Proc. of 1st Int. Conf. on *Microchannels and Minichannels*, Rochester, N. Y., USA.

Cheng, P. and Wu, H. Y., 2006, Mesoscale and microscale phase change heat transfer, *Adv. Heat Transfer,* 39, 461-573.

Cheng, P., Wu, H. Y. and Hong, F. J., 2007, Phase change heat transfer in microsystems, *ASME J Heat Transfer,* 129, 101-108.

Cheng, P., Wang, G. D. and Quan, X., J., 2009, Recent work on boiling and condensation in microchannels, *ASME J Heat Transfer,* 131, 043211-1-15.

Lee, J. D. and Pan, C., 1999, Dynamics of multiple parallel boiling channel systems with forced flows, *Nucl. Engg. Des.,* 192, 31-44.

Mehendale, S. S., Jacobi, A. M. and Shah, R. K., 2000, Fluid flow and heat transfer at micro and meso scales with application to heat exchanger design, *Apll. Mech. Review,* 53(7), 175-193.

Vlasie, C., Machi, H., Guilpart, J. and Agostini, B., 2004, Flow boiling in small diameter channels, *Int. J. Refrigeration,* 27, 191-201.

Zhang, L., Wang, E. N., Goodson, K. E. and Kenny, T. W. 2005, Phase change phenomena in silicon microchannels, *Int. J. Heat Mass Transfer*, 48, 1572-1582.

Ishibashi, E. and Nishikawa, K., 1969, Saturated boiling heat transfer in narrow spaces, *Int. J. Heat Mass Transfer*, 12, 863-894.

Kandlikar, S. G., 2001, *Fundamental issues related to flow boiling in minichannels and microchannels,* ExHFT-5, 1, Pisa, Italy, 129-146.

Kandlikar, S. G., 2002, Fundamental issues related to flow boiling in mini channels and micro channels, *Exp. Thermal Fluid Science*, 26, 389-407.

Kandlikar, S. G., 2002, Two phase flow patterns, pressure drop and heat transfer during boiling in minichannel flow passages of compact evaporators, *Heat Transfer Engineering*, 23(1), 5-23.

Thome, J. R., 2004, Boiling in microchannels: a review of experiment and theory, *Int. J. Heat Fluid Flow*, 25, 128-139.

Yarin, L. P., Mosyak, A., Hetsroni, G. , 2009, Boiling in microchannels, in Fluid Flow, *Heat Transfer and Boiling in Microchannels*, Springer, Berlin.

Bertsch, S. S., Groll, E. A. and Garimella, S. V., 2008, Review and comparative analysis of studies on saturated flow boiling in small channels, *Nanoscale and Microscale Thermophysical. Engg.*, 12, 187-227.

Consolini, L., 2008, Convective boiling heat transfer in a single microchannel, *Ph.D. thesis,* EPFL, Lausanne, Switzerland.

Bogojevic, D., Sefiane, K., Walton, A. J., Lin, H. and Cummins, G., 2009, Two phase flow instabilities in a silicon microchannels heat sink, *Int. J Heat Fluid Flow*, in press.

Boure, J. A., Bergles, A. E. and Tong, L. S., 1973, Review of two phase flow instability, *Nuclear Engg. and Design*, 25, 165-192.

Tadrist, L., 2005, Review on two phase flow instabilities in narrow spaces, Proc. ECI Int. Conf. *Heat Transfer and Fluid Flow in Microscale*, Italy, paper no. KL4.

Tadrist, L., 2007, Review on two phase flow instabilities in narrow spaces, *Int. J. Heat Fluid Flow*, 28, 54-62.

Lorenzini, E., 1981, A simplified method proposal for practical determination of aperiodic two phase flow instability, *Int. J. Multiphase Flow,* 7(6), 635-645.

Ribatski, G., Wojtan, L. and Thome, J. R., 2006, An analysis of experimental data and prediction methods for two phase frictional pressure drop and flow boiling heat transfer in micro-scale channels, *Exp. Thermal Fluid Sci,* 31, 1-19.

Kakac, S. and Bon, B., 2008, A review of two-phase flow dynamic instabilities in tube boiling systems, *Int. J. Heat Mass Transfer*, 51, 399-433.

Garimella, S. V. and Sobhan, C. B., 2003, Transport in microchannels---a critical review, *Ann. Rev. Heat Transfer,* 13, 1-50.

Agostini, B. and Bontemps, A., 2005, Vertical flow boiling of refrigerant R134a in small channels, *Int. J. Heat Fluid Flow*, 26, 296-306.

Carey, van P., 1992, *Liquid-vapor phase change phenomena, an introduction to the thermophysics of vaporization and condensation processes in heat transfer equipment,* Hemisphere, New York.

Blander, M. and Katz, J. L.,1975, Bubble nucleation in liquids, *AIChE J,* 21, 833-848.

Celata, G. P., Cumo, M. and Mariarini, A.,1997, Experimental evaluation of the onset of subcooled flow boiling at high liquid velocity and subcooling, *Int. J. Heat Mass Transfer*, 40, 2879-2885.

Cole, R., 1974, *Boiling nucleation, in Advances in Heat Transfer,* 10, Academic, N. Y., 86-166.

Davis, E. J. and Anderson, G. H., 1966, The incipience of nucleate boiling in forced convection flow, *AIChE J.*, 12, 774-780.

Dhir, V. K., 1998, Boiling heat transfer, *Ann Rev Fluid Mech.*, 30, 365-401.

Ghiaasiaan, S. M. and Chedester, R. C., 2002, Boiling incipience in microchannels, *Int. J. Heat Mass Transfer*, 45, 4599-4606.

Hsu, Y. Y., 1962, On size range of active nucleation cavities on a heating surface, *ASME J Heat Transfer,* 84, 207-216.

Li, J, and Cheng, P., 2004, Bubble cavitation in a microchannel, *Int. J. Heat Mass Transfer*, 47, 2689-2698.

Liu, D., Lee, P. S. and Garimella, S. V., 2005, Prediction of the onset of nucleate boiling in microchannel flow, *Int. J. Heat Mass Transfer*, 48, 5134-5149.

Qu, W. and Mudawar, I., 2002, Prediction and measurement of incipient boiling heat flux in microchannel heat sinks, *Int. J. Heat Mass Transfer*, 45, 3933-3945.

Bergles, A. E. and Rohsenow, W. M., 1964, The determination of forced convection surface boiling heat transfer, *ASME J Heat Transfer,* 86, 365-372.

Hapke, I., Boye, H. and Schmidt, J., 2000, Onset of nucleate boiling in microchannels, *Int. J. Thermal Sciences,* 39, 505-513.

Hetsroni, G., Mosyak, A., Pogrebnyak, E. and Segal, Z., 2005, Explosive boiling of water in parallel microchannels, *Int. J. Multiphase Flow*, 31, 371-392.

Hino, R. and Ueda, T., 1985, Studies on heat transfer and flow characteristics in subcooled flow boiling, Part 1: boiling characteristics, *Int. J. Multiphase Flow,* 11, 269-281.

Kennedy, J. E., Roach, G. M., Dowling, M. E., Abdel-Khalik, S. I., Ghiaasiaan, S. M., Jeter, S: M. and Qureshi, Z. H., 2000, The onset of flow instability in uniformly heated horizontal microchannels, *ASME J Heat Transfer,* 122, 118-125.

Hetsroni, G., Mosyak, A., Pogrebnyak, E. and Segal, Z., 2007, Natural convection boiling of water and surfactant in narrow horizontal annular channels, *Int. J. Multiphase Flow,* 33, 469-483.

Hetsroni, G., Gurevich, M., Mosyak, A., Rozenbilt, R. and Yarin, L. P., 2002, Subcooled boiling of surfactant solutions, *Int. J. Multiphase Flow*, 28, 347-361.

Hetsroni, G., Mosyak, A:, Segal, Z. and Pogrebnyak, E., 2003, Two-phase flow pattern in parallel microchannels, *Int. J. Multiphase Flow*, 29, 341-360.

Hetsroni, G., Gurevich, M., Mosyak, A. and Rozenbilt, R., 2003, Surface temperature measurement of a heated capillary tube by means of an infrared technique, *Measurement Science and Technology*, 14, 807-814.

Hetsroni, G., Klein, D., Mosyak, A., Segal, Z. and Pogrebnyak, E., 2003, Convective boiling in parallel microchannels, Proc. 1st Int. Conf. Microchannels and Minichannels, April 24-25, Rochester, N. Y., USA., 59-67, *also in Microscale Thermophysical Engineering,* 8, 403-421.

Hetsroni, G., Mosyak, A., Segal, Z. and Ziskind, G., 2002, A uniform temperature heat sink for cooling of electronic devices, *Int. J. Heat Mass Transfer*, 45, 3275-3286.

Klein, D., Hetsroni, G. and Mosyak, A., 2005, Heat transfer characteristics of water and APG surfactant solution in a microchannel heat sink, *Int. J. Multiphase Flow,* 31, 393-415.

Kosar, A., Kuo, C. J. and Peles, Y., 2005, Boiling heat transfer with reentrant cavities, *Int. J. Heat Mass Transfer*, 48, 4867-4886.

Kuo, C. J. and Peles, Y., 2007, Local measurements of flow boiling in structured surface microchannels, *Int. J. Heat Mass Transfer*, 50, 4513-4526.

Ory, E., Yuan, H., Prosperetti, A., Popinet, S, and Zaleski, S., 2000, Growth and collapse of vapor bubble in a narrow tube, *Phys. Fluids,* 12, 1268-1277.

Ozawa, M., Akagawea, K. and Sakaguchi, T., 1989, Flow instabilities in parallel channel flow systems of gas-liquid two phase mixtures, *Int. J. Multiphase Flow*, 15: 639-657.

Steinke, M. E. and Kandlikar, S. G., 2004, Control and effect of dissolved air in water during flow boiling in microchannels, *Int. J. Heat Mass Transfer*, 47, 1925-1935.

Steinke, M. E. and Kandlikar, S. G., 2004, An experimental investigation of flow boiling characteristics of water in microchannels, *ASME J Heat Transfer,* 126, 518-526.

Lee, P. C., Tseng, F. C. and Pan, C., 2004, Bubble dynamics in microchannels, Part I: single microchannel, *Int. J. Heat Mass Transfer*, 47, 5575-5589.

Agostini, B., Thome, J. R., Fabbri, M., Michel, B., Calmi, D. and Kloter, U., 2008, High heat flux flow boiling in silicon multi-microchannels, Part I: heat transfer characteristics of refrigerant 236fa, *Int. J. Heat Mass Transfer*, 51, 5400-5414.

Agostini, B., Thome, J. R., Fabbri, M., Michel, B., Calmi, D. and Kloter, U., 2008, High heat flux flow boiling in silicon multi-microchannels, Part II: heat transfer characteristics of refrigerant 245fa, *Int. J. Heat Mass Transfer*, 51, 5415-5425.

Cole, R. and Shulman, H. L., 1966, Bubble growth rates at high Jacob numbers, *Int. J. Heat Mass Transfer*, 9, 1377-1390.

Geld, C., 2004, Prediction of dynamic contact angle histories of a bubble growing at a wall, *Int. J. Heat Fluid Flow*, 25, 74-80.

Han, C. H. and Griffith, P., 1965, The mechanism of heat transfer in nucleate pool boiling, Part I: bubble initiation, growth and departure, *Int. J. Heat Mass Transfer*, 8, 887-904.

Helden, W., Geld, C. and Boot, P., 1995, Forces on bubbles growing and detaching in flow along a vertical wall, *Int. J. Heat Mass Transfer*, 38, 2075-2088.

Hetsroni, G., Mosyak, A., Pogrebnyak, E., Sher, I. and Segal, Z., 2006, Bubble growth in saturated pool boiling in water and surfactant solution, *Int. J. Multiphase Flow*, 22, 159-182.

Hetsroni, G., Mosyak, A., Pogrebnyak, E: and Segal, Z., 2006, Periodic boiling in parallel microchannels at low vapor quality, *Int. J. Multiphase Flow*, 32, 1141-1159.

Klausner, J. F., Mei, R., Bernard, D: and Zeng, L., 1993, Vapor bubble departure in forced convection boiling, *Int. J. Heat Mass Transfer*, 36, 651-661.

Li, H. Y., Tseng, F. C. and Pan, C., 2004, Bubble dynamics in microchannels, Part II: parallel microchannels, *Int. J. Heat Mass Transfer*, 47, 5591-5601.

Peng, X. F., Hu, H. Y. and Wang, B. X., 1998, Boiling nucleation during liquid flow in microchannels, *Int. J. Heat Mass Transfer*, 41, 101-106.

Revellin, R. and Thome, J. R., 2008, A theoretical model for the prediction of the critical heat flux in heated microchannel, *Int. J. Heat Mass Transfer*, 51, 1216-1225.

Wang, G. D., Cheng, P. and Bergles, A. E., 2008, Effects of inlet outlet configurations on flow boiling instability in parallel microchannels, *Int. J Heat Mass Transfer*, 51, 2267-2281.

Hetsroni, G., 2009, Boiling in Microchannels, ECI Int. Conf. on Boiling Heat Transfer, *Florianopolis*, Brazil, 3-7 May.

Hetsroni, G., Segal, Z. and Mosyak, A., 2000, Nonuniform temperature distribution in electronic devices cooled by flow in parallel microchannels, Packaging of Electronic and Photonic Devices, *EEP,* 28, 1-9.

Hetsroni, G., Mosyak, A. and Segal, Z., 2001, Nonuniform temperature distribution in electronic devices cooled by flow in parallel microchannels, IEEE Trans., *Components Packaging Technology,* 24(1), 17-23.

Bowers, M. B. and Mudawar, I, 1994, High heat flux boiling in low flow rate, low pressure drop mini channel and micro channel heat sinks, *Int. J. Heat Mass Transfer*, 37, 321-332.

Bowers, M. B. and Mudawar, I., 1994, Two phase electronic cooling using minichannel and microchannel heat sinks: Part 1—design criteria and heat diffusion constraints, *J. Electronic Packaging,* 116, 290-297.

Bowers, M. B. and Mudawar, I., 1994, Two phase electronic cooling using minichannel and microchannel heat sinks: Part 2—flow rate and pressure drop constraints, *J. Electronic Packaging,* 116, 298-305.

Brutin, B. and Tadrist, L., 2004, Pressure drop and heat transfer analysis of flow boiling in microchannel: influence of the inlet condition on two-phase flow stability, *Int. J Heat Mass Transfer*, 47, 2365-2377.

Grohmann, S., 2005, Measurement and modeling of single phase and flow boiling heat transfer in micro tubes, *Int. J. Heat Mass Transfer*, 48, 4072-4089.

Hall, D. D. and Mudawar, I., 2000, Critical heat flux (CHF) for water flow in tubes---II, subcooled CHF correlations, *Int. H. Heat Mass Transfer*, 43, 2605-2640.

Hwan, Y. W. and Kim, M. S:, 2006, The pressure drop in micro tubes and correlation development, *Int. J. Heat Mass Transfer*, 49, 1804-1812.

Katto, Y. and Ohno, H., 1984, An improved version of the generalized correlation of critical heat flux for the convective boiling in uniformly heated vertical tubes, *Int. J. Heat Mass Transfer*, 27, 1641-1648.

Lazarek, G. M. and Black, S. H., 1982, Evaporative heat transfer, pressure drop and critical heat flux in a small vertical tube with R-113, *Int. J. Heat Mass Transfer*, 25, 945-959.

Lee, H. J. and Lee, S. Y., 2001, Pressure Drop correlations for two-phase flow within horizontal rectangular channels with small heights, *Int. J. Multiphase Flow*, 27, 783-796.

Lee. H. J. and Lee, S. Y., 2001, Heat transfer correlation for boiling flows in small rectangular horizontal channels with low aspect ratios, *Int. J. Multiphase Flow*, 27, 2043-2062.

Lee, H. J. and Mudawar, I., 2005, Two-phase flow in high heat flux microchannel heat sink for refrigeration cooling applications, Part I: pressure drop characteristics, *Int. J Heat Mass Transfer*, 48, 928-940.

Lee, H. J. and Mudawar, I., 2005, Two-phase flow in high heat flux microchannel heat sink for refrigeration cooling applications, Part II, heat transfer characteristics, *Int. J. Heat Mass Transfer*, 48, 941-955.

Lockhart, R. W. and Martinelli, R. C., 1949, Proposed correlation of data for isothermal two phase two component flow in pipes, *Chem. Engg. Progress*, 45, 39-48.

Mishima, K. and Hibiki, T., 1996, Some characteristics of air-water two-phase flow in small diameter vertical tubes, *Int. J. Multiphase Flow*, 22, 703-712.

Qu, W. and Mudawar, I. 2003, Measurement and prediction of pressure drop in two phase microchannel heat sinks, *Int. J. Heat Mass Transfer*, 46, 2737-2753.

Qu, W. and Mudawar, I., 2003, Flow boiling heat transfer in two phase micro channel heat sink, I: experimental investigation and assessment of correlation methods, *Int. J. Heat Mass Transfer*, 46, 2755-2771.

Qu, W. and Mudawar, I., 2003, Flow boiling heat transfer in two phase micro channel heat sink, II: annular two phase flow model, *Int. J. Heat Mass Transfer*, 46, 2773-2784.

Qu, W. and Mudawar, I., 2004, Measurement and correlation of critical heat flux in two phase microchannel heat sinks, *Int. J. Heat Mass Transfer*, 47, 2045-2059.

Qu, W. and Mudawar, I., 2004, Transport phenomena in two phase microchannel heat sinks, *J. Electron. Pack.*, 126, 213-224.

Quiben, J. M. and Thome, J. R., 2007, Flow pattern based two phase pressure drop model for horizontal tubes, Part I: Diabatic and adiabatic experimental study, *Int. J. Heat and Fluid Flow*, 28(5), 1049-1059.

Quiben, J. M. and Thome, J. R., 2007, Flow pattern based two phase pressure drop model for horizontal tubes, Part II: New phenomenological model, *Int. J. Heat and Fluid Flow*, 28(5), 1060-1072.

Shuai, J., Kulenovic, R. and Droll, M., 2003, Heat transfer and pressure drop for flow boiling of water in narrow vertical rectangular channels, Proc. of 1st Int. Conf. on Microchannels, Rochester, N. Y., 24-25 April 2003, *ICMM* 2003, 1084.

Wojtan, L., Revellin, R. and Thome, J. R., 2006, Investigation of critical heat flux in single uniformly heated microchannels, *Exp. Therm. Fluid Science,* 30, 765-774.

Wu, H. Y. and Cheng, P., 2003, Liquid/two phase/vapor alternating flow during boiling in microchannels at high heat flux, Int. Comm. *Heat Mass Transfer*, 39, 295-302.

Wu, H.Y. and Cheng, P., 2003, Visualization and measurements of periodic boiling in silicon microchannels, *Int. J Heat Mass Transfer*, 46, 2603-2614.

Wu, H. Y. and Cheng, P., 2003, *Two large amplitude/long period oscillations boiling modes in silicon microchannels,* Proc. 1st Int. Conf. on Microchannels and Minichannels, Rochester, N. Y., 629-633.

Wu, H. Y. and Cheng, P., 2004, Boiling instability in parallel silicon microchannels at different heat flux, *Int. J. Heat Mass Transfer*, 47, 3631-3641.

Yen, T. H., Kasagi, N. and Suzuki, Y., 2003, Forced convective boiling heat transfer in micro tubes at low mass and heat fluxes, *Int. J. Multiphase Flow*, 29, 1771-1792.

Yen, T. H., Shji, M., Takemura, F., Suzuki, Y. and Kasagi, N., 2006, Visualization of convective boiling heat transfer in single microchannels with different shaped cross sections, *Int. J Heat Mass Transfer*, 49, 3884-3894.

Zhang, W., Hibiki, T., Mishima, K. and Mi, Y., 2006, Correlation of critical heat flux for flow boiling of water in minichannels, *Int. J. Heat Mass Transfer*, 49, 1058-1072.

Wu, H. Y., Cheng, P. and Wang, H., 2006, Pressure drop and flow boiling instabilities in silicon microchannel heat sinks, *J Micromechanics and Microengineering,* 16, 2138-2146.

Mudawar, I. and Bowers, M. B., 1999, Ultra high critical heat flux (CHF) for subcooled water flow boiling, I: CHF data and parametric effects for small diameter tubes, *Int. J. Heat Mass Transfer*, 42, 1405-1428.

Moriyama, K. and Inoue, 1996, Thickness of the liquid film formed by growing bubble in a narrow gap between two horizontal plates, *ASME J Heat Transfer,* 118, 132-139.

Thome, J. R., Dupont, V. and Jacobi, A. M., 2004, Heat transfer model for evaporation in microchannels, Part I: presentation of the model, *Int. J. Heat Mass Transfer*, 47, 3375-3385.

Thome, J. R., Dupont, V. and Jacobi, A. M., 2004, Heat transfer model for evaporation in microchannels, Part II: comparison with database, *Int. J. Heat Mass Transfer*, 47, 3387-3401.

Celata, G. P., Cumo, M. and Mariani, A., 1993, Burnout in highly subcooled water flow boiling in small diameter tubes, *Int. J. Heat Mass Transfer*, 36, 1269-1285.

Kandlikar, S. G., 2006, Nucleation characteristics and stability considerations during flow boiling in microchannels, *Exp. Thermal Fluid Science,* 30, 441-447.

Liu, Z. and Winterton, R. H. S., 1991, A general correlation for saturated and subcooled flow boiling in tubes and annuli, based on a nucleate pool boiling equation, *Int. J. Heat Mass Transfer*, 34, 2759-2766.

Fukano, T. and Kariyasaki, L., 1993, Characteristic of gas liquid two phase flow in a capillary tube, *Nucl. Engg. Des.*, 141, 58-68.

Triplett, K., Ghiaasiaan, S., Abdel Khalik, S:, LeMouel, A. and McCord, B., 1999, Gas liquid two phase flow in microchannels, Part I: two phase flow patterns, *Int. J. Multiphase Flow*, 25, 377-394.

Peng, X. F. and Wang, B. X., 1998, Forced convection and boiling characteristics in microchannels, Proc. 11th Int. Heat Transfer Conf., Korea, 1, 371-390.

Dupont, V., Thome, J. R. and Jacobi, A. M., 2004, Heat transfer model for evaporation in microchannels, Part II: comparison with database, *Int. J. Heat Mass Transfer*, 47, 3387-3401.

121. Revellin, R., Agostini, B. and Thome, J. R., 2008, Elongated bubbles in microchannels, Part II: experimental study and modeling of bubble collisions, *Int. J. Multiphase Flow*, 34, 602-613.

Sato, T. and Matsumura, H., 1964, On the conditions of incipient subcooled boiling with forced convection, Bull. *JSME*, 7, 392-398.

Peng, X. F. and Wang, B. X., 1993, Forced convection and flow boiling heat transfer for liquid flowing through microchannels, *Int. J. Heat Mass Transfer*, 36(14), 3421-3427.

Qi, S. L., Zhang, P., Wang, R. Z. and Xu, L. X., 2007, Flow boiling of liquid nitrogen in micro tubes: Part I: the onset of nucleate boiling, two phase flow instability and two phase flow pressure drop, *Int. J. Heat Mass Transfer*, 50, 4999-5016.

Arcanjo, A. A., Freitas, J. O., Tibirica, C. B. and Ribatski, G., 2009, Two-phase flow characteristics during flow boiling of halocarbon refrigerants in micro-scale channels, ECI Int. Conf. on Boiling Heat Transfer, Florianopolis-SC, Brazil, 3-7 May, Paper No. MC3.

Revellin, R. and Thome, J. R., 2007, Adiabatic two phase frictional pressure drops in microchannels, Exp. Therm. Fluid Sci., 31, 673-685.

Serizawa, A., Feng, Z. and Kawara, Z., 2002, Two phase flow in microchannels, *Exp. Thermal Fluid Science*, 26, 703-714.

Wang, G. D., Cheng, P. and Wu, H. Y., 2007, Unstable and stable flow boiling in parallel microchannels and in a single microchannel, *Int. J Heat Mass Transfer*, 50, 4297-4310.

Tabatabai, A. and Faghri, A., 2001, A new two phase flow map and transition boundary accounting for surface tension effects in horizontal miniature and micro tubes, *ASME J Heat Transfer,* 123, 958-968.

Zhang, L., Koo, J. M.., Jiang, L., Asheghi, M., Goodson, K. E. and Santiago, J. G., 2002, Measurements and modeling of two phase flow in microchannels with nearly constant heat flux boundary conditions, *J. Microelectromechanical Systems*, 11(1), 12-19.

Jiang, L., Wong, M. and Zohar, Y., 2001, Forced convection boiling in a microchannel heat sink, *J. Microelectromechanical Systems*, 10 (1), 80-87.

Wang, G. D. and Cheng, P., 2008, An experimental study of flow boiling instability in a single microchannel, Int. Comm. *Heat Mass Transfer*, 35, 1229-1234.

Balasubramanian, P. and Kandlikar, S. G., 2005, Experimental study of flow patterns, pressure drop and flow instabilities in parallel rectangular minichannels, *Heat Transfer Engineering*, 26(3), 20-27.

Zivi, S. M., 1964, Estimation of steady state steam void fraction by means of the principle of minimum entropy production, *ASME J Heat Transfer,* 86, 247-252.

Agostini, B., Revellin, R. and Thome, J. R., 2008, Elongated bubbles in micro channels, Part I: Experimental study and modeling of elongated bubble velocity, *Int. J. Multiphase Flow*, 34, 509-601.

Sobierska, E., Kulenovic, R. and Mertz, R., 2007, Heat transfer mechanism and flow pattern during flow boiling of water in a vertical narrow channel—experimental results, *Int. J. Thermal Sciences,* 46(11), 1172-1181.

Hapke, I., Boye, H. and Schmidt, J., 2002, Flow boiling of water and N-Heptane in microchannels, *Microscale Thermophysical Engineering,* 6, 99-115.

Lee, H. J., Oh, B. D., Bae, S. W. and Kim, M. H., 2003, Single bubble growth in saturated pool boiling on a constant wall temperature surface, *Int.. J. Multiphase Flow,* 29, 1857-1874.

Kandlikar, S. G., 1991, Development of a flow boiling map for subcooled and saturated flow boiling of different fluids in circular tubes, *ASME J Heat Transfer,* 113, 190-200.

Kandlikar, S. G., Steinke, M. E., Tian, S. and Campbell, L. A., 2001, High speed photographic observation of flow boiling of water in parallel minichannels, *ASME Nat. Heat Transfer Conf.*, CA, USA, Jun 10-12.

Kandlikar, S. G. and Bulut, M. , 2003, An experimental investigation on flow boiling of ethylene glycol/water mixtures, *ASME J Heat Transfer,* 125, 317-325.

Revellin, R. and Thome, J. R., 2007, A new type of diabatic flow pattern map for boiling heat transfer in microchannels, *J. Micromechanics and Microengineering*, 17(4), 788-796.

Revellin, R. and Thome, J. R., 2007, Experimental investigation of R-134a and R-245fa two phase flow in microchannels for different flow conditions, *Int. J. Heat Fluid Flow*, 28, 63-71.

Taitel, Y. and Dukler, A. E., 1976, A model for predicting flow regime transitions in horizontal and near horizontal gas liquid flow, *AIChE J.*, 22, 47-55.

Tibirica, C. B., Nascimento, J. N. and Ribatski, G., 2009, *Experimental Thermal Fluid Science* (in Press).

Kandlikar, S. G., 2004, Heat Transfer mechanisms during flow boiling in microchannels, *ASME J Heat Transfer,* 126, 8-16.

Kandlikar, S. G. and Balasubramanian, P., 2004, An extension of the flow boiling correlation to transition, laminar and deep laminar flows in mini channels and microchannels, *Heat Transfer Engineering*, 25, 86-93.

Yu, W., France, D. M., Wambsganss, M. W. and Hull, J. R., 2002, Two phase pressure drop, boiling heat transfer, and critical heat flux to water in a small diameter horizontal tube, *Int. J. Multiphase Flow*, 28, 927-941.

Chen, T. and Garimella, S., 2006, Measurements and high speed visualizations of flow boiling of a dielectric fluid in a silicon microchannel heat sink, *Int. J. Multiphase Flow*, 32, 957-971.

Kandlikar, S. G., 1990, A general correlation for saturated two phase boiling heat transfer inside horizontal and vertical tubes, *ASME J Heat Transfer,* 112, 219-228.

Steinke, M. E. and Kandlikar, S. G., 2003, Flow boiling and pressure drop in parallel flow microchannels, *ICMM2003-1070*, Proc. 1st Int. Conf. on Microchannels and Minichannels, N. Y., 567-579.

Lin, S., Kew, P. A. and Cornwell, K., 2001, Flow boiling of refrigerant R141B in small tubes, *IChemE,* 79, Part A, 417-424.

Wambsganss, M. W., France, D. M., Jendrzejczyk, J. A. and Tran, T. N., 1993, Boiling heat transfer in a horizontal small diameter tube, *ASME J Heat Transfer,* 115, 963-972.

Yan, Y. and Lin, T., 1998, Evaporation heat transfer and pressure drop of refrigerant R-134a in a small pipe, *Int. J. Heat Mass Transfer*, 41, 4183-4194.

Bao, Z. Y., Fletcher, D. F. and Haynes, B. S., 2000, Flow boiling heat transfer of Freon 11 and HCFC123 in narrow passages, *Int. J. Heat Mass Transfer*, 43, 3347-3358.

Kandlikar, S. G., Steinke, M. E. and Balasubramanian, P., 2002, Single phase flow characteristics and effect of dissolved gases on heat transfer near saturation conditions in microchannels, IMECE2002-32382, *Int. Mech. Engg. Conf. and Exposition,* New Onleans, USA, Nov. 17-21.

Kandlikar, S. G. and Steinke, M. E., 2003, Predicting heat transfer during flow boiling in minichannels and microchannels, *ASHRAE Trans.,* 109(1), CH-03-13-1.

Jacobi, A. M. and Thome, J. R., 2002, Heat transfer model for evaporation of elongated bubble flows in microchannels, *ASME J Heat Transfer,* 124, 1131-1136.

Diaz, A. C. and Schmidt, J., 2007, Experimental investigation of transient boiling heat transfer in microchannels, *Int. J. Heat Fluid Flow*, 28, 95-102.

Xu, J. L., Shen, S., Gan, Y., Li, Y., Zhang, W. and Su, Q, 2005, Transient flow pattern based microscale boiling heat transfer mechanisms, *J. Micromechanics and Microengineering*, 15, 1344-1361.

Kandlikar, S. G. and Balasubramanian, P., 2004, Effect of gravitational orientation on flow boiling of water in 1054 × 197 parallel minichannels, Proc. of the 2nd Int. *Conf. on Microchannels and Minichannels, R*ochester, N. Y., USA, 539-550.

Dupont, V. and Thome, J. R., 2005, Evaporation in microchannels: influence of the channel diameter on heat transfer, *Microfluidics and Nanofluidics,* 1, 119-127.

Tiselj, I., Hetsroni, G., Mavko, B., Mosyak, A., Pogrebnyak, E. and Segal, Z., 2004, Effect of axial conduction on the heat transfer in microchannels, *Int. J. Heat Mass Transfer*, 47, 2251-2265.

Harirchian, T. and Garimella, S. V., 2008, Microchannel size effects on local flow boiling heat transfer to a dielectric fluid, *Int. J. Heat Mass Transfer*, 51, 3274-3735.

Kandlikar, S. G., 2009, Scale effects on flow boiling in microchannels: a fundamental perspective, *ECI Conf. Boiling Heat Transfer,* Florianopolis, Brazil, 3-7 May.

Kandlikar, S. G. and Balasubramanian, P., 2003, Extending the applicability of the flow boiling correlation to low Reynolds number flows in microchannels, ICMM2003-1075, Proc. of 1st Int. Conf. on Microchannels and Minichannels, Rochester, N. Y., USA, 603-608.

Kandlikar, S. G. and Balasubramanian, P., 2005, An experimental study on the effect of gravitational orientation on flow boiling of water in 1054 × 197 μm parallel minichannels, *ASME J. Heat Transfer,* 127(8), 820-829.

Ravigururajan, T. S., 1998, Impact of channel geometry on two phase flow heat transfer characteristics of refrigerants in microchannel heat exchangers, *ASME J Heat Transfer,* 120, 485-491.

Wang, G. D. and Cheng, P., 2009, Subcooled flow boiling and microbubble emission boiling phenomena in a partially heated microchannel, *Int. J. Heat Mass Transfer*, 52, 79-91.

Xu, J. L., Gan, Y. H., Zhang, D. C. and Li, X. H., 2005, Microscale boiling heat transfer in a micro-timescale at high heat fluxes, *J. Micromechanics and Microengineering*, 15, 1362-1376.

Agostini, B. and Thome, J. R., 2005, Comparison of an extended database for flow boiling heat transfer coefficients in multi microchannel elements with the three-zone model, E*CI Conf. on Heat Transfer and Fluid Flow in Microscale,* Italy, Sept. 25-30.

Revellin, R., Dupont, V., Thome, J. R. and Zun, I., 2006, Characterization of diabatic two phase flows in microchannels: Flow parameter results for R-134a in a 0.5 mm Channel, *Int. J. Multiphase Flow*, 32, 755-774.

Cornwell, K. and Kew, P. A., 1993, Boiling in small parallel channels, Energy Efficiency in Process Technology, *Elsevier Appl.* Sc., London, 624-638.

Li, J. and Peterson, G. P., 2005, Boiling nucleation and two phase flow patterns in forced liquid flow in microchannels, *Int. J. Heat Mass Transfer*, 48, 4797-4810.

Revellin, R. and Thome, J. R., 2006, New diabatic flow pattern map for evaporating flows in microchannels, 13th *Int. Heat Transfer Conf.*, Sydney, Aug 14-18.

Chung, P. M. Y. and Kawaji, M., 2004, The effect of channel diameter on adiabatic two phase flow characteristics in small tubes, *Int. J Multiphase Flow,* 30, 735-761.

Zhao, T. S. and Bi, Q. C., 2001, Co-current air water two phase flow patterns in vertical triangular microchannels, *Int. J. Multiphase Flow*, 27, 765-782.

Cubaud, T. and Chih-Ming, H., 2004, Transport of bubbles in square microchannels, *Phys. Fluids,* 16(12), 4575-4585.

Lin, S., Kew, P. A. and Cornwell, K., 2001, Two phase heat transfer to a refrigerant in a 1 mm diameter tube, *Int. J. Refrigeration,* 24, 51-56.

Owhaib, W., Palm, B. and Martin-Callizo, C., 2006, Flow boiling visualizations in a vertical circular minichannel at high vapor quality, *Exp. Thermal Fluid Science*, 30, 755-763.

Tran, T. N., Wambsganss, M. W. and France, D. M., 1996, Small circular and rectangular channel boiling with two refrigerants, *Int. J. Multiphase Flow*, 22, 485-498.

Xu, J. L., Zhou, J. J. and Gan, Y. H., 2005, Static and dynamic flow instability of a parallel microchannel heat sink at high heat flux, *Energy Conversion and Management,* 46, 313-334.

Pettersen, J., 2004, Flow visualization of CO_2 in microchannel tubes, *Exp. Thermal Fluid Sci*, 28, 111-121.

Zhang, W., Hibiki, T. and Mishima, K., 2004, Correlation for boiling heat transfer in minichannels, *Int. J Heat Mass Transfer*, 47, 5749-5763.

Celata, G. P., 2005, Preface to the special issue dedicated to the ECI Int. Conf. Heat Trannsfer and Fluid Flow in Microscale, *Exp. Thermal Fluid Science*, Italy, 25-30 September.

Bertsch, S. S., Groll, E. A. and Garimella, S. V., 2009, A composite heat transfer correlation for saturated flow boiling in small channels, *Int. J. Heat Mass Transfer*, 52, 2110-2118.

Consolini, L. and Thome, J. R., 2009, Microchannel convective boiling heat transfer with flow instabilities, ECI Int. Conf. on Boiling Heat Transfer, Florianopolis, Brazil, 3-7 May.

Aritomi, M., Miyata, T., Horiguchi, M. and Sudi, S., 1993, Thermohydraulics of boiling two phase flow in high conversion light water reactors (thermohydraulics at low velocities), *Int. J. of Multiphase Flow,* 19(1), 51-63.

Zhang, W., Hibiki, T. and Mishima, K., 2005, Correlation for flow boiling heat transfer at low liquid Reynolds number in small diameter channels, *ASME J Heat Transfer,* 127, 1214-1221.

Bertsch, S. S., Groll, E. A. and Garimella, S. V., 2009, Effects of heat flux, mass flux, vapor quality and saturation temperature on flow boiling heat transfer in microchannels, *Int. J. Multiphase Flow*, 35, 142-154.

Tibirica, C. B. and Ribatski, G., 2009, An experimental study on microscale flow boiling heat transfer, ECI Int. Conf. on Boiling Heat Transfer, Florianopolis-SC-Brazil, 3-7 May, Paper No. MC7.

Saitoh, S., Daiguji, H. and Hihara, E., 2007, Correlation for boiling heat transfer of R-134a in horizontal tubes including effect of tube diameter, *Int. J. Heat Mass Transfer*, 50, 5215-5225.

Lee, P. S. and Garimella, S. V., 2007, Saturated flow boiling heat transfer and pressure drop in silicon microchannel arrays, *Int. J. Heat Mass Tranfer*, 51, 789-806.

Bertsch, S. S., Groll, E. A. and Garimella, S. V., 2008, Refrigerant flow boiling heat transfer in parallel microchannels as a function of local vapor quality, *Int. J. Heat Mass Transfer*, 51, 4775-4787.

Sun, L. and Mishima, K., 2009, An evaluation of prediction methods for saturated flow boiling heat transfer in mini channels, *Int. J. Heat Mass Transfer*, 52, 5323-5329.

Tran, T. N., Chiyu, M. C., Wambsganss, M. W. and France, D. M., 2000, Two phase pressure drop of refrigerants during flow boiling in small channels: an experimental investigation and correlation development, *Int. J. Multiphase Flow*, 26, 1739-1754.

Muller-Steinhagen, R. and Heck, K., 1986, A simple friction pressure drop correlation for two-phase flow in pipes, *Chem. Eng. And Processing,* 20(6), 297-308.

Wen, D. S. and Kenning, D. B. R., 2004, Two phase pressure drop of water during flow boiling in a vertical narrow channel, *Exp. Therm Fluid Sci.*, 28, 131-138.

Zhang, M. and Webb, R. L., 2001, Correlation of two phase friction for refrigerants in small diameter tubes, *Exp. Thermal Fluid Science*, 25, 131-139.

Agostini, B., Watel, B., Bontemps, A. and Thonon, B., 2002, Friction factor and heat transfer coefficient of R134a liquid flow in minichannels, *Appl. Thermal Engg.,* 22(16), 1821-1834.

Warrier, G. R., Pan, T. and Dhir, V. K., 2002, Heat transfer and pressure drop in narrow rectangular channels, *Exp. Thermal Fluid Science*, 26, 53-64.

Martinelli, R. C. and Nelson, D. B., 1948, Prediction of pressure drop during forced circulation boiling of water, *ASME J Heat Transfer,* 70, 695-702.

Cheol, H. and Moo, H.. K., 2007, Pressure drop, boiling heat transfer and flow patterns during flow boiling in a single microchannel, *Heat Transfer Engineering*, 28(8-9), 730-737.

Kandlikar, S. G. and Steinke, M. E., 2003, Flow boiling and pressure drop in parallel flow microchannels, Proc., 1st Int. Conf. on Microchannels and Minichannels, April 24-25, *ICMM*2003-1070, 567-579.

Chisholm, D., 1973, Pressure gradients due to friction during the flow of evaporation of two phase mixtures in smooth tubes and channels, *Int. J. Heat Mass Transfer*, 16, 347-358.

Yun, R., Heo, J. H. and Kim, Y., 2006, Evaporative heat transfer and pressure drop of R410A in microchannels, *Int. J. Refrigeration*, 29, 92-100.

Nair, S., Lele, S., Ishii, M. and Revankar, S. T., 1996, Analysis of flow instabilities and their role on critical heat flux for two phase downflow and low pressure systems, *Int. J. Heat Mass Transfer*, 39(1), 39-48.

Qi, S. L., Zhang, P., Wang, R. Z. and Xu, L. X., 2007, Flow boiling of liquid nitrogen in micro tubes: Part II: heat transfer characteristics and critical heat flux, *Int. J. Heat Mass Transfer*, 50, 5017-5030.

Bergles, A. E., Lopina, R. F. and Fiori, M. P., 1967, Critical heat flux and flow pattern observations for low pressure water flowing in tubes, *ASME J Heat Transfer,* 89, 69-74.

Roach, G. M., Abdel Khalik, S. I., Ghiaasiaan, S. M., Dowling, M. F. and Jeter, S. M., 1999, Low flow critical heat flux in heated microchannels, *Nuclear Science and Engineering,* 131, 411-425.

Roday, A. P. and Jensen, M. K., 2007, Experimental investigation of the CHF condition during flow boiling of water in microtubes, Proc. ASME/JSME Thermal Engg. *Summer Heat Transfer Conf.*, 1, 153-165.

Shah, M. M., 1979, A generalized graphical method for predicting CHF in uniformly heated vertical tubes, *Int. J. Heat Mass Transfer*, 22, 557-568.

Shah, M. M., 1987, Improved general correlation for critical heat flux during upflow in uniformly heated vertical tubes, *Int. J. Heat Fluid Flow*, 8, 326-335.

Yuncu, H., Yildrim, O. T. and Kakac, S., 1991, Two phase flow instabilities in a horizontal single boiling channel, *Appl. Sci. Res.*, 48, 83-104.

Ding, Y., Kakac, S. and Chen, X. J., 1995, Dynamic instability of boiling two phase flow in a single horizontal channel, *Exp. Thermal Fluid Science*, 11, 327-342.

Wang, Q., Chen, X. J., Kakac, S. and Ding, Y., 1996, Boiling onset oscillation: a new type of dynamic instability in a forced convection upflow boiling system, *Int. J Heat Fluid Flow*, 17(4), 418-423.

Xiao, M., 1991, Dynamic instabilities of liquid vapor two phase flow in high pressured parallel channels, *Ph.D. dissertation*, Xian Jiaotong University, Xian, China.

Xiao, M., Chen, X. J., Zhang, M. Y., Vezigorlu, T. N. and Kakac, S., 1993, A multivariable linear investigation of two phase flow instabilities in parallel boiling channels under high pressure, *Int. J. Multiphase Flow*, 19(1993), 65-77.

Kim, J. M. and Lee, S. Y., 2000, Experimental observation of flow instability in a semi closed two phase natural circulation loop, *Nucl. Engg. Des.,* 196, 359-367.

Jiang, L., Wong, M. and Zohar, Y., 1999, Phase change in microchannel heat sinks with integrated temperature sensors, *J Microelectromechanical Syst.*, 8, 358-365.

Liu, H. T. and Kakac, S., 1991, An experimental investigation of thermally induced flow instabilities in a convective boiling upflow system, *Waerme und Stoffuebertragung,* 26, 365-376.

Ding, Y., 1994, Experimental investigation of two phase flow phenomena in horizontal convective in-tube boiling system, *Ph.D. dissertation*, University of Miami, Coral Gables, Florida, USA.

Ledinegg, M., 1938, Instability of flow during natural and forced circulation, *Die Waerme,* 61(8), 891-898.

Stenning, A. H., 1964, Instabilities in the flow of a boiling liquid, *ASME J. Basic Engg.,* 86, 213-228.

Liu, H. T., 1993, Pressure drop type and thermal oscillations in convective boiling systems, *Ph.D. thesis,* University of Miami, Coral Gables, Florida, USA.

Achard, J. L., Drew, D. A. and Lahey, R. T., 1985, The analysis of non-linear density-wave oscillations in boiling channels, *J. Fluid Mechanics*, 155, 213-232.

Zhang, Z. Y., Gao, Z. Y. and Wang, D. Z., 1991, An explicit criterion for analyzing two phase flow density wave instability, in Chen, X. J., Veziroglu, T. N. and Tien, C. L. (Eds.), *Multiphase Flow and Heat Transfer*, Hemisphere, N. Y., USA.

Padki, M. M., Palmer, K., Kakac, S. and Veziroglu, T. N., 1992, Bifurcation analysis of pressure drop oscillations and the Ledinegg instability, *Int. J. Heat Mass Transfer*, 35(2), 525-532.

Liu, H. T., Kocak, H. and Kakac, S., 1995, Dynamical analysis of pressure drop type oscillations with a planar model, *Int. J. Multiphase Flow*, 21 (5), 851-859.

Bergles, A. E., 1976, *Review of instabilities in two phase systems,* Proc. NATO Advanced Study Institute, Istanbul, 1, 383-422.

Ishii, M., 1976, Study on flow instabilities in two phase flow mixtures, *Report ANL,* 76-23, Argonne National Laboratory, USA.

Quandt, E., 1961, Analysis and measurement of flow oscillation, *Chem. Engg. Progress Symposium Series,* 57, 111-117.

Agostini, B., Fabbri, M., Park, J. E:, Wojtan, L., Thome, J. R. and Michel, B., 2007, State of the art of high heat flux cooling technologies, *Heat Transfer Engineering*, 28, 258-281.

Wallis, G. G. and Heasley, J. H., 1961, Oscillations in two phase flow system, *ASME J Heat Transfer,* 83, 363-369.

Yadigaroglu, G. and Bergles, A. E., 1972, Fundamental and higher mode density wave oscillations in two phase flows, *ASME J Heat Transfer,* 94, 189-195.

Saha, P., Ishii, M. and Zuber, N., 1976, An experimental investigation of the thermally induced flow oscillations in two phase systems, *ASME J Heat Transfer,* 616-622.

France, D. M., Carlson, R. D. and Roy, R. P., 1986, Measurement and analysis of dynamic instabilities in fluid heated two phase flow, *Int. J. Heat Mass Transfer*, 29(12), 1919-1929.

Wang, Q., Kakac, S., Chen, X. J and Ding, Y., 1994, An experimental investigation of density wave type oscillations in a convective boiling upflow system, *Int. J. Heat Fluid Flow*, 15(3), 241-246.

Meyer, J. E. and Rose, R. P., 1963, Application of a momentum integral model to the study of parallel channel boiling flow oscillations, *ASME J Heat Transfer,* 85(1).

Zuber, N., 1958, On the stability of boiling heat transfer, *ASME J Heat Transfer,* 80, 711-720.

Takahashi, R. and Shindo, M., 1971, Theoretical study of two phase flow oscillation in a hot channel, (I), Theoretical study for interpreting the mechanism of hydrodynamic instability, *J. Nucl. Sci. Technology*, 8(11), 637-643.

Takahashi, R. and Shindo, M., 1971, Theoretical study of two phase flow oscillation in a hot channel, (II), Influence of heat transfer characteristics on the flow stability, *J. Nucl. Sci. Technology*, 8(12), 690-695.

Takitani, K., 1978, Density wave instability in once through boiling flow system—II, Lumped parameter model with moving boundaries, *J. Nucl Sci and Tech.*, 15(6), 389-399.

Akyuzlu, K., Veziroglu, T. N., Kakac, S. and Dogan, T., 1980, Finite difference analysis of two phase flow pressure drop and density wave oscillations, *Waerme and Stoffuebertragung,* 14, 253-267.

Fukuda, K. and Hasegawa, S., 1979, Analysis of two phase flow instability in parallel multichannels, *J. Nucl Sci and tech.*, 16(3), 190-199.

Achard, J. L., Drew, D. A. and Lahey, R. T. Jr., 1981, Effect of gravity and friction on the stability of the boiling flow in a channel, *Chem. Engg. Commun.,* 11(1-3), 59-79.

Yokomizo, O., 1986, Equivalent linearization technique for two phase flow instability analysis, *Nucl. Engg. Design*, 92(2), 141-151.

Khabenskii, V. B. and Kvetnyl, M. A., 1988, Simplified formulae for evaluating the boundary of oscillatory thermohydraulic instability in a two phase flow, *Thermal Engg*., 35(4), 208-212.

Clausse, A., Lahey, R. T. Jr. and Podowski, M., 1989, An analysis of stability and oscillation modes in boiling multichannel loops using parameter perturbation methods, *Int. J. Heat Mass Transfer*, 32, 2055-2064.

Delmastro, D., Juanico, L. and Clausse, A., 2001, A delay theory for boiling flow instability analysis, *Int. J. Multiphase Flow*, 27, 657-671.

Kakac, S., Veziroglu, T. N., Ozboya, N. and Lee, S. S., 1977, Transient boiling flow instabilities in a multichannel upflow system, *Waermi und Stoffuebertragung,* 10, 175-188.

Kakac, S., Veziroglu, T. N., Padki, M. M., Fu, L. Q. and Chen, X. J., 1990, Investigation of thermal instabilities in a forced convection upward boiling system, *Exp. Thermal Fluid Sci,* 3, 191-201.

Mentes, A., Yildirim, O. T., Gurgenci, H., Kakac, S. and Veziroglu, T. N., 1983, Effect of heat transfer augmentation on two phase flow instabilities in a vertical boiling channel, *Waerme und Stoffuebertragung,* 17, 161-169.

Gurgenci, H., Veziroglu, T. N. and Kakac, S., 1983, Simplified nonlinear descriptions of two phase flow instabilities in vertical boiling channel, *Int. J. Heat Mass Transfer*, 26(5), 671-679.

Lin, Z. H., Zhang, X., Chen, X. J., Veziroglu, T. N. and Kakac, S., 1988, An analytical study of the pressure drop type instabilities in a horizontal hairpin tube, *Int. J. Engg. Fluid Mech.,* 1(4), 427-444.

Padki, M. M., Liu, H. T. and Kakac, S., 1991, Two phase flow pressure drop type and thermal oscillations, *Int. J. Heat Fluid Flow*, 12, 240-248.

Cao, L., Kakac, S., Liu, H. T. and Sarma, P. K., 2000, The effects of thermal non-equilibrium and inlet temperature on two phase flow pressure drop type instabilities in an upflow boiling system, *Int. J Thermal Sci,* 39, 886-895.

Widman, F., Comakli, O., Gavrilescu, C. O., Ding, Y. and Kakac, S., 1995, The effect of augmented surfaces on two phase flow instabilities in a horizontal system, *J. Enhanced Heat Transfer,* 2(4), 263-267.

Karsli, S., Yilmaz, M. and Comalki, O., 2002, The effect of internal surface modifications of flow instabilities in forced convection boiling, *Int. J. Heat Fluid Flow*, 23, 776-791.

Mawaha, P. R. and Gross, R. J., 2001, Periodic oscillations in a horizontal single boiling channel with thermal wall capacity, *Int. J. Heat Fluid Flow*, 22, 643-649.

Guo, L. J., Fen, Z. P. and Chen, X. J., 2002, Transient convective heat transfer of steam water two phase flow in a helical tube under pressure drop type oscillations, *Int. J. Heat Mass Transfer*, 43, 533-542.

Gurgenci, H., Two phase flow instabilities, 1982, *Ph.D. thesis,* University of Miami, Coral Gables, Florida, USA.

Mentes, A., 1983, Effect of heat transfer augmentation on two phase flow instabilities in a single channel boiling system, *Ph.D. thesis,* University of Miami, Coral Gables,

Florida, USA.

Padki, M. M., 1990, Thermal oscillations in a forced convective single channel upward system, *Ph.D. dissertation*, University of Miami, Coral Gables, Florida, USA.

Dogan, T., Kakac, S. and Veziroglu, T. N., 1983, Analysis of forced convection boiling flow instabilities in a single channel upflow system, *Int. J. Heat Fluid Flow*, 4(3), 145-156.

Unal, H. C:, 1981, Density wave oscillations in sodium heated once through steam generator tubes, *ASME J Heat Transfer,* 103, 485-491.

Kandlikar, S. G., Kuan, W. K., Willistein, D. A. and Borrelli, J., 2006, Stabilization of flow boiling in microchannels using pressure drop elements and fabricated nucleation sites, *ASME J Heat Transfer,* 128, 389-396.

Kakac, S., 1985, Two phase flow instabilities in boiling systems: summary and review, *METU J. Pure Appl. Sci.,* 18(2), 171-252.

Bergles, A. E., Lienhard, J. H. V, Kendall, G. E. and Griffith, P., 2003, Boiling and evaporation in small diameter channels, *Heat Transfer Engineering*, 24, 18-40.

Brutin, D., Topin, F. and Tadrist, L., 2003, Experimental study of unsteady convective boiling in heated minichannels, *Int. J. Heat Mass Transfer*, 46, 2957-2965.

Wen, D. S., Kenning, D. B. R. and Yan, Y., 2002, *Flow boiling of water in a narrow vertical channel at low mass flux: observations of local phenomena*, Proc. 12th Int. Heat Transfer Conf., 773-778.

Brutin, D., Topin, F. and Tadrist, L., 2002, *Experimental study on two phase flow and heat transfer in capillaries,* Proc. of the 12th Int. Heat Transfer Conf., France, 784-789.

Suzuki, K., Saito, H., Matsumoto, K., Torikai, K. and Maki, H., 2002, *Subcooled flow boiling of water with microbubble emission in a horizontal channel,* Proc. of the 12th Int. Heat Transfer Conf., France, 719-724.

Jain, K. C., 1965, *Self sustained hydrodynamic oscillations in a natural circulation two phase flow boiling loop*, Ph.D. thesis, Northwestern University, Illinois, USA.

Wedekind, G. L., 1971, An experimental investigation into the oscillatory motion of the mixture—vapor transition point in horizontal evaporating flow, *ASME J Heat Transfer,* 93, 47-54.

Chang, K. H. and Pan, C., 2007, Two phase flow instability for boiling in a microchannel heat sink, *Int. J. Heat Mass Transfer*, 50, 2078-2088.

Li, H. Y., Lee, P. C., Tseng, F. C. and Pan, C., 2003, Two phase flow instability of boiling in a double microchannel system at high heating powers, Proc. 1st Int. Conf. on Microchannels and Minichannels, *ICMM*2003-1077, Rochester, N. Y., April 24-25, 615-621.

Stelling, R., McAssey, E. V., Dougherty, T. and Yang, B. W., 1996, The onset of flow instability for downward flow in vertical channels, *ASME J Heat Transfer,* 118(3), 709-714.

Kosar, A., Kuo, C. J. and Peles, Y., 2006, Suppression of boiling flow oscillations in parallel microchannels by inlet restrictors, *ASME J Heat Transfer,* 128, 251-260.

Kuo, C. J. and Peles, Y., 2009, Pressure effects on flow boiling instabilities in parallel microchannels, *Int. J Heat Mass Transfer*, 52 (1-2), 271-280.

Kandlikar, S. G., 2003, Heat transfer mechanisms during flow boiling in microchannels, 1st Int. Conf. Microchannels and Minichannels, April 24-25, Rochester, N. Y., USA, 33-46.

Kenning, D. B. R. and Yan, Y., 2001, Saturated flow boiling of water in a narrow channel: experimental investigation of local phenomena, *IChemE,* 79, part A, 425-436.

Jones, R. C. and Judd, R. L., 2003, An investigation of dryout/rewetting in subcooled two phase flow boiling, *Int. J. Heat Mass Transfer*, 46, 3143-3152.

Hamamura, Y. and Katto, Y., 1983, A new hydrodynamic model of critical heat flux, applicable widely to both pool and forced convective boiling on submerged bodies in saturated liquids, *Int. J. Heat Mass Transfer*, 26, 389-399.

Huh, C., Kim, J, and Kim, M. H., 2007, Flow pattern transition instability during flow boiling in a single microchannel, *Int. J. Heat Mass Transfer*, 50, 1049-1060.

Consolini, L., Ribatski, G., Thome, J. R., Zhang, W. and Xu, J., 2007, Heat transfer in confined forced flow boiling, *Heat Transfer Eng.*, 28(10), 826-833.

Collier, J. G. and Thome, J. R., 1994, Convective boiling and condensation, 3rd ed., Oxford University Press.

In: Frontier Research in Microscale and Nanoscale... ISBN 978-1-61324-665-8
Editor: Lixin Cheng

Chapter 8

EXPERIMENTAL INVESTIGATION OF NEW STRUCTURE MICROCHANNEL HEAT SINKS WITH LAMINAR FLOW HEAT TRANSFER

Guodong Xia**, Mingzheng Zhou, Jingzhi Qi, Zhenzhen Cui and Jian Li

College of Environment and Energy Engineering, Beijing University of Technology, Key Laboratory of Intensified Heat Transfer and Energy Saving, Ministry of Education, Beijing 100124, China

ABSTRACT

In this chapter, a new microchannel heat sink with staggered structure was designed based on thermal boundary layer technology for enhancing heat transfer performance and reducing flow resistance. Comparative experiments were carried out between a test section having the new staggered structure microchannel and a conventional straight microchannel keeping the values of three control parameters, inlet temperature, heating power, and flow rate, very close. The results indicated that there was an obvious difference between the friction resistance factor from experimental data and calculated values from theoretical formula. The staggered structure increased the partial friction resistance to a certain extent, but the advantage for heat transfer was more evident. The temperature distribution of the wall was more uniform along the flow direction.

Keywords: Microchannel heat sink; Staggered structure; Heat transfer; Resistance.

NOMENCLATURE

A_b	heat film area, m^2
A_c	microchannel cross-section area, m^2

---

* E-mail: xgd@bjut.edu.cn

A_f	fin convection area, m^2
A_h	heat virtual heat transfer area, m^2
A_l	fin cross-section area, m^2
c_p	constant pressure specific heat
D_h	hydraulic diameter, m
f	friction resistance factor
H	height of the microchannel m
L	total microchannel length, m
m	constant parameter
N	microchannel number
Nu	Nusselt number
p	perimeter of the fin, m
Pr	Prandtl number
q	heat flux, W/cm^2
Q	flow rate, m^3/s
Q_{loss}	heat loss of the device
Re	Reynolds number
R_t	total thermal resistance, /W
T_i	measurement temperature,
T_{in}	inlet temperature,
T_{out}	outlet temperature,
T_{ref}	reference temperature ,
T_{wi}	inside average temperature,
T_{wo}	outside average temperature,
$\overline{u}$	average velocity, m/s
V	volume flow rate, ml/min

Greek Symbols

δ	wafer bottom thickness, m
ΔT_m	average temperature difference,
Δp	pressure drop, Pa
Φ	heating power, W
ν	kinematic viscosity coefficient, m^2/s
η	fin efficiency
λ	wafer thermal conductivity, W/(m·K)
λ_f	water thermal conductivity, W/(m·K)
ρ	water density, Kg/m^3

1. INTRODUCTION

The heat removal problem has become an important factor in the advancement of microelectronics, due to both drastically increased integration density of chips in digital devices and an increased current and voltage handling capability of power electronic devices. The microchannel heat sink was first presented by Tuckerman [1] in the 1980's, with the aim to decrease the probability of damage of the electric device caused by superheat and to improve the capability and credibility of the electric device. It has been studied and tested here as a high performance and compact cooling scheme in microelectronics cooling applications.

Two characteristics limit the broad application of the traditional microchannel (TMC) heat sink. One is the large pressure drop caused by small diameters, and the other is large temperature change between the inlet and outlet of the channel. With the development of micro-electro-mechanical system (MEMS) technology, many kinds of microchannel heat sinks with new configurations were explored.

Jeng [2] experimentally studied the pressure drop and heat transfer of a square pin-fin array in a rectangular channel by using the transient single-blow technique. The optimal inter-fin pitches are provided based on the largest Nusselt number under the same pumping power. It was found that the optimal inter-fin pitches of square pin-fins are $X_T = 2$ (X_T, relative transverse pitch) and $X_L = 1.5$ (X_L, relative longitudinal pitch) for the arrays in in-line arrangements as well as $X_T = 1.5$ and $X_L = 1.5$ for the arrays in staggered arrangements. Kosar [3] experimentally studied the heat transfer and pressure drop of de-ionized water over a bank of shrouded staggered micro pin-fins 243μm long with hydraulic diameter of 99.5μm. Average heat transfer coefficients were obtained for effective heat fluxes ranging from 3.8 to 167 W/cm^2 and Reynolds numbers from 14 to 112. By applying the porous approach, Kim and Kuznetsov [4], and Kim *et al.* [5] modeled pin-fin heat sinks as porous media and examined their fluid flow and heat transfer performance in a jet impinging channel and a cross-flow channel, respectively. Chen [6] fabricated a new fractal tree-like microchannel heat sink for cooling of electronic chips on silicon wafers by advanced MEMS technology. It was experimentally confirmed that the thermal efficiency of the fractal tree-like microchannel net was much higher than the traditional parallel channel arrays for the sink.

Xia [7] numerically simulated the structure of a copper heat sink for cooling high-power diode laser bar. Their results showed that the structure of a microchannel heat sink is important to the thermal resistance, pumping power and temperature distribution on the surface of high-power diode laser bar. The most important effect on the pumping power was the width of the microchannel inlet and outlet. Liu [8] numerically and experimentally researched the fluid and heat transfer in V-shaped microchannel. It was concluded that a V-shaped microchannel can improve the heat dissipation ability of the diode laser arrays. The experimental data agreed well with the simulation result. Jeevan [9] optimized the two-layered microchannels heat sink with counter flow used for cooling diode laser bar.

Associated with intense heat transfer theory, breaking-up or decreasing thermal boundary layers is a significant means to enhance the heat transfer capacity of laminar flow. As reported in this paper, an extensive experimental investigation using de-ionized water as the testing fluid has been carried out on viscous dissipation, friction resistance factor and heat transfer with laminar flow in a test section which had new staggered structure microchannels with rectangular cross section. The comparisons between the staggered structure microchannels and the conventional structure microchannels are used to examine the capability of new style microchannels.

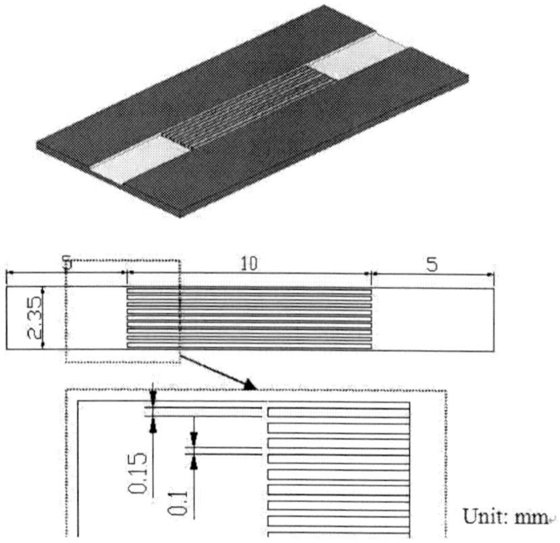

Figure 1. Schematic diagram of the heat sink # 1.

2. EXPERIMENTAL APPARATUS AND PROCEDURE

2.1. Test Section

Before investigating a new staggered structure microchannel, a microchannel(# 1) with traditional structure and a length of 10mm was investigated experimentally to illuminate that how we designed the new structure microchannel. The results indicated that the temperature gradient on the surface of the heat film along the flow direction gradually decreased, and the temperature reached the maximum at 2mm before the exit. The largest difference between maximum and minimum temperature on the surface of the heat film was 18 °C under a heat flux of 122W/cm2. In order to decrease the temperature difference of the surface and improve heat transfer efficiency, the temperature gradients of the inlet section should be decreased, and the maximum temperature of the exit section should be reduced. A new structure microchannel design was based on the above analysis. The fins are to be chopped up, and the staggered distributed in the inlet section and outlet sections in order to break up the thermal boundary layer and let the flow impact the fins. The staggered fins in the inlet section can decrease the temperature gradients of the wall, while the staggered fins in the outlet section can reduce the peak of wall temperature; both of them can obviously improve the heat transfer performance.

Table 1. Geometric parameters of two heat sinks

Device	structure	N	H/mm	D_h/mm	A_c/m^2	A_l/m^2	p/ mm
# 1	straight channel	10	0.2	0.133	0.02	0.03	20.30
# 2	staggered channel	10	0.2	0.133	0.02	0.03	5.5for long fins 2.4for short fins

In this chapter, two styles of microchannels were used: traditional straight microchannels (# 1) and the staggered structure microchannels (# 2). As shown in Figure 1, # 1was made up of 9 rectangle fins, each having a length of 10mm, a width of 150μm and a height of 200μm, and the width of the microchannel was 100μm. Microchannel # 2, as shown in Figure 2, had three sections of long fins with a length of 2.6mm and two sections of short fins with a length of 200μm. The distance between long and short fins was 350μm. The other geometric parameters were show in table 1.

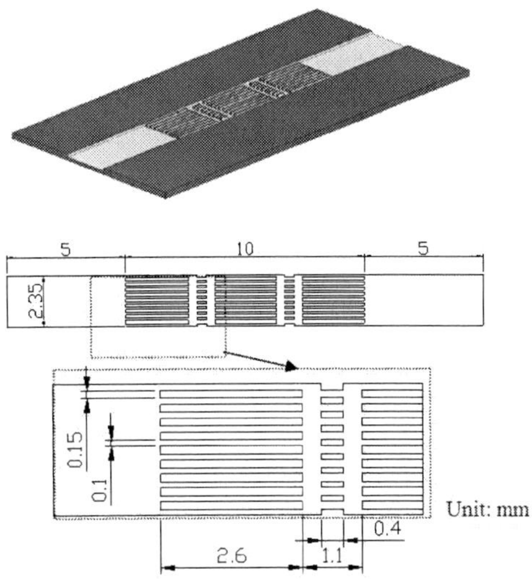

Figure 2. Schematic diagram of the heat sink # 2.

Both of the test sections were fabricated by the wet etching technique. Each of the silicon wafers was then bonded with a thin pyrex glass plate to close the microchannels. A 0.15μ thick layer of foil was uniformly deposited on the back of the silicon wafer by an electroplating method. The area of the heater was $2.25 \times 8mm^2$. A thin black lacquer was sputtered onto the heater to improve the emissivity, which was then measured to be 0.95.

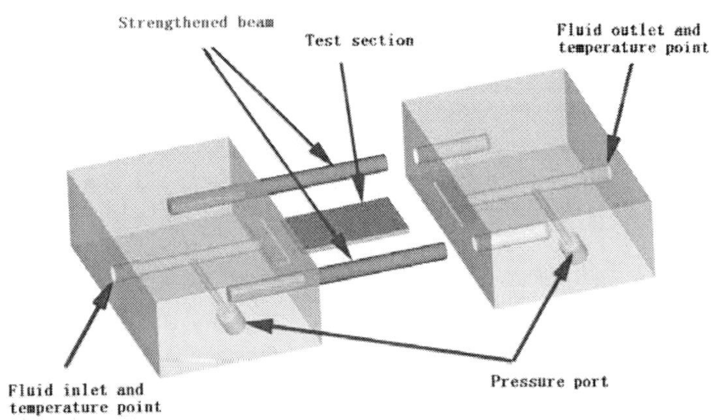

Figure 3. Device package.

As shown in Figure 3. The microchannel device is packaging by two Plexiglass block, which is convenient for observing during the experiment. The block and microchannel were conglutinated by epoxy resin glue, which can solidify rapidly in room temperature. The device was reinforced by two strengthened beam to avoiding twist.

2.2. Experimental Apparatus

The experimental apparatus used in this chapter (Figure 4) consisted of a tranquil flow pump, a filter, a test section, a collecting container and an infrared radiator image system. The de-ionized water, which was deposited in the constant temperature vessel, flows to the tranquil flow pump and is supercharged there. It then successively flows through the liquid valve, the filter, the silicon wafer test section and is collected in the liquid container. A 10μm filter was used to keep patency of the microchannels. While the working fluid flows steadily in the system, resistance, pressure, flow measurements, and temperature distribution of the wall are made at various locations.

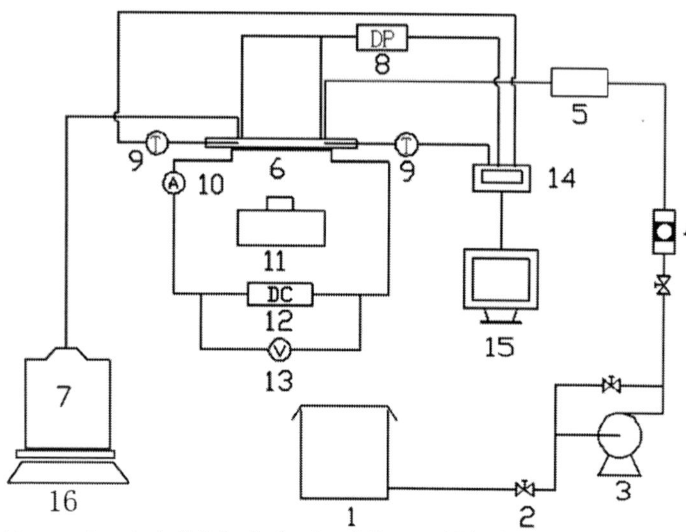

(1 Constant-temperature tank 2 Spherical valve 3 Pump 4 Liquid Filter 5 Liquid Manostat 6 Test section 7 Collect utensil 8 Pressure gauge 9 T-type thermocouple 10 Current gauge 11 Infrared thermal imaging system 12 Power supply 13 Voltage gauge 14 Date acquisition unit 15 Computer 16 Electronic balance)

Figure 4. Schematic diagram of the experimental set-up.

2.3. Parameter Measurements

The inlet and outlet temperatures of the microchannel heat sink were measured by two high-precision jacket thermocouples, which were located at the centerline of the connection tubes and about 2.0mm before and after the heat sink, respectively. T_{in} was about 20□, while the values of T_{out} were in the range of 20.2 to 47.4□ for different experiment condition. A smart differential pressure transducer was installed across the microchannel heat sink to measure the pressure drop, the pressure ports were about 1.0mm before and after the heat sink, respectively. The inlet pressure was measured by a Setra pressure transducer (model SBP800). The fluid temperature, pressure sensor and differential pressure transducer had the response times of 0.2s, 0.1 s and 0.01s, respectively. The signals were collected by an Agilent 34970A high speed data acquisition system, displayed by a PC monitor and stored in the PC memory for further analysis. The mass flow rate was measured from the increased mass in the liquid container over a long period of time using a high-precision electronic balance. The electrical power was supplied to the device with an HY1792-10S direct current power supply, which enabled the measurement of electrical current and voltage. All experiments were performed under ambient room temperature.

For all the tests, an IR camera (NEC 5104) was focused on the heating area of the silicon wafer ($2.25\times8mm^2$). The temperature gradient and the maximum wafer temperature were measured at different power levels. The variations in the maximum chip temperature and the transient response were also detected. The IR image was displayed on a PC screen and stored in the PC memory for further analysis. The uncertainties of measurement apparatus and parameters are shown in table 2.

Table 2. Experimental uncertainties

Uncertainty	Error
Thermocouple	±0.2□
Pressure transducer (model SBP800), kPa	±0.25%
Differential pressure transducer, kPa	±0.1%
electronic balance, *g*	±0.002g
IR camera, □	±0.1□
Direct current power, I and V	±0.01A,±0.1V
Flow rate, *Q*	±0.5%
Reynolds number, *Re*	±0.88%
friction resistance factor, *f*	±8.25%
Nusselt number, *Nu*	±6.21%

2.4. Experimental Procedure

The tranquil flow pump was started up first to the flow rate with the desired value, and then the power was opened to the right voltage to heat the silicon wafer. Experimental data were noted and the IR images of the heat film were saved after steady flow condition was reached under atmospheric exit pressure. Then, one of the three control parameters, inlet temperature T_{in}, heating power Φ, and flow rate Q, would be changed while the others were kept constant, in order to investigate other flow conditions. All experiments were performed under ambient temperature.

3. Data and Uncertainty Analysis

3.1. Flow Friction and Heat Transfer

The friction pressure drop for present study was taken to be:

$$f = 2D_h \Delta p / (\rho L \bar{u}^2) \tag{1}$$

Where, D_h is hydraulic diameter of the channel, Δp is pressure drop between inlet and outlet of the microchannel, ρ is density of the working liquid (de-ionized water), L is total length of microchannel (10mm). The average velocity of the flow, $\bar{u}$ was calculated from

$$\bar{u} = Q/(NA_c) \tag{2}$$

Where, Q is volume flow rate, N is number of the microchannels (10), A_c is cross-section area of microchannel.

The Reynolds number is calculated based on the hydraulic diameter, as:

$$\mathrm{Re} = \bar{u} D_h / \nu \tag{3}$$

Where ν is the kinematic viscosity coefficient of de-ionized water.

The average convection heat transfer coefficient between the working liquid and the wall of the microchannels, h, is given by

$$h = (\phi - Q_{loss}) / ((A_b + \eta A_f) \cdot \Delta T_m) \tag{4}$$

Where, ϕ is heating power, Q_{loss} is heat loss of the device, A_b is heat transfer area of the bottom of the channels, A_f is heat transfer area of the fin, η is efficiency of the fin, which will be calculated from

$$\eta = \frac{th(mH)}{mH} \tag{5}$$

The parameter m in the Eq. (5) is a constant, which is expressed as: $m = \sqrt{hp/(\lambda A_l)}$. where, A_l is cross-section of the fin, p is perimeter of the fin. The heat loss was investigated by the Kosar`s[10] method, and found that the heat loss in present condition was less than 3.0%. The efficiency of the fin under different flow rates and heat flux can be calculated from Eqs. (4 and 5), using an iterative method.

The average difference of heat transfer, ΔT_m, is given by

$$\Delta T_m = T_{wi} - \frac{1}{2}(T_{in} + T_{out}) \tag{6}$$

Where, T_{wi} is average temperature of the wafer inside surface, which is determined from

$$T_{wi} = T_{wo} - (\phi - Q_{loss}) \cdot \delta / (A_h \cdot \lambda) \tag{7}$$

Where, δ is thickness of the bottom of the wafer, λ is thermal conductivity of the wafer, A_h is virtual heat transfer area of the heater, T_{wo} is average temperature of the wafer outside surface and can be calculated from

$$T_{wo} = \frac{1}{20}\sum_{i=1}^{20} T_i \tag{8}$$

Where, T_i is a point temperature, determined from the infrared thermal imager, which are got in the center line of the heat film. The average of T_i is used as T_{wo}. The temperature data indicate that the difference between T_{wi} and T_{wo} is less than 2□.

The average Nusselt number (Nu) is:

$$Nu = hD_h / \lambda_f \tag{9}$$

Where, λ_f is thermal conductivity of the liquid.

The total thermal resistance, R_t, was calculated from:

$$R_t = \frac{\Delta T_{\max}}{\phi - Q_{loss}} = \frac{T_{wo} - T_{in}}{\phi - Q_{loss}} \tag{10}$$

Where, T_{in} is inlet temperature.

3.2. Error Analysis

The uncertainties of flow rate, Reynolds number, friction resistance factor and Nusselt number, given in Table 2, are obtained from the manufacturer`s specification sheets, while the uncertainties of derived parameters are calculated using the method developed by Kline and McClintock[11]. It can be seen that the errors of f and Nu due to measurement errors in this study were no more than 8.25% and 6.21%, respectively.

4. Results and Discussion

4.1. Flow Resistance Characteristics

For the flow conditions in the present study, test section □ 1 was first examined to determine its friction factor. The friction factor data were then compared with the conventional values for laminar flow in the microchannels. Figure 5 shows the friction factor, f, in the form of f versus Re log-log plot. A comparison of experiment with Ref. 12 data (steel tube) and Ref. 13 with MMC heat sink is presented in Figure 6.

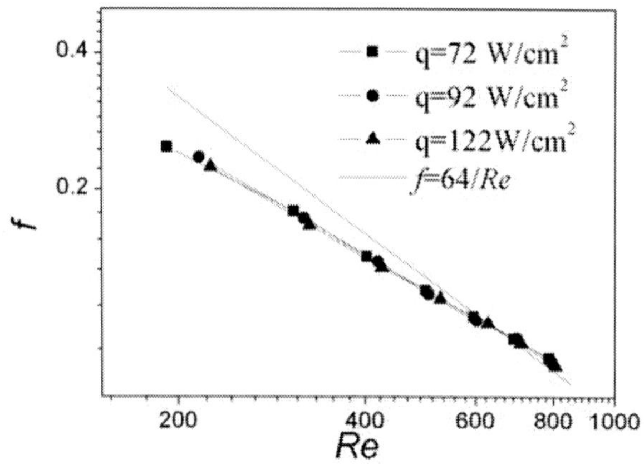

Figure 5. Relation of friction factor and Reynolds number for #1.

As shown in Figures 5 and 6, the friction factor of the experiment is lower than the theoretical value when the Reynolds number is below 600 with the maximal difference being 26 . However, the friction factor is higher than the theoretical value when the Reynolds number is over 600. The results from Ref. [12, 13] show that their experimental friction factor values agree well with the theoretical value for the low Reynolds numbers(Re≤800), which is different from present experimental data; a break is found, when the Reynolds number is larger than 800 for MMC heat sink, which is due to the transition from laminar to turbulent flow.

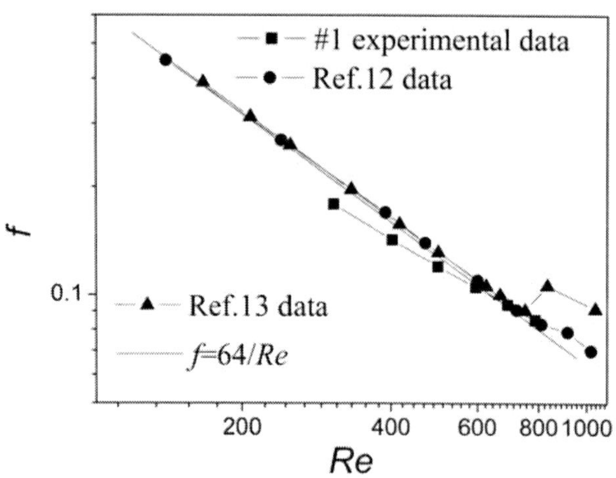

Figure 6. Comparison of friction factor at different test sections.

Effects of surface roughness on friction factor and flow pattern transition in circular tubes were numerically and experimentally studied by many researchers [14-18]. The channel processed by dry etching, which is used in the present experiment, has a very smooth surface. No obviously even at the at scan sizes of 1×1μm[19]. The material, used to comparing, are steel and copper, which are more rough relatively since the processing technics[20]. This may be the reason for the low values of friction factor in the present experiments.

The variation of friction factor with the Reynolds number and pumping power with flow rate for both test section are presented in Figures 7 and 8, respectively. The pumping power is calculated by flow rate times pressure drop (VΔp). The entity symbol is used for test section 1, while the void symbol is used for test section 2.

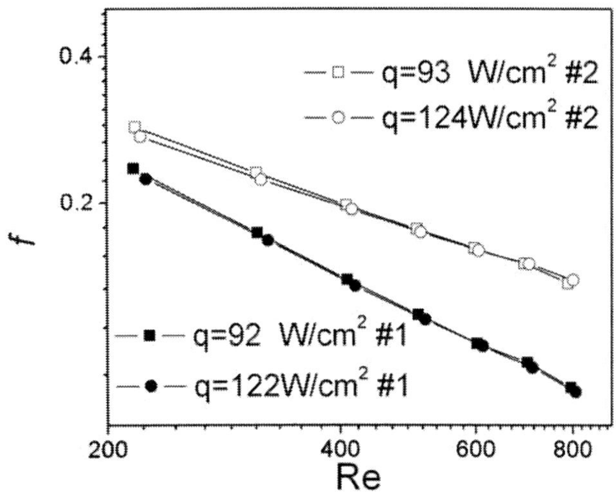

Figure 7. Relationship between friction factor and the Reynolds number.

The friction resistance factor and pumping power for #2 are higher than for #1, and the difference grows with increasing flow rate. The ratio of friction factor #1 and #2 grows from 1.2 to 1.7, and pumping power ratio grows from 1.2 to 1.6, as the flow rate increased from 20 to 60ml/min. The effects of heat flux on both friction factor and pumping power were negligible, as indicated in Figures 7 and 8.

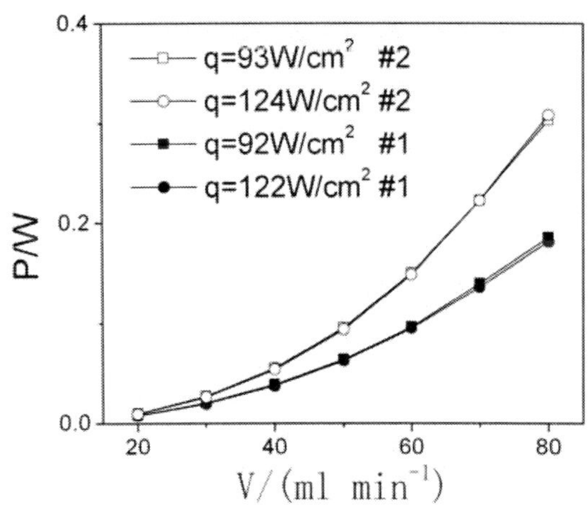

Figure 8.1 Relationship between pumping power and volume flow rate.

In the experiment with #2, the flow impinged the side wall of the fins, which enhanced the resistance and increased the pumping power. The higher velocity of flow created stronger disturbances, which created more resistance and larger energy loss.

4.2. Heat Transfer

The heat transfer performance of the two test sections was compared using Nusselt number Nu, total heat resistance R_t, average surface temperature T_{wo} and axial temperature in the central line of heat film $T_{wo\,x}$. As shown in Figures 9 and 10, Nusselt number of #2 were 2~2.5times larger than #1, while total heat resistances was 0.65~0.81times of #1. That result is proposed to be related to the fact that short fins of #2 breaking up the thermal boundary layers, produce more disturbances, and are impacted by the flow. The influences of the Reynolds number on both Nusselt number and total heat resistance are depressed with the growth of the Reynolds number, which indicates that the effect of convective heat transfer is weak in present microchannels.

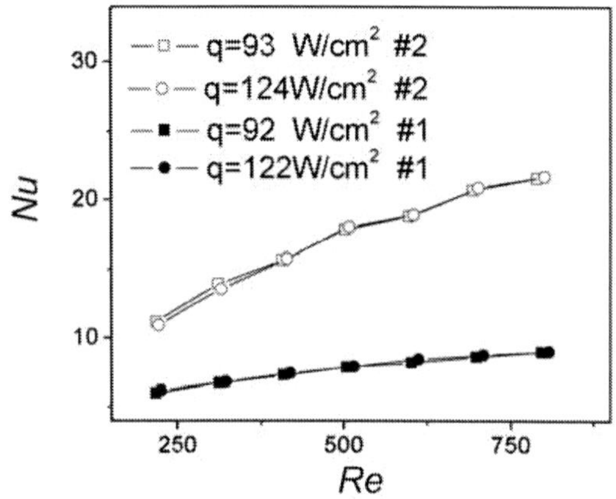

Figure 9. Relationship between average Nusselt number and the Reynolds number.

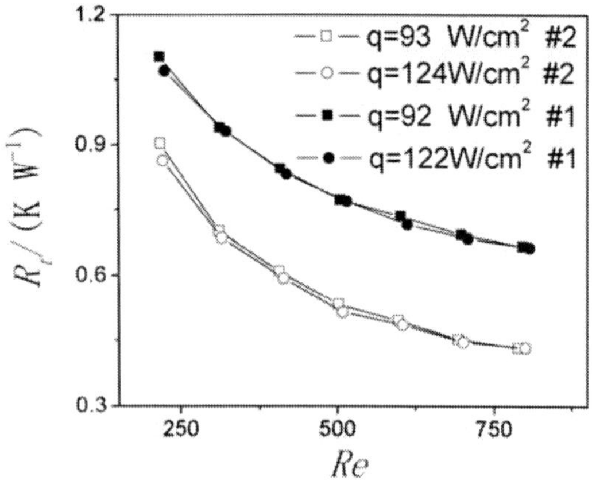

Figure 10. Relationship between total thermal resistance and the Reynolds number.

The surface average temperature of #2 is about 4~5□ lower than of #1 in the same flow condition as seen from Figure 11. Figure 12 presents axial temperature in the central line of heat film for both test sections under different flow rate. As shown in Figure 12, the variation of the axial temperature for #1 is smoother; the temperature decreased near the exit, which effect was caused by the heat conduction of the wafer. For #2, two inflexions of the axial temperature curves are found just at the position where the long and short fins were switched. The surface temperature of the #2, with the maximal temperature difference of about 3□, was more uniform than that of #1.

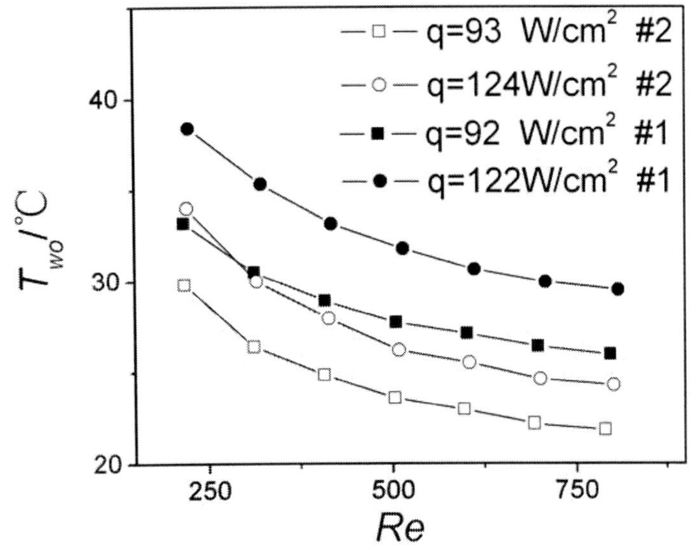

Figure 11. Average surface temperature of heat film on different heat flux.

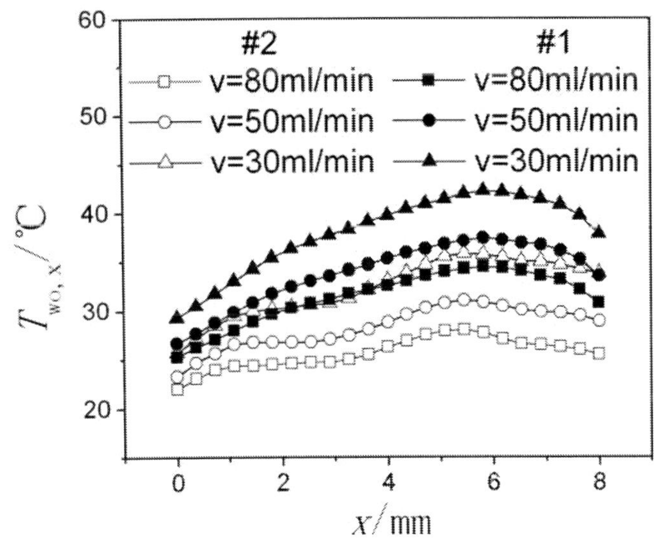

Figure 12. Average surface temperature of heat film on different flow rate（#1: q=122W/cm^2; #2: q=124W/cm^2）

Also shown in Figure 12, the inlet wall temperature of the two test sections are different even at the same inlet water temperature and equal heat flux. The wall temperature of #2 is lower than that of #1, which indicates a much higher heat transfer coefficient. The infrared thermal images of #1 and #2 with the same condition (q=124W/cm2, V=30ml/min) are shown in Figure 13. There has a large area in the infrared thermal image of #1, where the temperature is very high, while the temperature of #2 is more uniform.

On the basis of the experimental data and above analysis, the friction factor and pumping power for #1 are a little higher than #2, but the superiority of heat transfer performance for #2 is much stand out.

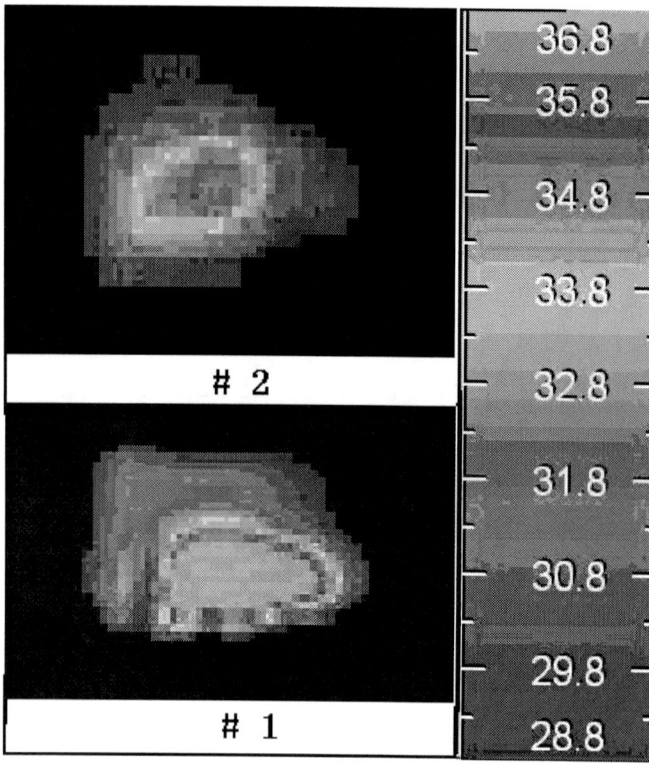

Figure 13. Infrared thermal image of heat film (q=124W/cm2, V=30ml/min).

CONCLUSION

An experimental investigation of a new configuration microchannel heat sink has been performed in this chapter. The main conclusions are as follows:

1. The friction factor of the experiment was different from the theoretical value and experimental data from other reports, which resulted from different wall roughness.
2. The new staggered structure heat sink increased the friction resistance factor and pumping power, but the heat transfer was outstanding. The average Nusselt number was higher; heat resistance was decreased; heat transfer coefficient was improved; and surface wall temperature was reduced and more uniform.
3. Attempting to improve the heat transfer efficiency by increasing the Reynolds numbers was not seen to be effective, rather, the structure and the material should be considered synthetically.

ACKNOWLEDGMENT

Thanks to Mary York for providing the article editing assistance. This chapter was supported by National Natural Science Foundation of China (No. 50876003); Beijing Municipality Natural Science Foundation (No. 3092009); Funding Project for Academic Human Resources Development in Institutions of Higher Learning Under the Jurisdiction of Beijing Municipality (200906104). The first session of the doctoral innovative projects of BJUT (No.bcx-2009-046)

REFERENCES

[1] Tuckerman, D. B. and Pease, R. F. W. High performance heat sinking for VLSI, *IEEE Electron Device Letter* 2(5) (1981) 126-129.

[2] Jeng, T. M. and Tzeng, S. C. Pressure drop and heat transfer of square pin-fin arrays in in-line and staggered arrangements, *Int. J. Heat and Mass Transfer* 50 (2006) 2364-2375.

[3] Kosar, A. and Peles, Y. Thermal-hydraulic performance of MEMS-based pin-fin heat sink, *J. Heat Transfer* 128 (2006) 121-131.

[4] Kim, S. Y. and Kuznetsov, A. V. Optimization of pin-fin heat sinks using anisotropic local thermal non equilibrium porous model in a jet impinging channel, *J. Heat Transfer* (Part A) 44 (2003) 771-787.

[5] Kim, D., Kim, S. J. and Ortega, A. Compact modeling of fluid flow and heat transfer in pin-fin heat sinks, *ASME J. Electron Packaging* 126 (2004) 342-350.

[6] Chen, Y. P. and Zheng, P. Development on fractal tree-like microchannel heat sink, *J. Engineering Thermophysics* 27(5) (2006) 853-855.

[7] Xia, G. D., Liu, Q., and Wang, M. Three-dimensional numerical optimization of the manifold microchannel heat sink, *J. Engineering Thermophysics* 27 (1) (2006) 145-147.

[8] Liu, T. T., Gao, Y. and Han, B. Research on fluid flow and heat transfer of the V-Shaped microchannel heat sink, *J. Sensors and Actuators* 19 (5) (2006)1683-1685.

[9] Jeevan, K., Azid, I. A. and Seetharamu, K. N. Optimization of double layer counter flow (DLCF) micro-channel heat sink used for cooling chips directly. Proceedings of 6th Electronics Packaging Technology Conference, *EPTC* 2004. 2004, 553-558.

[10] Kosar, Heat and fluid flow in MEMS-based pin fin heat sinks, *PhD thesis,* Rensselaer Polytechnic Institute, New York, 2006.

[11] Kline, S. and McClintock, F. A. Describing uncertainties in single-sample experiments, *Mech. Eng.* 75 (1) (1953) 3-8.

[12] Liu, Z. G. *Visual study and analysis on flow and heat transfer in microchannel,* PhD thesis, the Chinese Academy of Sciences. Beijing, China 2005.

[13] Liu, Q. M., Xia, G. D. and Qi, J. Z. Influence of surfactant on frictional pressure drop in manifold microchannel, *J. Chemical Industry and Engineering* 57(11) (2006) 2525-2530.

[14] Celata, G. P., Cumo, M. and Gulielmi, M. Experimental investigation of hydraulic and single phase heat transfer in 0. 130mm capillary tube, Proceeding of International

conference on *Heat Transfer and Transport Phenomena in Microscale*, G. P. Celata, Begell House, Inc, New York, Wallingford, U. K, (2000) 108-113.

[15] Wu, P. Y. and Little, W. A. Measurement of friction factors for the flow of cases in very fine channels used for microminiature, *Joule-Thomson Refrigerators Cryogenics* 23(1983) 273-277.

[16] Mala, G. M. and Li, D. Flow characteristics of water in microtubes, *Int. J Heat and Fluid Flow* 20 (1999) 142-148.

[17] Du, D. X. *Effect of compressibility and roughness on flow and heat transfer in microtubes,* Doctoral degree thesis, Tsinghua University. Beijing, 2000. (in Chinese)

[18] Li, Z. X., Du, D. X. and Guo, Z. Y. Experimental study of flow characteristics of liquid in circular microtubes, *Microscale Thermo-physical Engineering* 7 (2003) 253-265.

[19] Chandrasekaran, S. and Sundararajan, S. Effect of microfabrication processes on surface roughness parameters of silicon surfaces, *Surface & Coatings Technology* 188–189 (2004) 581– 587

[20] Kandlikar, S. G., Joshi, S. and Tian, S. Effect of surface roughness on heat transfer and fluid flow characteristics at low Reynolds numbers in small diameter tubes, *Heat Transfer Eng.* 24 (3) (2003) 4–16.

In: Frontier Research in Microscale and Nanoscale… ISBN 978-1-61324-665-8
Editor: Lixin Cheng

Chapter 9

THERMAL BEHAVIOR OF A TWO PHASE CLOSED THERMOSYPHON USING CUO/WATER NANOFLUID

M. Kahani, S. Zeinali Heris*, S. M. Nowee and S. H. Noie

Chemical Engineering Department, Engineering Faculty, Ferdowsi University of Mashhad, Mashhad, Iran

ABSTRACT

A two-phase closed thermosyphon (TPCT) is highly effective thermal spreader consisting of an evacuated closed tube filled with a certain amount of working fluid. In this study, nanofluids of aqueous CuO nanoparticles were prepared with volume concentrations varying from 1 to 3 percent. The effect of these different working fluids and the filling ratio on thermal performance of a TPCT was investigated and compared with pure water. Experimental results showed that the efficiency of TPCT significantly increases when the CuO/water nanofluid was used instead of pure water. Also, maximum efficiency of the TPCT was observed with filling ratio of 35%.

Keywords: CuO/water nanofluid; two-phase closed thermosyphon; filling ratio, heat transfer efficiency.

NOMENCLATURE

A_{cross}	:	Cross-sectional area of pipe (m^2)
A_{rad}	:	Radial area of pipe (m^2)
Cp	:	Specific heat of water $(J\ kg^{-1}K^{-1})$
D_i	:	Internal diameter of main tube (m)

* Corresponding author: Email: zeinali@ferdowsi.um.ac.ir, Fax: (+98)(511) 8816840, Tel: (+98) (511) 8816840

g	:	Gravitational acceleration (m/s^2)
$h_{conv.}$	:	Convective heat transfer coefficient $(Wm^{-2}\,°C^{-1})$
h_{fg}	:	Latent heat of vaporization (J/kg)
I	:	Current (A)
$k_{surr.}$	:	Thermal conductivity of surrounding air $(Wm^{-1}°C^{-1})$
L_e	:	Length of evaporator section (m)
L_t	:	Total length of tube (m)
m	:	Coolant water mass rate (kg/s)
Nu	:	Nusselt number ($= h_{conv.} L_t / k_{surr.}$)
Pr	:	Prandtl number $(= \vartheta / \alpha)$
$Q_{conv.}$	:	Convective heat transfer rate (W)
Q_{in}	:	Inlet heat by evaporation (W)
Q_{max}	:	Heat transfer rate limits at the TPCT (W)
Q_{out}	:	Outlet heat by condensation (W)
Q_{rad}	:	Radiation heat transfer rate (W)
Ra	:	Rayleigh number ($= g\beta(T_{ins.} - T_{surr})L_t^3 / \alpha\vartheta$)
T_a	:	Adiabatic temperature (°C)
T_c	:	Condenser temperature (°C)
T_e	:	Evaporator temperature (°C)
T_{in}	:	Inlet temperature of cooling water (K)
$T_{ins.}$	:	Temperature on external surface of insulation (K)
T_{out}	:	Outlet temperature of cooling water (K)
$T_{surr.}$	:	Surrounding temperature (K)
V	:	Voltage (V)

Greek letters

ρ_l	:	Density of liquid (kg/m^3)
ρ_v	:	Density of vapor (kg/m^3)
ε	:	Emissivity factor of insulator

δ	:	Boltzmann constant $(Wm^{-2}K^{-4})$
σ	:	Surface tension (N/m)
η	:	Efficiency of TPCT
α	:	Thermal diffusivity (m^2s^{-1})
ϑ:	:	Momentum diffusivity

Abbreviations

A.R	:	Aspect Ratio (L_e/D_i)
B.L	:	Boiling Limitation
CCFL	:	Counter Current Flow Limitation
F.R	:	Filling Ratio of the working fluid
TPCT	:	Two- Phase Closed Thermosyphon

1. Introduction

In order to cope with ever increasing demands from the electronic, automotive and aerospace industries, cooling devices have to be small in size and light-weighted with high performance. The level and reliability of heat rejection efficiency highly determine the optimal design of cooling devices.

Two-phase closed thermosyphons (TPCT) are highly effective thermal spreaders capable of transporting heat at a very high rate with a relatively small temperature drop, i.e. they present a low thermal resistance. In practice, the effective thermal conductivity of thermosyphon exceeds that of copper by a factor of 200 ~ 500 [1].

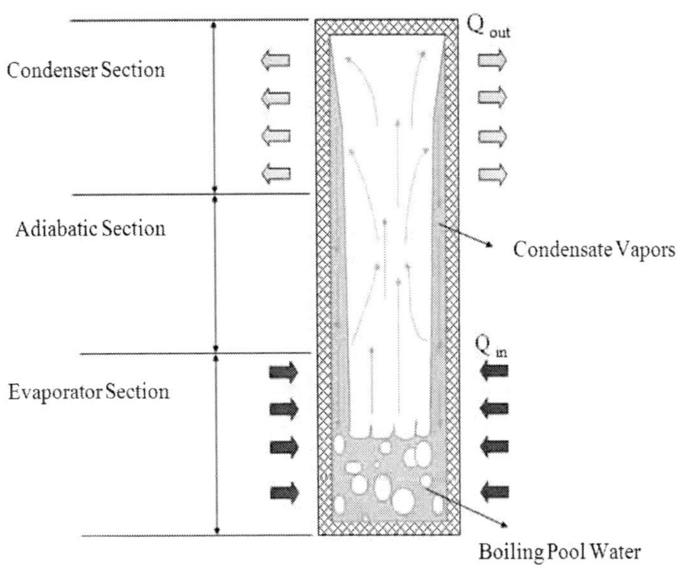

Figure 1. Schematic of a conventional thermosyphon.

A typical TPCT is shown in Figure 1. A small quantity of water is placed in a tube which air is then evacuated from and sealed. Heating the lower end of the tube causes the liquid to vaporize upward to the other end which is cold and where condensation occurs. The condensate returns to the hot end by gravity and this cycle is repeated continuously. Since the latent heat of evaporation is large, considerable quantities of heat can be transported with a very small temperature difference between two ends. Consequently, the system has a high effective thermal conductance which makes TPCTs a suitable device to be used in various applications by differing the size operating conditions, material and working fluid, such as industrial heat recovery, electronic components and solar heating systems and so on (Noie et al., [2], Akbarzadeh et al. [3], Xiau et al., [4], Liu et al., [5], Vasiliev et al., [6]). While Peterson et al. [7], among the other researchers, have reviewed the theory and the applications of TPCT technology.

Apart from featuring a low thermal resistance, another interesting characteristic of TPCT is, providing a very uniform temperature distribution along the condenser section while the heat transfer coefficient between the condenser and the surrounding environment is small, such as in baking oven applications.

As stated, the heat transfer performance of a TPCT is significantly affected by geometry, inclination angle, vapor temperature and pressure, filling ratio and thermo-physical properties of the working fluid.

Some experimental investigations have been reported the appropriate range of filling ratio (F.R), defined as the volume ratio of the charged liquid to the evaporator or the whole TPCT. Based on the former definition, Imura et al. [8] and Harada et al. [9] proposed that the filling ratio should be 1/5 ~1/3 and 0.25 ~0.30, respectively, whereas 30%, 60% and 90% are proposed by Noie [10] with different geometries. The experimental results of Feldman and Srinivasan [11] showed that the appropriate range of filling ratio was 18 ~22%. Hussein et al. [12] suggested the optimal value of 20%.

On the other hand, the processes of heating or cooling fluids are important in many industrial sectors, including energy supply and production, transportation and electronics. Thus, development of advanced heat transferring fluids is clearly essential to improve the effective heat transfer behavior of conventional fluids and various techniques have been proposed to reach this goal.

One of aforementioned methods is producing slurries from different types of powders such as metallic, non-metallic and polymeric particles added to the base heat transferring fluids. The thermal conductivities of fluids with suspended particles are expected to be higher than the pure ones.

By intuition, one expects that thermal conductivities of particle-fluid mixtures are higher than those of pure fluids, as pointed out by Maxwell [13] more than a century ago. However, due to the large size and high density of particle, there is no good way to prevent the solid particles from settling down. The lack of stability of such suspensions induces additional flow resistance and possible erosion.

Modern nanotechnology provides great opportunities to process and produce materials with average crystallite sizes below 50 nm. Recognizing an opportunity to apply this emerging nanotechnology to established thermal energy engineering, Argonne has developed the concept of a new class of heat transfer fluids called nanofluids, which transfer heat more efficiently than conventional fluids [14]. Nanofluids are engineered by suspending ultrafine metallic or nonmetallic particles of nanometer dimensions in traditional heat transfer fluids such as water, methanol and ethylene glycol. Particles are suspended by Brownian motions and when particles are in equilibrium with no flow, they are distributed in a balance between buoyant weight and thermal agitation (Brownian motion). The much larger relative surface area of nanoparticles, compared to those of conventional particles, should not only significantly improve heat transfer capabilities, but also should increase the stability of the suspensions. As a result nanofluid technology has succeeded in enhancing heat transfer properties of working fluids in the absence of aforementioned problems with ordinary suspensions [14].

Nanofluids are thought to be the next-generation heat transfer fluids, and they offer exciting possibilities due to their enhanced heat transfer performance compared to ordinary fluids. Nanofluids are proposed for various uses in important fields such as electronics, transportation and medicine [15].

Several existing published works mainly focus on prediction and measurement techniques of characterizing nanofluids. Copper oxide (CuO) and alumina (Al_2O_3) are the most common and inexpensive nanoparticles used by many researchers' experimental investigations.

Lee et al. [16] suspended CuO and Al_2O_3 with two different base fluids (water and ethylene glycol (EG)) and experimentally showed that nanofluids have substantially higher thermal conductivities than the original pure liquids. With 4 vol% CuO/EG mixture of nanoparticles, enhancement of more than 20% was observed. The effects of particle shape [17] and size [18,19] on the enhancement of the thermal conductivity of nanofluids were studies in other works. Xuan and Li [20] experimentally investigated flow and convective heat transfer characteristics of Cu/water nanofluids passing through a straight tube with a constant heat flux at wall. Results showed that the nanofluids provide substantial improvement of heat transfer rate compared to pure water. Maiga et al. [21] numerically studied the hydrodynamic and thermal characteristics of nanofluids flowing through a uniformly heated tube (L = 1m) in both laminar and turbulent regimes using the single phase model with adjusted properties. Observations showed that addition of nanoparticles can increase the heat transfer significantly compared to the base fluid alone. Zeinali et al. [22] investigated laminar flow of CuO/water and Al_2O_3/water nanofluids through a 1 m annular copper tube of 6 mm inner diameter and 0.5 mm thickness, and 32 mm diameter outer stainless steel tube. While constant wall temperature boundary condition was provided by circulating saturated steam, rather than constant heat flux condition studied by other researchers. The comparison of empirical results highlighted that the heat transfer coefficient improved with increasing volume fraction of nanoparticles as well as Peclet number while Al_2O_3/water showed greater improvement.

Heat transfer coefficients of Au/water, SiO_2/water, and SiO_2/ethylene glycol nanofluids under pool boiling conditions was measured and reported by Wang et al. [23]. From their study, Au/water nanofluids showed an increase in heat transfer rates compared to that of pure water while SiO_2/water and SiO_2/ethylene glycol nanofluids showed a reverse trend. Also, Tu et al. [24] found significant enhancement in pool boiling heat transfer with the Al_2O_3 nanofluid.

In contrast, one of the effective methods of enhancing the performance of TPCTs is to reduce thermal resistance by charging them with nanofluids, as proposed by Shung et al. [25] who used silver nanofluid in heat pipes. Noie et al. [26] prepared Nanofluids of aqueous Al_2O_3 nanoparticles suspensions in various volume concentration of 1~3% and used in a

TPCT as working media. They showed that for different input powers, the efficiency of the TPCT increases up to 14.7% when Al_2O_3/water nanofluid was used instead of pure water. Experimental investigations carried out by Baojin et al. [27] to study heat transfer characteristics of titanium /water and copper/water TPCT. Their experimental results showed that there are no remarkable differences of heat transfer coefficients in evaporator between the two kinds of TPCTs, whereas surprisingly the experimental results of heat transfer coefficient in condenser of Ti/H_2O TPCTs are about 2–3 times more than that of Cu/H_2O TPCTs. Tsai et al. [28] studied the effect of structural character of nanoparticles on heat pipe thermal performance and concluded that the thermal resistance of heat pipes with nanofluids was lower than those with distilled water. Naphon et al. [29] introduced a nanofluid in a heat pipe and emphasized that at the optimum conditions, the heat pipe with 0.1% nanoparticles concentration in refrigerant gives an efficiency of 1.40 times higher than that with a pure refrigerant. The effect of using a silver-nanofluid on thermal characteristics of a two-phase closed thermosyphons at the normal operating states was investigated by Pauramatethanuwat et al. [30]. They achieved approximately 20% better performance with nanofluid TPCT compared to the pure water one. Khandekar et al. [31] also studied thermal performance of a close two-phase thermosyphon charged with nanofluids and observed that nanofluids show inferior thermal performance than pure water. In addition, they claimed that an inferior heat transfer coefficient of nanofluids under pool boiling conquests high thermal conductivity of nanofluid. Kang et al. [32] employed aqueous suspension of 10 and 35 nm silver nanoparticles in a sintered circular heat pipe. At an input power of 30–50 W with the same charging volume, they demonstrated that the measured temperature difference for nanofluid filled heat pipes showed 0.56 ~0.65 °C declines, compared to DI-water filled ones. In addition, the input power for heat pipes with nanofluid as the working medium can be at the maximum of 70 W which is about 20 W higher than pure water loaded ones. For the heat pipe with 0.10% volume concentration of titanium nanoparticles, Naphon et al. [33] highlighted that the thermal efficiency was improved 10.60% higher than that with the base working fluid.

In the present study, CuO/water Nanofluid is employed as the working medium for conventional TPCT to investigate the efficiency improvement.

2. Experimental Setup

In order to study the thermal performance of a TPCT, a specific experimental setup has been designed and constructed and is presented schematically in Figure 2.

The thermosyphon consists of copper tube with 1000 mm length, 21 and 20 mm outer and inner diameters, respectively with evaporator section and condenser section 350 and 400 mm in length, respectively. To be ensured there are no remaining remnants of the earlier sample inside the device especially when there is a changeover from one nanofluid to another, it is necessary to have cleaned tube from inside prior starting any experiment. This was accomplished by three-cycle cleaning operation with trichloroethane, methanol and vacuum drying.

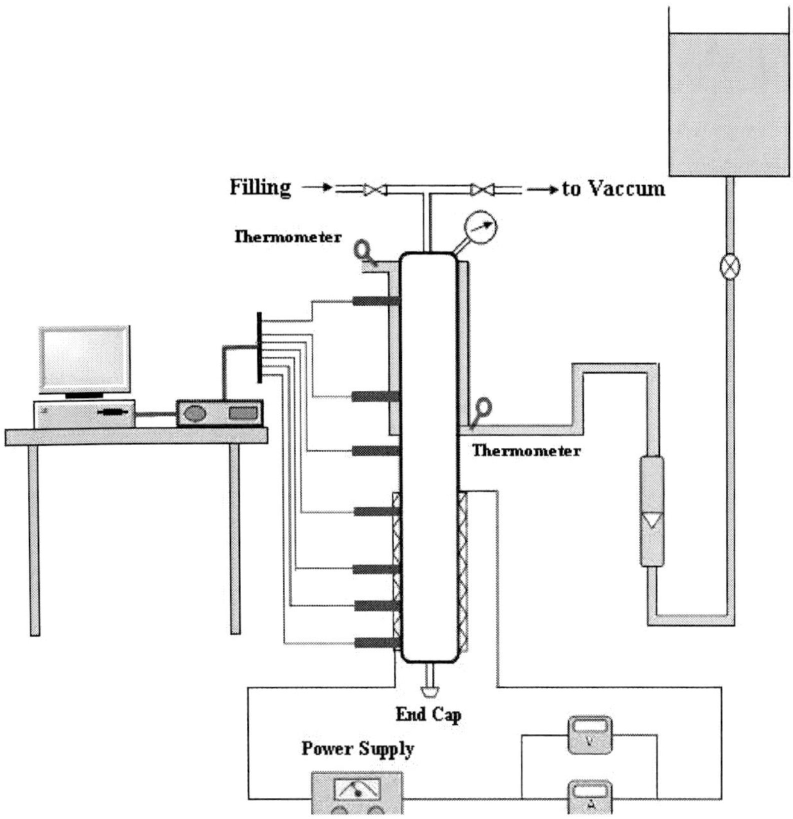

Figure 2. Schematic of experimental set-up.

The average diameter, thermal conductivity and actual density of CuO nanoparticles that used in this study are 30-50nm, 69 W/m.K and 6350 Kg/m3 respectively. The nanofluid in 1, 1.5, 2, 2.5 and 3 percent volume fraction were prepared. Preparation of nanoparticles suspensions is the first step in applying nanofluids in heat transfer enhancement. In the present study, CuO nanoparticles were dispersed in distillate water without using any dispersant or stabilizer, because of addition of any agent may change the chemical properties of the fluid. After preparing the required volume of the powder, using the equivalent weight of the solid, nanoparticles were mixed with distilled water in a flask. These solutions were subjected to ultrasonication for about four hours. No precipitation/settlement of nanoparticles was seen after 6 hours.

In order of reducing the heat loss to the surrounding, all sections of the thermosyphon were covered by rock wool. The local temperatures of the TPCT were measured by seven (type-LM35) thermocouples which were sandwiched in seven halved copper pipes and the TPCT. Data were monitored on a PC connected to a data-logger which was acquiring data from all the thermocouples. The temperatures of water inlet and outlet to the jacket were measured by two digital thermometers, "TESTO". The flow rate of cooling water was measured using a rotameter, while the operating pressure was read by a pressure gage connected to the upper part of the TPCT. All electrical equipments (such as thermocouples, thermometers, ammeter and voltmeter) and mechanical equipment (such as rotameter) were calibrated, initially. The thermocouples had a maximum precision of 0.1oC. Flow rates were measured using rotameter an uncertainty of 5.0%. The maximum precision of ammeter and voltmeter is 1 volts and 0.1 ampere, respectively. Thus the maximum uncertainty of efficiency evaluated is 5.41% [34].

There are various parameters that impose limitations and constraints to the steady and transient operation of a TPCT which the most important ones are the dry-out limitation, the Counter Current Flow Limitation (CCFL) or flooding and the boiling limitation. Dry-out limitation is usually not important unless the filling ratio is less than 15% [35]. If F.R is above this limit, dry-out phenomenon is not observed and other mechanisms like CCFL/flooding or Boiling Limitation (BL) limit the maximum heat flux [36].

Flooding is one of the most significant and a common limitation found in TPCTs and is evaluated based on a correlation proposed by Faghri [37] as follow:

$$Q_{\max} = Kh_{fg}A_{cross}\left[g\sigma(\rho_l - \rho_v)\right]^{0.25}\left[\rho_v^{-0.25} + \rho_l^{-0.25}\right]^{-2} \tag{1}$$

where:

$$K = \left(\frac{\rho_l}{\rho_v}\right)^{0.14} \tanh^2(Bo)^{0.25} \tag{2}$$

$$Bo = D_i\left[g(\rho_l - \rho_v)/\sigma\right]^{0.5} \tag{3}$$

The Boiling Limitation is observed in TPCT having large liquid filling ratio and high radial heat fluxes in the evaporator section. Under this limitation, at the critical heat flux, vapor bubbles coalesce near the pipe wall and prohibit contact of working liquid to wall surface resulting in rapid increase in evaporator wall temperature. For Boiling Limitation, the correlation suggested by Imura et al. [8] may be used in the following form:

$$Q_{max} = Ku\ h_{fg}A_{rad}\rho_v^{0.25}\left[g\sigma(\rho_l - \rho_v)\right]^{0.25} \tag{4}$$

The Equation (12) provides a good estimation for B.L. The Ku term in the above relation is calculated as follows:

$$Ku = 0.16[1 - exp\left\{\left(\frac{-D_i}{L_e}\right)\left(\frac{\rho_l}{\rho_v}\right)^{0.13}\right\}] \tag{5}$$

In our experiment for 3 vol% CuO/water mixture of nanoparticles, The CCFL/flooding and B.L calculated 2039W and 2310W respectively.

3. Experimental Results

The voltage and the current to the heater were measured by a voltmeter and an ampere meter respectively, and input power is calculated by the products as follow:

$$Q_{in} = VI \tag{6}$$

The amount of heat transferred to the coolant water can be calculated by the water temperature variation, taking into account the water flow rate and its specific heat, using following relation:

$$Q_{out} = mc_p(T_{out} - T_{in}) \tag{7}$$

The efficiency of TPCT can be expressed as a ratio of the output heat by condensation to the inlet heat by evaporation as follow:

$$\eta = \frac{Q_{out}}{Q_{in}} \tag{8}$$

To compare the efficiency improvement of the TPCT filled with nanofluids, the TPCT charged only by pure water at first and was also tested.

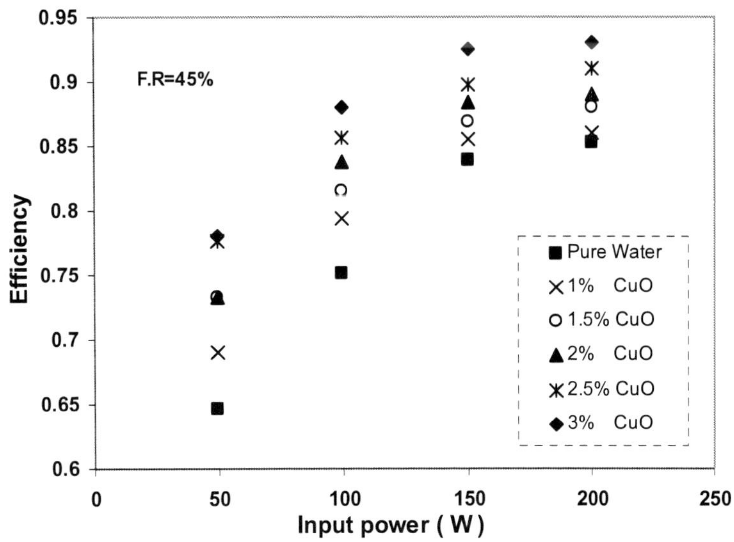

Figure 3. Efficiency of TPCT versus input power and concentration of nanofluid.

3.1. Effect of Nanofluid Concentration

Figure 3 shows the efficiency of TCPT versus the input power for pure water and different concentrations of nanofluids at F.R=45%. It was recognized that the efficiency improves as concentration of CuO nanoparticles increase. However, the efficiency difference between pure water and nanofluids is larger at lower input power. For example, at 50W, the efficiency of TPCT improved 13.36% by using 3 Vol. % of CuO nanoparticles in pure water, but at 200 W the efficiency improvement was only 7.78%. Also, there was significant enhancement on efficiency when the input power increased to 150W, but at higher input powers this was negligible comparably.

3.2. Effect of Filling Ratio

In order of examining the effect of filling ratio on efficiency improvement of the TPCT, the experimental results were plotted in Figure 4. The Figure presents the variation of efficiency against the filling ratio for pure water and three different concentrations of nanofluid. It was found that the efficiency for filling ratio 35% was the highest among other filling ratios for input power 100 W. Figure 4 shows that the efficiency improved as CuO nanoparticles concentration increased and maximum efficiency took place for 3 Vol. % concentration. For example, the efficiency difference of TPCT between pure water and nanofluid with 3% volume concentration at 15%, 23%, 35% and 45% filling ratios are 12.3%, 16.7%, 15.0% and13.6%, respectively.

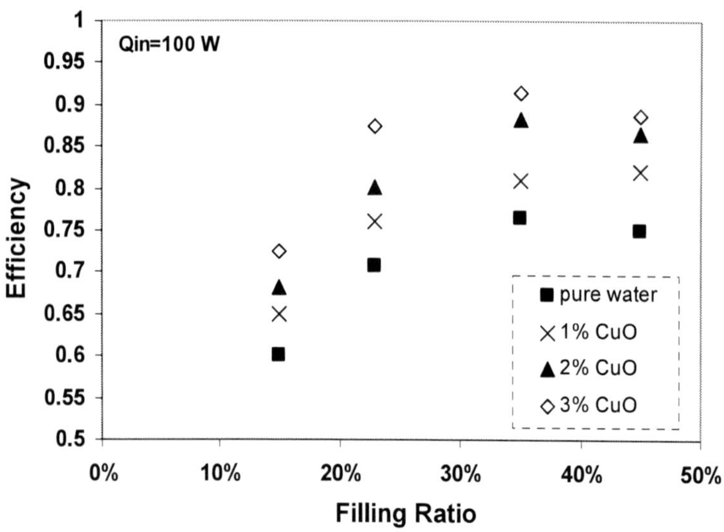

Figure 4. Efficiency of TPCT versus filling ratio and concentration of nanofluid.

3.3. Temperature Distribution along the Outside Wall of TPCT

In order to observe the temperature distribution over the entire length of the TPCT, the temperatures at four points on the evaporator section, one point on the adiabatic section, and two points on the condenser section, are simultaneously monitored. Figure 5(a-d) presents the temperature distribution along the TPCT contains pure water and nanofluids with different concentration for various input powers (50 W, 100 W, 150 W and 200 W).

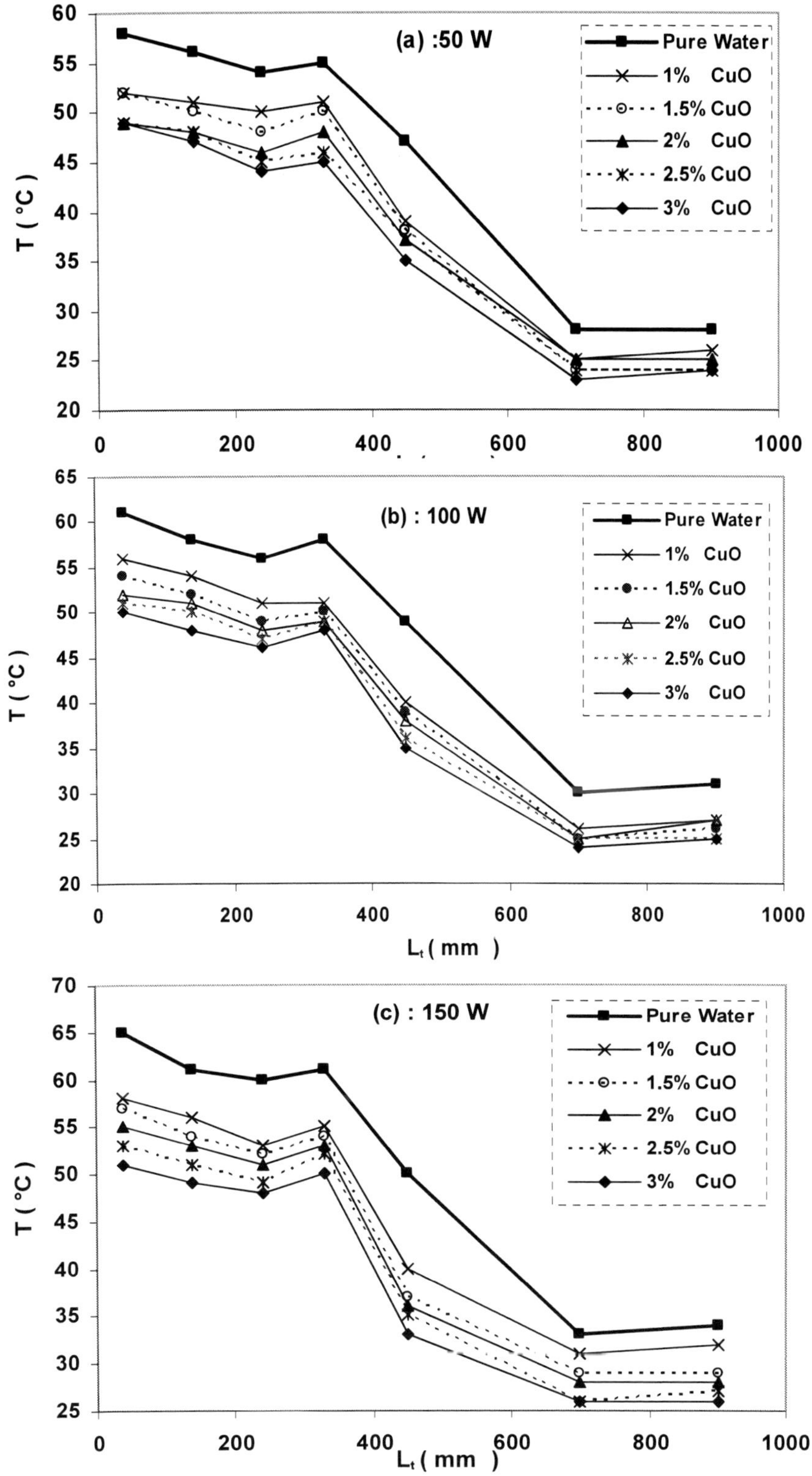
(a) :50 W
(b) : 100 W
(c) : 150 W
Pure Water
1% CuO
1.5% CuO
2% CuO
2.5% CuO
3% CuO
T (°C)
L_t (mm)

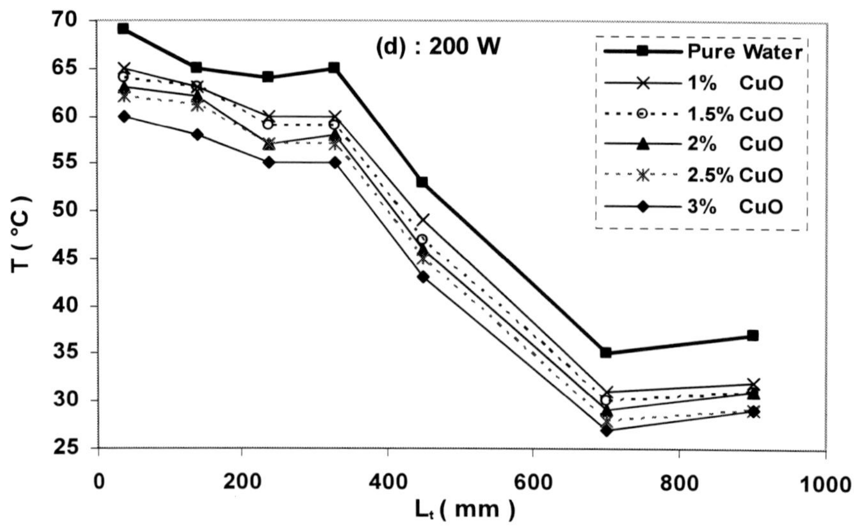

Figure 5(a-d). Average temperature of TPCT distribution versus input power and concentration of nanofluid.

At first the temperature of evaporator section decreases but there is a small growth in temperature in the end of evaporator section. When TPCT is charged with a nanofluid, temperature differences between evaporator and condenser are less than that of pure water. It confirms that thermal performance is better when a nanofluid is used in TPCT instead of pure water. It is concluded that a smaller rise in wall temperature is observed for TPCT with higher concentration of CuO nanoparticles in the working fluid compared to TPCT with pure water under various heat loadings. The trend of data for different concentration in evaporator section is not exactly the same because of the nanofluid boiling heat transfer performance. Vapor bubbles create in inside wall of evaporation section and the suspended nanoparticles tend to bombard the vapor bubbles during the bubble formation. Random movement and dispersion effect of nanoparticles inside evaporator section can affect the temperature profile along inside wall of TPCT.

The possible reason for improving the thermal performance of TPCT with the nanofluid as the working fluid can be explained as follows.

A major thermal resistance of TPCT is caused by the formation of vapor bubbles at the liquid–solid interface. A larger bubble nucleation size creates a higher thermal resistance that prevents the transfer of heat from the solid surface to the liquid [38]. The suspended nanoparticles tend to bombard the vapor bubbles during the bubble formation. Therefore, it is expected that the nucleation size of vapor bubble is much smaller for fluid with suspended nanoparticles than that without them. Also during nucleate boiling some nanoparticles deposit on the heater surface to form a porous layer. This layer improves the wettability of the surface considerably [39].

It is concluded that nanofluid has its challenge due to the limited research in this area. A comprehensive review in this new research frontier was carried out by Cheng et al. [40] which specially deals with the current research status of nanofluid two-phase flows and the research needs in this important area.

CONCLUSION

The main objective of the present investigation was to study the thermal performance of a two-phase closed thermosyphon (TPCT) with different concentrations of CuO/Water nanofluid (1~3%). In order to evaluate the heat transfer enhancement of nanofluid, if any, as compared to standard working fluid (degassed de-ionized water) an experimental setup of TPCT was designed, fabricated and examined. And the following statements are concluded:

1) Nanofluids in all concentrations that were studied showed better thermal performance than pure water. The efficiency of TPCT improved as concentration of CuO nanoparticles increased.
2) With filling ratio of 45%, the maximum efficiency improvement took place at 50W input power.
3) The best thermal performance of the TPCT was achieved at 35% filling ratio.
4) Temperature distributions along the TPCT were lower in magnitude with nanofluid than pure water. Temperature differences between evaporator and condenser section and thus thermal resistance of the thermosyphon loaded with nanofluids were less than that with pure water.

ACKNOWLEDGMENT

The authors thank the Iranian Fuel Conservation Company (IFCO) for financial supports of this project.

REFERENCE

Dunn, P. D. and Reay, D. A. *Heat Pipes*, third edition, Pergamon Press, 1994.

Noie, S. H. and Majideian, G. R. Waste heat recovery using heat pipe heat exchanger (HPHE) for surgery rooms in hospitals, *Applied Thermal Engineering* 20 (1999) 1271-1282.

Akbarzadeh, A., and Wadowski, T. Heat pipe-based cooling systems for photovoltaic cells under concentrated solar radiation, *Applied Thermal Engineering* 16 (1) (1997) 81-87.

Xiau, P., Johnson, P. and Akbarzaseh, A. A study of heat pipe heat exchanger effectiveness in an air conditioning application, 5th Int. *Heat Pipe Symposium,* Melbourne, Australia, 1996.

Liu, G. The application of heat pipe heat exchanger in exhaust gas heat recovery system and its thermodynamic analysis, 8th Int. *Heat Pipe Conference*, Beijing, China, (1992) 582-585.

Vasiliev, L., Lossouarn, D., Romestant, C., Alexandre, A., Bertin, Y., Piatsiushyk, Y. and Romanenkov, V. Loop heat pipe for cooling of high-power electronic components, *Int. J. of Heat and Mass Transfer* 52 (1-2) (2008) 301-308.

Peterson, G. P. *An Introduction to Heat Pipes, Modeling, Testing, and Applications,* John

Wiley & Sons, New York, 1994.

Imura, H., Ssasaguchi, K. and Kozai, H. Critical heat flux in a closed two phase thermosyphon, *Int. J. of Heat and Mass Transfer* 26 (8) (1983) 1181– 1188.

Harada, K., Inoue, S. and Fujita, J. Heat transfer characteristics of large heat pipe (in Japanese), *Hitachi Zosen Tech. Rev.* (41) 1980.

Noie, S. H. Heat transfer characteristics of a two-phase closed thermosyphon, *Applied Thermal Engineering* 25 (4) (2005) 495–506.

Feldman, K. T. and Srinivasan, R. *Investigation of heat transfer limits in two-phase closed thermosyphon*, 5th Int. Heat Pipe Conference, Tsukuba, Japan, 1984.

Hussein, H. M. S., El-Ghetany, H. H. and Nada, S. A. Performance of wickless heat pipe flat plate solar collectors having different pipes cross sections geometries and filling ratios, *Energy Conversation Management* 47 (2006) 1539–1549.

Maxwell, J. C. *A Treatise on Electricity and Magnetism*, second edition, Clarendon Press, Oxford, 1881.

S. U. S. Choi, Enhancing thermal conductivity of fluid with nanoparticles, Developments and Applications of Non-Newtonian Flows, *FED-V. 231*/ MD-V. 66, ASME, New York, (1995) 99-105.

Zussman, S. More about Argonne's stable, highly conductive nanofluids, Technology Transfer at Argonne, Public communication, Argonne National Laboratory, IL, USA, 2002.

Lee, S., Choi, S. U. S., Li, S. and Eastman, J. A. Measuring thermal conductivity of fluids containing oxide nanoparticles, *Int. J. of Heat Transfer* 121(1999) 280-289.

Hamilton, R. L. and Crosser, O. K. Thermal conductivity of heterogeneous two component systems, I *& Ec Fundamentals* (1) 1962.

Wang, X., Xu, X. and Choi, U. S. Thermal conductivity of nanoparticle-fluid mixture, *J. Thermophysics Heat Transfer* 13 (1999) 474-480.

Eastman, J. A., Choi, S. U. S., Li, S., Yu, W. and Thompson, L. J. Anomalously increased effective thermal conductivities of ethylene glycol-based nanofluids containing copper nanoparticles, *Applied Physics letters* 78 (6) (2001) 718-720.

Xuan, Y. and Li, Q. Investigation on convective heat transfer and flow features of nanofluids, *J. Heat Transfer* 125 (2003) 151–155.

Maiga, S. E. B., Nguyen, C. T., Galanis, N. and Roy, G. Heat transfer behaviours of nanofluids in a uniformly heated tube, *Superlattices and Microstructures* 35 (2004) 543–557.

S. Zeinali Heris, S. Gh Etemad. and M. Nasr Esfahany, Experimental investigation of oxide nanofluids laminar flow convective heat transfer, *Int. Communications Heat Mass Transfer* 33 (2006) 529–533.

Wang, X. and Mujumdar, A. S. Heat transfer characteristics of nanofluids: A review, *Int. J. of Thermal Science* 46 (2007) 1–19.

Tu, J. P., Dinh, N. and Theofanous, T. An experimental study of nanofluid boiling heat transfer, In: Proc. of 6th *Int. Symposium on Heat Transfer,* Beijing, China, 2004.

Shung, W. K. and Wei, W. C. Experimental Investigation of silver nanofluid on heat pipe thermal performance, Prooceedings of Tamkang University, *Int. Nano and MEMS Workshop*, Thaiwan, 2005.

Noei, S. H., Zeinali Heris, S., Kahani, M. and Nowee, S. M. Heat transfer enhancement using Al_2O_3/Water nanofluid in a two-phase closed thermosyphon, *Int. J. Heat and Fluid Flow*, 30 (4) (2009) 700-705.

Baojin, Q., Li, Z. and Yan, S. Heat transfer characteristics of titanium/water two-phase closed thermosyphon, *Energy Conversion and Management*, in press.

Tsai, C. Y., Chien, H. T., Ding, P. P., Chan, B., Luh, T. Y. and Chen, P. H. Effect of structural character of gold nanoparticles in nanofluid on heat pipe thermal performance, *Materials Letters* 58 (9) (2004) 1461-1465.

Naphon, P., Thongkum, D. and Assadamongkol, P. Heat pipe efficiency enhancement with refrigerant-nanoparticles mixtures, *J. of Energy Conversion Management.* in: Proc, 2008.

Pauramatethanuwat, T., Rittidech, S. and Booddachan, K. Effect of inclination angles and aspect ratios on heat transfer characteristics of a two- phase closed thermosyphon Using De Ionize Water Mix Silver Nano, 9th *Int. Heat Pipe Symposium,* Malaysia, 2008.

Khandekar, S., Joshi, Y. and Mehta, B. Thermal performance of closed two-phase thermosyphon using nanofluids, *Int. J. Thermal Science* 47 (2007) 59-667.

Kang, S. W., Wei, W. C., Tsai, S. H. and Huang, C. C. Experimental investigation of nanofluids on sintered heat pipe thermal performance, *Applied Thermal Engineering* in: Proc, 2008.

Naphon, P., Assadamongkol, P. and Borirak, T. Experimental investigation of titanium nanofluids on the heat pipe, *Int. Communications in Heat and Mass Transfer* 35 (10) (2008) 1316-1319.

Holman, J. D. *Experimental methods for engineers,* fifth edition, Mc. Grow-Hill, New York. 1989.

Park, Y. J., Kang, H. K. and Kim, C. J. Heat transfer characteristics of a TPCT to the fill charge ratio, *Int. J. of Heat and Mass Transfer* 45 (2002) 4655-4661.

Faghri, A. *Heat Pipe Science and Technology,* Taylor & Francis, London, 1995.

Faghri, A., Chen, M. M., and Morgan, M. Heat transfer in two-phase closed conventional and concentric annular thermosyphon, *Int. J. Heat and Mass Transfer* 1 (1) (1989) 611-618.

Collier, J. G. and Thome, J. R. *Convective boiling and condensation*, Clarendon Press, Oxford, 1996.

Kim, S. J., Bang, I. C., Buongiorno, J. and Hu, L. W. Surface Wettability Change during Pool Boiling of Nanofluids and its effect on Critical Heat Flux, *Int. J. Heat and Mass Transfer* 50 (2007) 4105-4116.

Cheng, L., Bandarra Filho Enio, P. and Thome John, R. Nanofluid two-phase flow and thermal physics: a new research frontier of nanotechnology and its challenges, *International Journal of Nanoscience and Nanotechnology* 8 (2008), 3315–3332.

ISBN 978-1-61324-665-8

Chapter 10

STUDY OF VOID FRACTION IN SUBCOOLED FLOW BOILING THROUGH VERTICAL NARROW ANNULAR CHANNELS

Shunyu Su[1*]***, Suyi Huang*** [2]***, Chunzhi Zhang***[1] ***and Jian Chen***[1]

[1]College of Urban Construction, Wuhan University of Science and Technology, No. 199 Xiongchu Avenue, Wuhan 430070, P.R.China

[2]School of Energy and Power Engineering, Huazhong University of Science and Technology, Wuhan 430074, P.R.China

ABSTRACT

Using distilled water as the working fluid, the experiments with dual tube heating were made under two types of test section. The values of the onset of nucleate boiling Z_{ONB} could be measured in the experiments. And the values of the start of fully developed subcooled boiling Z_{FDB} were calculated by Bowring method. Differential equations for the relevant parameters such as true mass quality x, liquid subcooled temperature θ_l, distance along the heated channel measured from the inlet Z and void fraction α were developed by applying Zuber-Findlay correlation and improved Rouhani-Axelsson mechanistic model which considered the influence of mass flux. The fourth-order Runge-Kutta method was used to integrate the relevant equations simultaneously along the heating section while the experimental data were the initial conditions. The calculation results of the improved correlation were compared with the data of numerical calculation and agreed well with it while the deviation was less than ±12%

Keywords: Void fraction; Subcooled flow boiling; Mechanistic model; Correlation.

* Corresponding author: Tel: +86-027-68893576; fax: +86-027-68893641, E-mail: shunyusu@163.com

Nomenclature

A	flow area in the channel (m^2)
c_p	specific heat (J kg^{-1} K^{-1})
D_h	hydraulic diameter (m)
d	diameter (m)
G	mass flux (kg m^{-2} s^{-1})
g	gravitational acceleration (m s^{-2})
h	heat transfer coefficient (W m^{-2} K^{-1})
i_{fg}	latent heat of vaporization (J kg^{-1})
k	thermal conductivity (W m^{-1} K^{-1})
$\dot{m}_s$	mass flux which converted into steam (kg m^{-2} s^{-1})
N_q	middle variable defined by Eq. (18)
P	system pressure (Pa)
P_h	heated perimeter (m)
Pr	Prandtl number
Q	amount of heat (W)
q	heat flux (W m^{-2})
Re	Reynolds number
T	absolute temperature (K)
U	wet perimeter (m)
u_l	liquid velocity (m s^{-1})
x	true mass quality
Z	distance from the tube inlet (m)

Greek Letters

α	void fraction
η	empirical factor
θ	subcooled temperature (K)
μ	dynamic viscosity (Pa s)
ρ	density (kg m^{-3})
σ	surface tension of liquid (N m^{-1})
ϕ	diameter (m)

Subscripts

b	boiling
c	condensation
FD	fully developed subcooled boiling
B	vapor
g	inner; iterative time
i	inlet
in	liquid
l	outer

	o	onset of nucleate boiling
	ON	saturation
B		subcooled boiling
	sat	wall
	sub	
	w	

1. INTRODUCTION

Boiling phenomena usually occur in heated liquid. Boiling bubbles can grow and detach freely in a large channel. But in a narrow channel, boiling bubbles suffer more restriction of space and surface force. Thome [1] gave an overview of experiment and theory which addressed the topics of macroscale versus microscale boiling heat transfer. Ghiaasiaan and Chedester [2] suggested that boiling incipience in microchannels was controlled by the thermocapillary force that tended to suppress the microbubbles which formed on wall cavities. Haynes and Fletcher [3] concluded that the heat transfer coefficient in narrow passages could be described accurately as a simple additive combination of single-phase liquid-only convection heat transfer and nucleate boiling. Steiner *et al.* [4] proposed a Chen-type superposition model to compute the effective wall heat flux in subcooled boiling flow. They suggested the focus should have to be put on the consideration of the bubble-bubble as well as the bubble-liquid interactions in the fully developed boiling (FDB) region. Ahmadi *et al* [5] developed a mechanistic model for the prediction of onset of nucleate boiling wall heat flux for subcooled flow boiling. They considered that all the wall energy is first transferred to the superheated liquid layer adjacent to the wall, from where a fraction of it goes to vaporization.

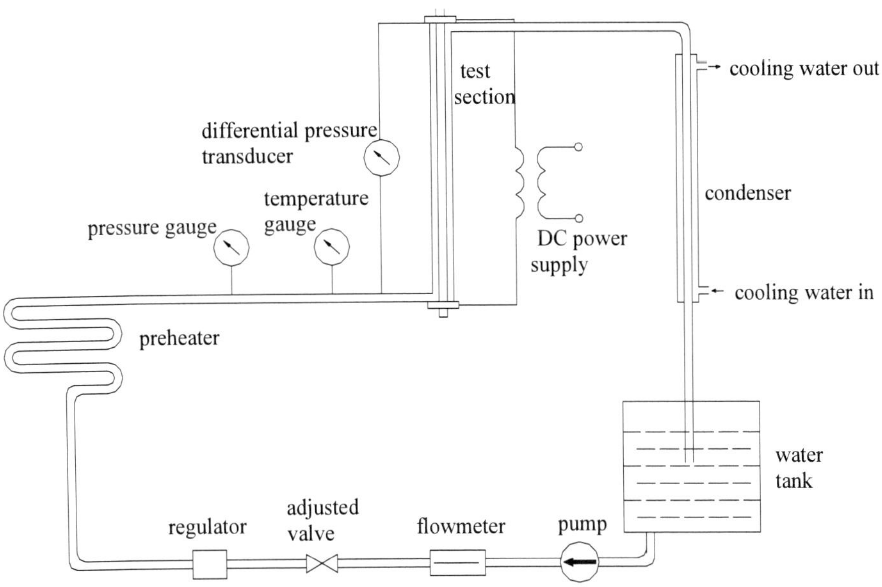

Figure 1. Schematic diagram of the experimental system.

Subcooled flow boiling exists in many industrial areas, and void fraction is the important parameter in subcooled boiling heat transfer [6,7]. Void growth models can be separated into two categories: profile-fitting models and mechanistic models. Profile-fitting models predict void growth successfully at high pressure and mechanistic models assume the variation of certain parameters introduced during the modeling of void growth [8]. The earlier models were identified in a review of literature [9]. To improve the prediction capabilities of void fraction models, Harms *et al* [10] developed a mechanistic void fraction model for annular flow in horizontal tubes. Zeitour and Shoukri [11] presented a mechanistic model to predict the axial void fraction profile. And the significance of the accurate estimation of condensation rate was emphasized in literatures [12-14].

The purpose of this paper is to analyze the characteristics of void fraction in subcooled flow boiling through narrow channels. Rouhani-Axelsson mechanistic model [15] was commonly used to calculated void fraction in subcooled boiling regions. But the correlation of void fraction of fully developed subcooled boiling didn't consider the influence of mass flux. In this paper, an improved Rouhani-Axelsson mechanistic model was presented while the effect of mass flux was taken account of. And the location of bubble detachment was determined by Bowring's correlations [16].

2. Experiments

2.1. Experimental System

A schematic diagram of the experimental system is presented in Figure 1. The distilled water was driven from the water tank by a pump and regulated by the adjusted valve and regulator. A standard hole-plate flowmeter, calibrated to an accuracy of ±0.5%, was employed to measure mass flux. After passing the preheater, subcooled water was led into the test section in which it gradually became saturated water, steam and superheated vapour. The temperature and pressure of inlet flow were measured by temperature and pressure gauge respectively. The pressure drop of two-phase flow was obtained by a CECC-650G type capacitive pressure transducer calibrated to an accuracy of ±0.5%. After passing through the test section, the vapour flew into a condenser in which it was condensed to liquid water. Measured data including mass flux, pressure and temperatures along the inner and outer tubes were acquired, displayed and processed by FLUKE-2645 data acquisition system and a PC computer.

The experiments with the loop of high temperature and high pressure were carried out within the following ranges: mass flux 45-180kg/m^{2}s, system pressure 2-3.5MPa, temperature of inlet water 50-180℃, heat flux 40-210kW/m^2.

2.2. Test Section

Two tubes with different diameters formed a vertical annular channel in the test section as Figure 2 shows. Table 1 shows test section dimensions. The inner tube and outer tube were made of stainless steel and heated simultaneously by a high-current-regulated DC power supply. The experimental electric current was measured by a T24-AV type amperemeter calibrated to an accuracy of ±0.2%. The experimental voltage was measured by a HP3428A type digital voltmeter calibrated to an accuracy of ±0.2%. The test tubes were composed of two types. Table 1 presents the test section dimensions. Precautions were taken to achieve uniform distribution of heat flux between the two tubes in each test section.

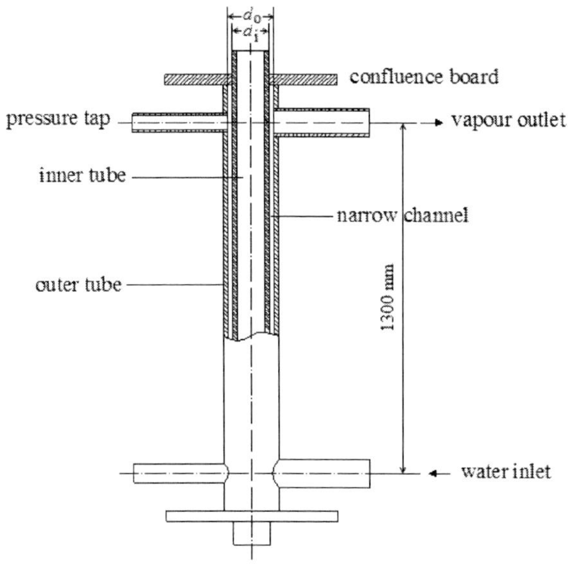

Figure 2. Schematic diagram of the test section.

The temperature of inlet water was measured by metal clad thermocouple (K-type) of ϕ 1mm. The inner wall temperatures of inner tube were measured by metal clad thermocouple (K-type) of ϕ 1mm, which was dragged along the inner wall. The outer wall temperatures of outer tube were measured by the chromel-alumel thermocouples (K type) of ϕ 0.3mm, which were shot welded in the outer wall along the tube and reinforced by inorganic adhesives. The accuracy of all thermocouples used in the experiments was ±0.2℃.

Table 1. Test section dimensions (mm)

Outer tube	Inner tube	Length	D_h
ϕ 12×1	ϕ 8×1.2	1300	2
ϕ 13×1.5	ϕ 7×1	1300	3

2.3. Measurement of ONB Point

The onset of nucleate boiling (ONB) point could be identified as the point where the wall temperature changed noticeably before the saturated section of the wall temperature curve. Figure 3 shows the experimental results of the wall temperature varying along the test tube. Since the heat transfer coefficient increased dramatically when subcooled boiling started, and that resulted in a sharp drop of wall temperature under the condition of a constant surface heat flux, the ONB point occurred at the first turning point in the wall temperature curve, which was shown in Figure 3.

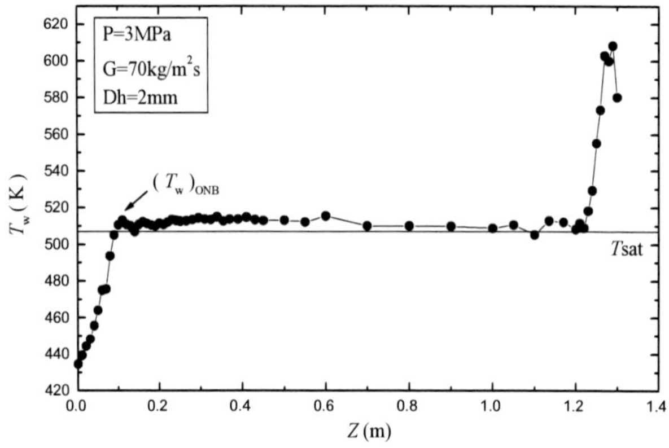

Figure 3. Wall temperature along the tubes.

3. Mechanistic Void Fraction Model

3.1. Subcooled Boiling Regions

Subcooled boiling starts after the heating wall temperature exceeds the saturation temperature enough to cause nucleation at a point called the onset of nucleate boiling (ONB). The length from the tube inlet to the point of ONB, Z_{ONB}, could be measured in the experiments. After Z_{ONB}, the fluid flow enters into highly subcooled region in which evaporation is essentially a wall effect. The generated vapour remains as discrete bubbles attached to the wall surface where bubbles growing and collapsing. At the fully developed subcooled boiling Z_{FDB}, bubbles begin to detach the wall surface. The fluid flow then enters into low subcooled region in which bubbles grow and detach, condensing only slowly as they pass through the slightly subcooled liquid. Figure 4 shows the subcooled boiling regions.

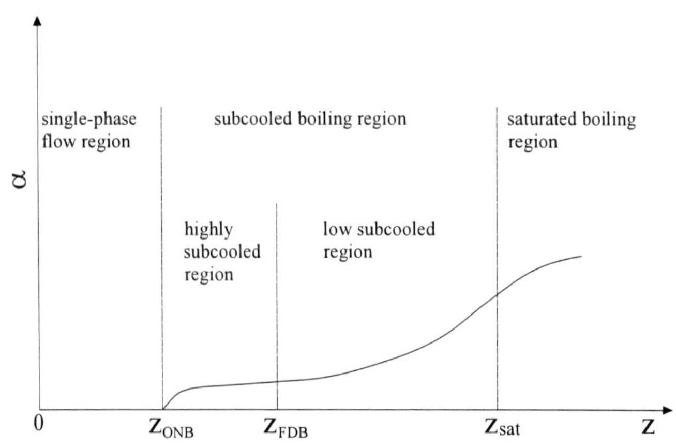

Figure 4. Schematic diagram of subcooled boiling regions.

The heat equilibrium up to distance Z from the test section inlet gives [17],

$$q\pi D_h Z = \frac{1}{4}\pi D_h^2 G c_{pl}\left(T_Z - T_{in}\right) \tag{1}$$

$$Z = \frac{G D_h c_{pl}}{4q}\left(T_Z - T_{in}\right) \tag{2}$$

or

$$Z = \frac{G D_h c_{pl}}{4q}\left[\left(T_{sat} - T_{in}\right) - \left(T_{sat} - T_Z\right)\right] \tag{3}$$

therefore,

$$Z = \frac{G D_h c_{pl}}{4q}\left[\left(\Delta T_{sub}\right)_{in} - \left(\Delta T_{sub}\right)_Z\right] \tag{4}$$

where

$$D_h = \frac{4A}{U} = \frac{\pi\left(d_o^2 - d_i^2\right)}{\pi\left(d_o + d_i\right)} = d_o - d_i \tag{5}$$

The length from the tube inlet to the start of fully developed subcooled boiling, Z_{FDB}, is given by Bowring method [16]:

$$\left(\Delta T_{sub}\right)_{FDB} = \eta q / (u_l)_{in} \tag{6}$$

or

$$\left(\Delta T_{sub}\right)_{FDB} = \eta q \rho_l / G \tag{7}$$

While the working fluid is water, then

$$\eta = \left(14 + 0.987P\right) \times 10^{-6} \tag{8}$$

According to equation (4),

$$Z_{FDB} = \frac{G D_h c_{pl}}{4q}\left[\left(\Delta T_{sub}\right)_{in} - \left(\Delta T_{sub}\right)_{FDB}\right] \tag{9}$$

According to equation (2),

$$Z_{sat} = \frac{GD_h c_{pl}}{4q}\left(T_{sat} - T_{in}\right) \tag{10}$$

3.2. Equations

For single-phase liquid, we use Dittus-Boelter correlation which gives

$$h = 0.023\,\mathrm{Re}^{0.8}\,\mathrm{Pr}^{0.4}\,\frac{k_l}{D_h} \tag{11}$$

In subcooled boiling region, the heat balance on the heated wall is [15]

$$q = h\theta_l\left(1 - \frac{\alpha}{\alpha_{FDB}}\right) + \dot{m}_s i_{fg} + \frac{\dot{m}_s}{\rho_g} c_p \rho_l \theta_l \tag{12}$$

Void fraction of subcooled boiling is calculated by Zuber-Findlay correlation [18]:

$$\alpha = \frac{x}{\rho_g}\left\{1.54\left[\frac{x}{\rho_g} + \frac{1-x}{\rho_l}\right] + \frac{1.18}{G}\left[\frac{g\sigma\left(\rho_l - \rho_g\right)}{\rho_l^2}\right]^{\frac{1}{4}}\right\}^{-1} \tag{13}$$

The amount of heat which goes to steam generation within dz along the channel will be:

$$\begin{aligned} dQ_b &= \dot{m}_s i_{fg} P_h dz \\ &= \frac{q - h\theta_l\left[1 - \left(\alpha / \alpha_{FDB}\right)\right]}{\rho_g i_{fg} + c_p \rho_l \theta_l} \rho_g i_{fg} P_h dz \end{aligned} \tag{14}$$

In low subcooled region, namely $\alpha > \alpha_{FDB}$, the term containing h must be eliminated as for equation (12) and equation (14).

The amount of heat which goes to the subcooled liquid through condensation of vapour bubbles within dz may be expressed as:

$$dQ_c = k_c \theta_l dz \tag{15}$$

According to literature[9], k_c could be expressed as

$$k_c = 30\frac{k_l}{\mathrm{Pr}}\left(\frac{\rho_g}{\rho_l}\right)^2 A^{\frac{2}{3}}\alpha^{\frac{2}{3}}.\frac{(\mathrm{Re})_l}{N_q^{0.5}} \tag{16}$$

where

$$(\mathrm{Re})_l = \frac{GD_h}{(1-\alpha)\mu_l} \tag{17}$$

$$N_q = \frac{q\mu_l}{i_{fg}(\rho_l - \rho_g)\sigma} \tag{18}$$

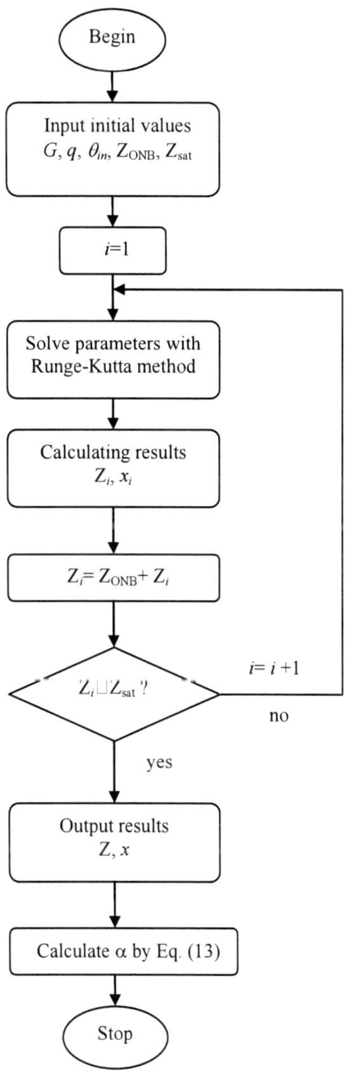

Figure 5. Flow chart of the simulation program for α in subcooled boiling.

The differential change of true mass quality gives as:

$$dx = \frac{dQ_b - dQ_c}{GAi_{fg}} \tag{19}$$

The differential change of subcooled temperature of liquid gives as:

$$-d\theta_l = \frac{qP_h dz - (dQ_b - dQ_c)}{GAc_p} \tag{20}$$

3.3. Solution

When the experimental data, the onset of nucleate boiling Z_{ONB} and the inlet liquid subcooled temperature θ_{in}, were the initial conditions, the fourth-order Runge-Kutta method was used to integrate the relevant equations (11)-(15) and (19)-(20) simultaneously along the Z axis. Since the calculation is started at Z=0 in equations (19)-(20), where is the point of ONB, the actual length from the inlet should add Z_{ONB} as it's shown in Figure 4. The fully developed subcooled boiling Z_{FDB} was calculated by equation (9). The void fraction α was calculated by equation (13) after the true mass quality x was obtained from above integrating calculation. Figure 5 shows the flow chart of the simulation program for void fraction α in subcooled flow boiling through narrow channels.

4. RESULTS

The fully developed subcooled boiling Z_{FDB} is the division between highly subcooled region and low subcooled region. For water as the boiling medium, the calculation of α_{FDB} was proposed by Rouhani [19]:

$$\alpha_{FDB} = 2.435 \times 10^{-3} P^{-0.237} \frac{P_h}{A} \tag{21}$$

From present calculated results, the value of α_{FDB} is also related to mass flux G. Figure 6 shows the influence of mass flux G on α_{FDB}. It indicates that, the fully developed subcooled boiling α_{FDB} decreases with increasing system pressure as it's shown in equation (21). But at a given pressure, α_{FDB} is higher for a higher mass flux G in both gap sizes of D_h=2mm and D_h=3mm. So it is reasonable to take account of the effect of mass flux G. From the experimental data, α_{FDB} can be expressed as

$$\alpha_{FDB} = 1.746 G^{0.469} P^{-0.237} \left(\frac{P_h}{A} \right)^{-0.066} \tag{22}$$

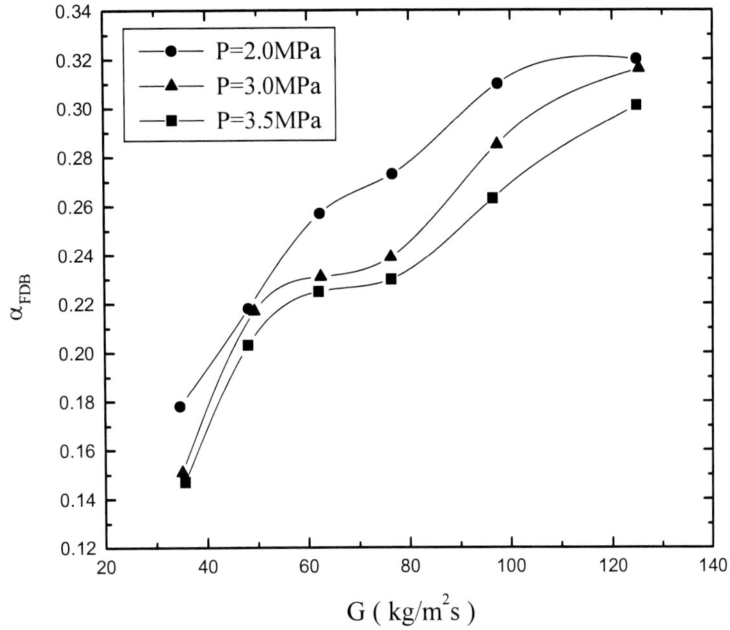

(a) D_h=2mm

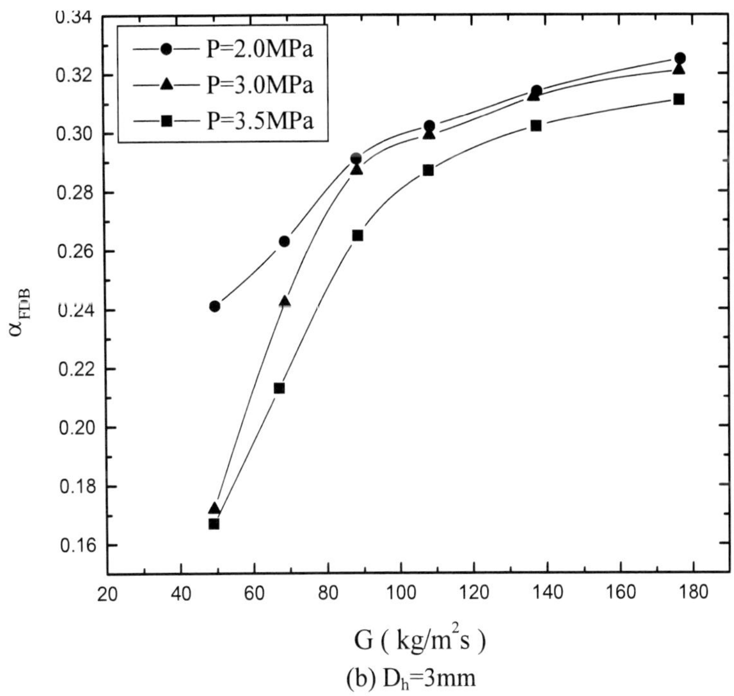

(b) D_h=3mm

Figure 6. The influence of mass flux G on α_{FDB}.

Figure 7 shows the comparison of α_{FDB} between improved correlation and numerical calculation with the deviation within±12%. The data from above two methods agree well with each other. It means equation (22) can be used to predict the value of fully developed subcooled boiling void fraction α_{FDB}.

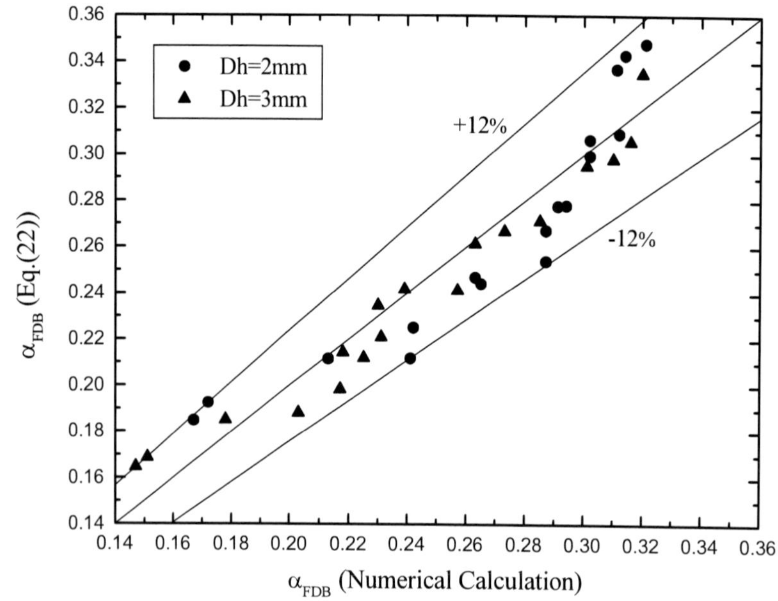

Figure 7. Comparison of α_{FDB} between improved correlation and numerical calculation.

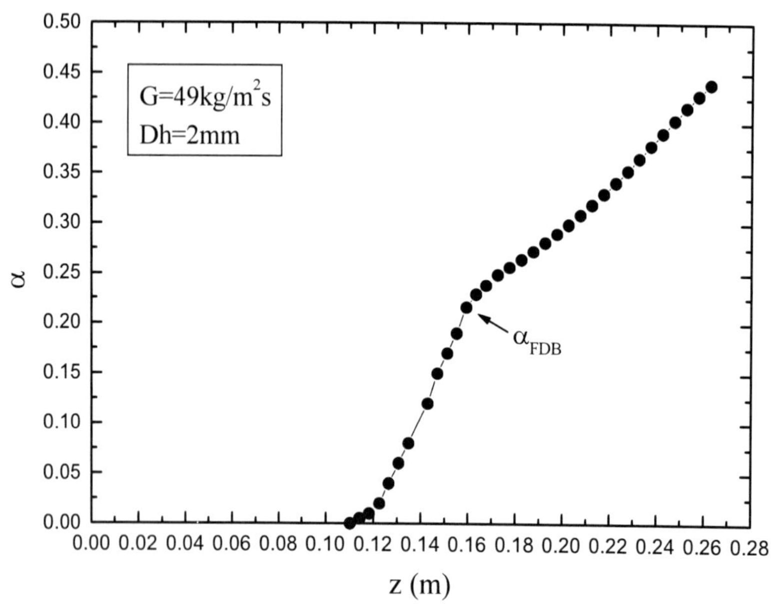

Figure 8. Curve of void fraction α along axial direction.

Figure 8 shows the curve of void fraction α along axial direction with system pressure P=2MPa. The first point of the curve which is in the Z axis is ONB point. Void fraction α turns at the point of fully developed subcooled boiling as it's shown in Figure 8. The corresponding value in the Z axis is Z_{FDB}. α_{FDB} is the turning point of subcooled boiling fluid from highly subcooled boiling region to low subcooled boiling region in the curve.

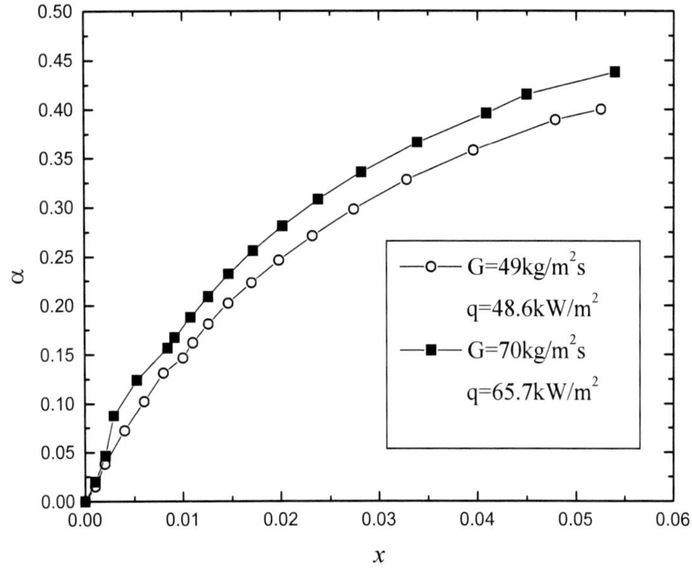

Figure 9. The influence of mass flux and heat flux on α.

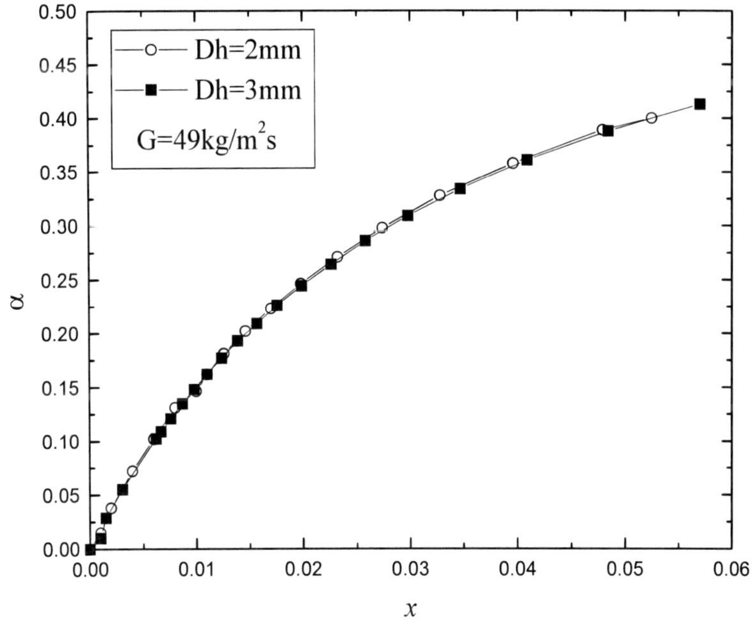

Figure 10. The influence of gap size on α.

Figure 9 shows the influence of mass flux and heat flux on void fraction of subcooled boiling. It indicates that the void fraction increases as mass flux and heat flux increase at a given gap size. In experimental process, heat flux always increases with increasing mass flux.

Otherwise, the boiling phenomena may not occur in test section while inputting amount of heat is too low. The results in Figure 9 are consistent with Zuber-Findlay correlation.

Figure 10 shows the influence of gap size on void fraction of subcooled boiling with mass flux G=49kg/m^{2}s. It indicates that the void fraction curves with different gap sizes are nearly overlapped. It is consistent with Zuber-Findlay correlation. Then it is concluded that gap size has little influence on void fraction of subcooled boiling flow at a given mass flux.

Conclusion

The improved Rouhani-Axelsson mechanistic model could be used to predict the void fraction in subcooled flow boiling through vertical narrow annular channels within the experimental ranges: heat flux, 40-210 kW/m^2; mass flux, 45-180 kg/m^{2}s; system pressure, 2-3.5MPa; gap size, D_h=2mm and D_h=3mm. The deviation is less than ±12%.

Void fraction increases as mass flux and heat flux increase at a given gap size. Gap size has little influence on void fraction of subcooled boiling flow at a given mass flux.

References

[1] Thome, J. R. Boiling in microchannels: a review of experiment and theory. *Int. J. Heat and Fluid Flow* 25 (2004) 128-139.

[2] Ghiaasiaan S. M. and Chedester, R. C. Boiling incipience in microchannels. *Int. J. Heat Mass Transfer* 45 (2002) 4599-4606.

[3] Haynes, B. S. and Fletcher, D. F. Subcooled flow boiling heat transfer a narrow passages. *Int. J. Heat Mass Transfer* 46 (2003) 3673-3682.

[4] Steiner, H., Kober, A. and Gebhard, L. A wall heat transfer model for subcooled boiling flow. *Int. J. Heat Mass Transfer* 48 (2005) 4161-4173.

[5] Ahmadi, R., Borujerdi, A. N., Jafari, J. *et al*, Experimental study of onset of subcooled annular flow boiling. *Progress in Nuclear Energy* 51 (2009) 361-365.

[6] Aoki, S., Inoue, A., Aritomi, M. and Sakamoto, Y. Experimental study on the boiling phenomena within a narrow gap. *Int. J. Heat Mass Transfer* 25 (7) (1982) 985-990.

[7] Zuber, N., Staub, F. W. and Bijwaard, G. Vapour void fraction in subcooled boiling and in saturated boiling systems, III International Heat Transfer Conference, *ASME*-153, 1966.

[8] Bibeau E. L. and Salcudean, M. Subcooled void growth mechanisms and prediction at low pressure and low velocity. *Int. J. Multiphase Flow* 20 (5) (1994) 837-863.

[9] Rice, C. K. The effect of void fraction correlation and heat flux assumption on refrigerant charge inventory predictions, *ASHRAE Trans.* 93 (1) (1987) 341-367.

[10] Harms, T. M., Li, D., Groll, E. A. *et al*, A void fraction model for annular flow in horizontal tubes, *Int. J. Heat Mass Transfer* 46 (2003) 4051-4057.

[11] Zeitour, O. and Shoukri, M. Axial void fraction profile in low pressure subcooled flow boiling, *Int. J. Heat Mass Transfer* 40 (4) (1997) 869-897.

[12] Zeitour, O., Shoukri, M. and Chatoorgoon, V. Measurement of interfacial area concentration of subcooled liquid-vapour flow, *Nuclear Engineering and Design* 152 (1994) 243-255.
[13] Zeitour, O., Shoukri M. and Chatoorgoon, V. Interfacial heat transfer between steam bubbles and subcooled water in vertical upward flow. *J. Heat Transfer* 117 (1995) 402-407.
[14] Lai, J. C. and Farouk, B. Numerical simulation of subcooled and low quality forced convection boiling flows. ANS Proceedings, *National Heat Transfer Conference,* vol. 6, 1992, pp. 12-20.
[15] Rouhani, S. Z. and Axelsson, E. Calculation of void volume fraction in the subcooled and quality boiling regions, *Int. J. Heat Mass Transfer* 13 (1970) 383-393.
[16] Bowring, R. W. Physical model based on bubble detachment and calculation of steam voidage in the subcooled region of a heated channel, *OECD Halden Reactor Project Report HPR*-10, 1962.
[17] Collier, J. G. *Convective boiling and condensation*, McGraw-Hill, New York, 1972, pp. 129-131.
[18] Zuber, N. and Findlay, Average volumetric concentration in two-phase flow systems, *J. Heat Transfer* 87 (Ser. C) (1965) 453-462.
[19] Rouhani, S. Z. Calculation of steam volume fraction in subcooled boiling, *J. Heat Transfer* 90 (1968) 158-164.

In: Frontier Research in Microscale and Nanoscale... ISBN 978-1-61324-665-8
Editor: Lixin Cheng

Chapter 11

TRSL LABORATORY DESCRIPTION

Mamoru Ishii and Takashi Hibiki*

Thermal-hydraulics and Reactor Safety Laboratory, School of Nulcear Engineering, Purdue University, West Lafayette, IN 47907-2017, U. S.

LABORATORY ORGANIZATION AND OVERVIEW

The Thermal-hydraulics and Reactor Safety Laboratory (TRSL) is a part of the Purdue University School of Nuclear Engineering and oversees several smaller laboratories. The research team at TRSL is led by two faculty members, Distinguished Professor Mamoru Ishii and Professor Takashi Hibiki. Working with these professors are a number of doctoral research associates and approximately 15 graduate research assistants working on specific research projects. Four full-time machinists work in the machine shop, which is attached to the laboratory. TRSL itself consists of extensive experimental loops, reactor safety systems, and advanced instrumentation systems supported by its own computer facility. The Institute of Thermal Hydraulics (ITH), which operates in collaboration with several other universities, was established in 1997 by the US Nuclear Regulatory Commission (NRC) to provide broad technical expertise and perform thermal hydraulics and reactor safety research in the areas of modern computer code architecture, advanced numerical methods, new two-phase flow models, constitutive models, graphical user interfaces, computational fluid dynamics (CFD), and fundamental experiments. The Purdue University Multidimensional integral test Assembly (PUMA) is the only existing US Government integral test facility for boiling water reactors (BWRs). It is designed to simulate the thermal hydraulic phenomena that may occur during design-basis accidents and other transients in innovative BWR designs that include various passive safety systems. Shown below is a chart that details the organizational structure of the laboratory.

* Email: hibiki@purdue.edu

Laboratory Capability I: Innovative Instrumentation

The laboratory's instrumentation capability permits very detailed experimental studies of complex multi-phase flow and heat transfer phenomena. TRSL is equipped with conventional instruments for thermal hydraulics experiments such as computerized high-speed data acquisition systems, high speed cameras, high resolution camcorders, and transient temperature, pressure, and flow rate measurement units. Some of these were developed at TRSL with advanced design, excellent functionality, and high accuracy results.

- 2-sensor conductivity probe to measure interfacial area concentration and void fraction
- 4-sensor conductivity probe to measure interfacial area concentration and void fraction
- Isokinetic droplet sampling system to measure droplet size distribution
- Ring-type void fraction sensor to measure void fraction
- Multi-electrode sensor to measure flow pattern identification
- Water film thickness probe to measure water film thickness
- Liquid metal film thickness probe to measure liquid metal film thickness
- Two-phase flow diagnostic system to measure flow regimes
- Gamma-densitometer to measure void fraction
- Laser Doppler anemometer to measure liquid velocity and turbulence
- Hot-film anemometer to measure liquid velocity and turbulence
- High-speed film camera to visualize flow
- High-speed digital camera to visualize flow
- Stereo microscope to calibrate probe geometry
- Image animation and processing system to measure interfacial area concentration, void fraction, bubble or droplet size
- Laser focus displacement meter to measure film thickness measurement
- Modified magnetic flow meter to measure two-phase flow liquid velocity

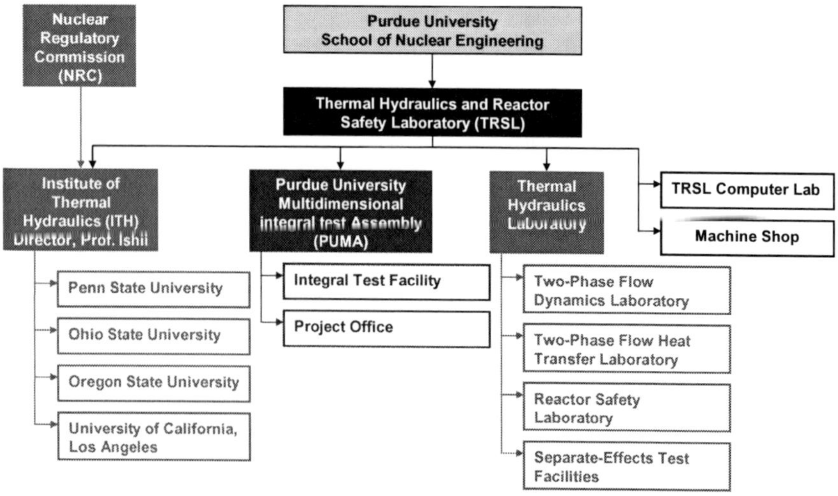

Laboratory Capability II: Thermal-Hydraulic Test Facilities

Ongoing projects cover a wide range of fundamental investigations. Some highlights of existing experimental capabilities are:

- Two-phase boiling and condensation (up to 10 bars pressure)
- Two-phase flow instability (nuclear coupling) (up to 10 bars)
- Freon 113 two-phase flow up to critical point
- Suppression pool separate-effects
- Annular flow entrainment and film dynamics
- Two-phase flow for confined channel sections
- Two-phase vertical upward flow (8 inch, 6 inch, 4 inch, 2 inch and 1 inch inner diameter)
- Two-phase downward flow (2 inch and 1 inch inner diameter)
- Two-phase Freon 113 natural circulation
- Inverted annular film boiling
- Microgravity simulation
- Boiling for nucleation and interfacial area study
- Flow-induced vibration effects
- Seismic vibration simulation

Laboratory Capability III: Integral Test Facility PUMA

Purdue University Multi-dimensional Test Assembly is a test facility designed to perform integral and separate effect tests, especially to assess the performance of natural circulation BWR with passive safety systems. Some of the features of the test facility are:

- Currently designed and scaled down from the GE-ESBWR based on the new three-level scaling approach. The major scaling ratios from PUMA to ESBWR are: (1) Pressure ratio: 1/1, (2) Area ratio: 1/128.8, (3) Length (height) ratio: 1/4.5, (4) Power ratio: 1/273, and (5) Time ratio: 1/2.12.
- Utilizes 38 electrical heater rods to simulate the history of core decay heat power with programmable power generation, maximum 650 kW.
- Monitored by the state-of-art measurement and display system. There are about 500 instruments in total installed, including thermocouples, P and DP cells, magnetic flow meters, vortex flow meters, in-house built impedance meters and in-house built conductivity meters.
- Equipped with a formal control room with capability to simulate all design-basis accidents (DBS) and many beyond DBS.

PUMA facility (Top view)

PUMA facility (Control room)

Laboratory Capability IV: Up-to-Date Computational Capability

TRSL has many computers for a variety of purposes, including computation, data acquisition, and data analysis. These computers have access to a number of different software packages for integral reactor safety analysis and three-dimensional two-phase flow computation. These codes are:

- Systems Analysis Codes
  - TRAC P, B, M
  - RELAP5
  - TRACE
  - CATHARE
  - RAMONA
- CFD Codes
  - CFX
  - FLUENT

Also, the computer hardware available at TRSL can be summarized as;

- About 50 personal computers in total
- 10 of the computers connected to the network for general usage purposes
- 30 computers used for acquisition of data from thermal-hydraulic and integral test facilities
- 2 computers for CFD codes
- 5~8 computers used by PUMA staff for system analysis codes and data analysis of PUMA experiments
- 1 Unix workstation for system analysis codes

Laboratory Capability V: Mechanistic Modeling

One of the main interests of TRSL is to develop and test mechanistic models. These models are developed using physical laws and advanced mathematical techniques, then implemented in system analysis or CFD codes. The results of these codes are then tested experimentally to verify that the model is accurate. The modeling activities include detailed 3-D drift-flux and two-fluid model formulation, interfacial area transport equation, and development of constitutive equations for thermo-fluid phenomena under various conditions such as adiabatic-to-boiling flow, bubbly-to-annular mist flow, atmospheric pressure-to-1 MPa condition, normal gravity-to-microgravity condition, transient-to-steady flow, stable-to-unstable flow, developing flow-to-fully-developed flow.

In: Frontier Research in Microscale and Nanoscale... ISBN 978-1-61324-665-8
Editor: Lixin Cheng

Chapter 12

EFFECT OF NANO-PARTICLES FOR POOL BOILING HEAT TRANSFER OF REFRIGERANT 141B ON HORIZONTAL TUBES

Chien-Yuh Yang*[*] *and Da-Wei Liu
Department of Mechanical Engineering/Institute of Energy Engineering
National Central University, 32054 Chung-Li, Taiwan

ABSTRACT

Fluids with nano-sized particles have been proved that may effectively enhance the single-phase convective heat transfer performance. For pool boiling heat transfer, the published test results seems conflicted to each other. Some measured heat transfer coefficient decreased with increasing particle concentration but some showed no apparent difference.This study provides an experimental investigation on pool boiling heat transfer performance of refrigerants R-141b with and without nano-sized Au particles on horizontal plain tubes. The change of surface roughness and fluid particles condition were also observed for further understanding of the mechanisms that affect the pool boiling heat transfer performance.

The test results show that the boiling heat transfer coefficients increase with increasing nano-particles concentration. At particles concentration of 1.0 %, the heat transfer coefficients are more than twice higher than those without nano-particles. The SPM investigation shows that the test tube surface roughness decreased from 0.317 μm before boiling test to 0.162 μm after test. Further investigation by TEM and Dynamic Light Scattering particle analyzer shows that the nano-particles aggregated from the average size of 3 nm before test to 110 nm after test. This results show that the nano-sized Au particles are able to significantly increase pool boiling heat transfer of refrigerant R-141b on plain tube surface. The tube surface roughness and particle size changed after boiling test. Both of these effects degrade the boiling heat transfer coefficients.

* E-mail address: cyyang@ncu.edu.tw Tel: 886-3-4267347. Corresponding author: Department of Mechanical Engineering/Institute of Energy EngineeringNational Central University, 32054 Chung-Li, Taiwan

Keywords: pool boiling, nano fluid, refrigerant

INTRODUCTION

Nucleate boiling is one of the most important modes of heat transfer. It is popularly applied on the shell side of tubular heat exchangers in the refrigeration and process industries. For improving the nucleate boiling heat transfer performance, numerous enhancement techniques have been developed on tube surface and fluid treatment.

Water with nano-sized particles have been tested that may effectively enhance the single-phase convective heat transfer performance. The enhancement may up to 24% for water containing only 1.0 vol. %nanoCu or Al_2O_3particles (Li and Xuan [1] and Pak and Cho [2]). For two-phase heat transfer, Das et al. [3] tested water-Al_2O_3nano-fluids with concentration up to 4.0 % pool boiling on plain and rough tubes. Bang and Chang [4] tested the same nano-fluids on plain surface. Theirmeasured boiling heat transfer coefficient decreased with increasing particle concentration. They concluded that the change of boiling performance was due to the trapped particles on surface and reduced the number of active nucleation sites.

However, You et al. [5] and Vassallo et al. [6] tested the boiling performance of water-Al_2O_3 and water-silica nano-fluids respectively. They found that the critical heat flux increases dramatically in comparing to pure water, but no appreciable difference in heat transfer coefficient for powers less than CHF.

Wen and Ding [7] carried out systematic experiments to formulate stable aqueous based nano-fluids containing α-alumina nano-particles with particle size from 10 to 50 nm, and investigated their pool boiling heat transfer behavior. Their results show that alumina nano-fluids can significantly enhance boiling heat transfer. The enhancement increases with increasing particle concentration and reach 40 % at a particle loading of 1.25 % by weight. They suggested that the reported controversies in the pool boiling heat transfer behavior be associated with the properties and behavior of both nano-fluids and boiling surface, as well as their interactions. The test results by the above researchers on pool boiling heat transfer performance of nano-fluids seems conflicted to each other. More testsare necessary for clarifying the effect of nano-particles on the pool boiling heat transfer performance.

All of the previous stated researches used water as the base fluid. Refrigerants HFCs and HCFCs have been popularly used in the air condition and refrigeration systems. But there is no any test of heat transfer performance for refrigerant with nano-particles has been published up-to-date. This study provides an experimental investigation by measuring the pool boiling heat transfer performance of refrigerant R-141b with nano-sized Au particles on plain tubes. It will be helpful for further understanding the pool boiling heat transfer enhancement mechanism of nano-fluids.

EXPERIMENTAL SETUP

A special fabricated thick wall plain copper tube with length of 100 mm and outside diameterof18.0 mm was tested in this study. Four thermocouple grooves with 2.5 mm depth and 90 degrees apart were made by EDM (electric discharging machining) inside the tube as shown in Figure 1. A cartridge heater with a diameter of 6.5 mm and an effective length of 100 mm was installed in the test tube. The power range of the cartridge heater can be controlled within 10 ~ 700 W and measured by using a precision power meter. Four T-type thermocouples were installed in the grooves located at 50mm from the flange to measure the temperature variations around the tube wall.

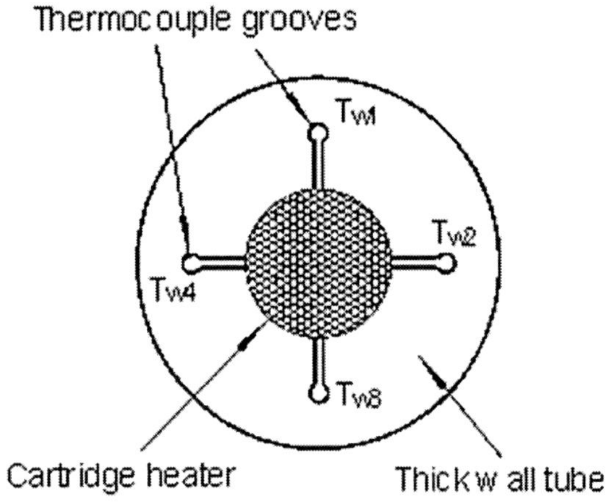

Figure 1.Cross-section view of the test tubes.

A schematic diagram of the test system is shown in Figure 2. It is a natural circulation system that includes a test chamber, a condenser, andrelevant connecting pipes.The test chamber is made of stainless steel with windows at the side and front for boiling phenomena observation. The test tube was immersed in a refrigerant bath in the test chamber. Heat was supplied to the test tube by an internal cartridge heater.Refrigerant evaporated on the surface of the test tube, through the vapor pipeline, and condensed in a water-cooled condenser. The heat flux is defined as the power input to the cartridge heaterdivided by the tube outside surface area. The range of heat flux tested was from 3.5 to 100 kW/m^2 with decreasing heat flux.

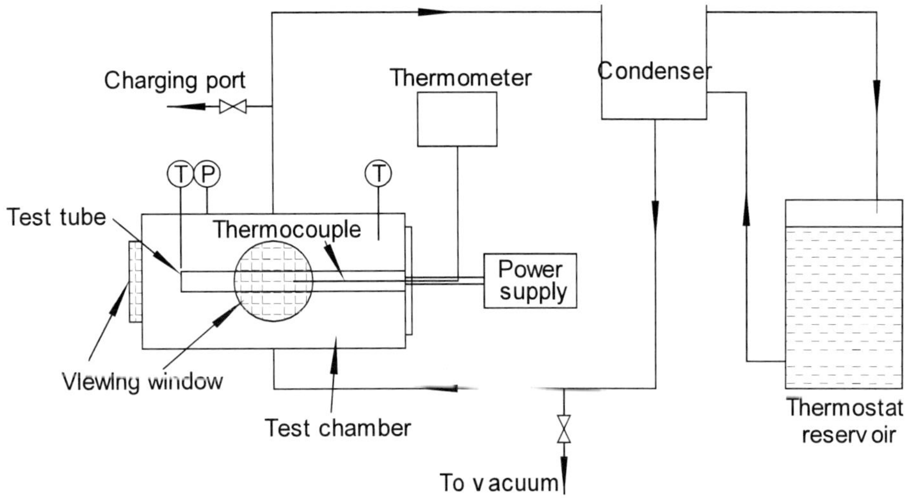

Figure 2.Schematic diagram of the experimental system.

A pressure transducer calibrated with an accuracy of ±0.2% of the measuring span was placed at the top of the test chamber to measure the system pressure. Two RTDs were installed in the liquid pool and the vapor space to measure the saturation liquid and vapor temperatures respectively. All of the thermocouples and RTDs were calibrated by a temperature calibrator having a calibrated accuracy of ±0.1°C.All data wererecordedand converted to digital signals by a data acquisition system. The converted signals were then transmitted through an IEEE-488 interface to a personal computer for further operation.

Because of the difficulties on nano-fluids production at high pressure condition, a low-vapor pressure refrigerant R-141b was selected as the working fluid in the present test. Mono-dispersed gold nano-particles were prepared by the reduction of aqueous hydrogen tetrachloroaurate (HAuCl4, Aldrich Chemical) with trisodium citrate and tannic acid (Aldrich Chemical) using a standard procedure (Hayat [8]) as described by Tsai et al. [9]. Three nano-particles volume concentrations, 0.09%, 0.4% and 1.0 %, of the nano-fluid were tested in this study.

DATA REDUCTION

The calculated wall temperatureT_{wall} is the average wall temperature at the outer surface of the tube, which can be calculated by conduction theory from the inside temperature T_{wi}, and is given by

$$T_{wall} = T_{wi} - q\frac{\ln(D_o/D_i)}{2\pi kL} \tag{1}$$

whereq is the electric heating power andT_{wi} is the arithmetic mean of the measured inside wall temperature T_{w1}, T_{w2}, T_{w3} and T_{w4}.

$$T_{wi} = \frac{T_{w1} + T_{w2} + T_{w3} + T_{w4}}{4} \tag{2}$$

The heat transfer coefficient for each power input was calculated as follows:

$$h = \frac{q}{A(T_{wall} - T_s)} \tag{3}$$

whereT_s is the saturation temperature based on the measured saturation pressure. The difference between T_s and the saturation vapor temperature measured by the RTD has been checked within ±0.1°C. The outside surface area, A, is evaluated as $\pi D_o L$ where D_o is the outside tube diameter, and L is the effective length of the cartridge heater. The experimental uncertainties of boiling heat transfer coefficients at various power input are listed in Table 1.

Table 1. Uncertainties of testapparatus and derived parameters

Parameters	Uncertainty
RTD	±0.1 °C
thermocouple	±0.1 °C
pressure transducer	±0.2%
heat transfer coefficient	
Power (W)	(%)
600	±3.93
450	±4.17
300	±4.40
200	±4.97
100	±5.35
40	±7.04
20	±12.22

EXPERIMENTAL RESULTS AND DISSCUSSIONS

Figure 3 shows the measuredboiling heat transfer coefficients versus heat flux for refrigerant R-141bwith and without nano-particles. The predictive results by Cooper [10] correlation given below are also plotted for comparison.

$$h = 90(q'')^{0.67} M^{-0.5} P_R{}^{m} (-\log_{10} P_R)^{-0.55} \tag{4}$$

where M is the working fluid molecular weight, P_R is the reduced pressure, and m = 0.12 – 0.2 $\log_{10} R_p$. The surface roughness, R_p, is measured to be 0.317μm. The Figure shows that the tested heat transfer coefficients for R141b agree very well with those predicted by Cooper [10] correlation. For R-141b with 0.09 % nano-particles, there is no significant effect on pool boiling heat transfer performance at such low concentration nano-particles. The boiling heat transfer coefficients increase with increasing nano-particles concentration. At particles concentration of 1.0 %, the heat transfer coefficient is more than twice higher than those without nano-particles. The addition of nano-particles significantly increases refrigerant R-141b boiling heat performance. These experimental results agree with those by Wen and Ding [7] but are in contradiction to those by Das et al. [3] and Bang and Chang [4] who observed decrease of boiling heat transfer coefficients due to the presence of nano-particles.

For assuring the previous nano-fluid heat transfer performance has been correctly measured, the same test was repeated for four times with the intervals of every five days. Figure 4 shows the measured boiling heat transfer coefficients decreased for each test. And finally close to those without nano-particles. Das et al. [3] concluded from their test that the change of boiling performance was due to the trapped particles on surface and reduced the number of active nucleation sites.

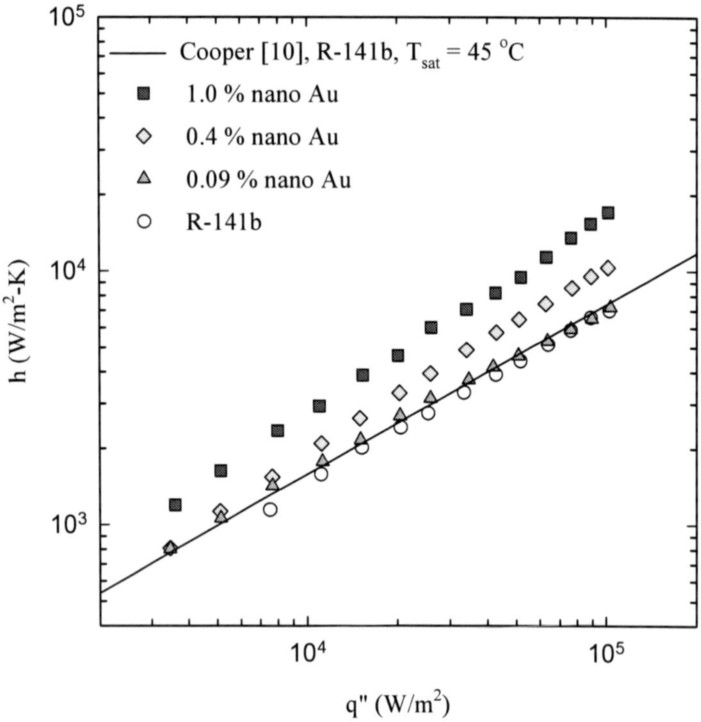

Figure 3. Heat transfer coefficients versus heat flux for various concentration nano-particles.

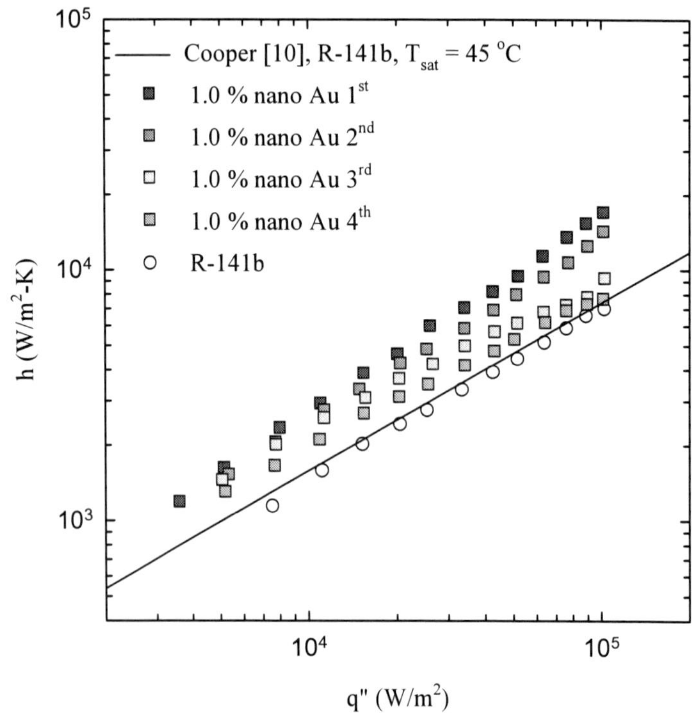

Figure 4. Heat transfer coefficients for each test after every five days.

To verify this postulation, the tube surface roughness was measure by SPM (Scanning Probe Microscope) before and after tests in the present study. The results are shown in Figures 5 and 6. Figure 5 shows that clear machining scratches can be observed on the tube surface before tests. Its surface roughness was measured as 0.317 μm. However, after boiling test as shown in Figure 6, the tube surface was covered by a layer of nano-particles sediment. The machining scratches can not be seen again and the measured surface roughness reduced to 0.162 μm. This surface condition change probably causes the deterioration of boiling heat transfer performance for each test after every five days as shown in Figure 4.

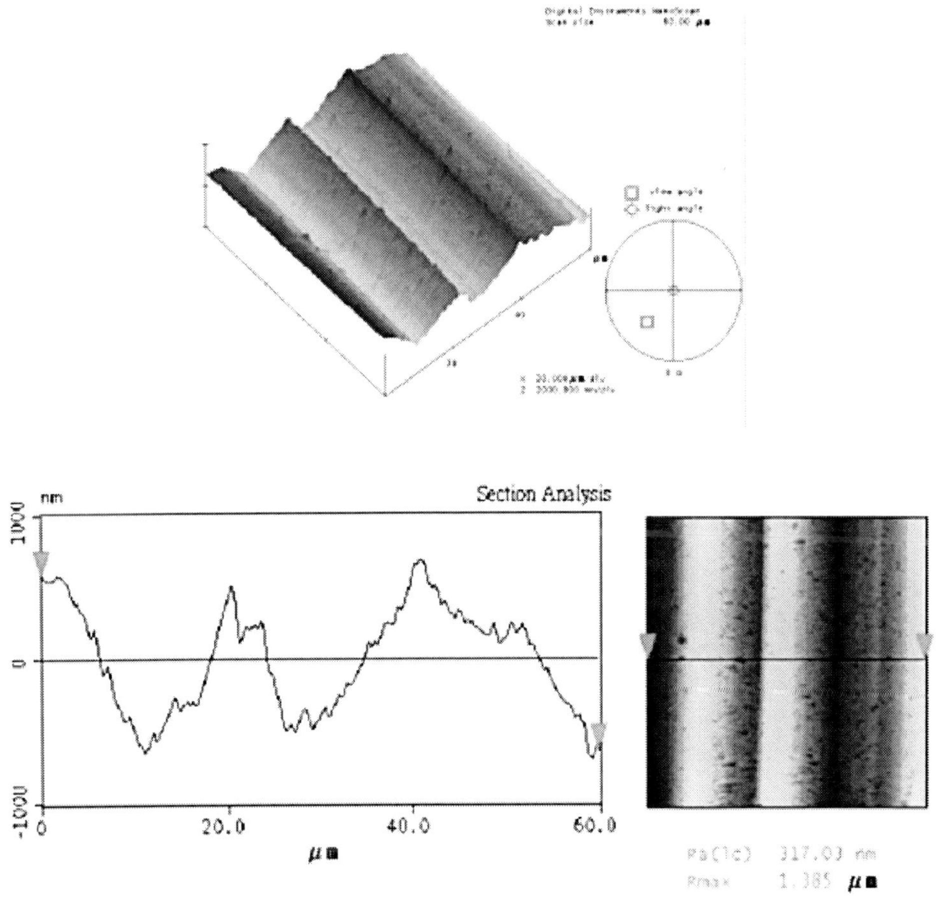

Figure 5. Tube surface roughness before test.

The tube surface was then cleaned thoroughly by ultrasonic cleaner and the same boiling test was repeated on the cleaned tube surface again. Figure 7 shows the boiling heat transfer coefficients on cleaned tube surface are higher than those on the tube before clean but still much lower than the first test results. There must be some other effects beside the tube surface conditions that degraded the boiling heat transfer performance.

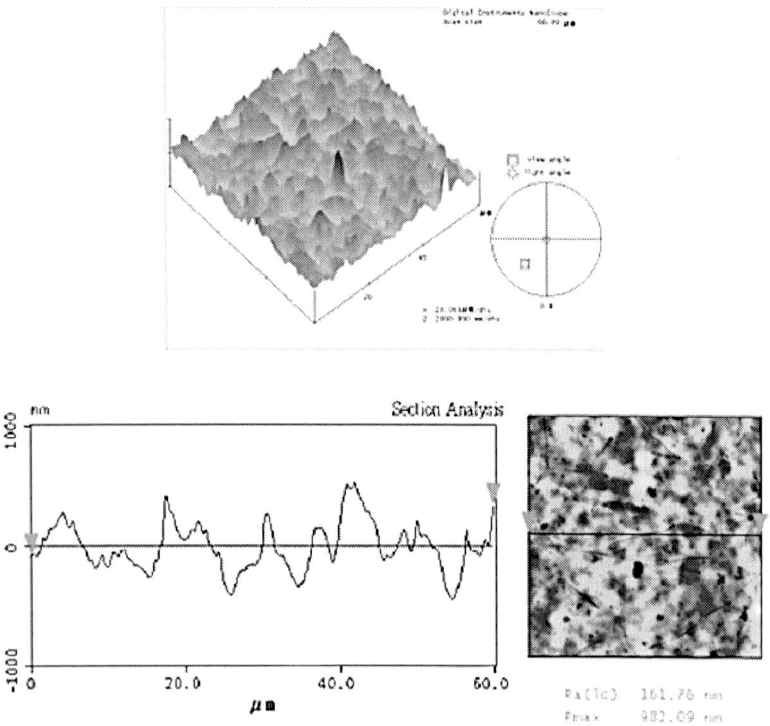

Figure 6. Tube surface roughness after test.

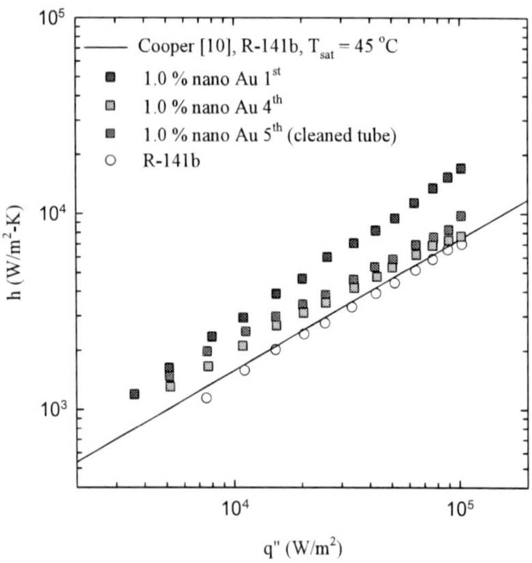

Figure 7. Heat transfer coefficients on cleaned tube.

Wen and Ding [7] suggested that the reported controversies in the pool boiling heat transfer behavior be associated with the properties and behavior of both nano-fluids and boiling surface, as well as their interactions. The effect of boiling surface in the present study has been verified and discussed in the previous section. For the nano-fluid itself in the present study, there is no any change after test can be observed with the naked eyes. However, while it is investigated with TEM and Dynamic Light Scattering particle size analyzer as shown in Figure 8, a significant particles aggregation can be observed. The average particles size changed from approximately 3 nm before tests to 110 nm after tests. The aggregation of nano-particles reduced the number of particles and therefore degrades the effect boiling heat transfer enhancement.

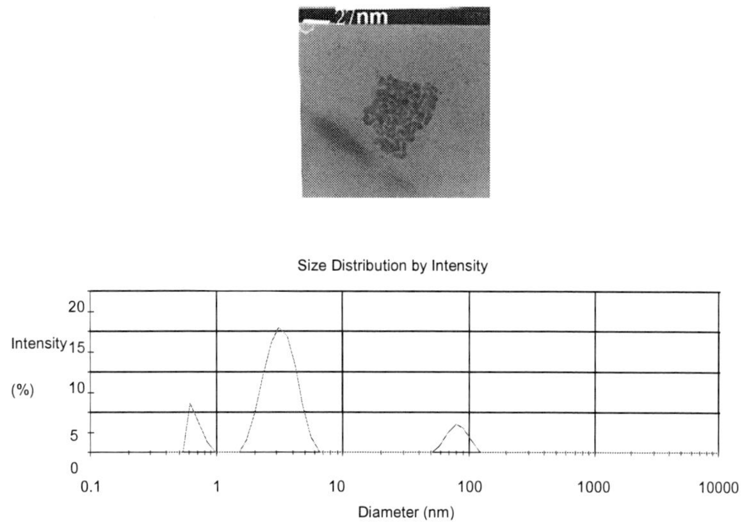

(a) Particles size before test

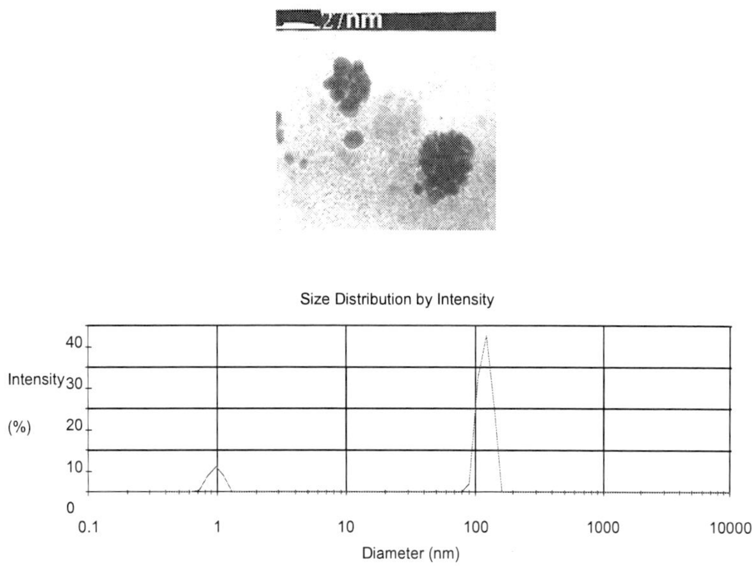

(b) Particles size after test

Figure 8. Particles size observed by TEM and Dynamic Light Scattering particle size analyzer.

The previous results show that the nano-sized Au particles are able to increase pool boiling heat transfer of refrigerant R-141b on plain tube surface. However, tube surface roughness and particle size changed after boiling tests. Both of these effects degrade the boiling heat transfer coefficients.

Conclusion

This study provides an experimental investigation on pool boiling heat transfer performance of refrigerants R-141b with and without nano-sized Au particles on horizontal plain tubes. The test results show that the heat transfer coefficients for R141b without nano Au particles agree very well with those predicted by Cooper [10] correlation. The boiling heat transfer coefficients increase with increasing nano-particles concentration. At particles concentration of 1.0 %, the heat transfer coefficient is more than twice higher than those without nano-particles.

The SPM investigation shows that thetest tube surface roughness decreased from 0.317 μm before boiling test to 0.162 μm after test.Further investigation by TEM and Dynamic Light Scattering particle analyzer shows that the nano-particles aggregated from the size of 3 nm before test to 110 nm after test. This results show that the nano-sized Au particles are able to significantly increase pool boiling heat transfer of refrigerant R-141b on plain tube surface. However, tube surface roughness and particle size changed after boiling test. Both of these effects degrade the boiling heat transfer coefficients.

Acknowledgments

The authors would like to express their thanks to Prof. B. -H. Chen and his graduate student Mr. S.-T. Chien in National Taiwan University for providing the nano-fluid and its related properties used in this research.

Nomenclature

D_i	Inside diameter	m
D_o	Outside diameter	m
h	Heat transfer coefficient	W/m^2K
k	Tube thermal conductivity	W/mK
L	Heating length	m
M	Molecular weight	kg/kmol
m	Index in equation (4)	dimensionless
P_R	Reduced pressure	dimensionless
q	Heatingpower	W
q”	Heat flux	kW/m^2
R_P	Surface roughness	μm
T_{wall}	Average walltemperature	^oC
T_s	Saturation temperature	^oC

T_{w1}	Top wall temperature	^oC
T_{w2}, T_{w4}	Side wall temperature	^oC
T_{w3}	Bottom wall temperature	^oC
T_{wi}	Arithmetic mean of the measured inside wall temperatures	^oC

REFERENCES

[1] Li, Q. and Xuan, Y. Convective heat transfer performance of fluids with nano-particles,*Proceedings of the 12th International Heat Transfer Conference*, 2002, Grenoble, France.

[2] Pak, B. C.and Cho, Y. I. Hydrodynamic and heat transfer study of dispersed fluids with submicron metallic oxide particles, *Experimental Heat Transfer***11**, 1998, pp. 151-170.

[3] Das, S. K, Putra, N. and Poetzel, W. Pool boiling characteristics of nano-fluids, *International Journal of Heat and Mass Transfer***46**, 2003, 851-862.

[4] Bang, I. C. and Chang, S. H. Boiling Heat Transfer Performance and Phenomena of Al_2O_3-Water Nano-fluids from a Plain Surface in a Pool, *International Journal of Heat and Mass Transfer***48**, 2005, pp. 2407-2419.

[5] You, S. M., Kim, J. H. and Kim, K. H. Effect of nanoparticles on critical heat flux of water in pool boiling heat transfer, *Applied Physics Letter***83**, 2003, pp. 3374-3376.

[6] Vassallo, P., Kumar, R. and D'Amico, S. Pooling boiling heat transfer experiments in silica-water nano-fluids, *International Journal of Heat and Mass Transfer***47**, 2004,pp. 407-411.

[7] Wen, D. and Ding, Y. Experimental Investigation into the Pool Boiling Heat Transfer of Aqueous Based α-alumina nanofluids, *Journal of nanoparticle research* **7**, 2005, pp. 265-274.

[8] Hayat, M. A. (Ed.), *Colloidal Gold: Principles, Methods and Applications,* vols. 1 and 2, Academic Press, San Diego, CA, 1989.

[9] Tsai, C.-Y., Chien, H.-T., Ding, P.-P., Chan, B., Luh, T.-Y. and Chen, P.-H. Effect of structural character of gold nanopaticles in nanofluid on heat pipe thermal performance, *Materials Letters*, **58**, 2004, pp. 1461-1465.

[10] Cooper, M.G. Saturation nucleate pool boiling - asimple correlation, International *Chemical Engineering Symposium Series***86**, 1984, pp. 785-792.

In: Frontier Research in Microscale and Nanoscale
Editor: Lixin Cheng
ISBN 978-1-61324-665-8

Chapter 13

LATTICE BOLTZMANN METHOD FOR MODELING MICROFLOW

Lajos Szalmás*
Department of Mechanical Engineering,
University of Thessaly,
Pedion Areos, 38344 Volos, Greece

Abstract

The lattice Boltzmann model is a novel method to simulate gaseous or liquid flows. It successfully fills the gap between the macroscopic description, the Navier-Stokes equation, and particle based methods, such as the direct simulation of Monte-Carlo.

In the last years, the lattice Boltzmann model has attracted increasing attention in modeling flows in microfluidic devices. This area has developed rapidly. Microelectromechanical systems (MEMS), gaseous sensors, lab-on-chips produce significant grow in various areas of engineering and technology.

In this paper, we give an overview of the lattice Boltzmann method and its key concepts. We review recent developments related to the field of microflows and microfluidics. Particular attention is paid for non-continuum behaviours observed at the micro-scale, such as the slip and jump of macroscopic variables in a micro-device.

Finally, we provide further ideas and directions which can serve as targets of future developments. Beyond the lattice Boltzmann model, these ideas can stimulate progress and investment in applications and technology.

Keywords: lattice Boltzmann method, microflow, computational kinetic theory

*E-mail address: lszalmas@gmail.com, Telephone: 00-30-2410-74337, Fax: 00-30-24210-74085

1. Nomenclature

Collision operator:	Q_i
Distribution function:	$f_i(\boldsymbol{r},t)$
Equilibrium distribution function:	$f_i^e(\boldsymbol{r},t)$
First relaxation time:	τ_1
Fluid velocity:	$\boldsymbol{u}$
Kinematic viscosity:	ν
Knudsen layer width:	l
Knudsen number:	Kn
Macroscopic size:	L
Magnitude of the speed vectors:	c
Mean free path:	l_0
Non-equilibrium part of the pressure tensor:	$p^n_{\alpha\beta}$
Particle density:	n
Pressure tensor:	$p_{\alpha\beta}$
Second relaxation time:	τ_2
Space step:	δx
Relaxation time:	τ
Time step:	δt
Transformed collision operator:	Q_i'
Transformed distribution function:	$f_i'(\boldsymbol{r},t)$
Velocity vector:	$\boldsymbol{c}_i$
Weight function:	w_i

2. Introduction

In the last few decades, the lattice Boltzmann method (LBM) has matured into a flexible computational platform in modeling complex hydrodynamic problems. The essence of the method is that the description of the media starts at the kinetic scale instead of the usual hydrodynamics which operates with macroscopic quantities. Due to this kinetic nature, the model is well-suited for modeling microflows.

In recent years, the field of microflows has undergone a rapid development [1,2]. These flows show special physical behaviors due to the increased role of the molecular interactions. The flow pattern and the profile of the fluid dynamic variables in a microfluidic device basically differ from the ones observed at the macro-scale. Understanding and modeling flows under such circumstances is challenging and important. There is a huge interest in microflows in engineering and technology.

In this paper, I give an introduction into the LBM and overview its main concepts. Then, I discuss recent developments in the LBM for modeling microflows. Finally, some challenges for further research in the LBM are presented.

3. Concepts of the LBM

Historically, the LBM was derived from its predecessor the lattice gas automata (LGA). McNamara and Zanetti replaced the Boolean particles travelling on lattice links by continuous distribution functions [3]. The new-born model was free from some practical hold-back factors of the LGA, noise and the so-called spurious invariants. The original version of the LBM was based on the multi-relaxation-time formalism [4, 5]. This approach inherited the heuristic collision model of the LGA. Later, subsequent developments were performed. The collision operator was simplified and linearized. The single-relaxation-time approach, the Bhatnagar-Gross-Krook (BGK) approximation [6], was also introduced. The BGK approach was applied for magnetohydrodynamics [7] or the recovery of the Navier-Stokes equations [8, 9]. This latter development made the LBM powerful for practical simulations. It was realized that the LBM has large capabilities in computational fluid dynamics [4, 10]. The theoretical background of the method was established. The method was successfully married with computational fluid dynamic techniques. An important result was that the discretization of the velocity and the coordinate spaces can be chosen independently [11]. This gave rise to the development of different realization techniques, such as the finite-difference (FDLBM) [12] or finite-volume based LBM (FVLBM) [13].

The starting point of the LBM is a discrete kinetic equation of the form

$$\frac{\partial f_i(\boldsymbol{r},t)}{\partial t} + \boldsymbol{c}_i \frac{\partial f_i(\boldsymbol{r},t)}{\partial \boldsymbol{r}} = Q_i, \tag{3.1}$$

which is the governing equation of the one-particle distribution function $f_i(\boldsymbol{r},t)$. The molecular interactions are encoded in the collision operator Q_i. The macroscopic fluid dynamic quantities, the particle density n, the velocity $\boldsymbol{u}$, the pressure tensor p_{ab} are defined as the moments of the distribution function $n = \sum_i f_i$, $\boldsymbol{u} = \sum_i \boldsymbol{c}_i f_i$, $p_{ab} = \sum_i c_{ia} c_{ib} f_i$.

The essential features of the model depend on the speed vectors and the collision operator. The speed vectors can be constructed in different ways. Here, we mention the *DnQm* family which is based on the roots of the Hermite polynomials [14]. n, m denote the dimension and the number of speed vectors, respectively. In several applications, the collision term is usually defined as the BGK operator. Note that the resulting dynamic system on the basis of the BGK model contains one-relaxation time. As a consequence, only one transport coefficient can be chosen freely, which is usually the viscosity. However, much can be gained by applying the so-called multi-relaxation-time (MRT) collision operator. This allow us choosing independently the relaxation times for different kinetic fields. We mention that the LBM was born in MRT form. However, it is crucial to realize that the idea of MRT is more general. One can show that MRT collision operators with variable Prandtl-number in the LBM are basically equivalent with the Shakov (S) model, which is often used in rarefied gas dynamics for describing thermal problems [15].

Up to this point, our kinetic model, Eq. (3.1), accounts for a set of partial differential equations. This immediately raises the following question. How we can develop a real computational method on the basis of Eq. (3.1)? This can be achieved by the so-called realization procedure. The coordinate space is discretized by choosing a coordinate grid comprising fluid nodes. In this paper, I show two basic realization procedures, the so-called top-down (or bottom-up) and the finite-difference based approaches.

In the top-down approach, the grid is constructed from the speed vectors. If one fluid node $\boldsymbol{r}$ resides on the grid, its shifted neighborhoods $\boldsymbol{r}' = \boldsymbol{r} + \boldsymbol{c}_i\delta_t$ also determine fluid nodes on the grid for all $\boldsymbol{c}_i$. This construction results in a lattice structure. This approach is usually applied for generating cubic lattices, and typical examples are the $D2Q9$ or the $D3Q15, D3Q27$ lattices. For a given lattice, one can also define the space step δ_x. After the lattice is constructed, Eq. (3.1) is integrated by using the so-called trapezoid integration scheme. The resulting dynamic system can be written by

$$f_i'(\boldsymbol{r} + \boldsymbol{c}_i\delta_t, t + \delta_t) - f_i'(\boldsymbol{r}, t) = Q_i', \tag{3.2}$$

where $f_i'(\boldsymbol{r}, t)$ and Q_i' are transformed quantities. Eq. (3.2) is called as the lattice Boltzmann equation. I established a general procedure for deriving the top-down LBM for the MRT operator.

In the finite-difference based LBM, the coordinate grid is independent of the speed vectors. The grid can be chosen freely, and one can use body-fitted, curvilinear grids as well. The time and space derivatives in Eq. (3.1) are replaced by finite-difference forms. In my works, I frequently used the combination of the Runge-Kutta time marching scheme and midpoint or upwind, downwind differential forms for the space derivatives. This choice ensures second-order accuracy in both time and space. The time and space discretizations introduce also the time and space steps δ_t, δ_x as in the top-down approach.

The next question is how to connect the LBM to real physical experiments. In our kinetic model, we encounter four important parameters, the magnitude of the speed vectors c, a main relaxation time τ which is usually connected to the viscosity and the time and space steps δ_t, δ_x. Having these quantities, we can dimensionally connect the model to real, physical processes. Furthermore, we can define the mean free path l_0, the average distance travelling a molecule between two subsequent collisions, and the Knudsen number $Kn = l_0/L$, the ratio of the mean free path and a relevant macroscopic size L of the problem, such as the channel width for microflows.

Finally, let us discuss the macro-dynamic behavior of the method. The governing equations of the fluid dynamic quantities are obtained from Eq. (3.1) as its moments. The derivation of the hydrodynamic equations is usually based on the Chapman-Enskog expansion. With an appropriately chosen collision term, the equations of continuum fluid mechanics can be recovered by the LBM in the small Kn number, hydrodynamic limit.

4. LBM for Microflows

The lattice Boltzmann model is especially well-suited for modeling microflows. This is because the method is based on micro-physical concepts instead of the continuum description. Microflows can significantly depart from macro-hydrodynamics usually characterized by the Navier-Stokes equations.

In recent years, it has been demonstrated that the LBM succeeds in simulating micro-gas flows. In this case, the main challenge is to capture the so-called rarefaction effects [16]. In contrast to the simple macro-hydrodynamics, the Knudsen number is significantly larger for micro-gas systems and mainly lies in the range of $Kn = 0.001 - 10$. Micro-gas flows can be basically classified into two regions. The slip flow assumption is used in the range of $0.001 < Kn < 0.1$. In this situation, the Navier-Stokes equation can be considered valid, but an appropriate slip or jump condition needs to be applied at the solid walls. The so-called first-order slip model characterizes the slip or jump of the macroscopic quantities, velocity or temperature, as a linear function of the Kn number. The slip-flow assumption can be extended for larger Kn numbers by using the second-order slip model, which applies a second-order expansion in the Kn number. The transient region lies in the range of $0.1 < Kn < 10$. In this case, the Navier-Stokes hydrodynamics loses its validity. The so-called Knudsen layer, a highly non-equilibrium gas region near the solid-walls, starts to overlap in the micro-device.

There is clear evidence that the LBM as a kinetic model describes well micro-gases in the slip region. The LBM with Navier-Stokes level velocity models, such as the $D1Q3$ class, captures the slip and jump conditions of the macroscopic quantities on the solid walls. In addition, the LBM permits to use realistic boundary conditions for the distribution function, as derivatives from kinetic theory. The diffuse-reflection boundary condition has been implemented into the LBM framework [17]. Such a boundary treatment does not use accommodation coefficients. However, these accommodation coefficients can be applied for the method in order to describe variable momentum transfer at the wall. They allow a sophisticated and better description of the solid-gas interface. The combination of the bounce-back and specular-reflection boundary conditions has been used for simple geometries [18, 19] and moving flat walls [20]. Boundary condition with arbitrary accommodation on curved surfaces has also been developed [21]. This has made the LBM applicable for modeling slip flow on complex geometries. It is mentioned that the diffuse-specular boundary condition has also been used in LBM simulations [22]. These kinetic boundary conditions capture important rarefaction effects, however, other standard boundary treatments were also used at the beginning of the development [23]. The slip models have been applied for thermal problems as well [24]. In addition, recently, the Langmuir slip model has been incorporated into the method [25].

In this direction an important result was that the LBM with the multi-relaxation-time collision operator captures the second slip model [26]. I showed that the relaxation times in the MRT operator can be chosen to recover tunable second-order slip coefficient in the slip model. This approach is especially useful for modeling pressure driven flows well beyond the slip flow assumption. For micro-channels, the mass flow rate is usually an important quantity from either theoretical or technological standpoints. In this case, one encounters the so-called Knudsen minimum phenomenon, that is the mass flow rate has a minimum point as a function of the Kn number in the transient region. In Fig. 1 we show the mass flow rate provided by different types of LBM models versus the Kn number. The results are compared to the second-order slip models of Cercignani [27] and Hadjiconstantinou [28]. Here, it is mentioned that Cercignani developed one of the first second-order slip boundary conditions. His approach is based on the Boltzmann equation with the BGK operator. Later, this model has been reformulated by Hadjiconstantinou for hard-sphere gases. In addition, a Knudsen layer correction has been implemented into the boundary condition. In the present study, LBM I and II stand for the MRT schemes which recovers either the boundary condition of Cercignani or Hadjiconstantinou. For comparative purposes, the result provided by the LBM BGK with the bounce back boundary condition, which is often used in hydrodynamic simulations, is also presented as indicated by LBM III. In addition, the Navier-Stokes results with the no-slip boundary condition is also shown in the figure. It is demonstrated that an excellent agreement between the MRT LBM and the sophisticated second-order slip models can be achieved. It can also be seen that the Navier-Stokes hydrodynamics with no-slip boundary condition can not reproduce the Knudsen minimum in the transient region.

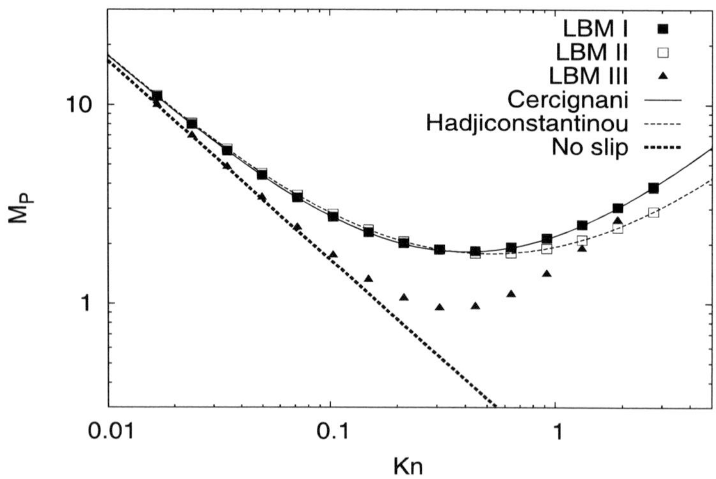

Figure 1. Normalized mass flow rate for Poiseuille flow. LBM I,II denote the results of the LBM MRT with kinetic boundary conditions. LBM III represents the result of the LBM BGK with bounce back boundary condition. The flow rate from Cercignani and Hadjiconstantinou second slip models [27, 28] are presented by lines. The result of the Navier-Stokes equation with no slip boundary condition is also plotted.

The remarkable results provided by the LBM in the slip flow region suggest that the method is extend-able into the transient region. Since the LBM is a derivative model from the Boltzmann transport equation, we can increase the number of velocity vectors used in the approach to get a better approximation. In this way, we obtain the so-called higher-order lattice Boltzmann models. These models formally recover some higher-order macroscopic moment equations than those provided by the Navier-Stokes description. As it has been shown, the higher-order LBM can be successfully used to capture the rarefaction effects in the transient region.

A main result was in the development the so-called Knudsen layer theory of the higher-order LBM [29]. The Knudsen layer characteristic of the method was elaborated. It was shown that the additional velocity vectors used in the higher-order LBM provides additional contribution to the hydrodynamic velocity profiles. These modes are in close relationships with the Knudsen layer functions appearing in rarefied gas theory. I also showed the advantageous features of the MRT collision operator in modeling the Knudsen layer. The MRT operator provides a boundary layer with a variable layer width. This finding is extremely useful for modeling micro-gas flows in the transient region.

Now, we turn our attention to the details of the MRT higher-order scheme and its relation to the Boltzmann equation. Following previous developments [29], the basic discrete

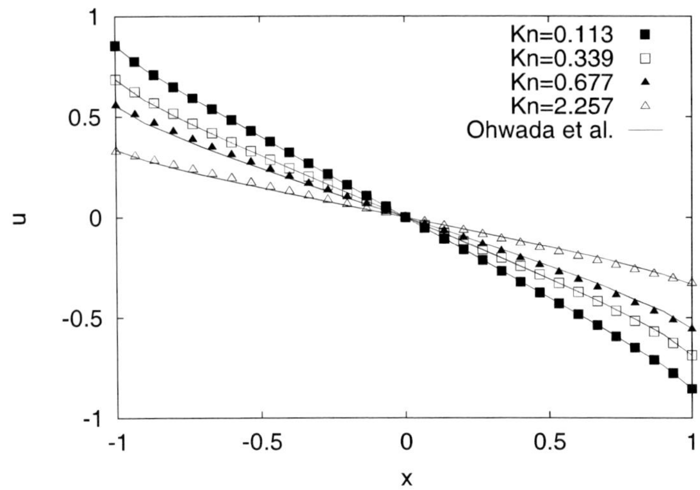

Figure 2. Normalized velocity profiles for Couette flow between two parallel plates. Symbols represent the results of the higher-order LBM MRT versus the *Kn* number. The corresponding velocity profiles on the basis of the linearized Boltzmann equation with hard-sphere interaction [30] are plotted with solid lines.

kinetic equation (3.1) can be written by

$$\frac{\partial f_i(\boldsymbol{r},t)}{\partial t} + \boldsymbol{c}_i \frac{\partial f_i(\boldsymbol{r},t)}{\partial \boldsymbol{r}} = -\frac{1}{\tau_2}(f_i(\boldsymbol{r},t) - f_i^e(\boldsymbol{r},t)) + \left(\frac{1}{\tau_2} - \frac{1}{\tau_1}\right) \frac{w_i}{2c^4}(c_{i\alpha}c_{i\beta} - c^2\delta_{\alpha\beta}) p^n_{\alpha\beta}(\boldsymbol{r},t), \tag{4.3}$$

The present model applies the $D2Q16$ higher-order velocity set with the $\boldsymbol{c}_i, w_i$ speed vectors and weights. The collision operator is a combination of a BGK term and a pressure term. This latter uses the non-equlibrium part of the pressure tensor $p^n_{\alpha\beta}(\boldsymbol{r},t)$. The model applies two relaxation times, τ_1 and τ_2. These quantities are connected to the Boltzmann equation and the Knudsen layer characteristics. At this point, it is to be mentioned that the LBM is a fluid or gas model, which should reproduce the correct hydrodynamics as follows from the Boltzmann equation. The hydrodynamic property is characterized by the viscosity of the gas. In the present model, the kinematic viscosity is given by $\nu = c^2\tau_1$. This relation fixes the first relaxation time. The second relaxation time is connected to the width of the Knudsen layer $l \sim c\tau_2\sqrt{3}$. By comparing the Knudsen layers provided by the present method and the Boltzmann equation, τ_2 can be determined. In this way, the relationship between the MRT higher-order method and the Boltzmann equation is established.

In order to demonstrate the promising results of the higher-order MRT LBM, we show the velocity profiles provided by the LBM for Couette flow in comparison with the accurate solution of the Boltzmann equation in Fig. 2. In the MRT LBM, the second relaxation time is chosen by $\tau_2 = \gamma\tau_1$ with $\gamma = 0.3$. Very good agreement can be achieved between the two approaches. Other results also suggested that the LBM models for transient micro-gas flows should be based on higher-order velocity sets.

The MRT higher-order method can be extended for non-isothermal problems. In that case, it is useful to connect the scheme to the S kinetic model. For non-isothermal problems, the thermal conductivity is an additional transport coefficients. The MRT LBM can be formulated for thermal problems as well. In that case, more velocity vectors are needed than in the case of isothermal problems. The pressure and the heat flux relaxation times need to be connected to the viscosity and thermal conductivity. The remaining relaxation parameters in the MRT LBM can be used the optimize the Knudsen layer characteristics. In this way, the hydrodynamic behaviors dictated by the S kinetic or Boltzmann equation can be recovered. This is the general treatment of relating the MRT LBM to the S kinetic model or the Boltzmann equation for non-isothermal problems.

Another approach has been also appeared to capture the Knudsen layer characteristics, the so-called wall-function concept [31]. In this case, the relaxation time is redefined in the boundary layer near the wall. This is a phenomenological approach and based on the idea that the effective mean free path may be smaller near the wall than in the bulk region due to the wall collisions. The method has attracted some interest in modeling microflows. The approach has been also used with the higher-order LBM [32].

Beyond single component gases, the LBM has been developed for modeling binary mixtures. With regard to this subject, it is important to capture the correct value of the slip coefficient at walls. The first approach in this direction used a modified diffuse-reflection boundary condition, which can reproduce variable slip coefficients in the method [33]. A different lattice Boltzmann method has also been developed for binary micro-gas flows [34]. In this latter case, the combination of the bounce-back and specular-reflection boundary condition is used for moving walls [20].

5. Outlook and Further Challenges

In the LBM development there are further directions and challenges regard to the microflow domain as targets for research. It was realized that the higher-order LBM provides reliable results for microflows. In principle, this method could be further developed. There are complex problems in microflows of which description requires larger moment systems than those provided by the standard LBM. A typical example is thermal problems, where a larger moment system should be used. Another example is gaseous mixtures, where the higher-order dynamics needs to be recovered in order to capture important rarefaction effects. From methodological viewpoint, the higher-order LBM could be developed by using optimized velocity sets. It needs to be realized that the Knudsen characteristics depend on the used velocity model, hence an optimization strategy could be useful. An important advantage of the LBM is that it is a real time model. As a consequence, it can be generally used for modeling time-dependent gaseous flows. A good research project would be the development of the higher-order LBM with larger velocity sets on parallel computers for fast computation. We remind the reader that the LBM can be easily parallelized as a result the parallel-performance can be adopted to the higher-order scheme.

6. Conclusion

In this paper, I have given a brief introduction into the lattice Boltzmann model. I have discussed the aspects of the method in modeling microflows. Recent developments clearly show that the LBM is a viable tool for modeling this area. The model with its kinetic origin has large capabilities to describe important rarefaction effects. I have also presented research topics for further developments in the field. It is straightforward to develop the LBM further along these lines.

References

[1] C.M. Ho, Y.C. Tai, *Annu. Rev. Fluid. Mech.* **30** (1998) 579-612.

[2] P. Tabeling, *Introduction to Microfluidics*; Oxford University Press, Oxford, 2005.

[3] G.R. McNamara, G. Zanetti, *Phys. Rev. Lett.* **61** (1988) 2332-2335.

[4] F.J. Higuera, S. Succi, R. Benzi, *Europhys. Lett.* **9** (1989) 345-349.

[5] F.J. Higuera, J. Jimenez *Europhys. Lett.* **9** (1989) 663-668.

[6] P. Bhatnagar, E.P. Gross, M. Krook, *Phys. Rev.* **94** (1954) 511-525.

[7] S. Chen, H. Chen, D. Martinez, *Phys. Rev. Lett.* **67** (1991) 3776-3779.

[8] J. Koelman, *Europhys. Lett.* **15** (1991) 603-607.

[9] H. Chen, S. Chen, W. Matthaeus, *Phys. Rev. A* **45** (1992) 5339-5342.

[10] R. Benzi, S. Succi, M. Vergassola, *Phys. Rep.* **222** (1992) 145-197.

[11] N.Z. Cao, S.Y. Shen, S. Jin, D. Martinez, *Phys. Rev. E* **55** (1997) R21-R24.

[12] R. Mei, W. Shyy, *J. Comp. Phys.* **143** (1998) 426-448.

[13] G.W. Peng, H.W. Xi, C. Duncan, S.H. Chou, *Phys. Rev. E* **59** (1999) 4675-4682.

[14] Y.H. Qian, D. dHumieres, P. Lallemand, *Europhys. Lett.* **17** (1992) 479-484.

[15] F. Sharipov, V. Seleznev, *J. Phys. Chem. Ref. Data.* **27** (1998) 657-706.

[16] S. Colin, *Microfluid. Nanofluid.* **1** (2005) 268-279.

[17] S. Ansumali, I.V. Karlin, *Phys. Rev. E* **66** (2002) 026311.1-026311.6.

[18] C.Y. Lim, C. Shu, X.D. Niu, Y.T. Chew, *Phys. Fluids* **14** (2002) 2299-2308.

[19] M. Sbragaglia, S. Succi, *Phys. Fluids* **17** (2005) 093602.1-093602.8.

[20] L. Szalmás, *Phys. Rev. E* **73** (2006) 066710.1-066710.5.

[21] L. Szalmás, *Int. J. Mod. Phys. C* **18** (2007) 15-24.

[22] G.H. Tang, W.Q. Tao, Y.L. He, *Phys. Fluids* **17** (2005) 058101.1-058101.4.

[23] X. Nie, G.D. Doolen, S. Chen, *J. Stat. Phys.* **107** (2002) 279-289.

[24] C. Shu, X.D. Niu, Y.T. Chew, *J. Stat. Phys.* **121** (2005) 239-255.

[25] S. Chen, Z. Tian, *Phys. A* **388** (2009) 4803-4810.

[26] L. Szalmás, *Phys. A* **379** (2007) 401-408.

[27] C. Cercignani, *Institute of Engineering Research Report*, AS-64-19, University of California, Berkeley, 1964.

[28] N.G. Hadjiconstantinou, *Phys. Fluids* **15** (2003) 2352-2354.

[29] L. Szalmás, *Europhys. Lett.* **80** (2007) 24003.p1-24003.p5.

[30] Y. Sone, S. Takata, T. Ohwada, *Eur. J. Mech. B/Fluids* **9** (1990) 273-288.

[31] Y.H. Zhang, X.J. Gu, R.W. Barber, D.R. Emerson, *Phys. Rev. E* **74** (2006) 046704.1-046704.7.

[32] G.H. Tang, X.J. Gu, R.W. Barber, D.R. Emerson, Y.H. Zhang, *Phys. Rev. E* **78** (2008) 026706.1-026706.8.

[33] L. Szalmás, *Cent. Eur. J. Phys.* **6** (2008) 786-791.

[34] Z. Guo, P. Asinari, C. Zheng, *Phys. Rev. E* **79** (2009) 026702.1-026702.9.

In: Frontier Research in Microscale and Nanoscale
Editor: Lixin Cheng
ISBN 978-1-61324-665-8

Chapter 14

HEAT TRANSFER IN ELECTRO-OSMOTIC FLOW OF POWER-LAW FLUIDS IN A SLIT MICROCHANNEL

Waqar A. Khan[a] ***and Rama Subba Reddy Gorla***[b]
[a]Department of Engineering Sciences,
PN Engineering College, National University of Science and Technology,
PNS Jauhar, Karachi, Pakistan.
[b]Department of Mechanical Engineering,
Cleveland State University, Cleveland, Ohio, U. S.

Abstract

Heat transfer in electro-osmotic flow of power-law fluids in a slit microchannel is investigated for uniform wall heat flux thermal boundary condition. Energy equation is solved analytically for $0 \leq \kappa H \leq 1$ and numerically for $\kappa H > 1$ using an implicit finite difference method. Previous analytical expressions for the velocity distribution are used to solve energy equation. The viscous dissipation effects are included in the analysis. The effects of electro kinetic parameter κH, flow behavior index n, and Brinkman number Br on the dimensionless local temperature and average heat transfer rate are examined and the results are presented graphically.

Keywords: Electro-osmotic Flow, Power-Law Fluids, Slit Microchannel, Brinkman number, Uniform heat flux, EDL.

Nomenclature

c_p	specific heat
E_0	external electric field
H	channel half height
m	flow consistency index
Nu	Nusselt number
n	power-law index
q	heat flux $[W/m^2]$
T	temperature $[°C]$
u	axial velocity $[m/s]$
v	transverse velocity $[m/s]$
x	axial coordinate $[m]$
y	transverse coordinate $[m]$

Subscripts

∞	free stream conditions
m	mean
w	wall

Greek Symbols

α	thermal diffusivity $[m^2/s]$
ε	dielectric constant
ε_0	permitivity of vacuum $[C/Vm]$
η	dimensionless tranverse coordinate
κ	inverse Debye length $[1/m]$
μ	absolute viscosity $[Ns/m^2]$
ν	kinematic viscosity $[m^2/s]$
ψ	electro kinetic potential $[V]$
θ	dimensionless temperature

1. Introduction

The electro-osmosis is the bulk movement of liquid relative to a stationary surface due to an externally applied electric field. Most solid substances will acquire a relative electric charge when in contact with an aqueous electrolytic solution, which in turn influences the charge distribution in the solution.

Ions of the opposite charge to that of the surfaces are attracted towards the surface and ions of the same charge are repelled from the surface as shown in Fig 1(a).The net effect is the formation of the region close to the charged surface called the electric double layer (EDL) in which there is an excess of counter ions over co-ions and which are disturbed in a diffused manner. The charge distribution in fluid therefore falls from its maximum near the wall to the charge in the fluid core. The thickness of EDL is characterized Debye length, which is the wall normal distance over which the net charge has decreased form a magni-

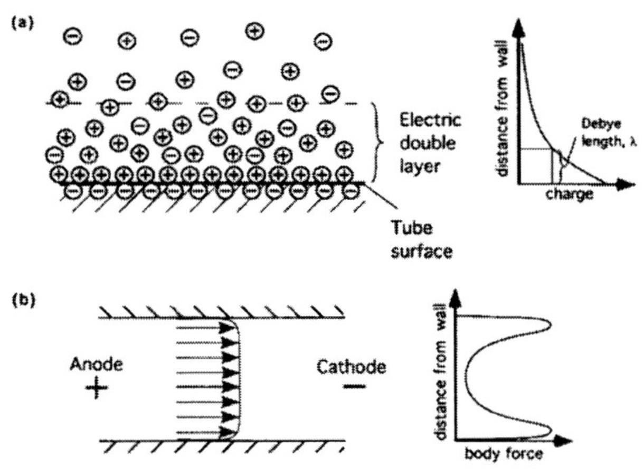

Figure 1. Schematic illustration of electro-osmotically generated flow.

tude near the tube surface to 1/e (37%) of the surface charge. The positively charged cations and solvent molecules strongly absorbed at the wall remain stationary under the influence of an electric potential in the streamwise direction, while the mobile cations in the EDL very near the tube walls will migrate toward the cathode due to the excess charge in the layer. This gives rise to a concentrated fluid body force near the channel or tube walls as illustrated in Fig 1(b). Viscous shear force transmitted from EDL to the channel or tube center pull the core fluid towards the cathode as well. The resulting electro-osmotic flow velocity distribution is a function of the ratio of the hydraulic radius to the Debye length.

Electro-osmotic flow has shown immense potential as a non-mechanical driving mechanism for Newtonian and non-Newtonian fluids in microchannels. It offers many advantages over other methods including plug-like velocity profile, high reliability, ease of fabrication and control, and no noise. The details of electro-osmotic flow are given by Masliyah and Bhattacharje [1]. Non-Newtonian fluids are generally divided in to shear thinning (or pseudoplastic, e. g. glues, blood, and polymer melts), viscoplastic (e. g. paints and concentrated suspensions) fluids , and shear thickening (or dilatant, e. g. wet sand, sugar and borax solutions) fluids. The properties and behavior of power-law fluids are discussed by Chhabra and Richardson [2]. Kang et al. [3] analyzed the electro-osmotic flow in a microchannel under both direct and alternating electric fields. They solved analytically Navier-Stokes and Poisson-Boltzmann equations to obtain interstitial tubular velocity and estimated the electroosmosis in the overall porous media. Park et al. [4] developed a reduced-order model for electro-osmotic flow in a microchannel of nonuniform cross section. They used the Karhunen-Loeve Galerkin (KLG) procedure and predicted electro-osmotic flows accurately for a wide range of parameters. Datta et al. [5] determined experimentally the electro-osmotic flow rate in fused silica microchannels. They found that the flow rate decreases when trace quantities of salts containing the divalent cations are added to an electrolytic solution flowing in a channel having negatively charged walls. Kwak et al. [6] conducted a numerical investigation on the electro-osmotic flow and heat transfer in a 2D microchannel. They explored a new conceptual idea to control electro-osmotic flow by using a thermal field effect through the temperature-dependent physical properties.

Marcos et al. [7] analyzed the dynamic aspects of the electro-osmotic flow in rectangular microchannels. They obtained an analytical solution for electrical potential distribution in the channel. Zhang et al. [8] presented computer simulations of electro-osmotic flow through rectangular microchannels. They investigated the effects of the channel cross-section, channel size, the ionic concentration and the strength of the applied electrical field on the volumetric flow rate and the average velocity. Craven et al. [9] presented a complete model for the ion concentrations, electric field, and fluid flow in complex microchannels. They proposed an empirical slip velocity boundary condition for electro-osmotic flow. Hu et al. [10] studied experimentally and theoretically the electro-osmotic flow in slit microchannels. They found that the average electro-osmotic velocity was lower than that in a smooth microchannel and that this velocity reduction increases with the size of the prisms.

Tang et al. [11] presented numerical analysis for transient temperature development in microchannels. They proposed a model based on the Poisson-Boltzmann equation, the modified Navier-Stokes equations, the conjugate energy equation, and the mass species transport equation. Horiuchi and Dutta [12] analyzed heat transfer characteristics of steady electro-osmotic flows in two-dimensional straight microchannels. They found that the Nusselt number of pure electro-osmotic flow is higher than that of plane Poiseulle flow and the presence of the electro-osmotic flow with pressure gradient makes the thermal entrance length smaller and enhances heat transfer rates. Monazarami et al. [13] developed a numerical model to study the heat transfer characteristics of pressure driven and electro-osmotic flow through a microchannel. Peng et al. [14] performed computer simulations of electro-osmotic flow through rectangular microchannels. They found significant influences of the channel aspect ratio on the volumetric flow rate. Teck et al. [15] studied theoretically the electrokinetic flow through star-point family microchannels. They developed a model for the steady, fully developed electro-osmotic flow in the closed-end star point family microchannels. Teck et al. [16] studied theoretically and experimentally the electro-osmotic displacing process between two solutions in a rectangular microchannel. They developed a reliable theoretical model to predict the electro-osmotic displacing process.

Chakraborty [17] developed closed form expressions for Nusselt number variation in a microtube. He considered the effects of both electro-osmotic forces and imposed pressure gradients and analyzed the results, as a function of the dimensionless tube radius. Liqing and Dongqing [18] investigated the characteristics of electro-osmotic flow in a cylindrical microchannel with nonuniform zeta potential. They developed models for the electrical double-layer and the flow fields using the Poisson-Boltzmann and momentum equations. Their numerical results show possible effects of bioadhesion on the electro-osmotic flow in biochip devices. Yang et al. [19] analyzed the electro-osmotic flow induced by an applied electrostatic potential field through microchannels. They solved Poisson-Boltzmann and Laplace equations numerically using a finite difference method. They discussed the effects of the electrical double-layer field and the applied electrostatic field on the velocity distribution, pressure drop, and skin friction. They found a small pressure drop along the parallel plates, which was neglected in the literature.

Bhattacharyya, and Nayak [20] discussed a two-dimensional model to analyze the electro-osmotic flow in a micro- or nanochannel. Their model can handle electrolyte of arbitrary ionic valence. They employed a pressure correction-based iterative algorithm (SIMPLE) for investigating both weak and strong electrolytes at different channel heights. Arun and Suman [21] obtained analytical solutions for the temperature and Nusselt number in the thermal entrance region of a parallel plate microchannel. They considered both pressure-driven and electro-osmotic transport mechanisms along with the the effects of viscous dissipation, Joule heating and axial conduction. Mayness and Webb [22] analyzed the influence of viscous dissipation on thermally fully-developed, electro-osmotically generated flow in a parallel plate microchannel and circular microtube under constant wall temperature and constant wall heat flux boundary conditions. They determined the solution for the dimensionless temperature distribution and Nusselt number for both geometries and thermal boundary conditions.

The above literature survey reveals that all the studies are concerned with electro-osmotic or pressure driven flows of Newtonian fluids in microchannels. The studies related to the electro-osmotic flow of non-Newtonian fluids in microchannels are very scarce.

Tang et al. [23] carried out a numerical study of electro-osmotic flow of non-Newtonian fluid in microchannels. They solved lattice Boltzmann equations to obtain the electric potential distribution in the electrolyte and the flow field for commonly used non-Newtonian fluids. They found that the fluid rheological behavior changes the electro-osmotic flow pattern significantly. Zhao et al. [24] analyzed electro-osmotic flow of power-law fluids in a slit channel. They determined analytical solution of the velocity distributions, shear stress and dynamic viscosity. They examined the effects of κH, flow behavior index, double layer thickness, and applied electric field on the velocity distribution, shear stress, dynamic viscosity, and flow rate of the electro-osmotic flow of power-law fluids. Philip and Wooding [25] developed an analytical approximation for the solution of the Poisson-Boltzmann equation. They treated a symmetrical electrolyte specifically, that applies also to nonsymmetrical ones. Their analysis preserves the mathematical character of the exact problem and gives results with negligible errors. Das and Chakraborty [26] obtained analytical solutions for the transport characteristics of a non-Newtonian fluid flowing a rectangular microchannel. They analyzed the flow behaviour of a blood sample and developed a model for the hematocrit fraction in the sample.

This survey reveals that no study investigates the heat transfer in electro-osmotic flows of power-law fluids through microchannels. In this study, the dimensionless heat transfer rate in electro-osmotic flow of power-law fluids is obtained for two different cases when $0 < \kappa H \leq 1$ and $\kappa H > 1$ and the effects of the dimensionless electrokinetic parameter, κH, flow behavior index n, and Brinkman number Br on the dimensionless local temperature and average heat transfer are examined for both cases.

2. Analysis

The geometry of a slit microchannel is shown in Figure 2. The height of the channel is $2H$. The surfaces of channel are assumed to be smooth. A liquid electrolyte (power-law fluid) of dielectric constant ε, the flow consistency index m, and the flow behavior index n is used in the channel. The fluid properties are assumed to be constant. The slit wall is assumed to be uniformly charged with a zeta-potential, ψ_w and an external electric field E_0

is imposed along x direction. It is assumed that the electric double layers formed at the slit microchannel wall do not overlap, and the Debye-Hückel linearization principle is valid. It is also assumed that no pressure is applied and there is no gravitational body force. The flow in the channel is assumed to be steady, laminar and fully developed. The temperature of the channel walls is assumed to be $T_w(x)$ with an inlet fluid temperature of T_∞. At the channel wall, a uniform heat flux is applied. Taking advantage of symmetry, the upper half of the channel is selected. The fluid sets in motion due to electroosmosis, and the fluid

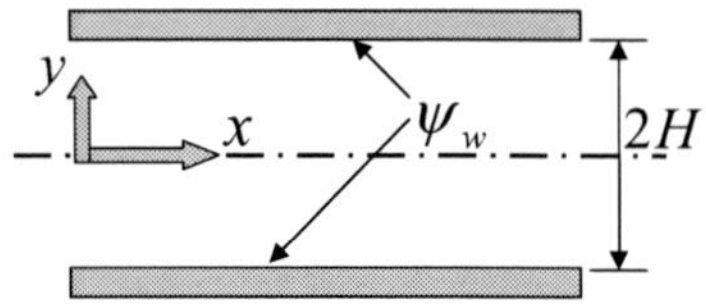

Figure 2. Geometry of a 2D slit microchannel.

flow is governed by continuity and Cauchy momentum equations, whereas heat transfer is governed by energy equation. Following [1] and [21], and considering the effect of the viscous dissipation, the governing equations can be written for fully developed flow as:

$$\frac{\partial u}{\partial x} + \frac{\partial v}{\partial y} = 0 \tag{2.1}$$

$$\frac{d}{dy}\left[\left(-\frac{du}{dy}\right)^n\right] = -\frac{\kappa^2 \varepsilon_0}{m}\psi \tag{2.2}$$

$$u\frac{\partial T}{\partial x} + v\frac{\partial T}{\partial y} = \alpha \frac{\partial^2 T}{\partial y^2} + \frac{\nu}{c_p}\left(\frac{du}{dy}\right)^{n+1} \tag{2.3}$$

with boundary conditions

$$\left.\begin{array}{lll} y = 0 & : \frac{du}{dy} = 0, & \frac{\partial T}{\partial y} = 0 \\ y = H & : u = 0, & T = T_w(x) \end{array}\right\} \tag{2.4}$$

where $\frac{\partial T}{\partial x} = \frac{dT_m}{dx}$=constant for a thermally fully developed flow with isoflux boundary condition.

According to the theory of electrostatics, the net charge density ρ_e in the diffuse layer of the wall EDL is given by the Poisson's equation:

$$\varepsilon \frac{d^2\psi}{dy^2} = -\rho_e \tag{2.5}$$

With the assumptions of the Boltzmann distribution and small zeta potentials, the electrical potential profile in the EDL is governed by the linearized poisson-Boltzmann equation expressed as:

$$\frac{d^2\psi}{dy^2} = \kappa^2\psi \tag{2.6}$$

The boundary conditions are given by:

$$\text{at} \quad y = H, \psi = \psi_w \quad \text{and} \quad \text{at } y = 0, d\psi/dy = 0$$

The solution for the electrical potential distribution is given by

$$\psi(y) = \psi_w \frac{\sinh(\kappa y)}{\cosh(\kappa H)} \tag{2.7}$$

The net charge density can be expressed as a function of the EDL potential

$$\rho_e(y) = -\kappa^2 \varepsilon \psi \tag{2.8}$$

Zhao et al. [21] integrated Eq. (2) twice and obtained the following velocity distribution

$$u(y) = \kappa^{(n-1)/n} \left(-\frac{\varepsilon \psi_w E_0}{m} \right)^{1/n} \frac{\int_{\kappa y}^{\kappa H} \sinh^{1/n}(\kappa y) d(\kappa y)}{\cosh^{1/n}(\kappa H)} \tag{2.9}$$

It is important to note that Zhao et al. [21] claimed analytical solutions for $n = 1, 1/2$, and $1/3$ only, whereas these authors have found analytical solutions for $n = 1/j$, where $j = 1 \cdots 10$. The velocity distribution for $j = 1, 3, 5, 7, 9, 10$ can be determined from

$$\begin{aligned} u(y) \;=\; & \frac{\left(-\frac{\varepsilon \psi_w E_0}{m} \right)^j}{\kappa^{j-1} \cosh^j(\kappa H)} \left[\frac{1}{j} (\sinh^{j-1}(\kappa H) \cosh(\kappa H) \right. \\ & - \sinh^{j-1}(\kappa y) \cosh(\kappa y))) - \frac{(j-1)}{j(j-2)} [\sinh^{j-3}(\kappa H) \\ & \cosh(\kappa H) - \sinh^{j-3}(\kappa y) \cosh(\kappa y)] + \frac{(j-1)(j-3)}{j(j-2)(j-4)} \left[\sinh^{j-5}(\kappa H) \cosh(\kappa H) \right. \\ & \left. - \sinh^{j-5}(\kappa y) \cosh(\kappa y) \right] \\ & - \frac{(j-1)(j-3)(j-5)}{j(j-2)(j-4)(j-6)} \left[\sinh^{j-7}(\kappa H) \cosh(\kappa H) - \sinh^{j-7}(\kappa y) \cosh(\kappa y) \right] \\ & + \frac{(j-1)(j-3)(j-5)(j-7)}{j(j-2)(j-4)(j-6)(j-8)} \\ & \left[\sinh^{j-9}(\kappa H) \cosh(\kappa H) \right. \\ & \left. - \sinh^{j-9}(\kappa y) \cosh(\kappa y) \right] \\ & - \frac{(j-1)(j-3)(j-5)(j-7)(j-9)}{j(j-2)(j-4)(j-6)(j-8)} [\kappa(H-y)] \end{aligned} \tag{2.10}$$

For other even values of j, the velocity expressions are given by

$$u(y) = \frac{\left(-\frac{\varepsilon \psi_w E_0}{m} \right)^2}{2\kappa \cosh^2(\kappa H)} [\sinh(\kappa H) \cosh(\kappa H) - \sinh(\kappa y) \cosh(\kappa y) - \kappa (H-y)] \tag{2.11}$$

for $j = 2$,

$$u(y) = \frac{\left(-\frac{\varepsilon\psi_w E_0}{m}\right)^4}{8\kappa^3 \cosh^4(\kappa H)} [2\sinh^3(\kappa H)\cosh(\kappa H) - 2\sinh^3(\kappa y)\cosh(\kappa y)$$
$$- \sinh(\kappa H)\cosh(\kappa H)$$
$$+ \sinh(\kappa y)\cosh(\kappa y) + \kappa(H - y)] \tag{2.12}$$

for $j = 4$,

$$u(y) = \frac{\left(-\frac{\varepsilon\psi_w E_0}{m}\right)^6}{48\kappa^5 \cosh^6(\kappa H)} [8\sinh^5(\kappa H)\cosh(\kappa H) - 8\sinh^5(\kappa y)\cosh(\kappa y)$$
$$-10\sinh^3(\kappa H)\cosh(\kappa H)$$
$$+10\sinh^3(\kappa y)\cosh(\kappa y) + 15\sinh(\kappa H)\cosh(\kappa H)$$
$$-15\sinh(\kappa y)\cosh(\kappa y) - 15\kappa(H - y)] \tag{2.13}$$

for $j = 6$, and

$$u(y) = \frac{\left(-\frac{\varepsilon\psi_w E_0}{m}\right)^8}{384\kappa^7 \cosh^8(\kappa H)} [48\sinh)^7(\kappa H)\cosh(\kappa H)$$
$$-48\sinh^7(\kappa y)\cosh(\kappa y) - 56\sinh^5(\kappa H)\cosh(\kappa H)$$
$$+56\sinh^5(\kappa y)\cosh(\kappa y) + 70\sinh^3(\kappa H)\cosh(\kappa H)$$
$$-70\sinh^3(\kappa y)\cosh(\kappa y) - 105\sinh(\kappa H)\cosh(\kappa H)$$
$$+105\sinh(\kappa y)\cosh(\kappa y) + 105\kappa(H - y)] \tag{2.14}$$

for $j = 8$.
Zhao et al. [21] approximated analytical velocity distributions for the following two cases:

2.1. Case I: $(0 < \kappa H \leq 1)$

Following Zhao et al. [21], the dimensionless velocity distribution for this case can be written as

$$\frac{u(y)}{u_{av}} = \frac{2n+1}{n+1}\left(1 - \frac{y}{H}\right)^{(n+1)/n} \tag{2.15}$$

where u_{av} is the average velocity given by

$$u_{av} = n\kappa^{(1-n)/n}\left(-\frac{\varepsilon\psi_w E_0}{m}\right)^{1/n} \frac{\frac{1}{1+2n}(\kappa H)^{(1+n)/n}}{\cosh^{1/n}(\kappa H)} \tag{2.16}$$

Using dimensionless variables

$$\eta = \frac{y}{H} \qquad \text{and} \qquad \theta(\eta) = \frac{T - T_w}{(q_w H/k)} \tag{2.17}$$

Equation (3) can be transformed in to dimensionless form

$$\theta''(\eta) + \frac{2n+1}{n+1}a\left(1-\eta^{\frac{n+1}{n}}\right) + \left(\frac{2n+1}{n}\right)^{n+1} Br\,\eta^{\frac{n+1}{n}} \tag{2.18}$$

where a is a constant and Br is known as the Brinkman number and are given by

$$\left.\begin{aligned} a &= -\frac{u_{av}H^2}{\alpha(q_w H/k)}\cdot\frac{dT_m}{dx} \\ Br &= \frac{\nu}{c_p}\frac{H^2}{\alpha(q_w H/k)}\left(\frac{u_{av}}{H}\right)^{n+1} \end{aligned}\right\} \tag{2.19}$$

The transformed boundary conditions can be written as

$$\theta'(0) = 0 \qquad \text{and} \qquad \theta(1) = 0 \tag{2.20}$$

The analytical solution of Eq. (19) with boundary conditions (20) can be written as

$$\theta(\eta) = \frac{1}{n(n+1)}\left[-\frac{na\eta^2(2n+1)}{2} + \frac{an^3\eta^{\frac{3n+1}{n}}}{3n+1} + \frac{\left(\frac{2n+1}{n}\right)^n Br\,n^3\eta^{\frac{3n+1}{n}}}{2n+1} - \frac{\left(\frac{2n+1}{n}\right)^n Br\,n^2\eta^{\frac{3n+1}{n}}}{(2n+1)(3n+1)}\right] + \frac{\left[4an + 2\left(\frac{2n+1}{n}\right)^n Br\,n + a\right]}{2(3n+1)} \tag{2.21}$$

For a fully developed flow and uniform wall heat flux boundary condition, the dimensionless mean or bulk temperature is defined as:

$$\theta_m = \frac{T_m - T_w}{\left(\frac{q_w H}{k}\right)} = \frac{\int_0^1 u(\eta)\theta(\eta)d\eta}{\int_0^1 u(\eta)d\eta} \tag{2.22}$$

which gives the Nusselt number

$$Nu = -\frac{2}{\theta_m} \tag{2.23}$$

2.2. Case II: $\kappa H > 1$

For this case, Zhao et al. [24] integrated Eq. (9) by employing approximation used by Philip and Wooding [25] and obtained the following dimensionless velocity distributions:

$$\frac{u(y)}{u_{av}} = \begin{cases} C\left\{\frac{2^{1/n}}{n+1}\left[1-(\kappa y)^{(n+1)/n}\right]+R\right\}, & 0 \le \kappa y \le 1 \\ C\left\{e^{\kappa H/n} - e^{\kappa y/n}\right\}, & \kappa y > 1 \end{cases} \tag{2.24}$$

where C and R are constants and given by

$$C = \left[\left(1-\frac{n}{\kappa H}\right)e^{\kappa H/n} + \frac{(n-1)\,e^{1/n}}{\kappa H} + \frac{2^{1/n}}{\kappa H(1+2n)}\right]^{-1} \tag{2.25}$$

and

$$R = e^{\kappa H/n} - e^{1/n} \tag{2.26}$$

Using these dimensionless velocities and variables defined by (17), Eq. (3) can be transformed in the dimensionless forms for both cases:

$$\theta''(\eta) + Br\left(\frac{C}{n}\right)^{n+1}(2\kappa H\eta)^{\frac{n+1}{n}} + aC\left\{\frac{2^{1/n}\left[1-(\kappa H\eta)^{\frac{n+1}{n}}\right]}{n+1}+R\right\} = 0 \tag{2.27}$$

for $0 \le \kappa y \le 1$ and

$$\theta''(\eta) + Br\left(\frac{C}{n}\right)^{n+1}\left[\left(\frac{\kappa H}{n}\right)e^{\kappa y/n}\right]^{n+1} + aC\left(e^{\kappa H/n} - e^{\kappa H\eta/n}\right) \tag{2.28}$$

for $\kappa y > 1$. The constant a and Brinkman number Br are given by Eq. (19). The Eqs. (27) and (28) are solved with (20) numerically using an implicit finite difference method and the dimensionless heat transfer rate Nu is obtained using Eq. (19).

3. Results and Discussion

Temperature distribution and dimensionless heat transfer rates in electro-osmotic flow of power-law fluids in a slit microchannel are determined by solving energy equation analytically for $0 \le \kappa H \le 1$ and numerically for $\kappa H > 1$ using an implicit finite difference method. The electrokinetic parameter, power-law index and the Brinkman number are the main parameters which govern heat transfer in a slit microchannel. The effects of these parameters on local dimensionless temperature and average heat transfer are investigated for both cases. These effects for the case I, $0 \le \kappa H \le 1$, are shown in Figs. 3 and 4. Figure 3 (a) shows the temperature distribution for $Br < 0$ whereas Fig. 3(b) shows temperature distribution for $Br > 0$. As expected, the temperature is maximum at the center of the channel and decreases up to the wall. When $Br < 0$, pseudoplastic fluids have higher temperature whereas dilatant fluids have higher temperature for $Br > 0$. This is due to the rheological properties of power-law fluids. The comparison shows that the effect of viscous dissipation on the dimensionless temperature distribution is greater in dilatant fluids than that in pseudoplastic fluids. Figures 4 (a) and (b) illustrate the variation of the Nusselt number with the

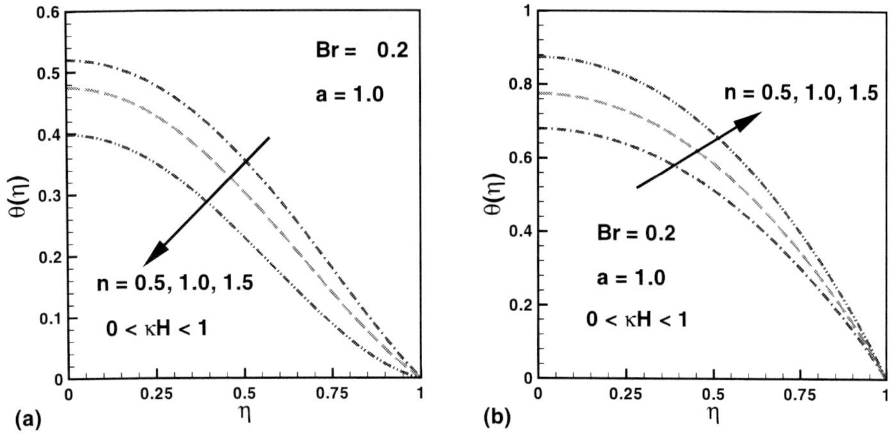

Figure 3. Temperature distribution for different power-law fluids in a slit microchannel when $0 < \kappa H < 1$.

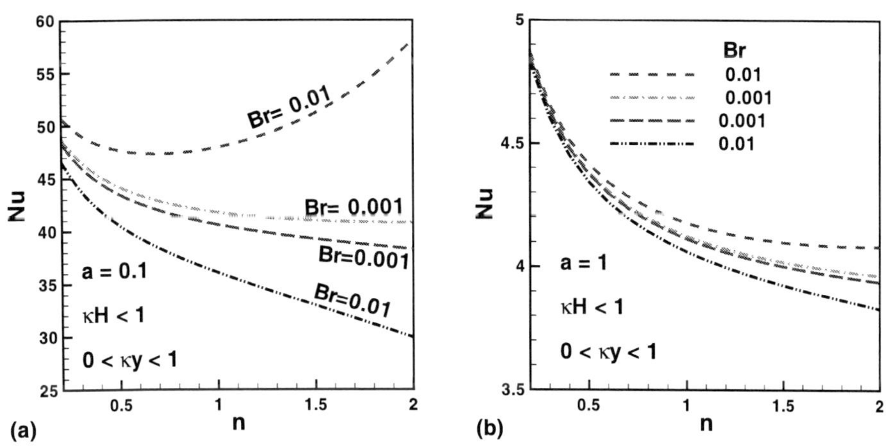

Figure 4. Effects of Brinkman number on dimensionless heat transfer rate in a slit microchannel.

power-law index for different Br. It is clear that the Nusselt number decreases as the power-law index increases for higher Br, but with the decrease in Br and specially for negative Br, Nusselt numbers starts increasing for dilatant fluids. This is because the positive values of Br correspond to the hot wall ($T_w > T_c$), while the negative values of Br correspond to the cold wall ($T_c > T_w$). It is also clear from Figs. 4 that Nu increases with the decrease in Br. This is because the viscous dissipation adds heat transfer for the wall cooling case. As seen from Fig. 4 (b), Nu decreases for all values of the n due to larger values of the constant a.

In addition, as observed from Fig. 4 (b), Br is more effective on Nu for pseudoplastic fluids than for dilatant fluids.

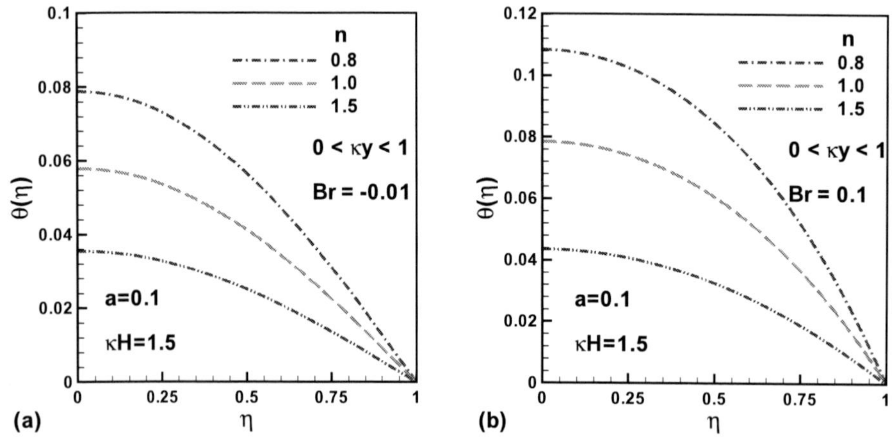

Figure 5. Effects of Br on dimensionless heat transfer rate for different power-law fluids.

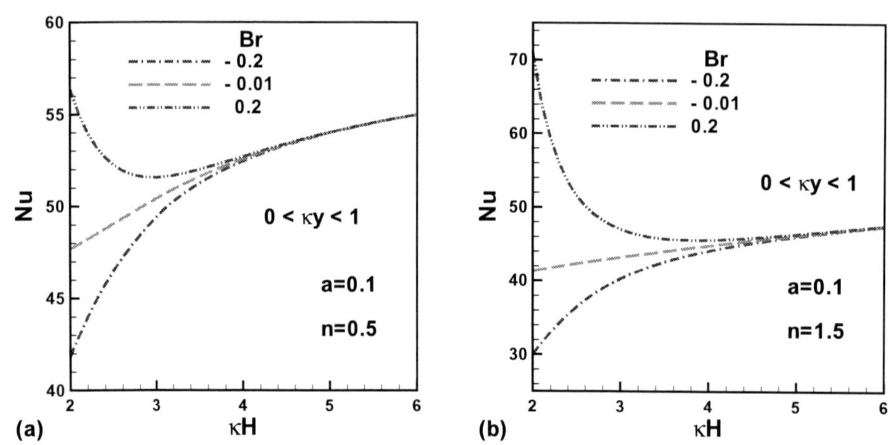

Figure 6. Effects of power-law index n and Br on dimensionless heat transfer rate when $0 < \kappa y < 1$.

The effects of main parameters on the local temperature and heat transfer rates are shown in Figs. 5 and 6 for the case II when $0 < \kappa y < 1$ and $\kappa H > 1$. Figures 5 (a) and (b) show the temperature distributions for different power-law fluids when $Br < 0$ and $Br > 0$. In both figures, the temperature for pseudoplastic fluids is found to be higher than dilatant fluids. Again, the temperature is maximum at the center and decreases up to the wall. It is also clear from Fig. 5 (b) that the temperatures of each fluid are higher for $Br > 0$. The effects of electro kinetic parameter κH and Br on dimensionless heat transfer Nu for pseudoplastic and dilatant fluids are shown in Fig. 6 (a) and (b). It is clear from both figures that the dimensionless heat transfer is independent of higher values of electro kinetic parameter κH for both fluids. However, for smaller values of κH, heat transfer

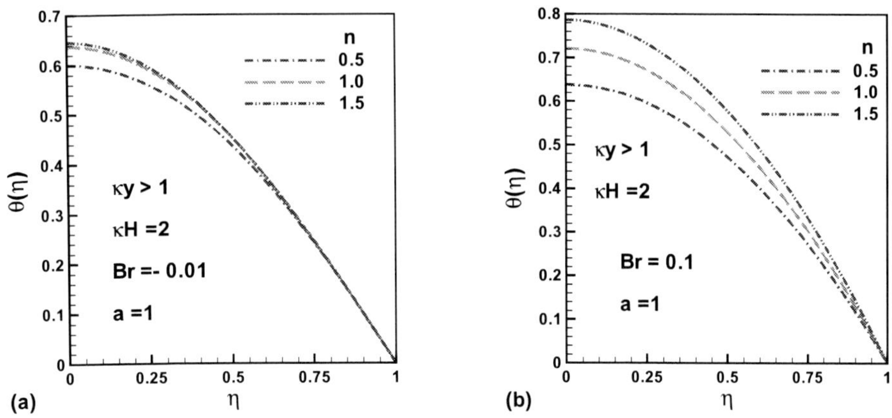

Figure 7. Effects of Br on temperature distribution for different power-law fluids when $\kappa y > 1$.

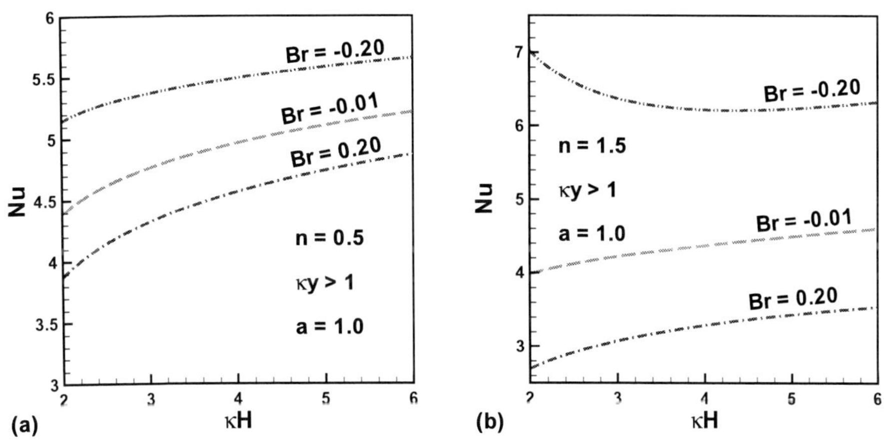

Figure 8. Effects of power-law index n and Br on dimensionless heat transfer rate when $\kappa y > 1$.

increases with the increase in Br. For $Br < 0$, heat transfer increases with κH, whereas for $Br > 0$, heat transfer decreases for smaller values of κH and increases for larger values of κH.

The local dimensionless temperature distribution and heat transfer from a slit microchannel for the case when $\kappa H > 1$ and $\kappa y > 1$ are shown in Figs. 7 and 8. Figure 7 shows the effects of power-law index and Br on the local dimensionless temperature distribution for $\kappa H = 2$. It is clear, from both figures, that in this case pseudoplastic fluids have lower temperature than dilatant fluids. As usual, temperature is maximum at the center and decreases up to the wall. The temperature also increases with Br. The variation of dimensionless heat transfer with the electro kinetic parameter is shown in Fig. 8 (a) for pseudoplastic fluid and in Fig. 8 (b) for dilatant fluid. It is clear that heat transfer of pseudoplastic fluid increases with κH for each Br, whereas heat transfer of dilatant fluid increases for $Br > 0$ and decreases for $Br < 0$.

4. Conclusion

Temperature distribution and heat transfer in electro-osmotic flow of power-law fluids in a slit microchannel are investigated for uniform wall heat flux thermal boundary condition. Energy equation with viscous dissipation term is solved for two cases, when $0 \leq \kappa H \leq 1$ and when $\kappa H > 1$. Analytical solutions are obtained for dimensionless velocity and temperature distribution for the first case $0 \leq \kappa H \leq 1$, whereas numerical solutions are obtained for the second case $\kappa H > 1$. The effects of electro kinetic parameter κH, flow behavior index n, and Brinkman number Br on the dimensionless local temperature and average heat transfer rate are examined for each case. It is found that for

1. $0 \leq \kappa H \leq 1$, pseudoplastic fluids have higher temperature for $Br < 0$, whereas dilatant fluids have higher temperature for $Br > 0$ under the same conditions. Br is found to be more effective on Nu for pseudoplastic fluids than for dilatant fluids.

2. $\kappa H > 1$ and $0 \leq \kappa y \leq 1$, the temperature for pseudoplastic fluids is found to be higher than dilatant fluids. The dimensionless heat transfer increases with the increase in Br for smaller values of κH,however, it is found independent of higher values of κH for both fluids.

3. $\kappa H > 1$ and $\kappa y > 1$, pseudoplastic fluids have lower temperature than dilatant fluids. The heat transfer of pseudoplastic fluid increases with κH for each Br, whereas heat transfer of dilatant fluid increases for $Br > 0$ and decreases for $Br < 0$.

References

[1] J. H. Masliyah and S. Bhattacharje, *Electrokinetic and Colloid Transport Phenomena,* JohnWiley & Sons, Inc., Hoboken, New Jersey, 2006.

[2] R.P. Chhabra, and J.F. Richardson, Non-Newtonian *Flow in the Process Industries:Butterworth-Heinemann,* Oxford, 1999.

[3] Kang, Y.J., Yang, C., and Huang, X.Y., Analysis of electro-osmotic flow in a microchannel packed with microspheres, *Microfluidics and Nanofluidics,* Vol. 1, No. 2, pp. 168-176, 2005.

[4] Park, H.M., Hong, S.M., and Lee, J.S., *Estimation of zeta potential of electro-osmotic flow in a microchannel using a reduced-order model, Biomedical Microdevices* , Vol. 9, No. 5, pp. 751-60, 2007.

[5] Datta, S., Conlisk, A.T., Li, H.F., and Yoda, M., *Effect of divalent ions on electro-osmotic flow in microchannels, Mechanics Research Communications,* Vol. 36, No. 1, pp. 65-74, 2009.

[6] Kwak, Ho Sang, Kim, H., Hyun, Jae Min, Song, Tae-Ho, Thermal control of electro-osmotic flow in a microchannel through temperature-dependent properties, *Journal of Colloid and Interface Science,* Vol. 335, No. 1, pp. 123-129, 2009.

[7] Marcos, Y. C., Wong, T. N., Ooi, K. T., Dynamic aspects of electro-osmotic flow in rectangular microchannels, *International Journal of Engineering Science,* Vol. 42, No. 13-14, pp. 1459-1481, 2004.

[8] Zhang, P., Zuo, C., Zhou, D., Chen, H., and Liu Y, Numerical simulation on electro-osmotic flow in a rectangular microchannel, *6th International Conference on Electronics Packaging Technology,* pp. 591-595, 2005.

[9] Craven, T.J., Rees, J.M., and Zimmerman, W.B., On slip velocity boundary conditions for electro-osmotic flow near sharp corners, *Physics of Fluids,* Vol. 20, No. 4, pp.1-14, 2008.

[10] Hu, Y., Xuan, X., Werner, C., and Li, D., Electro-osmotic flow in microchannels with prismatic elements, *Microfluidics and Nanofluidics,* Vol. 3, No. 2, pp. 151-160, 2007.

[11] G.Y. Tang, C. Yang, C.K. Chai, and H.Q. Gong, Numerical analysis of the thermal effect on electro-osmotic flow and electrokinetic mass transport in microchannels, *Analytica Chimica Acta,* Vol. 507, No. 1, pp. 27-37, 2004.

[12] K. Horiuchi, and P. Dutta, Heat transfer characteristics of mixed electro-osmotic and pressure driven micro-flows, *JSME International Journal, Series B (Fluids and Thermal Engineering),* Vol. 49, No. 3, pp. 812-19, 2006.

[13] R. Monazarami, A. Q. Zade, M. T. Manzari, Analysis of fluid flow and heat transfer in microchannels using combined pressure gradient and electro-osmotic pumping, *Proceedings of the 3rd International Conference on Microchannels and Minichannels* , 2005, Part A, pp. 503-510, 2005.

[14] Z. Peng, Z. Chuncheng, Z. Deyi, C. Hongli, and L. Yan, Numerical simulation on electro-osmotic flow in a rectangular microchannel, *6th International Conference on Electronics Packaging Technology*, 2005.

[15] N. W. Teck, Z. Yali, Y. Chun, and T. O. Kim, Electro-osmotic flow in irregular shape microchannels, *International Journal of Engineering Science,* Vol. 43, No. 19-20, pp. 1450-63, 2005.

[16] N. W. Teck, W. Cheng, Y. Chun, and T. O. Kim , Characterization of electro-osmotic flow in rectangular microchannels, *International Journal of Heat and Mass Transfer,* Vol. 50, No. 15-16, pp. 3115-21, 2007.

[17] S. Chakraborty, Analytical solutions of Nusselt number for thermally fully developed flow in microtubes under a combined action of electro-osmotic forces and imposed pressure gradients, *International Journal of Heat and Mass Transfer,* Vol. 49, pp. 810-813, 2006.

[18] Liqing Ren and Dongqing Li, *Electro-osmotic Flow in Heterogeneous Microchannels, Journal of Colloid and Interface Science*, Vol. 243, pp. 255-261, 2001.

[19] R. J. Yang, L. M. Fu, and Y. C. Lin, Electro-osmotic Flow in Microchannels, *Journal of Colloid and Interface Science,* Vol. 239, pp. 98-105, 2001.

[20] S. Bhattacharyya, A.K. Nayak, Electro-osmotic flow in micro/nanochannels with surface potential heterogeneity: An analysis through the Nernst-Planck model with convection effect, *Colloids and Surfaces A: Physicochem. Eng. Aspects,* Vol. 339, pp. 167-177, 2009.

[21] S. Arun, C. Suman, Semi-analytical solution of the extended Graetz problem for combined electro-osmotically and pressure-driven microchannel flows with step-change in wall temperature, *International Journal of Heat and Mass Transfer,* Vol. 51, pp. 4875-4885, 2008.

[22] D. Maynes, B.W. Webb, The effect of viscous dissipation in thermally fully-developed electro-osmotic heat transfer in microchannels, *International Journal of Heat and Mass Transfer,* Vol. 47, pp. 987-999, 2004.

[23] G.H. Tang, X.F. Li, Y.L. He, and W.Q. Tao, Electro-osmotic flow of non-Newtonian fluid in microchannels, *Journal of Non-Newtonian Fluid Mechanics,* Vol. 157, No. 1-2, pp. 133-137, 2009.

[24] C. Zhao, Z. Emilijk, H. M. Jacob, and Y. Chun, Analysis of electro-osmotic flow of power-law fluids in a slit microchannel, *Journal of Colloid and Interface Science*, Vol. 326, pp. 503-510, 2008.

[25] J. R. Philip, and R. A. Wooding, Solution of the Poisson-Boltzmann equation about a cylindrical particle, *Journal of Chemical Physics,* Vol. 52, No. 2, pp. 953-9, 1970.

[26] S. Das, S. Chakraborty, Analytical solutions for velocity, temperature and concentration distribution in electro-osmotic microchannel flows of a non-Newtonian bio-fluid, *Analytica Chimica Acta,* Vol. 559, pp. 15-24, 2006.

In: Frontier Research in Microscale and Nanoscale... ISBN 978-1-61324-665-8
Editor: Lixin Cheng

Chapter 15

SORPTION HEAT PUMPS AND HEAT PIPES TO INCREASE THE EFFICIENCY OF THE NEW POWER SOURCES

L. L. Vasiliev*, O. Filatova, M. Kovaleva, S. Tishkevich and A. A. Antukh
Luikov Heat and Mass Transfer Institute, National Academy of Science, Belarus

ABSTRACT

The modern tendency to improve the power sources efficiency is related to the fuel economy and nature protection. One of the promising solutions in the development of energy efficient technology of power sources is encompassed on the advanced solid sorption systems application (including co-generation and tri-generation, Photovoltaic/Thermal (PVT) solar collectors, fuel cells, Stierling and Diesel-generator) with heat pipe thermal management. Such technology provides the local heat, cold and electricity production. Some examples of heat pipes are shown oriented on its thermal control application in fuel cells and solid sorption heat pumps.

Keywords: fuel cell, Diesel engine, heat pump, heat pipe, sorption, alternative energy source.

1. INTRODUCTION

The 21st century will see the development of a wide range of active energy devices with application in energy management and power sources, electronic cooling, and bioengineering. Nowadays utilizing nature friendly biomass fuels and solar hydrogen can be considered as an ultimate solution to the fast growing energy demand and greenhouse in atmosphere. At the same time it is very important to ensure the efficient consuming of the natural gas in low temperature technologies, converting it to hydrogen. The new solid and liquid sorption

* E-mail address: Leonard_Vasiliev@rambler.ru

technology and sorption machines are a good way to combine the fossil fuel resources (exhausted gases, vapour and water) with alternative (solar, water, ground, and air) and autonomous sources of energy (transport) to reduce the energy consumption down to 15–20%, Figure 1.

It is known that for the most power generation systems the efficiency of the cycle is near 30–35 %. The energy losses to recover are: exhausted gases, vapour, water, and etc. The other nature friendly energy generating technologies are the photovoltaic system, wind energy and water resources. Sorption heat pump technology gives a possibility to apply the waste energy of traditional power generating devices as a booster to activate the alternative sources of energy. One of the keen low temperature technologies is the electrochemical power source, namely fuel cell that converts the chemical energy to transform of hydrogen in coupling with air into electricity and that generate hydrogen and oxygen when it is used reversibly as an electrolyzer. Unlike conventional power devices, i.e. steam turbines, gas turbines and internal combustion engines, which are based on certain thermal cycles, the maximum efficiency of fuel cells is not limited by the Carnot cycle (G.R. Simander, Fuel Cell Technology for Natural Gas, E.V.A. – the Austrian Energy Agency), Figures 2 – 4. The fuel cell efficiency is better to compare with the internal combustion engine (diesel), Figures 3 – 5. But even fuel cells need to improve their efficiency of coefficient of the amplification (COA) due to the heat dissipation inside on the level of 50 %.

The efficiency of the combination fuel cell + solid sorption heat pumps + heat pipe heat exchangers has COA = 1.6, Figure 4. The efficiency of the combination Diesel engine + power generator + solid sorption heat pumps + heat pipe heat exchangers ensures COA = 1.4, Figure 5.

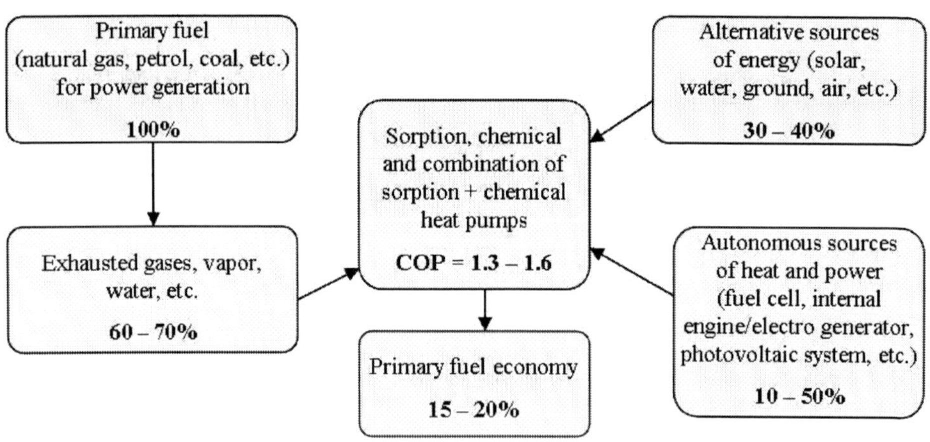

Figure 1.New possibilities to improve the efficiency of the fossil fuel, alternative and autonomous energy sources application using sorption and chemical heat pumps.

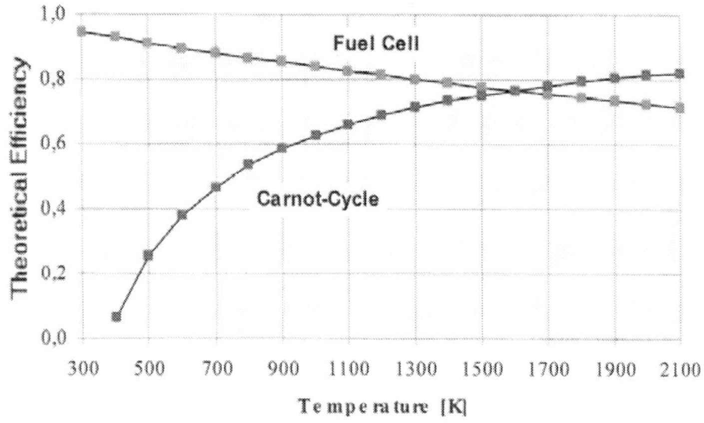

Figure 2.The efficiency of Carnot cycle and fuel cell cycle as a function of temperature.

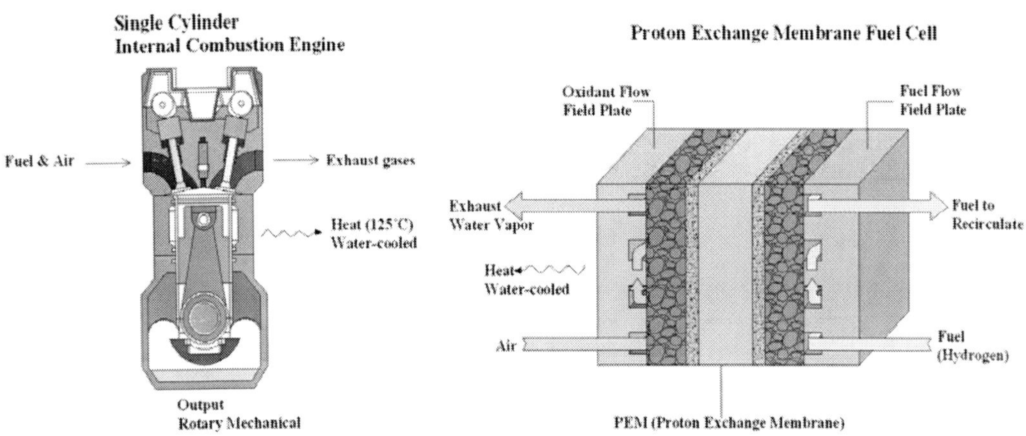

Figure 3. Carnot and fuel cells cycles schematic.

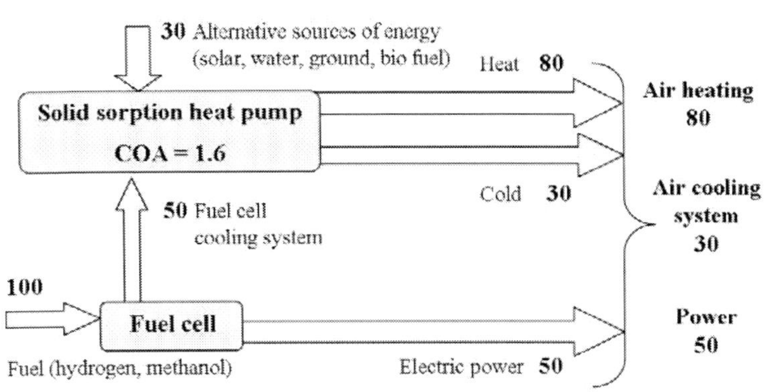

Figure 4. System of tri-generation based on fuel cell and solid sorption heat pump application.

Low temperature power systems are less expensive to build than high temperature ones. Since a major barrier to acceptance, this is a major concern for fuel cell technology, in general. For low-duty-cycle applications (such as private homes and many portable devices) the direct system cost is an important factor and efficiency is less important. For high-duty-cycle applications (such as hospital power and some telecommunications devices) the cost of the unit is offset by efficiency gains and reliability. Fuel cells, with unmatched efficiency and potentially greater reliability than other energy systems (because of greatly lower part counts, for example, and solid state construction in the case of solid oxide fuel cells (SOFCs)) are quite competitive in many applications, even at present cost. On the other hand, since the experience and familiarity with fuel cells systems is still small compared to many other energy systems, the risk cost is still considered to be high. In this respect, the low temperature systems enjoy an advantage over high temperature ones. We should expect to use many different technologies, often in combination. Fuel cells are well suited to this requirement, and operate with batteries, turbines, and exchangers as partners. Users of cars continually expect higher levels of comfort and ease of use. This leads to increasing requirements for on-board electrical power. The use of electrical equipment, such as electrical air conditioning during engine stand still is limited due to the batteries capacity. A fuel-based auxiliary power unit is one possible method of supplying this electrical power, Figure 4.

The thermodynamics of fuel cells and internal combustion engines is considered [1-3] as:

Fuel cell,

$$\Delta G = \Delta H - T\Delta S = -nF\Delta E \quad (1)$$

$$\mathrm{COA} = \frac{\Delta G}{\Delta H} = 1 - \frac{T\Delta S}{\Delta H}, \quad (2)$$

Carnot cycle engine,

$$\mathrm{COA} = \frac{W}{Q_1} = \frac{(Q_q - Q_2)}{Q_1} = \frac{(T_1 - T_2)}{T_1}, \quad (3)$$

The heat exchangers based on heat pipes are very welcome to connect thermally alternative sources of energy and fuel cells (or Carnot cycle engines) with solid/liquid sorption machines to increase its total efficiency. So, using sorption heat pumps in combination with power generating machines (co-generation and tri-generation), we have a real possibility to recover heat losses in fuel cell, or the Carnot cycle and transform them into heat, cold and power production, thus increase the cycle efficiency, Figure 1, Figures 4 – 5.

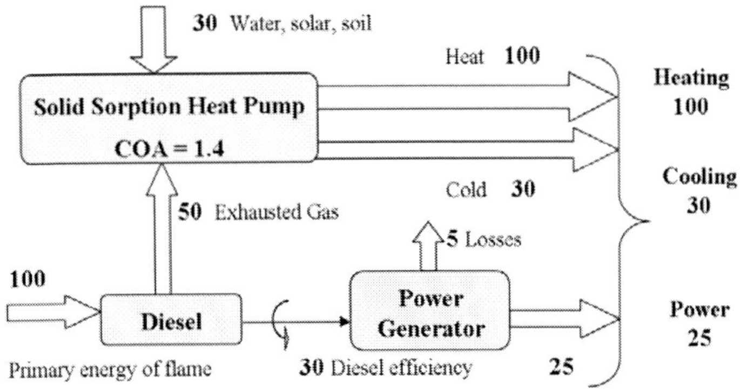

Figure 5. System of tri-generation based on Diesel engine and solid sorption heat pump application.

2. Solid Sorption Heat Pumps

Nowadays the sorption technology is steadily improving, and the increase at sorption market is strongly related to the energy policy in different countries. The solid sorption technology advantages at first are related to the nature friendly refrigerants such as water, ammonia, CO_2 (no CFC, HCFC, HFC) and at second they are thermally driven and can be coupled with a low temperature waste heat, solar heat, burning fossil fuel, or biomass. The solid sorption coolers demonstrate its possibility to be served as a very effective thermal compressor and accumulator of energy. Let us consider actual sorption technologies (liquid and solid sorption) as the potential source of energy recovery of primary fuels flames (including high temperature solid oxide fuel cells), which have different advantages and drawbacks with regard to their compactness, complexity, cost, the range of working temperature [2, 4, 5]. Sorption technologies represent the new stream and are divided on adsorption heat machines [3-9] and chemical heat machines [10-12]. Between adsorption and chemical machines are disposed special heat pumps and coolers with complex compound sorbent bed (complex compound sorption machines), in which adsorption and chemical reactions are available in the same space and are interacting at the same time of the cycle (ad/absorption) [13-18]. The sorbent bed of these complex compound sorption machines consists of adsorptive materials (for example, active carbon fibre) and micro/nano particles of chemicals (metal chlorides, metal hydrides, etc.) disposed on the surface of the carbon filaments. This complex compound has the advantages of the chemical heat pumps (high sorption capacity) and high speed of adsorption, typical for active carbons particles. There is a strong interaction (heat and mass transfer) between adsorptive materials and chemicals during the cycle of heating/cooling. The application of the activated carbon filaments and micro crystals on its surface increases the COP of the system. The advanced sorbent bed needs to have micro pores volume near 50 %, solid carbon near 40 % and meso/macro pores volume near 10 %. Activated carbon fibre "Busofit" used in the set of experiments as a fast sorbent material starts to react with ammonia in the early stage of heating/cooling time (up to 5 min) and accomplish its action after the chemical reaction of the salt is finished. Due to its high permeability "Busofit" is welcome for the cycles with a forced convection heating/cooling. Active carbon fibre

"Busofit" is a universal adsorbent and easily adsorbs different gases (H_2, N_2, O_2, CH_4, NH_3 etc.). It has such advantages as:

high rate of adsorption and desorption;
uniform surface pore distribution $(0.6–1.6)10^{-9}$ m;
small number of macropores $(100–200)10^{-9}$ m, with its specific surface 500–2000 m^2/g;
small number of mesopores with 50 m^2/g specific surface.

Therefore the pressure changing in the reactor (adsorber) is fast and starts before the salts are beginning to react. "Busofit" as a capillary-porous host material (binder) ensures the salt microcrystals distribution through the whole volume of a sorbent bed during the regeneration time (ammonia capillary condensation, liquid motion through the sorbent bed due to capillary forces action). This fibre is a typical active carbon structure, which has all advantages of the carbon nano-tubes technology.

2.1. Resorption Heat Pump with Two Reactors

In conventional resorption heat pumps the major entropy production in resorption systems is due to the superheating of the vapor during the cold production phase and de-superheating of the vapor during regeneration phase, the main part of reaction is used for pressurization and de-pressurization of the gas in reactors. "Busofit" action as a fast reacting material stimulates the pressure drop decrease between the phase of cold production and regeneration thus increasing the COP of the device. Microcrystals of such salts as $NiCl_2$, $MnCl_2$ and $BaCl_2$ disposed on the carbon fibre are suggested due to their stability, low cost and suitable temperature range. The ammonia was chosen as a working fluid to ensure the sufficient heat and cold output.

The resorption complex compound heat pump with two reactors and heat pipes is shown on Figure 6. If we have two reactors filled with different salts disposed on the carbon filaments, the cycle is separated in two main phases corresponding to two pressure levels. During the heat pump activation this pressure difference is quickly changing in time due to the effect of the ammonia adsorption/desorption on the carbon fibre.

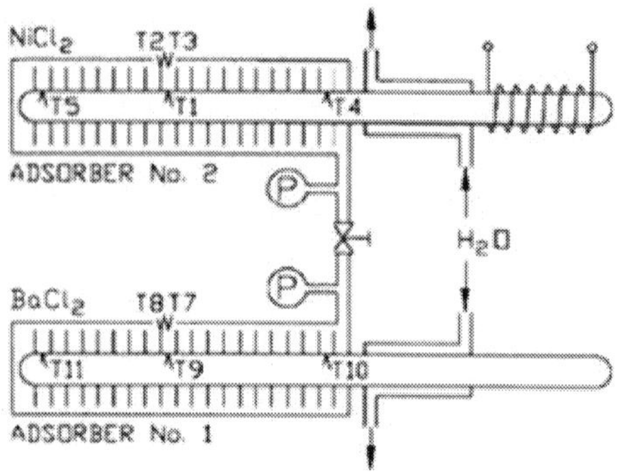

Figure 6. Sorption heat pump with two reactors (adsorbers) and heat pipe thermal control.

Such machines based on reversible solid-gas sorption cycles could have interesting application for space cooling, when a high temperature source of energy is available (Diesel, or SOFC) and/or the exigencies of the harsh external environment necessitates thermal control of an object. In the low temperature reactor the salt disposed on the "Busofit" surface is $BaCl_2$-8/0 NH_3 and its equilibrium curve is close to the ammonia saturation curve. The high temperature reactor is filled with "Busofit", and $NiCl_2$ microcrystals on its surface. The basic cycle is divided into two main reactions phases - (decomposition/desorption) at high temperature and (synthesis/adsorption) at low temperature, separated by two intermediate cycles (heating and cooling of the sorbent bed) allowing the reactor to reach the reaction phase conditions. The reactors procedure of heating and cooling is performed by heat pipes, Figure 6. The temperature of reactors is changing between 50 °C and 220 °C depending on the reaction phase (typical for $BaCl_2$ and $NiCl_2$ microcrystals). The time of cooling of the reactor is much faster than its heating. The reactor dimensions are: L = 1 m, D_{out}= 0.05 m, D_{in} = 0.049 m.

COP of the Resorption Cycle (two different salts) is:

Ideal:

$$COP^{i} = \frac{\Delta H_0}{\Delta H_1}, \tag{4}$$

Thermodynamic:

$$COP^{t} = \frac{[\Delta H_0 - C_p^{salt0}(T_0 - T_b)]}{[\Delta H_1 + C_p^{salt1}(T_h - T_0)]}, \tag{5}$$

Practical:

$$COP^{p} = \frac{\Delta H_0 - \Sigma C_p^{(salt0+carbon+steel)}(T_0 - T_b)}{\Delta H_1 + \Sigma C_p^{(salt0+carbon+steel)}(T_h - T_0)}, \tag{6}$$

For the combination of two reactors $BaCl_2$ / "Busofit" and $NiCl_2$ /"Busofit" COP = 1.15 – 1.2 is close to reality. Copper/water heat pipes with sintered powder wick ensure the efficient heating/cooling and the heat recovery between the reactors (adsorbers) 1 and 2. The particularities of temperature and pressure evolution in two reactors system are shown on Figures 7 – 8.

It is very important to note, that the temperature drop in the low temperature sorbent bed is negligibly small to compare with temperature drop in high temperature bed during the time of regeneration/desorption.

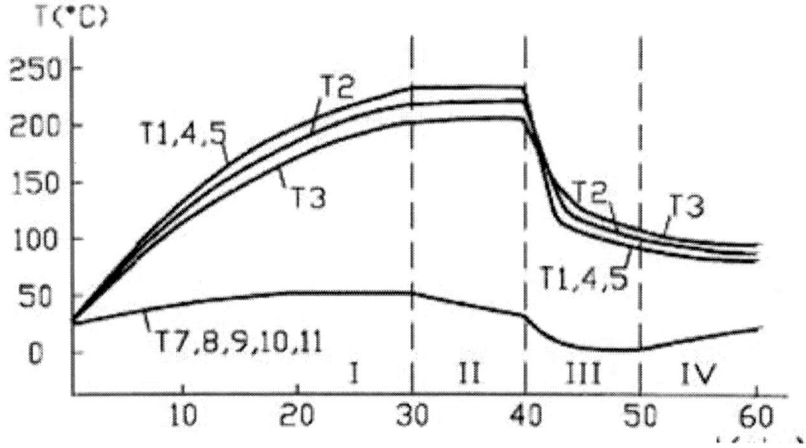

Figure 7. Temperature evolution in different parts of the resorption heat pump during its heating/cooling. T1–T5 – thermocouples in different parts of the high temperature sorbent bed, T7–T11 – thermocouples in different parts of low temperature sorbent bed.

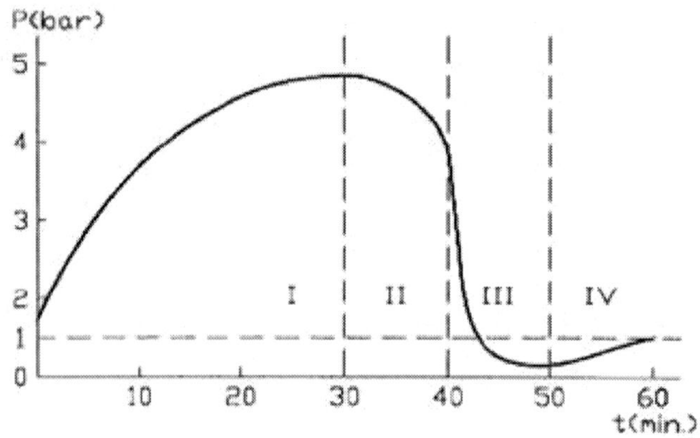

Figure 8.Pressure evolution in two reactors resorption heat pump during its heating/cooling.

2.2. Solid Sorption Cooler with Three Adsorbers + Condenser/Evaporator, Tri-Generation Approach

This cooler was suggested and tested in the Luikov Heat and Mass Transfer Institute and has original design of the evaporator/condenser. The cooler is used together with Diesel-generator engine for combined cold, heat and power (tri-generation) production. The cooler consists of two blocks of adsorbers (3+3) working out of phase. The selected sorbent materials are combinations of $BaCl_2$/"Busofit" – low temperature sample (LTS), $MnCl_2$/"Busofit"–medium temperature sample (MTS) and $NiCl_2$/"Busofit"– high temperature sample (HTS). The working fluid is ammonia. The total mass of the cooler is 9 kg, Figure 9.

Figure 9. Six adsorbers for three stage resorption cooler (active carbon fibre "Busofit" + salts) with heat pipe thermal control and internal heat recovery.

During the experiments with the cooler the ammonia vapour pressure is recorded as a function of temperature in each of three adsorbers. The operation of the cooler is based on the following reactions:

$$\begin{gathered} BaCl_2 + \cdot 8NH_3 \leftrightarrow Ba(NH_3)_8Cl_2 + 3766 \text{ J/mol} \\ Mn(NH_3)_2Cl_2 + 4NH_3 - 4741 \text{ J/mol} \leftrightarrow Mn(NH_3)_6Cl_2, \\ Ni(NH_3)_2Cl_2 + 4NH_3 - 5922 \text{ J/mol} \leftrightarrow Ni(NH_3)_6Cl_2 \end{gathered} \tag{7}$$

For simplicity we neglect the influence of "Busofit" on the ammonia chemical reactions with salts. This cooler enables near constant rate of the heating/cooling procedure. Two branches of the system are working out of phase and two sources of cold are active (1. $BaCl_2$ adsorber and 2. tubular evaporator with Al_2O_3 nano-porous coating). The evaporator has dimensions $L = 370$ mm and $D = 50$ mm. The cooler is ready to operate in two ways. It can be used as resorption cooler (working fluid - superheated ammonia) and as the adsorption cooler with evaporation and condensation phases (working fluid – liquid ammonia).

There are two possibilities to apply the waste heat of the diesel, or SOFC for the cold generation (air-conditioning) using such cooler with evaporator/condenser.

The cooler is oriented on the application of high temperature waste heat of the exhaust gas. This mode of cold generation is more efficient, because the heat and mass recovery of the energy of the high temperature adsorbers is used to preheat the low temperature adsorber.

It is convenient to generate the cold using the high temperature source of energy such as the exhaust gas (450-500 °C) of the engine. The Clapeyron (lnP vs 1/T) diagram, Figure 10, for such a case is more complicated to compare with resorption cooler, Figure 6., because the interaction between evaporator/condenser, active carbon fibre and three different salts is performed with heat recovery.

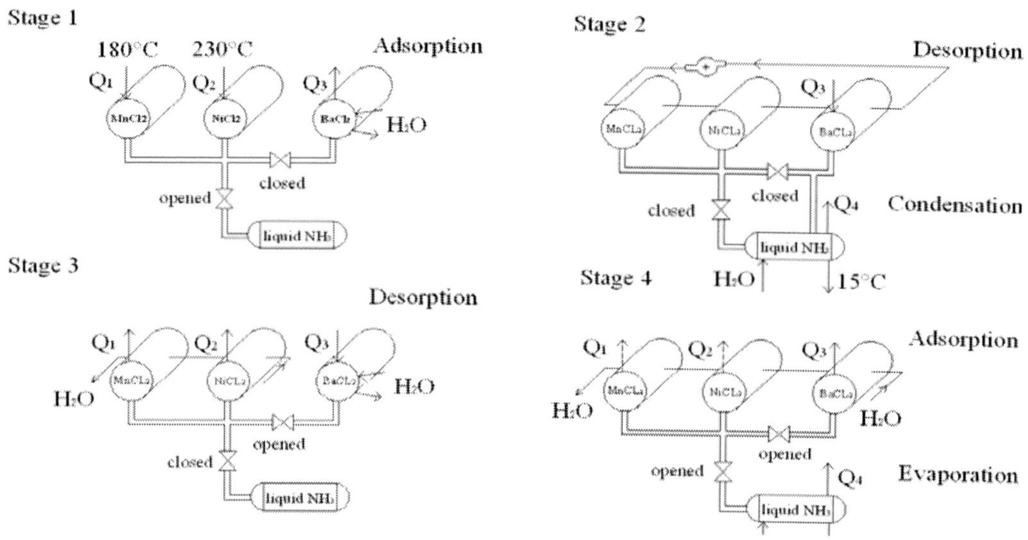

Figure 10. Solid sorption cooler ("Busofit" + salts) with heat recovery and two sources of cold – low temperature adsorber $BaCl_2$ and evaporator.

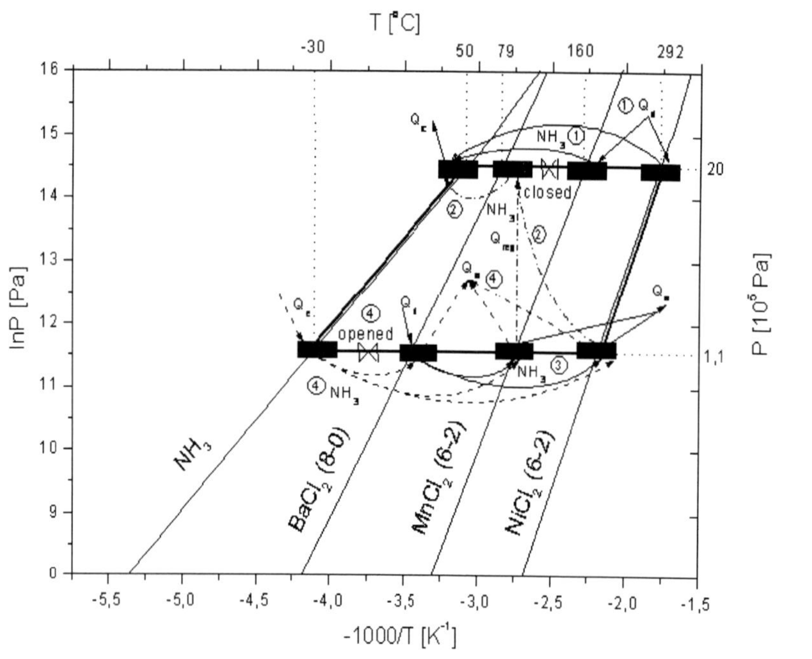

Figure 11. Clapeyron diagram of the solid sorption cooler with heat recovery. There are two cold generation sources: adsorber ($BaCl_2$/'Busofit') and the condenser/evaporator.

The volume for liquid ammonia in the evaporator exceeds the volume of ammonia in adsorbers. We use the recovered energy of the high temperature adsorbers $MnCl_2$/"Busofit", $NiCl_2$/"Busofit" (through the liquid heat exchanger)to heat the low temperature adsorber $BaCl_2$/"Busofit". This energy is used for desorbtion of ammonia inside the sorbent bed, Figure 10. In the experiments the total energy supply for two high temperature adsorbers

$MnCl_2$/"Busofit", $NiCl_2$/"Busofit"was simulated by the electric heaters (instead of the exhaust gas) through the heat pipes and was equal near 1400 kJ per cycle. The energy supply to the $BaCl_2$/"Busofit" adsorber was performed using the hot (90 °C) water heat exchanger (at the output of the $MnCl_2$/"Busofit" and $NiCl_2$/"Busofit" heat pipe heat exchangers) during the cycle of its cooling (wasted heat).

The cycle is devided by four stages:

1. At the first stage (time τ_1) $MnCl_2$/"Busofit" and $NiCl_2$/"Busofit" adsorbers are heated by the exhaust gas (electric heater) with further desorption of NH_3. The superheated vapor is condensing in the condenser/evaporator and is partly adsorbing by the low temperature sorbent bed of $BaCl_2$/"Busofit" adsorber.
2. At the second stage (time τ_2) $MnCl_2$/"Busofit" and $NiCl_2$/"Busofit" adsorbers are disconnected from the evaporator/condenser (the valve is closing) and cooled by the water circuit. The water flow (in heat exchanger) on the output of adsorbers with temperature of 90-95 °C enters the heat pipe heat exchanger of the low temperature $BaCl_2$/"Busofit" adsorber and heats the sorbent bed. This procedure is accompaning with ammonia desorption. Ammonia vapor is condensing in the evaporator/condenser.
3. At the third stage (time τ_3) all three adsorbers ($MnCl_2$, $NiCl_2$ and $BaCl_2$) are cooled down to the ambient temperature. Now adsorbers $MnCl_2$/"Busofit" and $NiCl_2$/"Busofit" are connecting to the low temperature adsorber $BaCl_2$/"Busofit" through the opened valve. The pass from the $BaCl_2$/"Busofit" adsorber to the condenser/evaporator is closed. More strong adsorbers $MnCl_2$/"Busofit", $NiCl_2$/"Busofit" suck the remaining part of ammonia from the adsorber $BaCl_2$/"Busofit". The desorption of the ammonia inside $BaCl_2$/"Busofit" adsorber stimulates the cold generation (resorption phenomena) inside the sorbent bed.
4. All three adsorbers now are connecting with the condenser/evaporator by the valve. The final stage (time τ_4) is responsible for the main cold generation in the evaporator. The temperature evolution of the liquid flow on the exit of heat exchanger of the evaporator (solid line 1) and $BaCl_2$/"Busofit" adsorber (dashed line 2) is shown on Figure 11.

In this experimental set-up the pressure and temperature sensors allow to check the dynamic of the pressure and temperature evolution of the sorbent bed, ambient temperature, the temperature of the vapour output and the temperature of the chilling water. The mass flow meters were used for the calculation of the degree of chemical reactions and physical adsorption.An example of high temperature adsorbers power charge-discharge procedure and heat input/output in the evaporator/condenser (1) and the $BaCl_2$/"Busofit" adsorber (2) via time of the cycle is presented on Figure 12. The max charging power of $MnCl_2$/"Busofit", $NiCl_2$/"Busofit" adsorbers is around 400 W each. Evaporation in the range of 10 °C results in cooling power of 200 W.

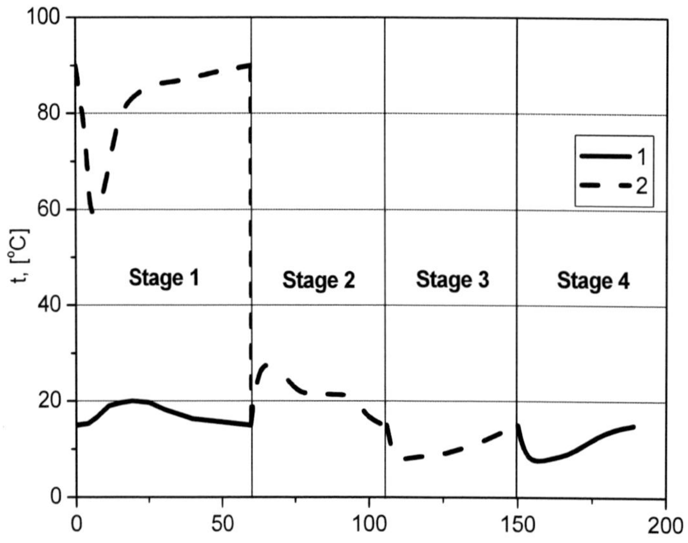

Figure 12. Solid sorption cooler ("Busofit" + salts) with heat recovery and two sources of cold – low temperature adsorber $BaCl_2$/"Busofit" and evaporator. The temperature evolution of the water flow on the exit of water heat exchanger of the evaporator (solid line 1) and adsorber $BaCl_2$/"Busofit" (dashed line 2).

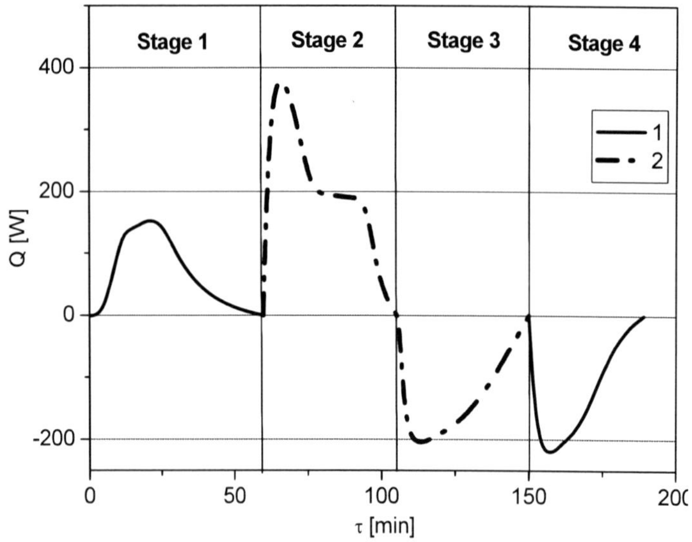

Figure 13. Solid sorption cooler ("Busofit" + salts) with heat recovery and two sources of cold – low temperature adsorber $BaCl_2$/"Busofit" and evaporator. The heat input/output in the evaporator/condenser (1) and the low temperature adsorber (2) as a function of the cycle.

The above mentioned experimental data validate the possibility to apply the waste energy of diesel engine, or SOFC for the air-conditioning and chilled water production.

Therefore, the solid sorption heat pumping and cooling technology allow to combine the low temperature alternative sources of energy together with the high temperature sources,

The loop thermosyphon evaporator needs to have a good thermal contact with the stack; the condenser ought to be cooled by air, or water, Figure 15. The loop thermosyphon transports thermal energy from a heat source to a sink by natural two-phase convective circulation without any external power supply (no electric pump). Thermosyphon evaporator and condenser are installed separately, but connected to each other by small diameter bendable (flexible) pipes (liquid and vapour lines). Due to the coupling between momentum and energy transport theoretical analysis of the loop performance is very complicate, therefore it is necessary that these problems be solved by experimental investigation before applying the loop thermosyphon to heat exchanger design. In the loop thermosyphon the heat transfer is considered to be affected by many factors, such as type and quantity of working fluid, pipe diameter, pipe length, and ratio of cooled surface to heated surface, the length of the adiabatic zone between heated and cooled sections, heat flux and operating temperature. The evaporator and condenser of the loop thermosyphon can be made of carbon-steel, cupper, aluminium and recently reinforced polymer tubes. Propane, R 134a, R 600, ammonia or water can be used as working fluid. When the copper is allowed to be applied water is the best working fluid. In order to establish heat transfer correlations for the application in the design program for the loop thermosyphon heat exchanger, regression analysis could be applied to experimental data for heat transfer coefficients in evaporator and condenser.

Typical loop thermosyphon with the flat evaporator (3 mm thickness) is shown on Figure 16.

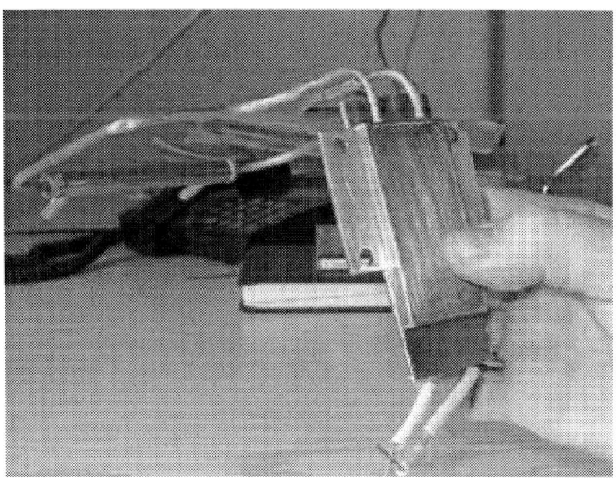

Figure 16. Copper/water loop thermosyphon for the fuel cell stack thermal management.

This thermosyphon is capable to transport Q_{max} near 100 W at the temperature of the adiabatic zone less 100 ^{0}C. The condenser is cooled by water circulation. The thermal resistance of thermosyphon R is 0.03 K/W.

Loop heat pipes are more flexible to compare with loop thermosyphons, due to its insincerity to the gravity field. Typical LHP for PEMFC with optimal heat flow rate 800 W at the working temperature near 80 ^{0}C is shown on Figure 17.

LHP with the non inverted meniscus of the evaporation designed and tested in the Luikov Institute is made from copper and has the wick performed from copper sintered powder. The working fluid is water. The typical maximum heat flow rate of such evaporator is near 1500 W, the thermal resistance of the evaporator R_e = 0.06 K/W.

Figure 17. Loop heat pipe with copper flat evaporator. The wick is made from sintered powder porous structure and has mini grooves for liquid suction.

3.3.Pulsating Heat Pipe Panels

Another alternative to the conventional heat pipe is an aluminium (multi-channel) heat pipe panel (Figure 18) with propane as a working fluid to cool FC stack [25]. Pulsating heat pipe (PHP) is one of several oscillatory thermal transport cycles under development that are obtaining attention as a potential semi-passive, high-power, high flux heat transport device. The PHP is unique in that it is capable of generating driving pressures in excess of many mechanically pumped loops. Capillary forces do not limit the PHP and it is capable of transferring high heat loads over long distances and against significant resistance (i.e. gravity, small tube diameters, etc.).

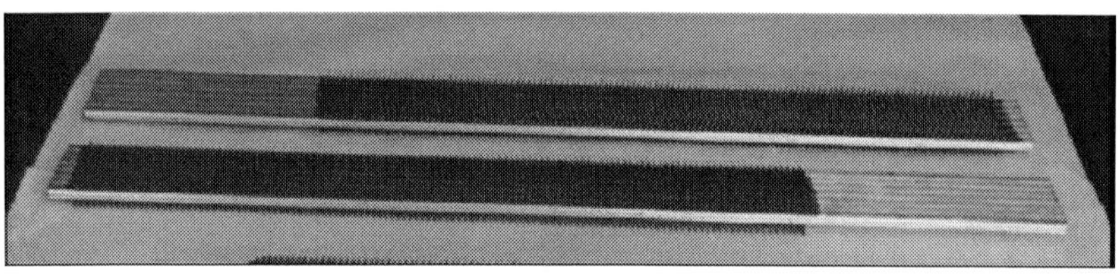

Figure 18. Aluminum pulsating heat pipe panel with mini channels inside and mini fins on the outer surface.

The main parameters of flat heat pipe panels developed in the Luikov Institute, Minsk are: HP width -70mm, HP height - 7 mm, HP length - 700 mm, evaporator length - 98 mm, condenser length - 500 mm, mass - 0, 43 kg. HP thermal resistance R = 0.05 K/W, evaporator heat transfer coefficient α = 8500 W/m^{2}K, condenser heat transfer coefficient α = 2500 W/m^{2}K. The working fluid (hydrocarbons) dynamic movement is stable with liquid filling ratio near 0.6 of the heat pipe volume.

3.4. Grooved Heat Pipes with Nano Porous Deposit on the Surface of Mini Channels

The thermal control of some new power sources (fuel cells) is interesting to realize by heat pipes with longitudinal grooves of triangular, rectangular, trapezoid, re-entrant, or Ω-form (GHPs), Figure 19, [26].

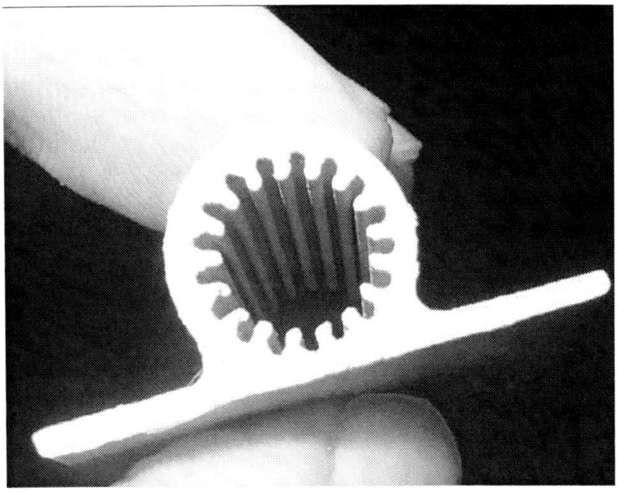

Figure 19. Cross section of the grooved heat pipe with nano porous deposit on mini grooves.

Usually, the geometric grooved heat pipes optimization is based on two contradictory objectives (the maximum of the power and the minimum of the external pipe diameter) and several objectives and constraints [27-31]. The liquid friction losses in GHPs are higher than the vapor ones; the main increase is performed on the hydraulic diameter parameter. The goal of the new groove configuration elaboration for the heat pipe thermal control of sorption machines and fuel cells is to reduce the friction losses and to enhance the capillary pumping action in the same time [32]. Proportional relation between the number of grooves and its transport capacity need to be determined. From the other hand, the heat transfer capability of grooved heat pipes is mainly limited by the liquid pressure drop in grooves and the evaporative surface of liquid meniscus. In order to improve the operating performance and heat transfer capacity of the grooved heat pipes especially for high heat flux applications it is necessary to enhance evaporation heat transfer within the grooves. The main aim is to improve the cooling capability of the GHP with Ω shape grooves, applying thin (nano particles) deposit (25-100μm) on the surface of the mini fins inside the heat pipe evaporator. In the present study two similar GHPs with Ω shape grooves, were tested simultaneously. The working fluid is ammonia. One GHP has grooves with smooth surface. Another GHP has porous layer on the grooves surface. The experimental part was devoted to validate by experimentally the role of porous layer in the improvement of evaporation efficiency within grooves. Therefore the present experimental plan was related with the determination of $\Delta Te_{/v}(Q)$, $R_{e/v}(Q)$, $R_{v/c}(Q)$ at the fixed temperatures in adiabatic zones for both GHPs. Here $\Delta T_{e/v}$ is the temperature difference between mean temperature of evaporator and the vapor temperature in the middle of the adiabatic zone, while $R_{e/v}=\Delta T_{e/v}/Q$ and $R_{v/c}=\Delta T_{v/c}/Q$. Seven different adiabatic temperature levels $T_v$were chosen namely -30, -10, 0, 10, 40, 50 and 80 °C for the comparable tests. As one can see from the Figure 20, at the given design of porous layer thermal resistance evaporator of GHP with porous depositis essential less to compare with conventional GHP (in 1.3 -1.8 times).

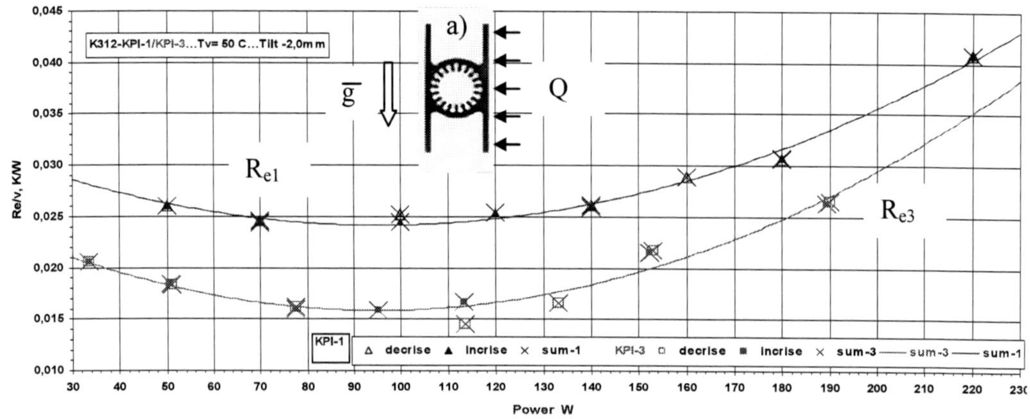

Figure 20. Thermal resistance $R_{e/v} = (T_e - T_v)/Q$ of evaporators. R_{e1}- thermal resistance of classical GHP evaporator, R_{e3} - thermal resistance of GHP evaporator with porous layer 50μm.Vapour temperature is 50 °C.L_e=0,2 m, L_{HP}=1 m; a – GHP cross section.

Both heat pipes were tested at the same time on the same experimental set-up in Luikov Heat and Mass Transfer Institute, Minsk. GHPs with porous layer and classical one were tested in the vapor temperatures diapason between -30 …+70 °C. It was found that for all vapor temperatures the evaporator thermal resistances ($R_{e/v}$) of GHP with porous layer is low to compare with $R_{e/v}$ for classical GHP. The difference in thermal resistances is between 1.3 and 1.8 times (or 0.015-0.02 K/W for GHP with porous layer and 0.025-0.035 K/W for classical one). The practical application of GHP with porous layer in the evaporator and classical condenser is alternative to conventional GHPs. Proposed method of evaporator heat transfer enhancing is perspective for mini grooves heat exchangers.

CONCLUSION

1. New power sources efficiency (co-generation and tri-generation systems, fuel cells, photovoltaic systems) have a good perspective to be increased using solid sorption heat pumps, refrigerators, accumulators of the heat and cold, heat transformers thermally connected with heat pipes and thermosyphons.
2. Experiments with the experimental set-up based on complex compound sorbent bed ($NiCl_2$, $MnCl_2$, $BaCl_2$ micro/nano crystals on the an active carbon fibre surface) have demonstrated a possibility to design a cooler with two independent sources of cold (low temperature $BaCl_2$adsorber and evaporator). The heat generation and chilled water production ensured by the cooler with $COP_{cooling}$ equal 0.6 and the heat pump COP near 1.63 was demonstrated.
3. Sorption heat pipes, loop heat pipes, thermosyphons and pulsating heat pipes, grooved heat pipes heat exchangers are suggested as an advanced thermal control system for solid sorption heat pumps and portable fuel cells with power generation > 100 W.

Nomenclature

C	–	heat capacity rate, W/K
E	–	thermal internal energy, J
F	–	surface, m^2
G	–	Gibbs energy, J
H	–	enthalpy, J
P	–	pressure, bar
Q	–	heat flow, W
S	–	entropy, J
T	–	temperature, K
W	–	electric power, Js^{-1}

Greek Symbols

Δ	–	increment

Subscripts

h	–	heat
i	–	ideal
p	–	practical
t	–	thermodynamic
AS	–	alternative source
FC	–	fuel cell
HE	–	exhausted heat
SK	–	sink
SR		source
TR	–	tri-generation

References

[11] Spinner, B., Stitou., D., Grini, P. G. Cascading Sorption Machines: New Concepts for the Power Control of Solid-Gas Thermochemical Systems Towards Sustainable Technologies, Absorption Heat Pump Conference, *Montreal,* 1996.

[12] Lepinasse, E., Marion, M., Goetz, V. Cooling storage with resorption process. Application to a box temperature control, *Appl. Therm. Eng.,* 21 (2001), pp. 1251-1263.

[13] Meunier, F. Second law analysis of a solid adsorption heat pump operating on reversible cascade cycles: application to the zeolite-water pair, *Heat Recovery Systems,* 1985, Vol. 5, pp. 133–141.

[14] Tozer, R.M. and James, I.W. Fundamental thermodynamics of ideal absorption cycles, *Int. Journal of Refrigeration*, 1997, Vol. 20, No. 2, pp. 120–135.

[15] El-Sharkwary, I. I., Kuwahara, K., Saha, B. B., Koyama, S. and Ng, K.C. Experimental investigation on adsorption of ethanol onto activated carbon fibers for possible application in adsorption cooling system, *Appl. Therm. Eng.,* 26 2006, 859-865.

[16] Sharkawy, I. I., Saha, B. B., Koyama, S., He, L., Ng, K. C. and Yap, C. Experimental investigation on activated carbon-ethanol pair for solar powered adsorption cooling applications, *Int. Journal of Refrigeration,* 31(2008), 1407-1413.

[17] Lu, Y. Z., Wang, R. Z., Zhang, M. and Jiangzhou, S. Adsorption cold storage system with zeolite – water working pair used for locomotive air conditioning, *Energy Conversion and Management,* 2003, Vol. 44, pp. 1733–1743.

[18] Wang, L. W., Wang, R. Z., Wu, J. Y. and Wang, K. Compound adsorbent for adsorption ice maker on fishing boats, *Int. Journal of Refrigeration,* 2004, Vol. 27, pp. 401-408.

[19] Lambert, M. A. and Jones, B. J. Review of Regenerative Adsorption Heat Pumps, *J. Thermophysics and Heat Transfer*, 2005, Vol. 19, No. 4, pp. 471-493.

[20] Raldow, W. M. and Wentworth, W. E. Chemical heat pumps – a basic thermodynamic analysis, *Solar Energy,* 1979, Vol. 23, pp. 75–79.

[21] Wonggsuwan, W., Kumar, S., Neveu, P. and Meunier, F. Review of chemical heat pump technology and applications, *Appl. Therm. Eng.,* 21 (2001), pp. 1489–1519.

[22] Poelstra, S., Haije, W. G. and Dijkstra, J. W. Techno-economic feasibility of high-temperature high – lift chemical heat pumps for upgrading industrial waste heat, *Appl. Therm. Eng.,* 22 (2002), pp. 1619–1630.

[23] Vasiliev, L., Kanonchik, L., Antukh, A., Kulakov, A. and Rosin, I. Waste Heat Driven Solid Sorption Coolers, *SAE Technical Paper Series* 941580, 1994.

[24] Vasiliev, L. L., Mishkinis, D. A. and Vasiliev, L. L. Jr., Complex Compound/Ammonia Coolers, SAE Technical Paper Series 961462, 26th Int. Conf. on Environmental Systems, *Monterey,* 1996.

[25] Vasiliev, L., Kanonchik, L., Antukh, A. and Kulakov, A. *NAX Zeolite, Carbon Fiber and $CaCl_2$ Ammonia Reactors for Heat Pumps and Refrigerators, Adsorption,* 2 (1996), pp. 311–316.

[26] Vasiliev, L. L., Gulko, N. V. and Khaustov, V. M. Solid adsorption refrigerators with active carbon – acetone and carbon- ethanol pairs, *Solid sorption refrigeration Symposium,* Paris, France, 1992.

[27] Vasiliev, L. L., Mishkinis, D. A., Antuh, A. A. and Vasiliev, L. L. Jr., Resorption heat pump, *Appl. Therm. Eng.,* 24 (2004), pp. 1893-1903.

[28] Amir Faghri and Zhen Guo, Challenges and opportunities of thermal management issues related to fuel cell technology and modeling, *International Journal of Heat and Mass Transfer* 48 (2005) 3891-3920.

[29] Aidoun, Z. and Ternan, M. Salt impregnated carbon fibres as the reactive medium in a chemical heat pump: the NH_3-$CoCl_2$ system, *Appl. Therm. Eng.,* 22 (2002), pp. 1163–1173.

[30] Aidoun, Z. and Ternan, M. The synthesis reaction in a chemical heat pump reactor filled with chloride salt impregnated carbon fibers: the NH_3-$CoCl_2$ system, *Appl. Therm. Eng.,* 22 (2002), pp. 1943–1954.

[31] Vasiliev, L. Solid sorption heat pumps for tri-generation, A*rchives of thermodynamics,* Vol. 28 (2007), No. 3, 1-4.

[32] US Patent 2005/0026015 A1, Feb. 3, 2005 Amir Faghri, Mansfilled, CT (US) *"Micro heat pipe embedded bipolar plate for fuel cell stacks"*.

[33] Vasiliev, L. L., Filatova, O. S., Kulakov, A. G., Vasiliev, L. L. Jr. and Tsitovich, A. P. Heat Pipes in new power sources technology, Proceedings of the VI Minsk *International Heat and Mass Transfer Forum,* 19-23 May, 2008, Minsk, Belarus, Vol. 2, pp. 21 -24.

[34] Vasiliev, L. L. and Vasiliev, L. L., Jr., Sorption heat pipe –a new thermal control device for space and ground application, *International Journal of Heat and Mass Transfer* 48 (2005) 2464-2472.

[35] Vasiliev, L. L. *Solid sorption heat pumps in tri-generation, in Proceedings of Heat transfer and renewable sources of energy,* HTRSE – 2006, Ed. by J. Mikielewicz, W. Nowak and A. A. Stachel, Szczecin, (2006), pp. 371-382.

[36] Vasiliev, L. Jr., Vasiliev, L., Romanenkov, V., Rabetsky, M. and Jean Claude Legros, Advance grooved heat pipe for space satellite thermal control system, *SAE Paper Series* 091SEC-0158, 2009.

[37] Schlitt, K., Kirkpatric, J. and Brennan, P. Parametric performance of extruded axial grooved heat pipes from 100 to 300K, AIAA-Paper N74-724, 1974, p. 9.

[38] Schneider, G. and Devos, R. Nondimensional analysis for heat transport capability of axially-grooved heat pipe including liquid/vapor interaction. AIAA-Paper. 1980. N 80-0214, P9.

[39] Goffaux, C., Pierret, S., Rossomme, S., Kelner, V., Van Oost, S. and Barremaecker, L. Geometric *Optimization of Grooved Heat Pipes by a Genetic Algorithm Technique*, Proceedings of the 6th International Conference on Heat Pipes, Heat Pumps and Refrigerators, Minsk, 2005.

[40] Kim, S. J., Ki Seo, J. and Hyung Do, K. Analytical and experimental investigation on the operational characteristics and the thermal optimization of a miniature heat pipe with a grooved wick structure, *International Journal of Heat and Mass Transfer* 46 (2003) 2051–2063.

[41] Suman, B,. De, S., and DasGupta, S. Transient modeling of micro-grooved heat pipe, *International Journal of Heat and Mass Transfer* 48 (2005) 1633–1646.

[42] Grakovich, L., Rabetsky, M., Zhuravlyov, A., Romanenkov, A., Vasiliev, L., Vasiliev, L. Jr., Baturkin, V., Rassamakin, B., Kharnasov, S. and Shevel, E. Grooved heat pipes with porous deposit to enhance heat transfer in the evaporator, International Conference "Heat pipes for space application", Moscow, Russia, 15-18 September 2009, p. 14-21.

In: Frontier Research in Microscale and Nanoscale… ISBN 978-1-61324-665-8
Editor: Lixin Cheng

Chapter 16

MICROGRAVITY AND MICRO-SCALE FLUID MECHANICS AND HEAT TRANSFER LABORATORY

J. N. Chung
Department of Mechanical and Aerospace Engineering,
University of Florida, U. S.

GENERAL RESEARCH INTERESTS

Dr. Chung's research activities for the past thirty years have been in the general areas of fluid mechanics and heat transfer with a special focus on phase change heat transfer and multiphase flows. Specifically, he has worked in the following areas : Numerical and analytical fluid mechanics and heat transfer associated with droplets and bubbles, boiling and condensation, heat and mass transfer in dehydration and porous media, two-phase flow and heat transfer, packed bed thermal storage system, particle and droplet dispersion in turbulent free shear flows, fluid mechanics and heat transfer in micro-gravity boiling, large-scale direct numerical simulation of transition to turbulence for heated flows by spectral methods, nuclear reactor thermal-hydraulic analysis, transport processes in manufacturing and materials synthesis for nano-cluster material. Recently his research is focused on nano to micro-scale transport phenomena in fundamental boiling, cryogenic two-phase flow and heat transfer, electro-thermal transport in fuel cell systems, and biomass and waste conversion to energy by thermal-chemical gasification.

RESEARCH TOPICS AND FACILITIES IN THE LABORATORY

I. Microgravity Boiling and Cryogenic Two-Phase Transport

I.1. Microgravity Boiling

Boiling is one of the most efficient heat transfer modes on earth. In microgravity, nucleate boiling will not function properly or can not even exist due to the loss of gravity. The main objective of this research in our lab is to study the physics and feasibility of using an externally applied force field to replace the terrestrial gravity for the purpose of maintaining nucleate boiling in microgravity. The research involves the investigations of effects of an external force field on the nucleation, bubble detachment from the heater surface, two-phase bubble dynamics, bubble-bubble interactions and boiling heat and mass transfer in a microgravity environment. Dielectrophoretic force, acoustic radiation force, and velocity shear force were identified and proven to be feasible for maintaining nucleate boiling in microgravity. This research has provided fundamental knowledge on the interaction of a boiling system with an external force field in the absence of gravity and suggested methods for engineering application in space.

I.2. Microgravity Cryogenic Two-Phase Transport

Our lab has been investigating the cryogenic chilldown process and two-phase heat transport under both terrestrial and microgravity conditions. For space explorations, cryogenic fluids are involved in propulsion, solid state cooling and life support systems. In addition to space applications, in a hydrogen economy, chilldown must be accomplished with a minimum consumption of cryogen for the overall energy efficiency to be within tolerable limits. Chilldown process under the microgravity condition is also of particular interest for space explorations. Within this effort, we obtained the chilldown boiling curve of a two-phase nitrogen flow under low flow rates. According to the experimental observations, a two-fluid chilldown model was developed that is applicable for both 1-g and microgravity conditions. This method accounts for the thermal non-equilibrium effects, which has long been believed to be essential for an accurate prediction of post-CHF heat transfer from previous experiments. Moreover, it successfully covers all the flow and heat transfer regimes and has been considered to have great engineering potentials. Currently, direct numerical simulations are under the development for cryogenic thermal-fluid transport and management in microgravity.

I.3. Research Facilities

(a) Microgravity Cryogenic Boiling and Two-Phase Flow Loop

Figure 1 shows the schematic of the boiling and two-phase flow loop and Figure 2 is a photographic view of this apparatus located in two side-by-side cubicles for the drop tower microgravity experiment. The experimental loop mainly consists of a bellows-controlled reservoir, a cryogenic valve, test section inlet portion, test section, test section outlet portion, liquid separator, lighting and video system. A vacuum jacket is provided to insulate the glass tube test section. Two different test sections were used in the experiment. Two test sections were built. The first is an unheated Pyrex glass tube which was used for the chilldown study while the second test section is a heated quartz tube coated with a thin film of Tin Oxide to provide a constant heat flux.

(b) Drop Tower Apparatus

The microgravity environment was provided by the University of Florida drop tower. The drop tower is 5-story high and is equipped with an airbag for deceleration. Fig. 3 shows the overall drop tower schematic and a picture of drop box inside a drag shield. At the current drop height of 15.25m, UF's drop tower produces 1.7 seconds of free fall. The experiment is actually located in a drag shield which prevents the experiment from experiencing the aerodynamic drag. As a result, the microgravity level is measured between 10^{-5} to 10^{-4} g.

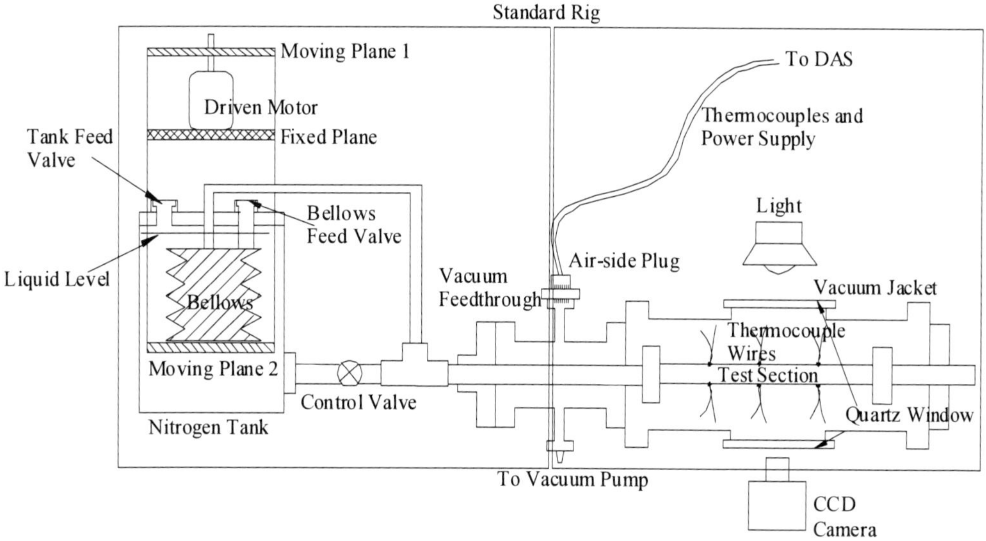

Figure 1. Schematic of the cryogenic boiling and two-phase loop.

Figure 2. Picture of microgravity boiling and two-phase loop in a drop tower box.

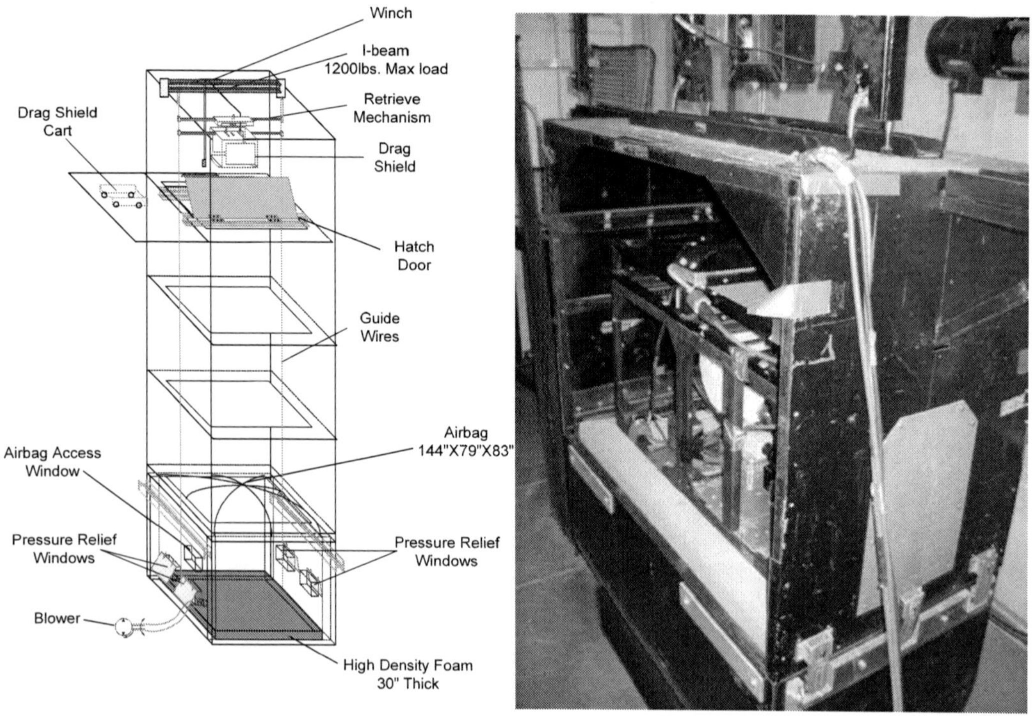

Figure 3. A schematic of the drop tower system and a picture of drop box inside a drag shield.

I.4. Representative Publication

1. T.J. Snyder and J.N. Chung, "Terrestrial and Microgravity Boiling Heat Transfer in a Dielectrophoretic Force Field," ***Int. J. Heat Mass Transfer***, Vol.43, 1547-1562, (2000).
2. Yue Ma and J.N. Chung, "An Experimental Study of Forced Convection Boiling in Microgravity," ***Int. J. Heat Mass Transfer,*** Vol. 41, pp. 2371-2382, (1998).
3. J.S. Sitter, T.J. Snyder, J.N. Chung and P.L. Marston, "Terrestrial and Microgravity Pool Boiling Heat Transfer From a Wire in an Acoustic Field", ***Int. J. Heat Mass Transfer***, Vol. 41, pp. 2143-2155,(1998).
4. Kun Yuan, Yan Ji and J. N. Chung, "Cryogenic Chilldown Process under Low Flow Rates", ***Int. J. Heat and Mass Transfer***, Vol. 50, pp. 4011-4022, (2007).
5. Kun Yuan, Yan, Ji, J. N. Chung and Wei Shyy, "Cryogenic Boiling and Two-Phase Flow During Pipe Chilldown in Earth and Reduced Gravity", ***J. Low Temperature Physics***, Vol. 150, pp. 101-122, (2008).

II. Micro-Scale Boiling and Microchannel Two-Phase Transport

II.1. Micro-Scale Boiling

Specifically in boiling on a micro heater array, our group has reported for the first time that the heat transfer rate as a function of time with a resolution of ten thousand hertz during the entire lifetime of a single bubble that clearly elucidates the respective contributions of bubble embryo nucleation, growth and detachment to the heat transfer characteristics. This research also reports the first quantitative measurements of the heat transfer enhancement due to the coalescence of bubbles in nucleate boiling that suggests a plausible scenario that the critical heat flux corresponds to a condition of maximum number of simultaneous individual coalescences, beyond that a vapor film will take over. In summary, this research contributes to the quantitative and fundamental understanding of the nucleate boiling mechanisms and also provides an explanation for the steep nucleate boiling curve and the critical heat flux.

On the numerical modeling and simulation, we employ a combination of continuum and molecular dynamics methods to predict the fundamental boiling physics. A finite-volume method with a sharp interface cut-cell method for the moving interface is developed for the continuum portion and a molecular dynamics method has been completed for the moving contact line among bubble vapor, liquid layer and solid heater surface.

II.2. Microchannel Two-Phase Transport

Applications of microtechnology must utilize components or systems with microscale fluid flow, heat and mass transfer. As the size of individual components shrinks and the surface-to-volume ratio increases drastically, the transport mechanisms involved also go beyond those covered by the traditional theories and understanding. The development of the new theories and modeling, and the fostering of up-to-date physical understanding have fallen behind the progress of micro machining and manufacturing. We have built a micro-PIV based experimental system and have performed single-phase and two-phase study such as single-phase fluid flows in serpentine and T-junction channels, micro-bubble dispensers, and gas-liquid flows for PEM fuel cells. Due to some limitations faced in the experiments at the microscale, we are also developing numerical simulation capabilities.

II.3. Research facilities

(a) Micro-Heaters Array Boiling Experimental System

This system, intended for microscopic boiling heat transfer study, includes a square array of 96 micro-heaters as shown in the middle of Fig. 4. Each microheater with a size of 0.3 mm x 0.3 mm is independently controlled. Each microheater as shown in on the left side of Fig. 4 is made of platinum filaments of 2 μm in width and submicron in thickness deposited on quartz substrate with a serpentine pattern. The control unit is capable of setting and maintaining either the constant temperature or the constant heat flux level for each microheater and then monitoring the other during the experiment. Because of the microscale infrastructure, the thermal inertia is negligibly small. With a sampling frequency up to 10,000 Hertz, the time and space resolved measured data are highly accurate. This system as shown in Fig. 5 has been used for quantitative measurement of micro-scale boiling heat transfer characteristics for a single bubble and also for coalescing duel and multiple bubbles on the microheater array surface.

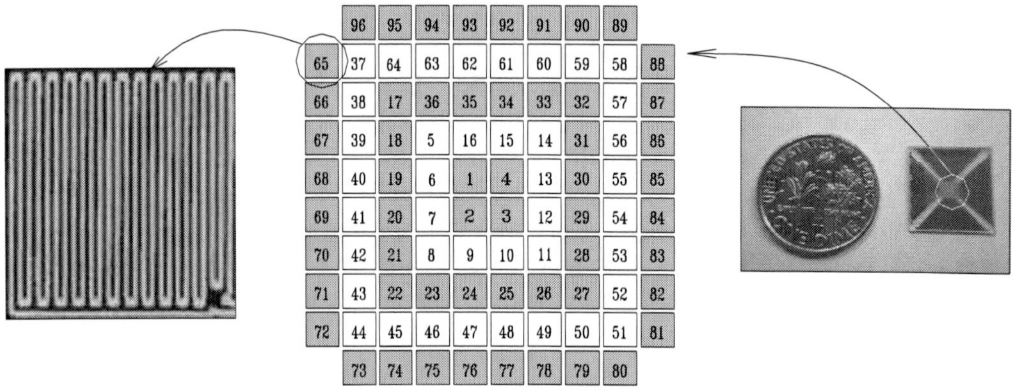

Figure 4. Micro-heaters array.

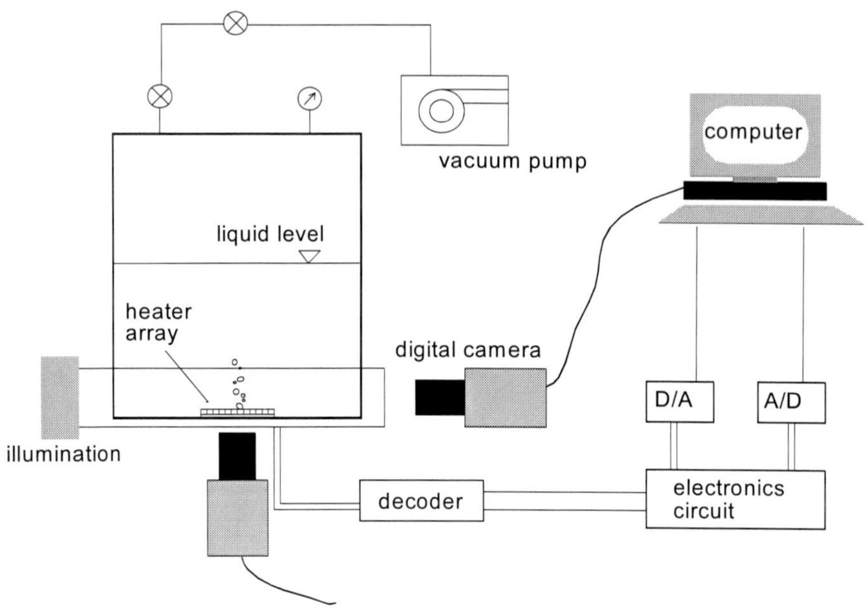

Figure 5. Micro-heaters boiling experimental system.

(b) Micro-PIV(particle image velocimetry) and Microchannel Experimental System

Figure 6 shows the schematic of the micro-PIV and microchannel experimental system. It includes two Nd:YAG lasers (Continuum), a timing controller (Labsmith), a CCD camera (Cooke), a filter cube (Olympus), a 0.5× magnification lens (Olympus), a microscope (Olympus), some optical lenses (Edmund), a syringe infusion pump (Cole-Parmer Instrument), a 60ml syringe (Mcmaster), a micro-filter (Swagelok), pressure transducers (Kavlico), straight and serpentine micro-channel test sections and computerized data acquisition system.

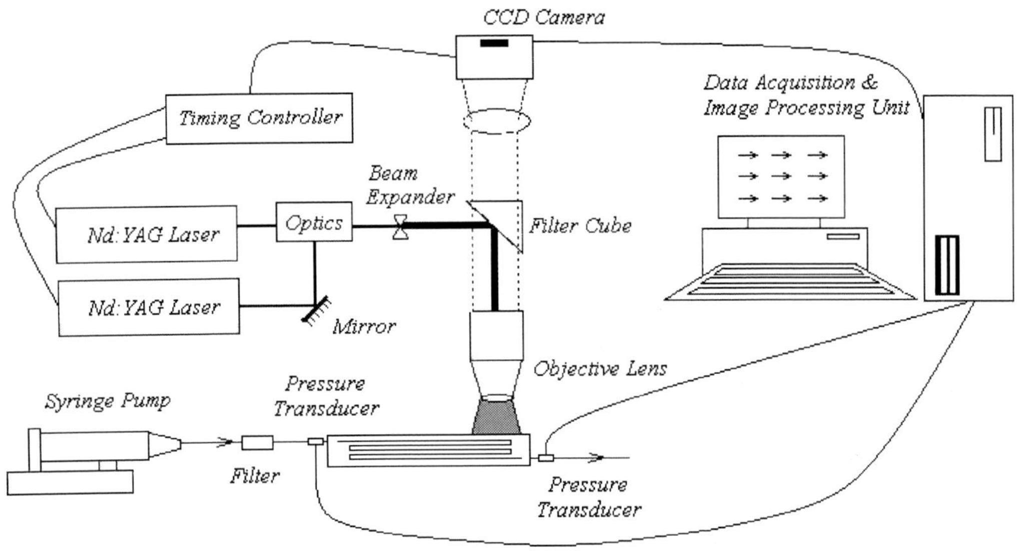

Figure 6. Micro-PIV and microchannel two-phase flow experimental system.

Figure 7. Photo of the optical subsystem (left) and the laser beam alignment component (right) of the micro-PIV apparatus.

For example, to visualize the flow structure around the corner of the serpentine micro-channel, a micro particle image velocimetry (micro-PIV) as shown in Fig. 7 was developed where fluorescent particles (Duke Scientific) were seeded into the de-ionized water flow. Two Nd:YAG lasers were directed to the same optical path by optical lenses and expanded by a beam expander made up of a concave and a convex lens. The 0.69 μm particles absorb green light (~542nm) and emit red light (~612nm). The emitted light is imaged through a 10× objectives lens (NA=0.3) and passed to the fluorescent filter cube, where the green light from background reflection is filtered out and the red fluorescence from the sub-micron particles is passed to the 0.5× lens and recorded on the CCD camera. Fig. 8 provides the fluorescent images in the serpentine micro-channel and the micro-PIV results for velocity vectors in the serpentine microchannel.

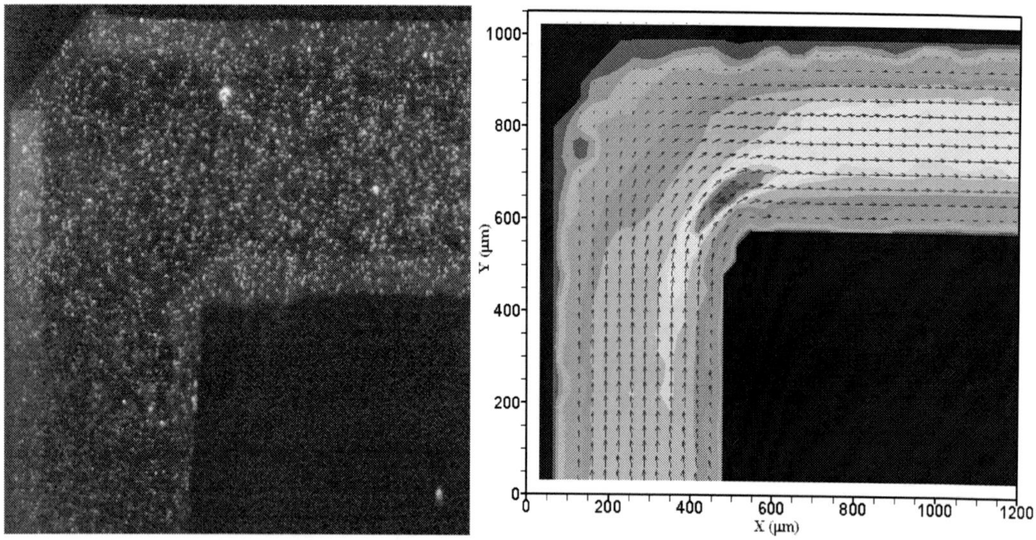

Figure 8. Fluorescent images in the serpentine micro-channel (left) and Micro-PIV results for velocity vectors in the serpentine microchannel (right).

II.4. Representative Publication

1. Tailian Chen and J. N. Chung, "An experimental study of miniature-scale pool boiling," ASME *J. Heat Transfer,* Vol. 125, pp. 1074-1086, (2003).
2. Tailian Chen and J.N. Chung, "Heat Transfer Effects of Coalescence of Bubbles from Various Site Distributions," *Proceedings of the Royal Society, London, A : Mathematical, Physical and Engineering Sciences,* Vol. 459, 2497-2527, (2003).
3. R. Xiong and J. N. Chung, An experimental study of adiabatic gas-liquid two-phase flow patterns and void fraction in micro-channels", *Physics of Fluids*, Vol. 19(3), Article 033301, (2007).
4. R. Xiong, M. Bai and J. N. Chung, "Formation of bubbles in a simple co-flowing microchannel", *Int. J. Micromechanics and Microengineering,* Vol. 17, 1002-1011, (2007).
5. A. K. Robinson and J. N. Chung, "The Feasibility Study of A Micro-Scale Vibration-Induced Droplet Atomization and Dropwise Condensation Membrane Device, Accepted for publication, *Int. J. of Microscale and Nanoscale Thermal and Fluid Transport Phenomena,* (2009).

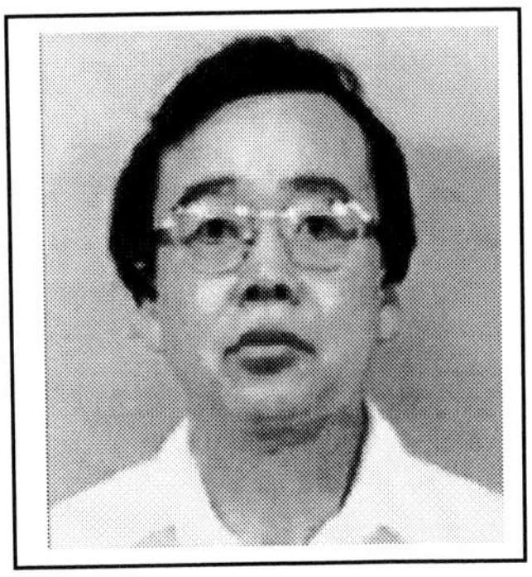

Jacob N. Chung

Jacob N. Chung is the Andrew H. Hines, Jr./Progress Energy Eminent Scholar Chair Professor in the Department of Mechanical and Aerospace Engineering at the University of Florida. He is also the Director of Micro-Scale and Microgravity Fluid Mechanics and Heat Transfer Lab at the University of Florida. He is a fellow of the American Society of Mechanical Engineers (ASME). From 2002-2005, he served as an associated editor for the ASME Journal of Heat Transfer in the areas of phase-change heat transfer and multiphase flows. Dr. Chung's research activities for the past thirty years have been in the general areas of fluid mechanics and heat transfer with a special focus on bubble and droplet dynamics and heat transfer, phase change heat transfer, multiphase flows, microgravity boiling, laminar-turbulent transition, turbulence in heated flows, nano- and micro-scale thermal transport phenomena, cryogenic two-phase heat transfer and thermal-chemical conversion of biomass to energy. He has authored and co-authored a total of 140 archival journal papers in the above research areas. Dr. Chung and two co-authors published a book entitled "Transport Phenomena with Drops and Bubbles".

In: Frontier Research in Microscale and Nanoscale...
Editor: Lixin Cheng ISBN 978-1-61324-665-8

Chapter 17

CLASSIFICATION OF SURFACE WETTABILITY AND FABRICATION METHODS IN SURFACE MODIFICATION: A REVIEW

Chin-Chi Hsu, Long-Sheng Kuo and Ping-Hei Chen*
Department of Mechanical Engineering
National Taiwan University
Taipei 10617, Taiwan, R.O.C.

ABSTRACT

This paper reviews the current status of research on surface wettability, and classifies wettability into five categories: isotropic wettability, anisotropic wettability, interlaced wettability, gradient wettability, and tunable wettability. Examples from nature include lotus leaves (isotropic), bamboo leaves (anisotropic), Namibia desert beetles (interlaced), and butterfly wings (gradient wettabilty). The tunable wettability is primarily formed in laboratories. A droplet can move freely on a horizontal surface with wettability gradients due to differences in surface potential energies. As such, the wettability of a surface can be tuned using various methods, such as applying external electrical fields or mechanical stresses. This paper also reviews the surface modification and fabrication methods currently used to achieve different wettability. Five main fabrication methods are considered in detail: etching, lithography, deposition, sol-gel processing, and templation. The transparency of the modified surface is also discussed since it is an important factor in some applications, such as solar cells. This review should be helpful in the design of surface modification, and future developments in this field.

Keywords: Wettability, Hydrophilic, Hydrophobic, Hierarchical, Micro/Nano structure.

* Corresponding author. E-mail: phchen@ntu.edu.tw. Tel and Fax:+886-2-2367-0781. Postal Address: No. 1, Sec. 4, Roosevelt Road, Taipei, 10617 Taiwan (R.O.C). Rm. 728, Engineering Bldg., Department of Mechanical Engineering. National Taiwan University

1. INTRODUCTION

The wettability of a solid surface is an important property in materials research [1-3]. The contact angle (CA) of water on a solid surface is usually used for classification: when CA is more than 90°, such the feature is called hydrophobic. Conversely, it is called hydrophilic when CA is less than 90°. A superhydrophobic surface has a CA greater than 150° (Figure 1), while a superhydrophilic surface has a nearly 0° CA. Liquid (water) tends to form a droplet on a hydrophobic surface, while it tends to form a film on a hydrophilic surface. These two characteristics are useful in various industrial applications [4-6]. For instance, a superhydrophobic surface can act as a self cleaner for use on windows [7] or vehicle windshields [8]. On the other hand, TiO_2 is a well-known material with a hydrophilic surface that can function as a defogger when exposed to UV light. Recently many researchers have focused on surfaces with specific wettability by mimicking from nature [9-16].

Lotus leaves are a good example of a superhydrophobic surface. There are two key issues for a surface to be superhydrophobic. The first is a de-wetting material, such as wax, on the surface. The second one is a rough surface with a micro/nano hierarchical structure [2, 17-19]. A lotus leaf has the binary scale structure, micro sized bumps, and nano hair-like structures. These two-fold structures hold air within irregular indentations, repelling water. Water will be repelled because of these two surface characteristics, which create a self-cleaning effect [9]. Recently, many researchers have tried to mimic lotus leaves and similar structures. Different methods have been developed to construct a micro/nano hierarchical surface [20,21]. There are also such the hierarchical examples from the creatures. The skin of a shark efficiently reduces resistance in water, saving more energy as the shark moves [22]. The water strider's legs have hierarchical micro/nano arrays [23] that allow them to walk on the water surface. The wings of the cicada orni [24] and grasshopper [25] have a superhydrophobic surface for flying purpose. Beetles in the Namib Desert have hierarchical structures on their back shells [26]. These micro-sized bumps are for collecting water from the mist in the air by the nano hydrophobic structure, allowing the beetles to survive in a desert environment.

The shape of a water drop on a lotus leaf is just like a spherical ball (Figure 2), while a water droplet on a bamboo leaf has an olive shape (Figure 3).

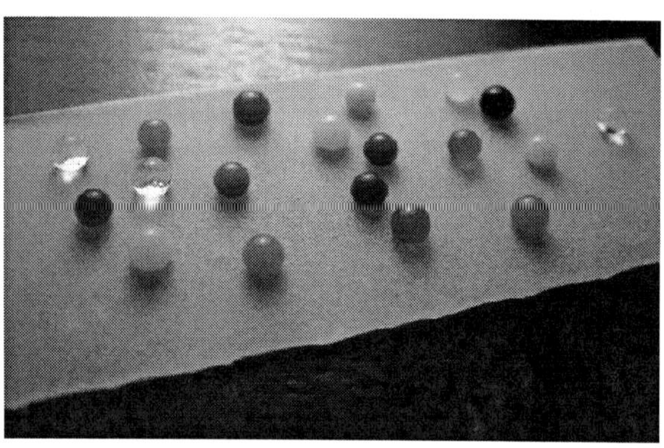

Figure 1. Superhydrophobic surface of a modified glass can make the dyed water droplets as spherical beads. (Courtesy to Li-De Liu for the photograph.).

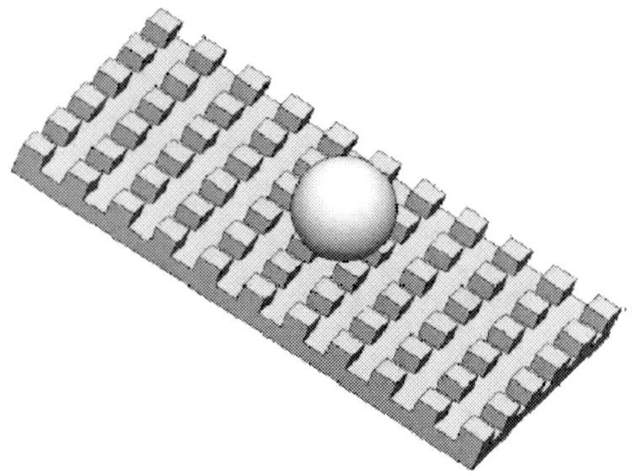

Figure 2. Isotropic wettability: uniform rough surface can achieve isotropic wettability.

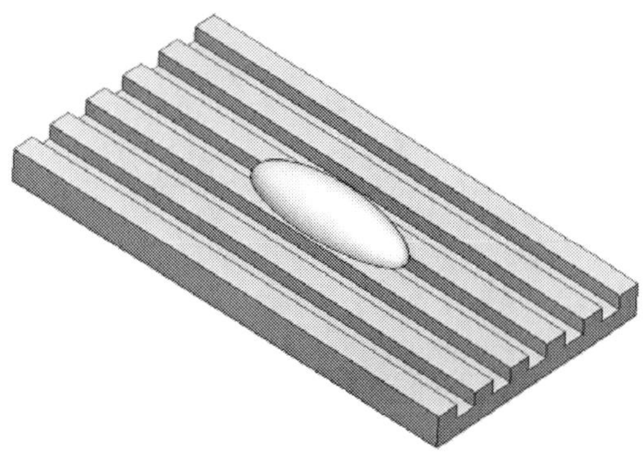

Figure 3. Anisotropic wettability: one dimensional structures can achieve anisotropic wettability.

The water droplets on a lotus leaf can roll freely in all directions, while there is a preferred direction of motion on a bamboo leaf. To distinguish between these two different surfaces, we call the lotus leaf an example of "isotropic wettability," while the bamboo leaf exhibits "anisotropic wettability." In other words, the CA of isotropic wetting is the same in every direction ($\theta_x = \theta_y$), but anisotropic wetting has different CAs at different directions ($\theta_x \neq \theta_y$, where θ_x and θ_y are CAs measured from two orthogonal directions). In addition to the two kinds of wetting phenomena mentioned above, researchers have also discovered that the Stenocara beetle has two kinds of structures on its body at the same time (Figure 4). One is the hydrophilic property, while the other is the hydrophobic property [26]. Because of these interlaced structures, the beetles are able to collect water mist in the air with the hydrophilic surfaces and transport effectively the water to their mouths with the help of the hydrophobic structures.

This type of surface is classified as "interlaced wettability." There is another type of wettability, which can be changed by external fields such as electrical fields, mechanical stresses, and other ways. The surface wettability is then to be able to vary from hydrophobic to hydrophilic or vice versa. Such the property is classified as "tunable wettability" (Figure 5).

Finally, there is the "gradient wettability" (Figure 6), which means that the wettability is not homogeneous on the surface. When water drops onto such the surface, the droplet will move from rough to smooth places automatically.

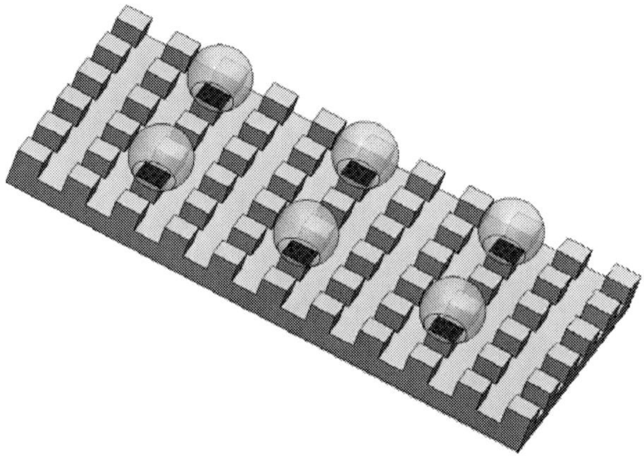

Figure 4. Interlaced wettability: hydrophilic materials (under droplet surface) pattern on a hydrophobic surface.

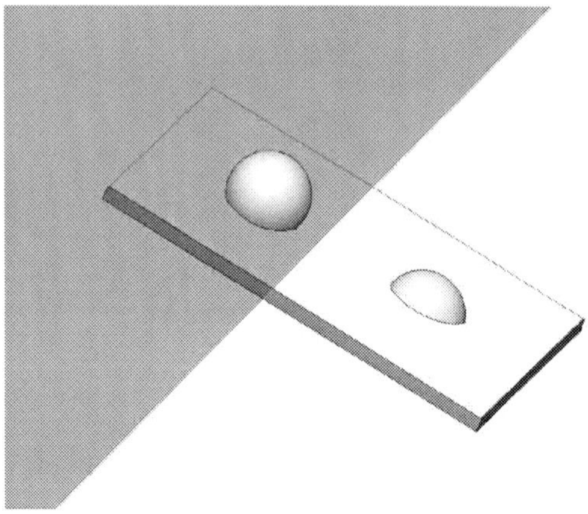

Figure 5. Tunable wettability: wettability can be tuned during some kinds of potential (electronic, temperature, light, mechanical stress, pH). For example, TiO_2 can be changed as superhydrophilic if exposed under UV light (right side). Surface without exposure remains larger CA (left side).

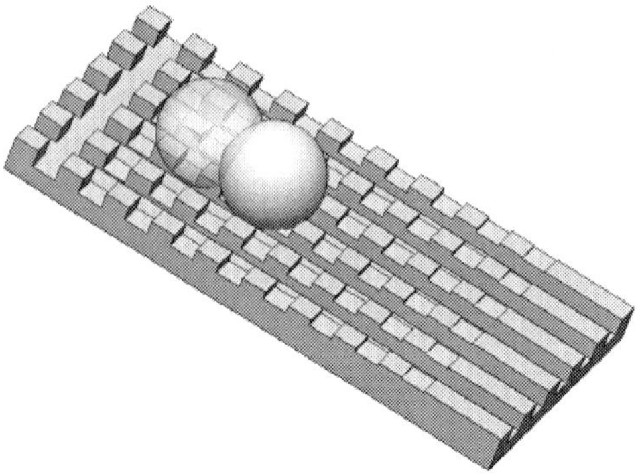

Figure 6. Gradient wettability: a water droplet moves from low surface energy (hydrophobic) region (left side) to high surface (hydrophilic) energy region (right side).

This paper gives a brief review of the current status of surface modification based on the characteristics of surface wettability. Section 2 gives detailed descriptions of the five kinds of surface wettability. Section 3 reviews fabrication methods for surface modification to achieve different wettabilities. Five main fabrication methods are considered in detail: etching, lithography, deposition, sol-gel processing, and templation. The transparency of the modified surface is also discussed since it is an important factor in some applications, such as solar cells. Section 4 provides conclusions. This study should be helpful in the design of surface modification and future developments in this field.

2. Classification of Surface Wettability

This section discusses five types of surface wettability, including isotropic wettability, anisotropic wettability, interlaced wettability, tunable wettability, and gradient wettability.

2.1. Isotropic Wettability

For isotropic wettability, the CA is the same in all directions. This may be the most common feature for material surfaces. In 1997, Barthlott [9] found that there are hierarchical micro-sized bumps and nano hair-like structures on the lotus leaf surface. These two scale structures create a rough surface. Researchers have tried to mimic the hierarchical structures of the lotus leaf to form superhydrophobic surfaces using either chemical etching [27], laser etching [28] nanoimprinting lithography (NIL) [29], or combined photolithography with reactive ion etching (RIE) [30]. By coating hydrophobic materials on glass, manufacturers can produce self-cleaning house windows, car windows, or solar cell cover glasses.

2.2. Anisotropic Wettability

Two examples of anisotropic wettability in nature are bamboo [1] and rice [2,31] leaves. This feature is the natural consequence of surface morphology. In these cases, long parallel periodic grooves exist on the leaf surface, creating smaller CAs along the longitudinal direction than those along the transverse direction. These patterns can be fabricated easily using lithography techniques. Sommers and Jacobi [32] have successfully made one-dimensional patterns on an aluminum surface to mimic anisotropic wettability. With this surface modification this approach enhances surface drainage, and can be applied on air cooler devices.

The depth and pitch of a one-dimensional period pattern are two key parameters for anisotropic wettability [33]. Fengxiang et al. [34] successfully built different scale parallel grooves, and discussed how to imprint different directions of nano scale parallel grooves (250nm) on micro scale parallel grooves (2μm). In stead of modifying the structures on the same materials, Bliznyuk et al. [35] and Morita et al. [36] coated two kinds of hydrophobic and hydrophilic materials with alternative stripe patterns on a silicon surface, resulting in anisotropic wettability.

2.3. Interlaced Wettability

After Parker and Lawrence [26] first published a paper discovering the interlaced wettability of Namibia desert beetle's back, many researchers began to mimic such the interlaced structures for various applications, like fog harvesting [37], and biomolecular immobilization [38]. Lei Zhai et al. [39] created hydrophilic dot arrays on a superhydrophobic surface and showed the function of collecting water on such the surface. Ihrig et al. [40] produced a partially hydrophilic surface of low density polyethylene (LDPE) using a plasma etching process. Choi et al. [41] deposited hydrophilic SiO2 dots on hydrophobic nanoscale structures. Garrod et al. [42] showed that with their modified interlaced surface, they can collect water on the surface of the area 5 x 10-8 m2 and reach 1.6g/cm2 in 2 hours.

2.4. Tunable Wettability

Tunable wettability means that surface wettability can be varied by changing the surface conditions. These so-called "smart" or "intelligent" surfaces are widely applied in sensing, manipulation of bioanalytical substances, intelligent drug release, controllable separation systems, and more [43-48]. In addition, such the surfaces can be used for the effective control of micro or nano-fluid motion on fluidic devices [49-54]. The switch functions of tunable wettability are discussed below in terms of electric potential, temperature variation, light irradiation, mechanical stress, and pH variation of solvent.

2.4.1. Electric Potential

Ahuja et al. [55] applied an electrical voltage to transform the wettability of a surface from superhydrophobic to the hydrophilic condition. They demonstrated that when the voltage is not applied, the CA of surface is 150°, but it becomes lower than 30° (strong wetting) when the voltage is applied. Lahann et al. [56] made MHA (16-Mercapto hexadecanoic acid) on the gold surface. MHA has a hydrophobic chain capped by a negatively-charged hydrophilic carboxylate group, forming a hydrophilic surface. The hydrophilic carboxylate groups will be attracted to the surface after the surface is charged positively. At this time, two ends of the chains are connected to the surface and the middle hydrophobic groups will be exposed on the surface, and the surface will become hydrophobic. Using static electricity to change the wetting behavior of dielectric film [57] is called electrowetting on dielectric (EWOD). This approach makes it possible to move water droplets across various surfaces [58], control the adhesion force of a water droplet [59], and develop lab-on-a-chip systems [60].

2.4.2. Temperature Variation

The wettability of poly(N-isopropylacrylamide) (PNIPAAm) [43,61-64] is temperature sensitive. There is a sharp wettability change of PNIPAAm at annealing temperature. When the temperature is below the annealing temperature, it is superhydrophilic (CA near 0°), while it becomes superhydrophobic if the temperature is higher than 30°C. This feature is due to the formation of C=O and N-H groups under heat. These two groups behave hydrophobically. Therefore, heating results in a nearly superhydrophobic surface (CA=149.3°)

The surface of poly (N-isopropylacrylamide-co-acrylic acid) (PNIPAAm-co-AAc) [64] can also be changed from superhydrophobic to superhydrophilic in response to both temperature and pH. The wettability of gold film hierarchical structures on a poly(dimethylsiloxane) (PDMS) surface has been successfully changed by plasma and heat treatments, switching from superhydrophobic (CA=160°) to superhydrophilic (CA= 0°) [65].

2.4.3. Light Irradiation

Materials like TiO_2 [18, 66-68] and Zinc oxide (ZnO) [69-73] can become superhydrophilic under exposure of ultraviolet (UV) light exposure and restore back to its original state if the UV light is removed. In 1997, Wang et al. [15] found that when polycrystalline TiO_2 was coated on glass irradiated by UV light, the CA of the surface was about 0°, but if the sample was placed in the dark for few days, the CA became 72°. Liu et al. [69] showed that the CA of the ZnO surface with nanostructures can change from 164° to blow 5° after UV exposure. A variety of ZnO and TiO_2 film fabrication methods have been reported, including CVD [67,69] thermal evaporation [74], pulsed laser deposition [75,76]. A hierarchical structure of micro-size Si spikes with nano-size ZnO protrusions was developed by pulsed laser deposition [75]. The wettability of the surface can even change from nearly superhydrophobic (CA~150°) to superhydrophilic (CA~0°) after UV light illumination.

2.4.4. Mechanical Stress

Jilin et al. [77] constructed a spring micro-structure with a triangular net-like structure of elastic polyamide film, which can be pulled by force. Therefore, they can change the surface roughness by extension. Before extension, the side length of this triangular structure is 200 μm and the surface CA is 150°. After applying the extension, the side length of the triangular can be extended to 450 μm and the surface becomes smoother. They showed that the CA can reduce to nearly 0°.

2.4.5. pH Variation of Solvent

Some polymers can change their chemical structures by adding different solvents with different pH values [78-85]. And different chemical structures have different wettability. It is then possible to use solvent to change wettability of a polymer coated surface. A variety of amphiphilic polymers can be used, such as polystyrene-poly(acrylic acid) (PS-PAA) [78]. Zhu et al. [44] demonstrated that the CA of a PS-PAA coated surface can change alternatively from 20° (by adding basic 0.1M NaOH) to more than 120° (by adding toluene). In addition to polymers, Yu et al. [86] coated a mixed monolayer containing both alkyl and carboxylic acid groups on gold surface with hierarchical structures, and showed that when pH value of solvent varies from 1 to 13, the CA of the surface can change from nearly 0 to nearly 160°.

2.5. Gradient Wettability

When the wettability of a surface is not homogeneous, it is classified as "gradient wettability." One example in nature is the wings of the butterfly called Morpho aega [87]. The morphology of the wing surface is like a ladder structure and extends to the rim of the wing, allowing droplets to easily roll off along the grooves. The gradient wettability of a surface implies that there exists difference of surface energy among different locations. Such the difference can be the driving force to move a droplet. In 1992, Chaudhury and Whitesides [88] showed that a water droplet (1 μL) can conquer gravitation and move up a slanted surface (tilted from a horizontal plane by 15°), revealing a new method of droplet motion. Lu et al. [89] made a poly(tetrafluoroethylene) (LDPE) surface with gradient roughness, resulting in that CA varies from 90° to over 150°. They demonstrated that a water droplet can move "automatically" from a hydrophobic location to a hydrophilic place. When applying non-uniform external fields to a surface with tunable wettability, people can get a surface with wettability gradient. Examples include electrowetting on dielectric (EWOD) [90-92], temperature gradient [93-95], and light potential [96].

3. Fabrication Methods for Surface Modification

One way to modify a surface wettability is to put different materials with desired wettability properties on the surface. For example, the common materials selected to enhance hydrophobic wettability include octadecyltrichlorosilane (OTS) [97], PTFE [98], fluoroalkylsilane [99,100], fluorocarcon [101], organosilanes [70], hexamethyldisiloxane (HMDSO) [102], tetramethylsilane (TMS) [99] dimethyldichlorosilane (DMDCS) [5] , *n*-octyldimethylchlorosilane (ODMCS) [5] and heptadecafluoro-1,1,2,2-tetrahydrodecyl dimethylchlorosilane (FDDCS) [5]. As for getting the hydrophilic property, 4-vinyl pyridine (4-VP), maleic anhydride (MA), glycidyl methacrylate (GMA), bromoethyl-acrylate (BEA), 4-vinyl aniline(4-VA) and 3-vinylbenzaldehyde (3-VBAL) [50], TiO_2 [66-68], or ZnO [67,70] has been used. In addition, the wettability of a surface can be governed not only by the wettability of the materials, but also surface roughness. Qian and Shen [103] demonstrated that the original hydrophilic copper surface (CA=75°) can even become hydrophoblic (CA=150°) by forming hierarchical structures on the surface. Making roughness surface is an important issue to change wettability on surface.

Two scale of hierarchical structure like lotus leaves usually can make a surface prone to hydrophobic wettability. Many methods have been developed to achieve various types of wettability. These methods include etching, deposition, sol-gel process, lithography, and templation processes. Lithography techniques can easily build one-dimensional structures, resulting in anisotropic wettability.

Patterning hydrophilic materials on a hydrophobic surface through deposition methods can achieve interlaced wettability. The following subsections give detailed descriptions for each fabrication methods.

3.1. Etching

Etching can easily remove materials from a surface, especially compared with other methods. Surface modification by etching can be divided into three categories: laser etching, chemical etching, and plasma etching. These techniques can easily make a rough surface.

3.1.1. Laser Etching

Laser etching can form a rough structure through a simple one-step process, and it is applicable to a wide variety of materials [104-106]. Gao and Yao applied laser etching to make hierarchical micro/nano structures on a silicon surface [107], resulting in a superhydrophobic surface with the CA greater than that of 175°.

3.1.2. Chemical Etching

Chemical etching uses a chemical solution to erode materials by chemical reaction. This method can be generally easily to make superhydrophobic surfaces [27,108-110]. Nitric acid and hydrogen peroxide solutions are used for steel and copper alloys [111], while the treatment of silicon involves ethanoic hydrofluoric acid [112] and hydrofluoric acid/hydrogen peroxide is used for titanium alloys [111]. Hierarchical flower-like structures were made on a Cu-Zn alloy surface using dodecanoic acid (DA) etching, resulting in CA of 161° [113]. Hong et al. [114] modified the copper surface with hierarchical flower-like structure, showing a CA of 158°. Li et al. [97] used multiple chemical etching steps to fabricate micro-pyramid structures and nano-wire hierarchical structures on a silicon surface. The first process produced micro-scale pyramid structures by using NaOH anisotropic etching, and the second one used silver-catalyzed chemical etching to produce nano wire structures. Finally, modified OTS (octadecyltrichlorosilane) was used to obtain superhydrophobicity on the silicon surface. The measurements get CA = 162°and sliding angle (SA) < 2°.

3.1.3. Plasma Etching

Plasma etching is a dry etching technique in which reactive atoms or ions with high momentum are used to shot the sample, and the splashed atoms are removed via vacuum systems. Different materials may require specific type of gases in plasma etching. For example, ion plasma etching is suitable for TEP-Teflon surfaces [68], argon plasma etching for polypropylene [115], domain-selective oxygen plasma etching for poly(ethylene terephthalate) [69].

Liu et al. [69] made a superhydrophobic poly (methyl methacrylate) (PMMA) surface with high transparency. They used oxygen plasma etching to produce rough surface and chemical treatment of fluorocarbon material to decrease surface energy of the solid. Shiu et al. [116] used plasma etching to form deep grooves on the surface, resulting in a superhydrophobic surface (CA = 171°). Kang et al. [117] used deep reactive ion etching (RIE) process to generate micro and nano structures on the silicon surface. If with only micro structures, they found that CA is 146.6°, while CA can increase to 158.3° with hierarchical (micro and nano) structures. Kim et al. [118] modified silicon surface with nanoscale structure by using CF_4 plasma etching, ending up with CA =137.8°. They further modified the silicon surface with larger copper dots (400 nm) to form the hierarchical structures. Finally they can get a superhydrophobic surface with CA =160° and hysteresis angle (HA) < 5°.

Table 1 shows the superhydrophobic surface fabricated by the etching methods. The CAs measured are all larger than 150° with hierarchical structures. When surface just has single micro scale structures, CA will be decreased by 12°- 20° compared with hierarchical structures [97,117]. This result demonstrates that fabrication of hierarchical structures is an effective strategy if higher CA is demanded. Table 1 also shows that the etching methods are applicable to either polymeric or metal substrates.

Table 1. Rough surface fabricated by etching methods

Fabriction Approach	Substrate	Contact angle CA (^{0})	Sliding angle SA (^{0})	Hysteresis angle HA (^{0})	Transmission T%	Chemical coating	references
femtosecond laser etching	Silicon	154±1(a)		5±2			[28]
Laser etching	Silicon	150.7(a)	<5			FAS	[107]
Chemical etching	Al	168(a)	<5			perfluorononane (C9F20)	[27]
	Al alloy	161(a)	<5			poly (dimethy lsiloxane) vinyl terminated (PDMSVT)	[27]
	Cu-Zn alloy	161(a)					[113]
One-step chemical immersion	copper	>150(a)	<5				[114]
NaOH anisotropic etching and chemical etching	Silicon	~160(a) ~140(b) ~150(c)	<2			OTS	[97]
plasma treatment	polystyrene	170(a)				Teflon	[116]
	PET	>150(a)			>90%	FAS	[99]
	Silicon	158.3(a) 146.6(b)					[117]

(a) Hierarchical structure (b)Micro scale structure (c) Nano scale structure.

3.2. Lithography

Lithography is extensively used in MEMS and semiconductor industry. At first, photoresist is coated on the surface. Depending on the methods to decide the amount of photon energy at different locations during exposure, which can end up different patterns on the surface, there are different lithography techniques, including photolithography (mask), laser interference lithography (interference), nano-imprinting lithography (mold pressing), and capillary force lithography (capillary force). These four techniques are descried as follows.

3.2.1. Photolithography

Kawai and Hitoshi [119] is the first group to change wettability on modification surface by using photolithography. They showed that aspect ratio (Ar) from 0 to1.2 of micro structures would result in CA from 60° to 100° on modified silicon surface. Bhushan et al. [120] used photolithography to form the micro-pillars (Ar = 2.1). They measured CA to be

154°. It seems that higher aspect ratio of structures would result in higher CA on modified surface.

The structure of anisotropic wettability can be easily produced by lithography techniques. Sommers et al. [32] combined photolithography with plasma etching to form one-dimensional structures on an aluminum alloy surface. They measured two different CAs along two mutual perpendicular directions, one is hydrophobic (CA=129°), the other is hydrophilic (CA=75°). Bliznyuk et al. [35] patterned hydrophobic perfluorodecyltrichlorosilane(PFDTS) material (CA =110°) and hydrophilic SiO_2 material (CA = 31°) to produce anisotropic wettability of stripe-patterned surface on silicon wafer by photolithography and vapor deposition. Wetting anisotropy is 25° (θx =105°, θy = 80°) when SiO_2 and PFDTS stripe width are 8μm.

Yasuda et al. [121] used photolithography to pattern hydrophilic SiO_2 and hydrophobic Cytop (amorphous fluoropolymer) material like wedge shape on silicon surface, resulting in gradient wettability. They found that a DI water droplet (5μL) can move automatically. Yang et al. [122] fabricated radial gradient surface by standard photolithography techniques and inductively coupled plasma etching on silicon substrates. They also found automatic motion of a droplet and the moving velocity is 62.5 mm/s from superhydrophobic location (CA=160°) to hydrophobic location (CA =120°). The wettability of OTS-SAM [124] and ODS-SAM [123,124] materials will be changed after exposure to UV light. Therefore, these two kinds of materials are usually used to fabricate gradient surface on silicon surface if a gradient photomask is applied.

3.2.2. Laser Interference Lithography (LIL)

Laser interference lithography is an effective way to fabricate periodic patterns over a wide area at a time without loss of focus [33,125-127]. Berendsen et al. [126] used LIL to make the mold and applied thermal imprinting to create an array of bumps on the surface, then coated with fluorocarbon to get a superhydrophobic surface (advancing CA =167°, HA< 5°). Chang et al. [128] used similar methods of LIL and nanoimprint to form hydrophobic (CA=133°) PET surface with nano-structures and high transmittance (87%). Instead of using molds, Yang [129] used LIL to generate interference patterns (photoresist FH6400) on a silicon surface (Figure 7a). Figure 7b shows the CA will increase with increased linewidth of nano structure. If coated with hydrophobic materials of hexamethyl-disilazane (HMDS), the CA of nanostructure with feature size of 300 nm are transformed from 17° (Figure 8a) up to 135° (Figure 8b). Anisotropic structures and then anisotropic wettability are specifically easy to be fabricated by using LIL because LIL can naturally generate one-dimensional interference patterns. Xia and Brueck [33] produced 1500 nm period of one dimensional pattern on silicon surface. Such the surface shows strong anisotropic wetting feature. Depending on the substrate materials, the wetting anisotropy can be 79° (θx =130°, θy = 51°) on Si substrate, and 82° (θx =125°, θy = 43°) on SiO_2 substrate.

3.2.3. Nanoimprinting Lithography (NIL)

In 1996, S. Y. Chou et al. [130] used nanoimprinting lithography to fabricate nanoholes and nanodots of patterns on a PMMA structure. Nanoimprinting lithography (NIL) is a good solution to mass production for nano scale structures on surface. NIL can also pattern different levels of hierarchical structures [29,34,131] on polymer surfaces. Zhang et al. [29] built hierarchical structures on PS and PMMA surface by nanoimprinting lithography.

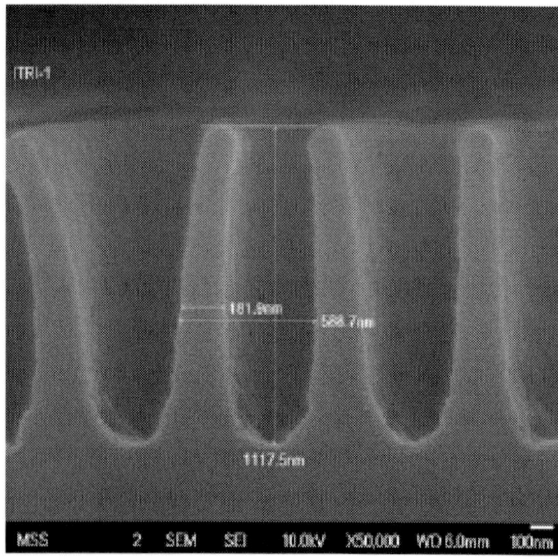

a

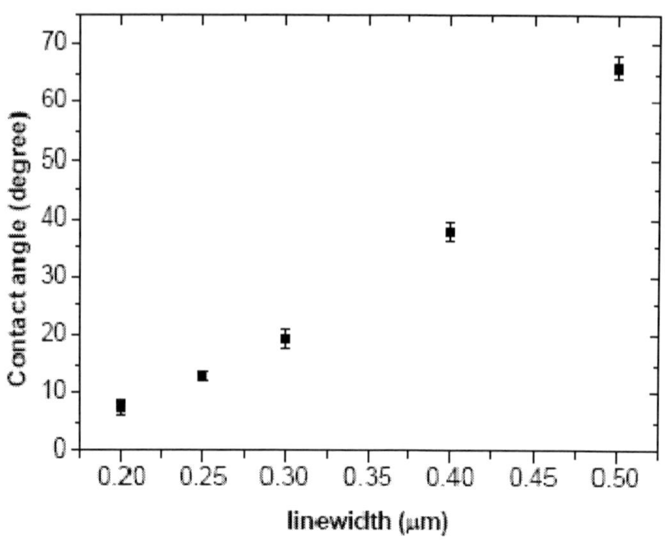

b

Figure 7. SEM images of nano structure formed by LIL, linewidth of 200 nm and the depth of 1200 nm as shown in Figure 7a. Nano structures of the various pitches as a function of structure geometry with contact angle as shown in Figure 7b. Reprinted with permission from [129].

They modified the surface such that the CA of the PS surface with nano pillars/micro-hemisphere structures is 128°, and 104° for PMMA surface. Pai et al. [131] coated one layer of PDMS on silicon substrate to imprint nanostructures on the silicon wafer. The CA of the modified surface increases to 111°, compared to 33° of a bare silicon surface. Zhang and Hong [34] fabricated hierarchical of micro (2 μm)/nano (250 nm) parallel grooves on PS and PMMA surface via two steps of NIL processes. The wetting anisotropy is 54° (θx =121°, θy = 67°) on Si substrate, 32° (θx =133°, θy = 101°) on SiO_2 substrate.

3.2.4. Capillary Force Lithography

When a patterned mold is placed on polymeric surface, and heated above to the polymer's glass-transition temperature, the capillarity force will drive polymeric melt into void space of mold and to form the structure.

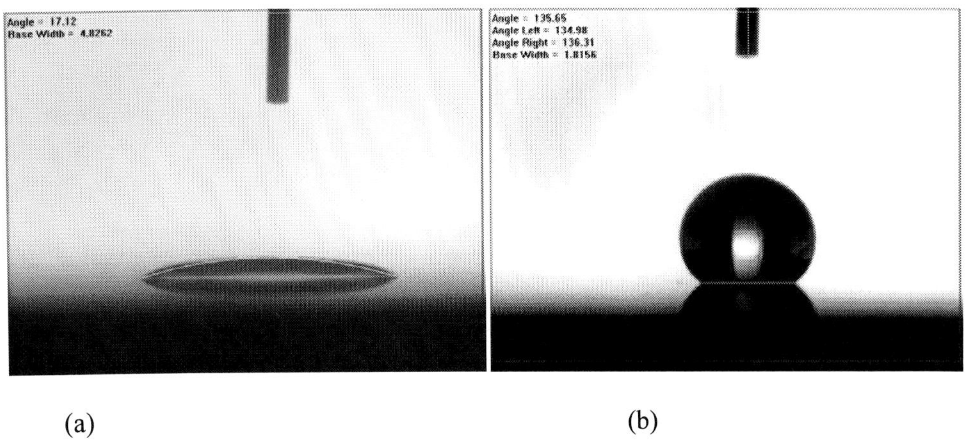

Figure 8. Images show the contact angles of a 4-μL drop of water on (a) nanowire without HMDS coating, and (b) nanowires with HMDS coating as a function of structure geometry. Reprinted with permission from [129].

In recent years, many researchers have used capillary force lithography (CFL) [132-134] to pattern structures on polymeric surface. Jeong et al. [134] used two kinds of mold (polyurethane acrylate (PUA) and PDMS) via two steps of CFL processes to build hierarchical surface. The CA is almost 170°. Suh et al. [135] used CFL to imprint structures on different materials (PMMA, PS ZnO). Suh and Lee [136] used the CFL method to generate anisotropic hydrophobic surface. Table 2 shows a list of some work done by lithography techniques, and polymeric substrates are usually used in this method. One can see that PET polymeric is a good material if high transmittance is required [127].

Table 2. Rough surface fabricated by lithography techniques

Fabriction Approach	Substrate	Contact angle CA (⁰)	Sliding angle SA (⁰)	Hysteresis angle HA (⁰)	Transmission T%	Chemicall coating	references
Nanoimpriting and LIL	PS	162.3(c)		4.7		fluorocarbon	[126]
	PET	131(c)			90%		[127]
Nanoimpriting	PS	128(a) 107(b) 115(c)					[29]
	PMMA	104(a) 93(b) 95(c)					[29]
Capillary force lithography	PET	170(a)		3~5			[134]

(a) Hierarchical structure (b)Micro scale structure (c) Nano scale structure.

3.3. Deposition

Deposition is a process to deposit layers at low pressure on a solid surface. People can coat various hydrophobic materials on a surface via deposition processes, such as PTFE [98], OTS [137], polymerized perfluoroalkyl ethyl methacrylate (PPFEMA) [138], methyloctyldimethoxysilane (MODMS) [139], fluorooctylmethyldimethoxysilane (FODMS) [113]. As for getting a hydrophilic surface, hydrophilic materials can be coated, like Ti [88], SiO_2 [41], and ZnO [67,70].

To mimic the interlaced wettability of the Stenocara beetle's back, Garrod et al. [42] deposited hydrophilic 4-vinyl pyridine (4-VP), maleic anhydride (MA), glycidyl methacrylate (GMA), bromoethyl-acrylate (BEA), 4-vinyl aniline(4-VA) materials on hydrophobic poly(tetrafluoroethylene) surface by using deposition processes.

Rough surface can be formed easily by depositing two kinds of molecules with different sizes using deposition processes. Liu et al. [140] created large rough area of superhydrophobic surface (CA=151°) by depositing binary cadmium sulfide (CdS) films on glass surface through microwave assisted chemical bath deposition (MA-CBD) method.

Liu et al. [141] deposited nano-scale (100-200 nm) Zn and ZnO molecules on glass surface by the novel plasma assisted thermal vapor deposition, and resulting superhydrophobic (CA=157°) and high transmittance (over 80%) glass surface. Hsieh et al. [142] fabricated a two-tier roughness with fluorinating carbon fibers (20 nm) by catalytic chemical vapor deposition (CCVD) to modify the carbon surface to reach superhydrophobic surface (CA=167°). Shirtcliffe et al. [143] first used photolithography and selectivity etching to dig array of micro structures (5-40 μm) on copper surface, then applied electrodeposition to deposit small scale of globular textures (cooper) to form the hierarchical structure. They measured the CA is 155°.

Table 3 lists researches using the deposition methods. It shows that the deposition method can apply to various substrates.

Table 3. Rough surface fabricated by deposition methods

Fabriction Approach	Substrate	Contact angle CA (^{0})	Sliding angle SA (^{0})	Hysteresis angle HA (^{0})	Transmission T%	Chemical coating	references
thermal vapor deposition	ZnO/Zn	157(c)	9		85%		[141]
chemical bath deposition	glass	151(a)					[140]
chemical vapor deposition	carbon fabric	169.7(a)				perfluoroalkyl methacrylic	[142]
Photolithography and chemical vapor deposition	copper	>150(a) 122(b) 152(c)		22		persulfate	[143]
Photolithography and chemical vapor deposition	silicon	169(a), 154(b), 158(c)					[120]

(a) Hierarchical structure (b)Micro scale structure (c) Nano scale structure.

3.4. Sol-Gel Processing

The sol-gel process is a wet chemical solution deposition commonly used in the field of material science. Recently, the sol-gel processing is usually used in surface modification to change the surface wettability. One way is to construct a rough glass surface by depositing particles on the glass substrate with the help of silanes as a coupling agent. The particles used include SiO_2 [144-147], TiO_2 [144], ZrO_2 [144] and ZnO [148,149]. For example, Lee [150] used two scale of SiO_2 particle sizes with methyltriethoxysilane (METES) and decyltrimethoxysilane (DECYS) by sol-gel processes to form hierarchical structures on glass surface (Figure 9a).

Modified surface has a high transmittance of over 85% and a high water CA over 160° (Figure 9b) on glass surfaces. Another strategy for surface modification is to deposit desired materials with particular functional groups on the surface to modify the wettability of the original surface.

The CAs can be increased either using fluorinated silanes [151-155], fluoroalkyl silane [147,153,156] or alkylated silanes [157-161] because all of the functional groups exposed are highly hdrophobic. Bravo et al. [162] used SiO_2 nano-particles with the layer-by-layer assembly technique to produce superhydrophobic surface (CA>160°) with high transmittance (>90%) in the visible light range. Yang et al. [163] made mulberry-like hierarchical structures by micro scale $CaCO_3$ particles (diameter: 0.8 - 1 μm) and nano scale SiO_2 particles (diameter: 30 nm).

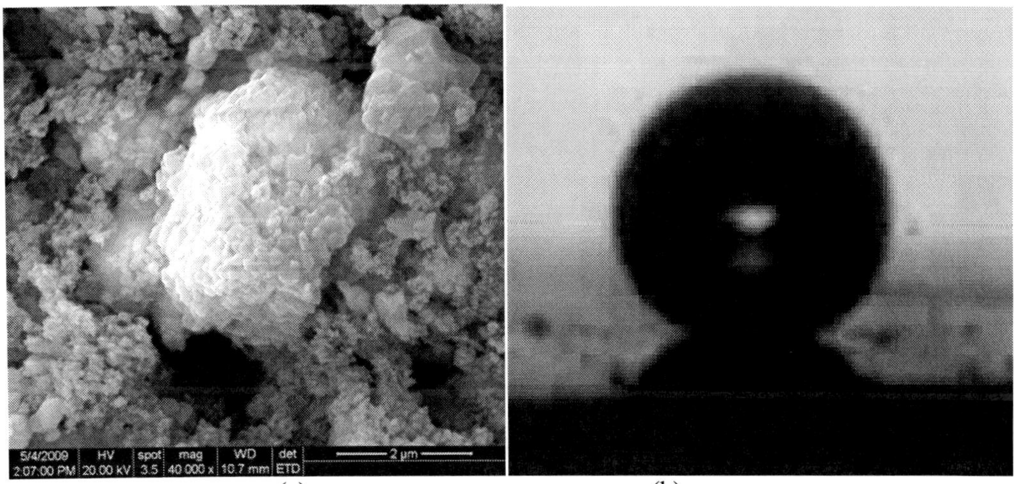

(a) (b)

Figure 9. Figure 9a shows multiple of lenses (40000X) of SEM picture with hierarchical structures on a glass surface. Water dropped on modified surface by using sol-gel process. Figure 9b shows that the measured CA can reach as high as 160°. Reprinted with permission from [150]

Table 4 lists the work done by sol-gel processes. Glass substrates are commonly used in sol-gel processes. The best advantage is its high transmittance of a glass substrate. It shows that most of transmittance can still be over 90% after sol-gel processes.

Table 4. Rough surface fabricated by sol-gel processes

Fabriction Approach	Substrate	Contact angle CA (^{0})	Sliding angle SA (^{0})	Hysteresis angle HA (^{0})	Transmission T%	Chemically coating	references
Sol gel	glass	~170(a)		<10	90%	fluoroalkyl silane	[156]
	glass	>160(a)	<1			FAS	[153]
Sol gel	PDMS	>160(a)	<10				[163]
Layer-by-layer assembly technique	glass	>160(a)		<10	>90%		[145]
	glass	~150(a)			>95%	fluoroalkylsilane	[146]

(a) Hierarchical structure (b)Micro scale structure (c) Nano scale structure.

3.5. Templation

Templation is a good method for mass production and the templates required can be used repeatedly to save cost. The mold prepared by a photolithography process for polymeric materials are widely used for templating surface, such as PDMS [163], PS [24,164], PMMA [165] and PTFE [164]. Kim et al. [147] used the nanodimpled (< 1μm) aluminum mold to form hierarchical of PDMS structures. The measured CA can reach as high as 141°. Huang [166] used a microstructured PDMS mold to pile up 85nm PS spheres to build hierarchical structures on glass substrates by templation method (Figure 10a). Figure 10b shows the SEM images of the hierarchical structures composed of PS spheres. The CA of the hierarchical structure is 140° and the transmittance can reach over 80%. Anodic aluminum oxide (AAO) is frequently used as a template for surface modification because of its easy preparation of either microscale or multiple scale structures on the aluminum oxide surface [24,164,165,167,168]. The measured CA of the templated surface by AAO can be as high as 169° [164]. Table 5 shows a list of work using the templation method. Although usually generating surfaces with less CA compared with other fabrication methods, the templation method is still an efficient way for surface modification because of its power of mass production.

Table 5. Rough surface fabricated by templation process

Fabriction Approach	Substrate	Contact angle CA (^{0})	Sliding angle SA (^{0})	Hysteresis angle HA (^{0})	Transmission T%	references
Templaion with AAO membranes	PTFE	169(a)	<2			[164]
	Ploycar carbouate	139(a),120(b), 116(c)			>80%	[167]
	PMMA	138(a)		<10		[165]
Photolithography and Templaion	PS	148-156(c)				[124]
	PS	140(b)			>80%	[166]
	PDMS	140(a),132.5(b), 118.5(c)	51	<10		[163]

(a) Hierarchical structure (b)Micro scale structure (c) Nano scale structure.

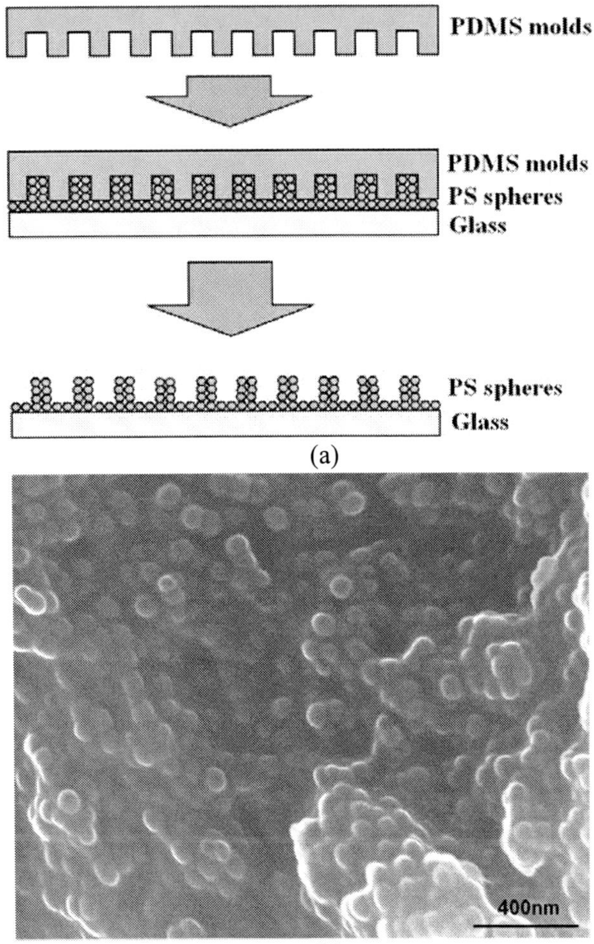

Figure 10. Figure 10a shows that by conformal lithography and replication to build PDMS mold, hierarchical structures of PS particle via thermal imprinting methods on glass substrate. SEM figure shows the different scale of hierarchical structures of PS particle in Figure 10b. Reprinted with permission from [167].

CONCLUSION

This work provides a classification of five kinds of surface wettability and an overall review of the main fabrication methods for surface modification. Isotropic wettability is the most commonly found in nature. With anisotropic structures, the surface can own the anisotropic wettability. The interlaced wettability has been found very useful to grab water from the mist by hydrophilic surface and driving them to specific location by hydrophobic surface. With the help of external sources like electrical field, temperature variation, or UV irradiation, the researches have produced modified surface with tunable wettability. Thanks to the gradient wettability, it has been demonstrated that a water droplet can move automatically without other external powers. In addition to the material properties, it is worthy to point out that, the surface structure is one of the key parameter to surface wettability. Rough surface is usually more hydrophobic than that of smooth surface. With two scales of micro and nano structures, it can become hydrophobic even when the original smooth surface is hydrophilic.

It is easy to modify a large surface area via deposition and laser interference lithography. Chemical etching and sol-gel processes have the advantage of low cost. Templation is suitable for mass production. The lithography process is a suitable way to produce a variety of patterns. When considering the materials of the substrates or the coating components, etching can be easily utilized with metals and nonmetals. The lithography process is most applicable to silicon substrates. Templation works well with polymers. The glass substrates are mostly modified by sol-gel processing. With the classification of surface wettability and the fabrication methods in mind, people should get clever insight into the subject of surface modification.

REFERENCES

[1] H. J. Li, X. B. Wang, Y. L. Song, Y. Q. Liu, Q. S. Li, L. Jiang, D. B. Zhu, Super-amphiphobic aligned carbon nanotube films, *Angew. Chem. Int. Ed.* 40 (2001) 1743-1746.

[2] L. Feng, S. H. Li, Y. S. Li, H. J. Li, L. J. Zhong, J. Zhai, Y. L. Song, B. Q. Liu, L. Jiang, D. B. Zhu, Super-hydrophobic surfaces: from natural to artificial, *Adv. Mater.* 14 (2002) 1857-1860.

[3] T. P. Russell, Surface-responsive Materials, *Science* 297 (2002) 964-967.

[4] T. Onda, S. Shibuichi, N. Satoh, K. Tsujii, Super-water-repellent fractal surfaces, *Langmuir* 12 (1996) 2125-2127.

[5] D. Öner and T. J. McCarthy, Ultrahydrophobic surfaces effects of topography length scales on wettability, *Langmuir* 16 (2000) 7777-7782.

[6] H. Y. Erbil, A. L. Demirel, Y. Avcı, O. Mert, Transformation of a simple plastic into a superhydrophobic surface, *Science* 299 (2003) 1377-1380.

[7] R. Blossey, Self-cleaning surfaces–virtual realities, *Nat. Mater.* 2 (2003) 301-306.

[8] H. M. Shang, Y. Wang, S. J. Limmer, T. P. Chou, K. Takahashi, G. Z. Cao, Optically transparent superhydrophobic silica-based films, *Thin Solid Films* 472 (2005) 37-43.

[9] W. Barthlott, C. Neinhuis, Purity of the sacred lotus, or escape from contamination in biological surfaces, *Planta* 202 (1997) 1-8.

[10] X. J. Feng, L. Feng, M. H. Jin, J. Zhai, L. Jiang, D. B. Zhu, Reversible super-hydrophobicity to super-hydrophilicity transition of aligned ZnO nanorod films, *J. Am. Chem. Soc.* 126 (2004) 62-63.

[11] T. L. Sun, G. J. Wang, L. Feng, B. Q. Liu, Y. M. Ma, L. Jiang, D. B. Zhu, Reversible Switching between superhydrophilicity and superhydrophobicity, *Angew. Chem. Int. Ed.* 116 (2004) 357-360.

[12] Q. Fu, G. V. R. Rao, S. B. Basame, D. J. Keller, K. Artyushkova, J. E. Fulghum, G. P. Lopez, Reversible control of free energy and topography of nanostructured surfaces, *J. Am. Chem. Soc.* 126 (2004) 8904-8905.

[13] J. Bico, C. Tordeux, D. Quéré, Rough wetting, *Europhys. Lett.* 55 (2001) 214-220.

[14] J. Bico, U. Thiele, D. Quere, Wetting of textured surfaces, *Colloids Surf.* A 206 (2002) 41-46.

[15] R. Wang, K. Hashimoto, A. Fujishima, M. Chikuni, E. Kojima, A. Kitamura, M. Shimohigoshi, T. Watanabe, Light-induced amphiphilic surfaces, *Nature* 388 (1997) 431-432.

[16] R. Wang, N. Sakai, A. Fujishima, T. Watanabe, K. Hashimoto, Rutile formation in hydrothermally crystallized nanosized titania, *J. Phys. Chem.* B103 (1999) 2188-2194.

[17] M. J. Hostetler, J. J. Stokes, R. W. Murray, Infrared spectroscopy of three-dimensional self-assembled monolayers: *n*-alkanethiolate monolayers on gold cluster compounds, *Langmuir* 12 (1996) 3604-3612.

[18] M. Sun, C. Luo, L. Xu, H. Ji, Q. Ouyang, D. Yu, Y. Chen, Artificial Lotus Leaf by Nanocasting, *Langmuir* 21 (2005) 8978-8981.

[19] D. H. Jung, I. J. Park, Y. K. Choi, S. B. Lee, H. S. Park, J. Rühe, Perfluorinated polymer monolayers on porous silica for materials with super liquid repellent properties, *Langmuir* 18 (2002) 6133-6139.

[20] M. Lehocky, P. F. F. Amaral, P. Stahel, M. A. Z. Coelho, A. M. B. Timmons, J. A. P. Coutinho, Preparation and characterization of organosilicon thin films for selective adhesion of Yarrowia lipolytica yeast cells, *J. Chem. Technol. Biotechnol.* 82 (2007) 360-366.

[21] M. Lehocky, P. F. F. Amaral, P. Stahel, M. A. Z. Coelho, A. M. B. Timmons, J. A. P. Coutinho, Deposition of yarrowia lipolitica on plasma prepared teflon-like thin films, *Surf. Eng.* 24 (2008) 23-27.

[22] P. Ball, Engineering shark skin and other solutions, *Nature* 400 (1999) 507-509.

[23] P. P. Goodwyn, E. D. Souza, K. Fujisaki, S. Gorb, Moulding technique demonstrates the contribution of surface geometry to the super-hydrophobic properties of the surface of a water strider, *Acta Biomater.* 4 (2008) 766-770.

[24] W. Lee, M. K. Jin, W. C. Yoo, J. K. Lee, Nanostructuring of a polymeric substrate with well-defined nanometer-scale topography and tailored surface wettability, *Langmuir* 20 (2004) 7665-7669.

[25] D. Byun, J. Hong, Saputra, J. H. Ko, Y. J. Lee, H. C. Park, B. K. Byun, J. R. Lukes, Wetting characteristics of insect wing surfaces, *J. Bionic Eng.* 6 (2009) 63-70.

[26] A. R. Parker, C. R. Lawrence, Water capture by a desert beetle, *Nature* 414 (2001) 33-34.

[27] Z. G. Guo, F. Zhou, J. C. Hao, W. M. Liu, Stable biomimetic super-hydrophobic engineering materials, *J. Am. Chem. Soc.* 127 (2005) 15670-15671.

[28] V. Zorba, E. Stratakis, M. Barberoglou, E. Spanakis, P. Tzanetakis, S. H. Anastasiadis, C. Fotakis, Biomimetic artificial surfaces quantitatively reproduce the water repellency of a lotus leaf, *Adv. Mater.* 20 (2008) 4049-4054.

[29] F. Zhang, J. Chan, H. Y. Low, Biomimetic, hierarchical via sequential imprint lithography, *Appl. Surf. Sci.* 254 (2008) 2975-2979.

[30] M. H. Chen, T. H. Hsu, Y. J. Chuang, F. G. Tseng, Dual hierarchical biomimic superhydrophobic surface with three energy states, *Appl. Phys. Lett.* 95 (2009) 023702-023703.

[31] X. M. Zhang, J. H. Zhang, Z. Y. Ren, X. Li, X. Zhang, D. F. Zhu, T. Wang, T. Q. Tian, B. Yang, Morphology and wettability control of silicon cone arrays using colloidal lithography, *Langmuir* 25 (2009) 7375-7383.

[32] A. D. Sommers, A. M. Jacobi, Creating micro-scale surface topology to achieve anisotropic wettability on an aluminum surface, *J. Micromech. Microeng.* 16 (2006) 1571-1578.
[33] Deying Xia and S. R. J. Brueck, Strongly anisotropic wetting on one-dimensional nanopatterned surfaces, *Nano Lett.* 8 (2008) 2819-2824.
[34] F. Zhang, H. Y. Low, Anisotropic wettability on imprinted hierarchical structures, *Langmuir* 23 (2007) 7793-7798.
[35] O. Bliznyuk, E. Vereshchagina, E. S. Kooij, B. Poelsema, Scaling of anisotropic droplet shapes on chemically stripe-patterned surfaces, *Phys. Rev.* E 79 (2009) 041601 1-6.
[36] M. Morita, T. Koga, H. Otsuka, A. Takahara, Macroscopic-wetting anisotropy on the line-patterned surface of fluoroalkylsilane monolayers, *Langmuir* 21 (2005) 911-918.
[37] http://www.fogquest.org/
[38] G. Macbeath, S. L. Schreiber, Printing proteins as microarrays for high-throughput function determination, *Science* 289 (2000) 1760-1763.
[39] L. Zhai, M. C. Berg, F. C. Cebeci, Y. Kim, J. M. Milwid, M. F. Rubner, R. E. Cohen, Patterned superhydrophobic surfaces: toward a synthetic mimic of the Namib desert beetle, *Nano Lett.* 6 (2006) 1213-1217.
[40] D. F. Ihrig, M. Licht, U. Brunert, J. Eggemann, Winning drinking water using radiation exchange, *Phys. Chem. Earth* 33 (2008) 86-91.
[41] J. H. Choi, Y. M. Kim, Y. W. Park, T. H. Park, K. Y. Dong, B. K. Ju, Hydrophilic dots on hydrophobic nanopatterned surfaces as a flexible gas barrier, L*angmuir* 25 (2009) 7156-7160.
[42] R. P. Garrod, L. G. Harris, W. C. E. Schofield, J. McGettrick, L. J. Ward, D. O. H. Teare, J. P. S. Badyal, Mimicking a stenocara beetle's back for microcondensation using plasmachemical patterned superhydrophobic-superhydrophilic surfaces, *Langmuir* 23 (2007) 689-693.
[43] T. L. Sun, G. J. Wang, L. Feng, B. Q. Liu, Y. M. Ma, L. Jiang, D. B. Zhu, Reversible switching between superhydrophilicity and superhydrophobicity, *Angew. Chem., Int.* Ed. 43 (2004) 357-360.
[44] Y. Zhu, M. H. Shi, X. D. Wu, S. R. Yang, Amphiphilic copolymer grafted "smart surface"enhanced by surface roughness, *J. Colloid Interface Sci.* 315 (2007) 580.
[45] L. B. Xu, Z. W. Chen, W. Chen, A. Mulchandani, Y. S. Yan, Electrochemical synthesis of perfluorinated ion doped conducting polyaniline films consisting of helical fibers and their reversible switching between superhydrophobicity and superhydrophilicity, *Macromol. Rapid Commun.* 29 (2008) 832-838.
[46] Y. H. Lin, H. W. Ren, Y. H. Wu, S. T. Wu, Y. Zhao, J. Y. Fang, H. C. Lin, Electrically tunable wettability of liquid crystal/polymer composite films, *Opt. Express* 16 (2008) 17591-17598.
[47] W. T. Sun, S. Y. Zhou, P. Chen, L. M. Peng, Reversible switching on superhydrophobic TiO_2 nano-strawberry films fabricated at low temperature, *Chem. Commun.* 5 (2008) 603-605.
[48] W. X. Hou, Q. Wang, UV-driven reversible switching of a polystyrene/titania nanocomposite coating between superhydrophobicity and superhydrophilicity, *Langmuir* 25 (2009) 6875-6879.
[49] Y. T. Cheng, D. E. Rodak, Is the lotus leaf superhydrophobic, *Appl. Phys. Lett.* 86 (2005) 144101 1-3.

[50] J. Xi, L. Feng, L. Jiang, A general approach for fabrication of superhydrophobic and superamphiphobic surfaces, *Appl. Phys. Lett.* 92 (2008) 053102 1-3.

[51] Z. Wang, N. Koratkar, L. Ci, P. M. Ajayan, Combined micro-/nanoscale surface roughness for enhanced hydrophobic stability in carbon nanotube arrays, *Appl. Phys. Lett.* 90 (2007) 143117-143119.

[52] C. Yang, U. Tartaglinoa, B. N. J. Persson, An access service scheme with anonymity for ubiquitous computing based on mobile IPv6, *Eur. Phys. J. E.* 25 (2008) 139-144.

[53] Y. Zhang, S. Sundararajan, Superhydrophobic engineering surfaces with tunable air-trapping ability, *J. Micromech. Microeng.* 18 (2008) 035024 1-7.

[54] W. Xiao, Z. Huang, Z. He, Tuning the wettability on $La_{0.7}Sr_{0.3}MnO_3$ coatings from superhydrophilicity to superhydrophobicity by hierarchical microstructure, *Appl. Phys. Lett.* 89 (2006) 083101 1-3.

[55] A. Ahuja, J. A. Taylor, V. Lifton, A. A. Sidorenko, T. R. Salamon, E. J. Lobaton, P. Kolodner, T. N. Krupenkin, Nanonails: a simple geometrical approach to electrically tunable superlyophobic surfaces, *Langmuir* 24 (2008) 9-14.

[56] J. Lahann, S. Mitragotri, T. N. Tran, H. Kaido, J. Sundaram, I. S. Choi, S. Hoffer, G. A. Somorjai, R. Langer, A reversibly switching surface, *Science* 299 (2003) 371-374.

[57] J. Lee, H. Moon, J. Fowler, T. Schoellhammer, C. J. Kim, Electrowetting and electrowetting-on-dielectric for microscale liquid handling, *Sens. Actuators, Phys.* A 95 (2002) 259-268.

[58] M. Abdelgawad, S. L. S. Freire, H. Yang, A. R. Wheeler, All-terrain droplet actuation, *Lab Chip* 8 (2008) 672-677.

[59] W. Shen, J. Kim, C. J. Kim, Controlling the Adhesion Force for Electrostatic ... Physical Surface Modification, *15th IEEE International Micro Electro Mechanical Systems Conference*, Las Vegas, Nevada, USA, Jan. (2002) 52-55.

[60] Y. Fouillet, D. Jary, C. Chabrol, P. Claustre, C. Peponnet, Digital microfluidic design and optimization of classic and new fluidic functions for lab on a chip systems, *Microfluid. Nanofluid.*4 (2008) 159-165.

[61] H. G. Schild, Poly(*N*-isopropylacrylamide): experiment, theory and application, *Prog. Polym. Sci.* 17 (1992) 163-249.

[62] Z. Hu, Y. Chen, C.Wang, Y. Zheng, Y. Li, Polymer gels with engineered environmentally responsive surface patterns, *Nature* 393 (1998) 149-152.

[63] L. Liang, P. C. Rieke, J. Liu, G. E. Fryxell, J. S. Young, M. H. Engelhard, K. L. Alford, Surfaces with reversible hydrophilic/hydrophobic characteristics on cross-linked poly(N-Isopropylacrylamide) hydrogels, *Langmuir 1*6 (2000) 8016-8023.

[64] F. Xia, L. Feng, S. T. Wang, T. L. Sun, W. L. Song, W. H. Jiang, L. Jiang, Dual–responsive surfaces that switch between superhydrophilicity and superhydrophobicity, *Adv. Mater.* 18 (2006) 432-436.

[65] J. Wu, H. J. Bai, X. B. Zhang, J. J. Xu, H. Y. Chen, Thermal/plasma-driven reversible wettability switching of a bare gold film on a poly(dimethylsiloxane) Surface by Electroless Plating, *Langmuir* 26 (2010) 1191 1198.

[66] R. D. Sun, A. Nakajima, A. Fujishima, T. Watanabe, K. Hashimoto, Photoinduced surface wettability conversion of ZnO and TiO_2 thin films, *J. Phys. Chem.*, B 105 (2001) 1984-1990.

[67] M. Miyauchi, N. Kieda, S. Hishita, T. Misuhashi, A. Nakajima, T. Watanabe, K. Hashimoto, Reversible wettability control of TiO_2 surface by light irradiation, *Surf. Sci.* 511 (2002) 401-407.

[68] P. Zeman, S. Takabayashi, Self-cleaning and antifogging effects of TiO_2 films prepared by radio frequency magnetron sputtering, J. Vac. Sci. Technol., A, *Vac. Surf. Films* 20 (2002) 388-393.

[69] H. Liu, L. Feng, J. Zhai, L. Jiang, D. B. Zhu, Reversible wettability of a chemical vapor deposition prepared ZnO film between superhydrophobicity and super–hydrophilicity, *Langmuir* 20 (2004) 5659-5661.

[70] S. N. Das, J. H. Choi, J. P. Kar, J. M. Myoung, Tunable and reversible surface wettability transition of vertically aligned ZnO nanorod arrays, *Appl. Surf. Sci.* 255 (2009) 7319-7322.

[71] C. Badre, T. Pauporte, M. Turmine, D. Lincot, A ZnO nanowire array film with stable highly water-repellent properties, *Nanotechnol.* 18 (2007) 365705 1-4.

[72] C. S. Lao, Y. Li, C. P. Wong, Z. L. Wang, Enhancing the electrical and optoelectronic performance of nanobelt devices by molecular surface functionalization, *Nano Lett.* 7 (2007) 1323-1328.

[73] G. Kenanakis, E. Stratakis, K. Vlachou, D. Vernardou, E. Koudoumas, N. Katsarakis, Light-induced reversible hydrophilicity of ZnO structures grown by aqueous chemical growth, *Appl. Surf. Sci.* 254 (2008) 5695-5699.

[74] Z. W. Pan, Z. R. Dai, Z. L. Wang, Nanobelts of semiconducting oxides, *Science* 291 (2001) 1947-1949.

[75] E. L. Papadopoulou, M. Barberoglou, V. Zorba, A. Manousaki, A. Pagkozidis, E. Stratakis, C. Fotakis, Reversible photoinduced wettability transition of hierarchical ZnO structures, *J. Phys. Chem.* C 113 (2009) 2891-2895.

[76] V. Zorba, L. Persano, D. Pisignano, A. Athanassiou, E. Stratakis, R. Cingolani, P. Tzanetakis, C. Fotakis, Making silicon hydrophobic: wettability control by two-lengthscale simultaneous patterning with femtosecond laser irradiation, *Nanotechnol.* 17 (2006) 3234-3238.

[77] J. Zhang, X. Lu, W. Huang, Y. Han, Reversible superhydrophobicity to superhydrophilicity transition by extending and unloading an elastic polyamide film, *Macromol. Rapid Commun.* 26 (2005) 477-480.

[78] D. Julthongpiput, Y. H. Lin, J. Teng, E. R. Zubarev, V. V. Tsukruk, Y-shaped polymer brushes: nanoscale switchable surfaces, *Langmuir* 19 (2003) 7832-7836.

[79] J. F. Hester, S. C. Olugebefola, A.M. Mayes, Preparation of pH-responsive membranes by self-organization, *J. Membr. Sci.* 208 (2002) 375-388.

[80] E. Tjipto, J.F. Quinn, F. Caruso, Assembly of multilayer films from polyelectrolytes containing weak and strong acid moieties, *Langmuir* 21 (2005) 8785-8792.

[81] Y. G. Jiang, Z. Q. Wang, X. Yu, F. Shi, H. P. Xu, X. Zhang, Self–assembled monolayers of dendronthiols for electrodeposition of gold nanostructures, *Langmuir* 21 (2005) 1986-1990.

[82] M. D. Wilson, G. M. Whitesides, The anthranilate amide of "polyethylene carboxylic acid" shows an exceptionally large change with pH in its wettability by water, *J. Am. Chem. Soc.* 110 (1988) 8718-8719.

[83] S. Minko, D. Usov, E. Goreshnik, M. Stamm, Environment-adopting surfaces with reversibly switchable morphology, *Macromol. Rapid Commun.* 22 (2001) 206-211.

[84] B. Zhao, W.J. Brittain, Synthesis, characterization, and properties of tethered polystyrene-b-polyacrylate brushes on flat silicate substrates, *Macromol.* 33 (2000) 8813-8820.

[85] R. Lupitskyy, Y. Roiter, C. Tsitsilianis, S. Minko, From smart polymer molecules to responsive nanostructured surfaces, *Langmuir* 21 (2005) 8591-8593.

[86] X. Yu, Z. wang, Y. Jiang, F. Shi, X. Zhang, Reversible pH-responsive from superhydrophobicity to superhydrophilicity, *Adv. Mater.* 17 (2005) 1289-1293.

[87] Y. Zheng, X. Gao, L. Jiang, Directional adhesion on superhydrophobic butterflies wings, *Soft Matter* 3 (2007) 178-182.

[88] M. K. Chaudhury, G. M. Whitesides, How to make water run uphill, *Science* 256 (1992) 1539-1541.

[89] X. Lu, S. Tan, N. Zhao, S, Yang, J. Xu, A unique behavior of water drops induced by low-density polyethylene surface with a sharp wettability transition, *J. Colloid Interface* Sci. 311 (2007) 186-193.

[90] S. K. Cho, H. Moon, C. J. Kim, Creating, transporting, cutting, and merging liquid droplets by electrowetting-based actuation for digital microfluidic circuits, *J. Microelectromech. Syst.* 12 (2003) 70-80.

[91] J. Lee, C. J. Kim, Surface-tension-driven microactuation based on continuous electrowetting, *J. Microelectromech. Syst.* 9 (2000) 171-180.

[92] P. Paik, V. K. Pamula, R. B. Fair, Rapid droplet mixers for digital microfluidic systems, *Lab Chip*, 3 (2003) 253-259.

[93] S. Daniel, M. K. Chaudhury, J. C. Chen, Fast drop movements resulting from the phase change on a gradient surface, *Science* 291 (2001) 633-636.

[94] D. T.Wasan, A. D. Nikolov, H. Brenner, Droplet speeding on surface, *Science* 291 (2001) 605-606.

[95] R. H. Farahi, A. Passian, T. L. Ferrell and T. Thundat, Microfluidic manipulation via Marangoni forces, *Appl. Phys. Lett.* 85 (2004) 4237-4239.

[96] K. Ichimura, S. K. Oh, M. Nakagawa, Light-driven motion of liquids on a photoresponsive surface, *Science* 288 (2000) 1624-1626.

[97] X. C. Li, B. K. Tay, P. Miele, A. Brioude, D. Cornu, Fabrication of silicon pyramid/nanowire binary structure with superhydrophobicity, *Appl. Surf. Sci.* 255 (2009) 7147-7152.

[98] K. K. S. Lau, J. Bico, K. B. K. Teo, M. Chhowalla, G. A. J. Amaratunga, W. I. Milne, G. H. McKinley, K. K. Gleason, Superhydrophobic carbon nanotube forests, *Nano Lett.* 3 (2003) 1701-1705.

[99] K. Teshima, H. Sugimura, Y. Inoue, O. Takai, A. Takano, Transparent ultra water-repellent poly(Ethylene Terephthalate) substrates fabricated by oxygen Plasma Treatment and Subsequent Hydrophobic Coating, *Appl. Surf. Sci.* 244 (2005) 619-622.

[100] T. Baldacchini, J. E. Carey, M. Zhou, E. Mazur, Superhydrophobic surfaces prepared by microstructuring of silicon using a femtosecond laser, *Langmuir* 22 (2006) 4917-4919.

[101] C. W. J. Berendsen, M. S. Keren, D. Najdek, F. Cerny, Superhydrophobic surface structures in thermoplastic polymers by interference lithography and thermal imprinting, *Appl. Surf. Sci.* 255 (2009) 9305-9310.

[102] C. K. Kang, S. M. Lee, I. D. Jung, P. G. Jung, S. J. Hwang, J. S. Ko, The fabrication of patternable silicon nanotips using deep reactive ion etching, *J. Micromech. Microeng.* 18 (2008) 075007 1-7.
[103] B. Qian, Z. Shen, Fabrication of superhydrophobic surfaces by dislocation-selective chemical etching on aluminum, copper, and zinc substrates, *Langmuir* 21 (2005) 9007-9009.
[104] D. Bäuerle, *Laser Processing and Chemistry*, Springer, Berlin, 2000.
[105] N. S. Murthy, R. D. Prabhu, J. J. Martin, L. Zhou, R. L. J. Headrick, Self-assembled and etched cones on laser ablated polymer surfaces, *Appl. Phys.* A 100 (2006) 023528 1-12.
[106] A. Bensaoula, C. Boney, A. Pillai, G. A. Shafeev, A. V. Simakin, D. Starikov, Arrays of 3D micro-columns generated by laser ablation of Ta and steel: modelling of a black body emitter, *Appl. Phys*. A 79 (2004) 973-975.
[107] X. Gao, X. Yao, L. Jiang, Effects of rugged nanoprotrusions on the surface hydrophobicity and water adhesion of anisotropic micropatterns, *Langmuir* 23 (2007) 4886-4891.
[108] J. Sui, Y. Zhang, S. Ren, M. Rinke, J. Lu, J. Jia, Carbon coating with combined super-hydrophobic and self-lubricating properties on titanium silicon carbide, *Carbon* 47 (2009) 629-634.
[109] Y. Xiu, L. Zhu, D. W. Hess, C. P. Wong, Hierarchical silicon etched structures for controlled hydrophobicity/superhydrophobicity, *Nano Lett*. 7 (2007) 3388-3393.
[110] X. Y. Fu, X. H. He, Fabrication of super-hydrophobic surfaces on aluminum alloy substrates, *Appl. Surf. Sci*. 255 (2008) 1776-1781.
[111] M. N. Qu, B. W. Zhang, S. Y. Song, L. Chen, J. Y. Zhang, X. P. Cao, Fabrication of Superhydrophobic Surfaces on Engineering Materials by a Solution-Immersion Process, *Adv. Funct. Mater*. 17 (2007) 593-596.
[112] M. Wang, N. Raghunathun, B. Ziaie, A nonlithographic top-down electrochemical approach for creating hierarchical (micro–nano) superhydrophobic silicon surfaces, *Langmuir* 23 (2007) 2300-2303.
[113] W. Wu, M. Chen, S. Liang, X. Wang, J. Chen, F. Zhou, Superhydrophobic surface from Cu-Zn alloy by one step O_2 concentration dependent etching, *J. Colloid Interface Sci*. 326 (2008) 478-482.
[114] Y. C. Hong, S. C. Cho, D. H. Shin, S. H. Lee, H, S, A facile method for the fabrication of super-hydrophobic surfaces and their resulting wettability, *Scr. Mater*. 59 (2008) 776-779.
[115] W. Chen, A. Y. Fadeev, M. C. Hsieh, D. Oner, J. Youngblood, T. J. McCarthy, Ultrahydrophobic and ultralyophobic surfaces: Some comments and examples, *Langmuir* 15 (1999) 3395-3399.
[116] J. Y. Shiu, C. W. Kuo, P. Chen, Fabrication of tunable superhydrophobic surfaces, *Proc. of SPIE* 5648 (2004) 325-332.
[117] C. K. Kang, S. M. Lee, I. D. Jung, P. G. Jung, S. J. Hwang, J. S. Ko, The fabrication of patternable silicon nanotips using deep reactive ion etching, *J. Micromech. Microeng.* 18 (2008) 075007 1-7.
[118] T. Y. Kim, B. Ingmar, K. Bewilogua, K. H. Oh, K. R. Lee, Wetting behaviours of a-C:H:Si:O film coated nano-scale dual rough surface, *Chem. Phys. Lett*. 436 (2007) 199-203.

[119] A. Kawai, H. Nagata, Wetting behavior of liquid on geometrical rough surface formed by photolithography, *Jpn. J. Appl. Phys*. 33L (1994) 1283-1285.

[120] B. Bhushan, K. Koch, Y. C. Jung, Fabrication and characterization of the hierarchical structure for superhydrophobicity and self-cleaning, *Ultramicroscopy* 109 (2009) 1029-1034.

[121] T. Yasuda, K. Suzuki, I. Shimoyama, Automatic transportation of a droplet on a wettability gradient surface, 7th lnternational Conference on Miniaturized Chemical and Blochemlcal Analysts Systems, Squaw Valley, Callfornla USA (2003) 1129-1132.

[122] J. T. Yang, J. C. Chen, K. J. Huang, J. A. Yeh, Droplet manipulation on a hydrophobic textured surface with roughened patterns, *J. Microelectromech. Syst*. 15 (2006) 679-707.

[123] S. V. Roberson, A. J. Fahey, A. Sehgal and A. Karim, Multifunctional ToF-SIMS: combinatorial mapping of gradient energy substrates, *Appl. Surf. Sci*. 100 (2002) 150-164.

[124] Y. Ito, M. Heydari, A. Hashimoto, T. Konno, A. Hirasawa, S. Hori, K. Kurita, A. Nakajima, The movement of a water droplet on a gradient surface prepared by photodegradation, *Langmuir* 23 (2007) 1845-1850.

[125] N. Feth, C. Enkrich, M. Wegener, S. Linden, Large-area magnetic metamaterials via compact interference lithography, *Opt. Express* 15 (2007) 501-507.

[126] C. W. J. Berendsen, M. S. Keren, D. Najdek, F. Cerny, Superhydrophobic surface structures in thermoplastic polymers by interference lithography and thermal imprinting, *Appl. Surf. Sci*. 255 (2009) 9305-9310.

[127] C. H. Choi, S. H. Hagvall, B. M. Wu, J. C. Y. Dunn, R. E. Beygui, C. J. Kim, Cell interaction with three-dimensional sharp-tip nanotopography, *Biomaterials* 28 (2007) 1672-1679.

[128] T. L. Chang, K. Y. Cheng, T. H. Chou, C. C. Su, H. P. Yang, S. W. Luo, Hybrid-polymer nanostructures forming an anti-reflection film using two-beam interference and ultraviolet nanoimprint lithography, *Microelectron. Eng*. 86 (2009) 874-877.

[129] Y. L. Yang, Study on surface wettability of periodical micro/nanostructures, Master thesis, National Taiwan University, Taipei, Taiwan, 2009.

[130] S. Y. Chou, P. R. Krauss, P. J. Renstrom, Imprint of sub-25 nm vias and trenches in polymer, *Appl. Phys. Lett*. 67 (1995) 3114-3116.

[131] I. T. Pai, I. C. Leu, M. H. Hon, Hierarchical structure through imprinting of polyimide without residual layers, *J. Micromech. Microeng*. 18 (2008) 105005 1-4.

[132] K. Y. Suh, Y. S. Kim, H. H. Lee, Capillary force lithography, *Adv. Mater*. 13 (2001) 1386-1389.

[133] D. Y. Khang, H. H. Lee, Pressure-assisted capillary force lithog- raphy, *Adv. Mater*, 16 (2004) 176-179.

[134] H. E. Jeong, M. K. Kwak, C. I. Park, K. Y. Suh, Wettability of nanoengineered dual-roughness surfaces fabricated by UV-assisted capillary force lithography, *J. Colloid Interface Sci*. 339 (2009) 202-207.

[135] K. Y. Suh, H. E. Jeong, J. W. Park, S. H. Lee, J. K. Kim, Fabrication of high aspect ratio nanostructures using capillary force lithography, Korean *J. Chem. Eng*. 23 (2006) 678-682.

[136] K. Y. Suh, H. H. Lee, Capillary force lithography: Large-area patterning, self-organization, and anisotropic dewetting, *Adv. Funct. Mater*. 12 (2002) 405-413.

[137] S. H. Choi and B. Z. Newby, Micrometer-scaled gradient surfaces generated using contact printing of octadecyltrichlorosilane, *Langmuir* 19 (2003) 7427-7435.

[138] M. Ma, Y. Mao, M. Gupta, K. K. Gleason, G. C. Rutledge, Superhydrophobic fabrics produced by electrospinning and chemical vapor deposition, *Macromol.* 38 (2005) 9742-9748.

[139] Z. Y. Song, J. Zhai, Y. L. Wang, L. Jiang, Fabrication of superhydrophobic surfaces by self-assembly and their water-adhesion properties, *J. Phys. Chem.* B 109 (2005) 4048-4052.

[140] Y. Liu, T. Tan, B. Wang, R. Zhai, X. Song, E. Li, H. Wang, H. Yan, Fabrication of CdS films with superhydrophobicity by the microwave assisted chemical bath deposition, *J. Colloid Interface Sci.* 320 (2008) 540-547.

[141] G. Li, B. Wang, Y. Liu, T. Tan, X. Song, H. Yan, Fabrication of superhydrophobic ZnO/Zn surface with nanowires and nanobelts structures using novel plasma assisted thermal vapor deposition, *Appl. Surf. Sci.* 255 (2008) 3112-3116.

[142] C. T. Hsieh, W. Y. Chen, F. L. Wu, Fabrication and superhydrophobicity of fluorinated carbon fabrics with micro/nanoscaled two-tier roughness, *Carbon* 46 (2008) 1218-1224.

[143] N. J. Shirtcliffe, G. McHale, M. I. Newton, C. C. Perry, Wetting and wetting transitions on copper-based super-hydrophobic surfaces, *Langmuir* 21 (2005) 937-943.

[144] R. Taurino, E. Fabbri, M. Messori, F. Pilati, D. Pospiech, A. Synytska, Facile preparation of superhydrophobic coatings by sol–gel processes, *J. Colloid Interface Sci.* 325 (2008) 149-156.

[145] J. Bravo, L. Zhai, Z. Wu, R. E. Cohen, M. F. Rubner, Transparent superhydrophobic films based on silica nanoparticles, *Langmuir* 23 (2007) 7293-7298.

[146] L. Zhang, Y. Li, J. Sun, J. Shen, Layer-by-layer fabrication of broad-band superhydrophobic antireflection coatings in near-infrared region, *J. Colloid Interface* Sci. 319 (2008) 302-308.

[147] J. Yang, P. Pi, X. Wen, D. Zheng, M. Xu, J. Cheng, Z. Yang, A novel method to fabricate superhydrophobic surfaces based on well-defined mulberry-like particles and self-assembly of polydimethylsiloxane, *Appl. Surf. Sci.* 255 (2009) 3507-3512.

[148] G. Goncalves, P. A. A. P. Marques, T. Trindade, C. P. Neto, A. Gandini, Superhydrophobic cellulose nanocomposites, *J. Colloid Interface Sci.* 324 (2008) 42-46.

[149] B. Xu, Z. Cai, Fabrication of a superhydrophobic ZnO nanorod array film on cotton fabrics via a wet chemical route and hydrophobic modification, *Appl. Surf. Sci.* 254 (2008) 5899-5904.

[150] P. S. Lee, A superhydrophobic glass surface with self-assembled silica particles, Master Thesis, National Taiwan University, Taipei, Taiwan, 2009.

[151] H. Okudera, A. Hozumi, The formation and growth mechanisms of silica thin film and spherical particles through the stober process, *Thin Solid Films* 434 (2003) 62-68.

[152] L. Yan, K. Wang, J. Wu, Lin Ye, Colloids and surfaces a: physicochemical and engineering aspects, *Colloid Surf. A: Physicochem. Eng. Aspects* 296 (2007) 123-131.

[153] C. Sun, L. Ge, Z. Gu, Fabrication of super-hydrophobic film with dual-size roughness by silica sphere Assembly, *Thin Solid Films* 515 (2007) 4686-4690.

[154] S. Pilotek, H. K. Schmidt, Wettability of microstructured hydrophobic sol-gel coatings, *J. Sol-Gel Sci. Technol.* 26 (2003) 789-792.

[155] H. Yang, Y. Deng, Preparation and physical properties of superhydrophobic papers, *J. Colloid Interface Sci.* 325 (2008) 588-593.

[156] Y. Xiu, F. Xiao, D. W. Hess, C. P. Wong, Superhydrophobic optically transparent silica films formed with a eutectic liquid, *Thin Solid Films* 517 (2009) 1610-1615.

[157] S. D. Bhagat, Y. H. Kim, Y. S. Ahn, Room temperature synthesis of water repellent silica coatings by the dip coat technique, *Appl. Surf. Sci.* 253 (2006) 2217-2221.

[158] L. J. Chen, M. Chen, H. D. Zhou, J. M. Chen, Preparation of super-hydrophobic surface on stainless steel, *Appl. Surf. Sci.* 255 (2008) 3459-3462.

[159] Y. Xiu, D. E. Hess, C. P. Wong, UV and thermally stable superhydrophobic coatings from sol–gel processing, *J. Colloid Interface Sci.* 326 (2008) 465-470.

[160] S. Smitha, P. Shajesh, P. Mukundan, T. D. R. Nair, K. G. K. Warrier, Synthesis of biocompatible hydrophobic silica–gelatin nano-hybrid by sol–gel process, *Colloid Surf. B: Biointerface* 55 (2007) 38-43.

[161] W. A. Daoud, J. H. Xin, X. M. Tao, Superhydrophobic silica nanocomposite coating by a low-temperature process, *J. Am. Cream. Soc.* 87 (2004) 1782-1784.

[162] J. Bravo, L. Zhai, Z. Wu, R. E. Cohen, M. F. Rubner, Transparent superhydrophobic films based on silica nanoparticles, *Langmuir* 23 (2007) 7293-7298.

[163] D. S. Kim, B. K. Lee, J. Yeo, M. J. Choi, W. Yang, T. H. Kwon, Fabrication of PDMS micro/nano hybrid surface for increasing hydrophobicity, *Microelectron. Eng.* 86 (2009) 1375-1378.

[164] L. Zhang, Z. Zhou, B. Cheng, J. M. DeSimone, E. T. Samulski, Superhydrophobic behavior of a perfluoropolyether lotus-leaf-like topography, *Langmuir* 22 (2006) 8576-8580.

[165] J. T. Kwon, H. G. Shin a, Y. H. Seo, B. H. Kim a, H. G. Lee, J. S. Lee, Simple fabrication method of hierarchical nano-pillars using aluminum anodizing processes, *Curr. Appl. Phys.* 9 (2009) e81-e85.

[166] W. T. Huang, Study of micro/nano structure effects on hydrophobic surfaces with high transparency, Master Thesis, National Taiwan University, Taipei, Taiwan, 2008.

[167] L. Feng, Y. L. Song, J. Zhai, B. Q. Liu, J. Xu, L. Jiang, D. B. Zhu, Creation of a superhydrophobic surface from an amphiphilic polymer, *Angew. Chem., Int. Ed.* 42 (2003) 800-802.

[168] I. Saarikoski, M. Suvanto, T. A. Pakkanen, Modification of polycarbonate surface properties by nano-, micro-, and hierarchical micro–nanostructuring, *Appl. Surf. Sci.* 255 (2009) 9000-9005.

In: Frontier Research in Microscale and Nanoscale... ISBN 978-1-61324-665-8
Editor: Lixin Cheng

Chapter 18

ENTROPY MINIMIZATION IN A MICRO-SCALE EVAPORATING THIN FILM HEAT TRANSFER DEVICE

***Rama Subba Reddy Gorla*[1*] *and David M. Pratt*[2]**

[1]Fenn Distinguished Research Professor
Department of Mechanical Engineering
Cleveland State University
Cleveland, Ohio 44115-2425, U. S.
[2]Thermal Structures Branch
Air Vehicles Directorate
Air Force Research Laboratory
Wright Patterson Air Force Base, Ohio 45433, U. S.

ABSTRACT

An analysis is provided for the entropy generated in a capillary pumped device for the combined heat and mass transfer in a circular tube and micro/nano scale heat and mass transfer in a capillary tube in terms of the gradients of velocity, temperature and concentration as well as the physical properties of the fluid. In phase change transport devices, capillary forces drive the overall circulation of working fluid from an evaporator to a condenser section. The heat and mass transfer rates are assumed to be uniform on the surface of the tube. The optimum geometric configuration that corresponds to the minimization of entropy generated and minimization of the fluid flow resistances is identified. Capillary pumped heat transfer devices do not achieve the maximum predicted heat flux because of interfacial thermocapillary stresses arising from liquid vapor interfacial temperature gradients. This can be counteracted by introducing naturally occurring concentration gradients using binary fluid mixtures.

Keywords: Entropy minimization, Capillary forces, microthin film.

---

* E-mail: r.gorla@csuohio.edu

Nomenclature

Be	Bejan number
C	concentration [mol m^{-3}]
C_p	specific heat at constant pressure [J kg^{-1} K^{-1}]
h	Film Thickness [m]
h_{fg}	Latent heat of vaporization of the liquid [J/kg]
k	thermal conductivity [W.m^{-1}.K^{-1}]
L	Length of plate [m]
$\dot{m}$	mass flux [kg.m^{-2}.s^{-1}]
$\dot{m}_d$	mass diffusion rate [kg.m^{-2}.s^{-1}]
M	molecular weight [g mol^{-1}]
N	entropy number
p	Pressure [Pa]
$\dot{q}''$	Heat flux [W.m^{-2}]
R	tube radius [m]
$\overline{R}$	gas constant
r	radial coordinate [m]
S	entropy generated [J m^{-1} K^{-1}]
T	Temperature [K]
u	velocity [m s^{-1}]
ρ	density [Kg/m^3]
μ	Absolute viscosity of liquid [kg.m^{-1}.s^{-1}]
ν	Kinematic Viscosity of liquid [m^2.s^{-1}]
ε	Dimensionless Film Thickness(h/R)

Subscripts

Evap	evaporative
G	generated
L	liquid phase
x	axial derivative
V	vapor phase

Introduction

Heat dissipation is always an important issue among machines. From computers, cars, to aircraft, scientists keep trying to design a better heat transport device which can dissipate heat energy. For hypersonic aircraft, though scientists have invented materials with high heat endurance, it is still not enough for the heat generated by the friction between air and leading edges. Air Force Research Laboratory (AFRL) has explored the use of capillary pumped heat transfer devices (CPHTD) for energy transport without the use of pumps or fans. Early systems encountered problems during start-up, were unable to withstand the large adverse acceleration and could inadvertently reach their transport limits. To be able to use these techniques, the AFRL must understand the physics associated with these devices.

Since Derjaguin [1] first analyzed the thermo-fluid characteristics of an evaporating thin film region, a number of excellent numerical and experimental investigations [2-4] have been introduced to predict the flow and heat transfer for various evaporator specifications. Das Gupta et. al. [5] studied the transport processes occurring in an evaporating extended meniscus. They suggested the augmented Young-Laplace equation with Kelvin-Clapeyron equation and kinetic theory that can be used to model fluid flow and evaporative heat transfer in the contact line region. Longtin et. al. [6] proposed a mathematical model to predict the heat transfer in a micro heat pipe having a triangular cross section. They concluded that the radius of curvature increases linearly along the flow direction. Sartre et. al. [7] presented a three dimensional model for predicting heat transfer in an extended meniscus region of a micro heat pipe array. They reported that the apparent contact angle and the heat transfer rate increase with increasing wall superheat. Qu et. al. [8] proposed a model to account for the formation of evaporating thin film and meniscus in capillary tubes. They investigated the effects of radius and heat transfer on the profile if evaporating extended meniscus region. Gorla et. al. [9-13] analyzed the augmentation of heat rates in a thin film.

The principle that generates geometric structure is the pursuit of a global objective subject to specified global and local constraints [14]. The global objective function can be the maximization of thermodynamic performance or the maximization of thermodynamic irreversibility. This design method is known as entropy generation minimization [15-17]. The generation of architecture on the basis of the objective and constraints principle is known as the constructal design [14].

Exergy analysis is pure thermodynamics. It relies on the laws of thermodynamics to establish the theoretical limit of ideal or reversible operation and the extent to which the operation of the given system departs from the ideal. The departure is measured by the calculated quantity called destroyed energy or irreversibility. This quantity is proportional to the generated entropy. Energy is the thermodynamic property that describes the useful energy content or the work producing potential of substances and streams. In real systems energy is always destroyed partially or totally when components and streams interact. The minimization of entropy generation requires the use of more than thermodynamics: fluid mechanics, heat and mass transfer, materials, constraints, and geometry are also needed in order to establish the relationships between the physical configuration and the destruction of energy. Reduction in energy destruction is pursued through changes in configuration. In the field of heat transfer, the entropy generation minimization method brings out the inherent competition between heat transfer and fluid flow irreversibilities in the optimization of devices subjected to overall constraints.

Hypersonic aircraft need to dissipate heat from leading edges to stay within material limits. Air Force Research Laboratory (AFRL) has explored the use of capillary pumped heat transfer devices for energy transport without the use of pumps or fans. Early systems encountered problems during start-up, were unable to withstand the large adverse accelerations and could inadvertently reach their transport limits. To be able to use these techniques the AFRL must understand the physics associated with these devices.

A study of thermodynamic irreversibilities is an important factor in the design of thermal systems. Entropy is a measure of molecular disorder within a system, which translates into the amount of energy not available to be converted into work. Therefore, the main objective is to determine possible ways of minimizing it. The method utilized to achieve this purpose is known as Entropy Generation Minimization or thermodynamic optimization. This method is a combination of the basic principles of thermodynamics, heat transfer and fluid mechanics in order to analyze the irreversibilities present in real devices concentrating special attention in finding the physical quantities that can be modified to optimize their specific performance.

Applications of second law analysis for channel and pipe may be found in references [18-21]. Entropy generation in a vertical concentric channel with variable viscosity was studied by Tasnim and Mahmud [22]. Haddad et. al. [23] investigated the entropy generation in a laminar forced convection in the entrance region of a concentric annulus. Nag and Kumar [24] performed second law analysis for convective heat transfer in rectangular ducts. San et. al. investigated the heat and mass in rectangular ducts by second law. Sahin [25] studied the effect of variable viscosity on entropy generation in tube flow. Pratt, Brown, Hallinan and Kihm [26-27] presented binary fluid mixture and thermocapillary effects on the wetting characteristics of a heated curved meniscus.

Thin liquid films have several practical applications in engineering, geology and biophysics. These films are subjected to gradients in surface tension and concentration in the presence of non-uniform interfacial temperature or concentration of a binary fluid. These gradients form Marangoni stresses and lead to flow generation and considerable deformation of the liquid film. The thin liquid film layer may be driven to rupture in the presence of evaporation.

The present work has been undertaken in order to study the entropy generation for the forced convection heat and mass transfer in a capillary pumped heat transfer device which includes the flow of vapor and liquid and the micro/nano scale heat and mass transfer in a circular tube. The optimum geometric configuration that corresponds to the minimization of entropy generated and minimization of fluid resistance is identified.

EXPERIMENTAL APPARATUS

Figures 1a and b show the details of the thermocapillary stress research vessel used at the Wright Research Laboratory. It consists of a single capillary pore evaporator with large pore condensers and vapor channels. Heat is introduced via an electric resistance heating element mounted on the outer diameter of the tube. All lengths were measured using a cathetometer having a bias limit of $\pm 0.0005\ cm$. The inner and outer diameter measurements have an error of $\pm 0.0005\ mm$ and $\pm 0.05\ mm$, respectively. The device is placed in a vacuum chamber to minimize convection heat losses. The chamber is equipped with feed throughs for thermocouples, pressure transducers, heat power connections, and cooling fluid lines. A roughing vacuum pump is used to evacuate the system. The system is filled to the appropriate level with the working fluid. To control the temperature of the condenser region, a refrigerated circulator is used to pump cooling fluid through a heat exchanger placed on the exterior of the condenser region.

The capillary tube is heated using a fine nichrome heater wire (spirally wrapped around the tube at the top) having a resistance of 1 ohm/cm. It is connected to a dc power supply capable of producing up to three amps at thirty volts. A shunt resistor connected in series with the heating element allowed for the current flowing through the heater to be determined by measuring the voltage drop across it. Calibrated thermocouples were used to measure the wall temperature of the capillary tube and to measure the temperature of the vapor space and liquid reservoir. To determine the temperature distribution of the evaporator region, thermocouples were positioned at 0.5 mm intervals along the capillary pore. This was accomplished using an A/D board interfaced to a Pentium III -500 PC. The associated error in the temperature readings are $\pm 0.3^{o}C$ with a 95% confidence interval. The refrigerated circulator temperature was used to set the liquid reservoir temperature at 5, 15 and 25^{0}C.

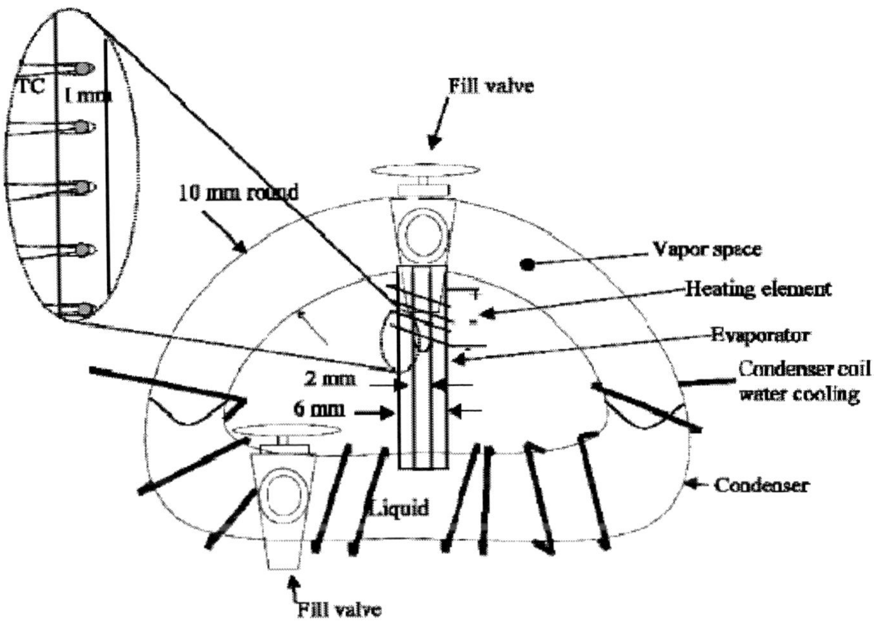

Figure 1a. Single Pore Capillary Pumped Loop Geometry.

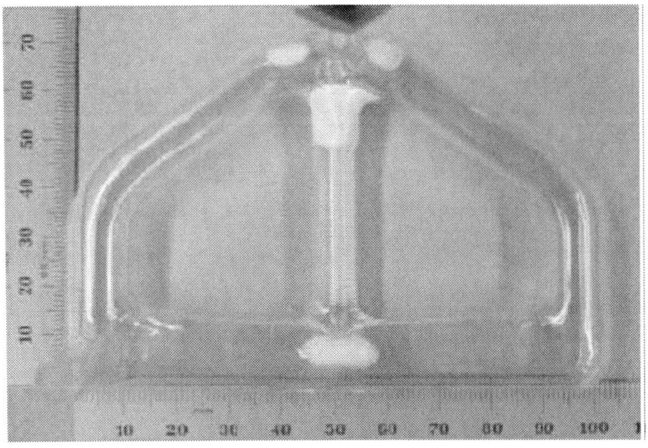

Figure 1b. Prototype Test Cell.

The heater section corresponds to the central tube L_1 where the thin film meniscus exists. The condenser section corresponds to the bottom horizontal tube (aL_2). We will now perform an analysis for optimization of diameters and lengths in the inverse T-shaped assembly shown in Figure 2.

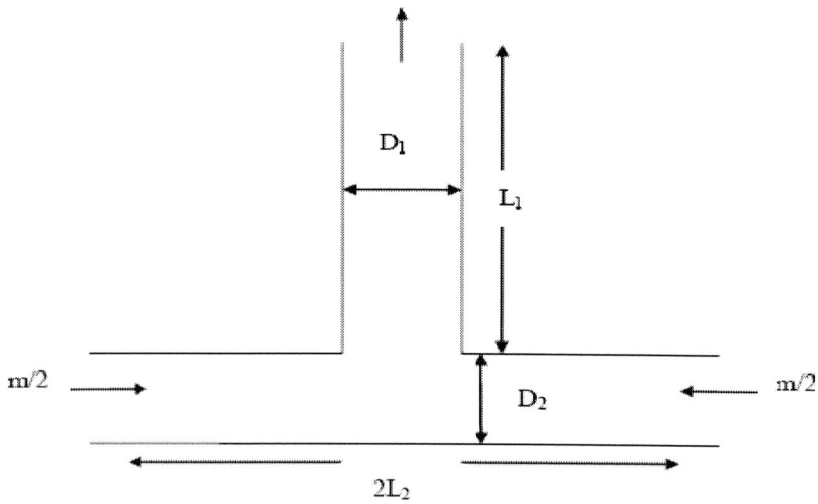

Figure 2. T-shaped Assembly.

Analysis

To optimize the diameters:

We assume that flow through the tubes is laminar and fully developed. It is further assumed that the tubes have both liquid and vapor. Figures 1 and 2 show the system under study. The total volume occupied by the tubes is given by:

$$\frac{\pi}{4}\left(D_1^2 L_1 + 2D_2^2 L_2\right) = cons\tan t \tag{1}$$

For Poiseuille flow, we have

$$\frac{\Delta p}{\dot{m}} = \frac{128}{\pi}\frac{\nu L}{D^4} \tag{2}$$

$$\text{Flow resistance} = \frac{\Delta p}{\dot{m}} \approx \frac{\nu L}{D^4} \tag{3}$$

$$\text{In section 1, resistance } R_1 = \frac{\nu_L L_{1L}}{D_1^4} + \frac{\nu_V L_{1V}}{D_1^4} \tag{4}$$

$$\text{In section 2, resistance } R_2 = \frac{\nu_L L_{2L}}{D_2^4} + \frac{\nu_V L_{2V}}{D_2^4} \tag{5}$$

The R_2 resistance are in parallel arrangement. Their overall effect is the resistance R_3

$$\frac{1}{R_3} = \frac{1}{R_2} + \frac{1}{R_2} \tag{6}$$

Therefore, $R_3 = \frac{R_2}{2}$ (7)

Resistance R_1 and R_3 are in series. The resistance is given by:

$$R = R_1 + R_3 = \frac{\nu_L L_{1L} + \nu_V L_{1V}}{D_1^4} + \frac{\nu_L L_{2L} + \nu_V L_{2V}}{2D_2^4} \tag{8}$$

Since the total volume is constant, we have:

$$D_1^2 (L_{1L} + L_{1V}) + 2D_2^2 (L_{2L} + L_{2V}) = V = constant \tag{9}$$

Let
$$x = D_1^2, \quad y = D_2^2 \tag{10}$$

Therefore, we have:

$$R = \frac{\nu_L L_{1L} + \nu_V L_{1V}}{x^2} + \frac{\nu_L L_{2L} + \nu_V L_{2V}}{2y^2} \tag{11}$$

is to be minimized subject to the condition

$$(L_{1L} + L_{1V})x + 2(L_{2L} + L_{2V})y - V = 0 \tag{12}$$

Minimize and Maximize Problem with Constraints:

Let us consider the problem of determining the stationary values of a function f(x, y) when two variables x and y are connected by a subsidiary condition g(x, y)=0. The optimization requires that the following equations must be solved:

$$\frac{\partial f}{\partial x} + \lambda \frac{\partial g}{\partial x} = 0 \tag{13}$$

$$\frac{\partial f}{\partial y} + \lambda \frac{\partial g}{\partial y} = 0 \tag{14}$$

$$g(x, y) = 0 \tag{15}$$

From equations (11) and (12) we have the following solutions:

$$\lambda = \frac{2}{x^3}\left[\frac{\nu_L L_{1L} + \nu_V L_{1V}}{L_{1L} + L_{1V}}\right] = \frac{1}{2y^3}\left[\frac{\nu_L L_{2L} + \nu_V L_{2V}}{L_{2L} + L_{2V}}\right] \tag{16}$$

$$\frac{x}{y} = 4^{1/3}\left[\frac{(\nu_L L_{1L} + \nu_V L_{1V})\,(L_{2L} + L_{2V})}{(\nu_L L_{2L} + \nu_V L_{2V})\,(L_{1L} + L_{1V})}\right]^{1/3} \tag{17}$$

$$\text{or,}\ \frac{D_1}{D_2} = 2^{1/3}\left[\frac{(\nu_L L_{1L} + \nu_V L_{1V})\,(L_{2L} + L_{2V})}{(\nu_L L_{2L} + \nu_V L_{2V})\,(L_{1L} + L_{1V})}\right]^{1/6} \tag{18}$$

Here, λ has no physical meaning.

To optimize the lengths:

We have the minimum resistance given by

$$R_{\min} = \frac{1}{x^2}\left[\left(\nu_L L_{1L} + \nu_V L_{1V}\right) + 2^{1/3}\left(L_{2L} + L_{2V}\right) \left[\frac{\left(\nu_L L_{1L} + \nu_V L_{1V}\right)\left(\nu_L L_{2L} + \nu_V L_{2V}\right)}{\left(L_{1L} + L_{1V}\right)\left(L_{2L} + L_{2V}\right)} \right]^{2/3} \right]$$

(19)

Let

$$R_{\min} = \xi = \frac{1}{x^2}\left[\left(\nu_L L_{1L} + \nu_V L_{1V}\right) + 2^{1/3}\left(L_{2L} + L_{2V}\right) \left[\frac{\left(\nu_L L_{1L} + \nu_V L_{1V}\right)\left(\nu_L L_{2L} + \nu_V L_{2V}\right)}{\left(L_{1L} + L_{1V}\right)\left(L_{2L} + L_{2V}\right)} \right]^{2/3} \right] \tag{20}$$

$$Area = aL_2^2(1-a) + aL_1L_2$$

We note that
$$L_1 = L_{1L} + L_{1V}$$
$$L_2 = L_{2L} + L_{2V}$$

Let $L_{1L} = \alpha L_1$ and $L_{2l} = \beta L_2$ (21)

We need to extremize ξsubject to the constraint condition given by equation (20). Using the Lagrangian multiplier method, we have:

$$\lambda = \frac{2^{1/3}\left(\nu_L\beta + \nu_V(1-\beta)\right)\left[\frac{\left(\nu_L L_{1L} + \nu_V L_{1V}\right)\left(L_{2L} + L_{2V}\right)}{\left(\nu_L L_{2L} + \nu_V L_{2V}\right)\left(L_{1L} + L_{1V}\right)} \right]^{2/3}}{\left[2L_2 a(1-a) + aL_1\right]} = -\frac{\left(\nu_L\alpha + \nu_V(1-\alpha)\right)}{\alpha L_2} \tag{22}$$

Therefore, the ratio of lengths is given by:

$$\frac{L_1}{L_2} = 2^{1/3}\left[\frac{\left(\nu_L\beta + \nu_V(1-\beta)\right)}{\left(\nu_L\alpha + \nu_V(1-\alpha)\right)} \right]^{1/3/} - 2(1-a) \tag{23}$$

ANALYSIS FOR HEAT TRANSFER IN CIRCULAR TUBE

Consider the fully developed laminar flow through a tube. The flow model and the co-ordinate system are shown in Figure 3.

Governing Equation:

Momentum balance:

$$\frac{\mu}{r}\frac{d}{dr}\left(r\frac{du}{dr}\right) = \frac{dp}{dx} \tag{24}$$

The boundary conditions are:

$$r = 0; \frac{du}{dr} = 0 \tag{25}$$
$$r = R; u = 0$$

From equations (24) and (25) we can obtain the velocity distribution in the following form:

$$u = \frac{R^2}{4\mu}\left(-\frac{dp}{dx}\right)\left[1-\left(\frac{r}{R}\right)^2\right]$$

$$= u_{\max}\left[1-\left(\frac{r}{R}\right)^2\right] \tag{26}$$

where

$$u_{\max} = \frac{R^2}{4\mu}\left(-\frac{dp}{dx}\right) \tag{27}$$

Energy Equation

Considering the constant wall temperature condition, we have

$$\frac{1}{r}\frac{\partial}{\partial r}\left(r\frac{\partial T}{\partial r}\right) = \frac{u}{\alpha}\frac{T_w - T}{T_w - T_m}\frac{dT_m}{dx} \tag{28}$$

The boundary conditions are

$$r = 0; \frac{\partial T}{\partial r} = 0 \qquad r = R; T = T_w \tag{29}$$

Combine equation (3), and (5), we have

$$2u_{MAX}\left(1-\frac{r^2}{R^2}\right)\frac{T_w - T}{T_w - T_m}\frac{dT_m}{dx} = \frac{1}{r}\frac{\partial}{\partial r}\left(r\frac{\partial T}{\partial r}\right) \tag{30}$$

Therefore, applied the boundary condition, we have

$$\frac{T_w - T}{T_w - T_m} = \sum_{n=0}^{\infty} C_{2n}\left(\frac{r}{R}\right)^{2n} \tag{31}$$

where C_{2n} are given as:

$C_0=1$

$$C_2 = \frac{\lambda_0^2}{4}$$

$$C_{2n} = \frac{\lambda_0^2}{(2n)^2}(C_{2n-4} - C_{2n-2}) \tag{32}$$

$$\lambda_0 = 2.704364 \tag{33}$$

Second Law and Entropy Generation:

Flow and heat transfer process inside the tube are irreversible. The non-equilibrium conditions arise due to the exchange of energy, mass and momentum within the fluid and at solid boundaries, thus resulting in entropy generation. A part of the entropy production is due to the heat and mass transfer in the direction of finite temperature gradients and the other part of entropy production arises due to the fluid friction. The general equation for the entropy generation per unit volume is given by:

$$S_G''' = \frac{\mu}{T}\left(\frac{\partial u}{\partial r}\right)^2 + \frac{k}{T^2}\left(\frac{\partial T}{\partial r}\right)^2 \tag{34}$$

In the above equation,

$\frac{\mu}{T}\left(\frac{\partial u}{\partial r}\right)^2$ =Friction component

$\frac{k}{T^2}\left(\frac{\partial T}{\partial r}\right)^2$ =Heat Flux component

(35)

The entropy generated per unit length is given by

$$S_G' = 2\frac{\mu}{T}u_{\max}R^2 + 1.99431\left(-\frac{1}{T_w - T_m}\right)\frac{1}{R^2} \tag{36}$$

Analysis in Micro-Scale Evaporating Thin Film Region

Figures 4 and 5 show a working fluid in a micro capillary tube evaporating due to the applied heat, resulting in the thin liquid film and an increase in the radius of curvature at the liquid/vapor interface. The capillary forces are controlled by surface tension and adsorption forces by van der Waals forces. Liquid flows in the steady film as a result of capillary and disjoining pressure gradients caused by evaporation due to superheating. Heat flows from the tube surface through the film to the liquid-vapor interface where evaporation occurs.

We assume that the flow of liquid and vapor is incompressible, laminar and steady. The surface tension is assumed to be temperature dependent. The inertial terms in the momentum equation are assumed to be negligible. The behavior of the liquid/vapor interface is governed by the pressure difference in the liquid and vapor phases. The pressure difference may be expressed by the augmented Young-Laplace equation.

$$p_v - p_l = \sigma K + \Psi \tag{37}$$

where p_v is the vapor phase pressure, p_l is the liquid pressure, K is the curvature, σ is the surface tension and Ψ is disjoining pressure. We assume that $\Psi = -\frac{A}{h^3}$ where A is the dispersion constant and h is the film thickness.

Following Das Gupta et. al. [18], based on Kelvin-Clapeyron equations, the evaporative mass flux as a function of temperature and pressure at the interface is given by,

$$m_{evap} = a(T_{lv} - T_v) + b(p_l - p_v) = b\Pi_0\left[\theta - \Pi_d - \eta''\right] \tag{38}$$

where $\Pi_0 = \frac{a}{b}\Delta T_0$ and $\Pi_d = \Psi / \Pi_0$.

Here, m_{evap} is the evaporative mass flux, defined as the amount of liquid evaporating from the liquid film per unit area per unit time in the direction away from film, T_{lv} is the temperature at the liquid-vapor interface, T_v is the temperature of vapor, a and b are functions of the system temperature and physical properties defined as

$$a = c_1\left(\frac{M}{2\pi R_g T_{lv}}\right)^{1/2}\frac{p_v M \Delta h_m}{R_g T_v T_{lv}};\ b = c_1\left(\frac{M}{2\pi R_g T_{lv}}\right)^{1/2}\frac{V_l p_v}{R_g T_{lv}} \tag{39}$$

where C_l is the accommodation coefficient, defined as equal to $2\gamma/(2-\gamma)$, where γ is the evaporation coefficient, M is the molecular weight, R_g is the universal gas constant, p_v is the bulk vapor pressure of liquid at T_v, V_l is the molar volume of the liquid at T_{lv} and Δh_m is the liquid enthalpy of vaporization per unit mass at T_{lv}.

Surface tension is related to liquid-vapor interfacial temperature:

(40)

$\sigma = \sigma_o + \gamma T_i$ where

σ_o=reference surface tension

$\gamma = \dfrac{d\sigma}{dT}$ (usually < 0)

T_i=interface temperature

Mean curvature

$$K = \left(\frac{1}{R-h}\right)\left(1+h_x^2\right)^{-0.5} + h_{xx}\left(1+h_x^2\right)^{-1.5} \tag{41}$$

where

R=Radius of pipe

Subscript x= $\dfrac{d}{dx}$

Lubrication Approximation:

$$\frac{1}{r}\frac{\partial}{\partial r}\left(r\frac{\partial u}{\partial r}\right) = \frac{1}{\mu_l}\frac{dp_l}{dx} \tag{42}$$

Boundary conditions

r = R; $u = 0$ (43)

$$r = R-h; -\mu\frac{\partial u}{\partial r} = \frac{d\sigma}{dx} \tag{44}$$

Solution for the velocity u(r):

Solution equation (23) along with (24) and (25)

$$u(r) = \frac{1}{4\mu_l}\frac{dp_l}{dx}r^2 + c_1 \ln r + c_2 \tag{45}$$

Where

$$c_1 = -\frac{1}{\mu_l}\frac{d\sigma}{dx}(R-h) - \frac{1}{2\mu_l}\frac{dp_l}{dx}(R-h)^2$$

$$c_2 = -\frac{1}{4\mu_l}\frac{dp_l}{dx}R^2 - c_1 \ln R$$

Heat Transfer Modeling:

$$\frac{d}{dr}\left(r\frac{dT}{dr}\right) = 0 \tag{46}$$

Boundary Conditions

$$r = R; T = T_w$$

$$r = R - h;\ K_l \frac{\partial T}{\partial r} = \dot{m}_{eva}.h_{fg} \tag{47}$$

From equations (46) and (47) we have:

$$T = \frac{\dot{m}h_{fg}}{K_l}(R-h)\ln\left(\frac{r}{R}\right) + T_w \tag{48}$$

Entropy generated per unit volume is given by

$$S''' = \frac{\mu}{T}\left(\frac{\partial u}{\partial r}\right)^2 + \frac{K}{T^2}\left(\frac{\partial T}{\partial r}\right)^2 \tag{49}$$

where u and T are given by equations (45) and (48) respectively.
Entropy generated per unit length of central capillary tube is given by

$$S' = \int_{R-h}^{R} S''' 2\pi r dr \tag{50}$$

For optimum design of the system, we can calculate the radius of the pipe for minimum entropy condition $\frac{d(S'_{total} L)}{dR} = 0$.

Here, S_{total} includes entropy generated in the central capillary tube, vapor flow tube and liquid flow tube. In dimensionless form, we may write:

$$N_s = \frac{S'_{total} K_L T_0^2}{(\dot{m}h_{fg})^2} = N_R + N_F \tag{51}$$

where the first term on the right hand side relates to radial conduction and the second term for fluid friction dissipation.

The irreversibility distribution ratio Φ is defined as

$$\Phi = \frac{N_F}{N_R} \tag{52}$$

Bejan number (Be) which is the ratio of entropy generation due to heat transfer to the total entropy generation may be written as:

$$Be = \frac{N_R}{N_S} = \frac{1}{1+\Phi} \tag{53}$$

Bejan number ranges from 0 to 1. Be = 0 is the limit where the irreversibility is dominated by fluid frictional effect and be = 1 corresponds to the limit where the irreversibility due to heat transfer by virtue of finite temperature differences dominates.

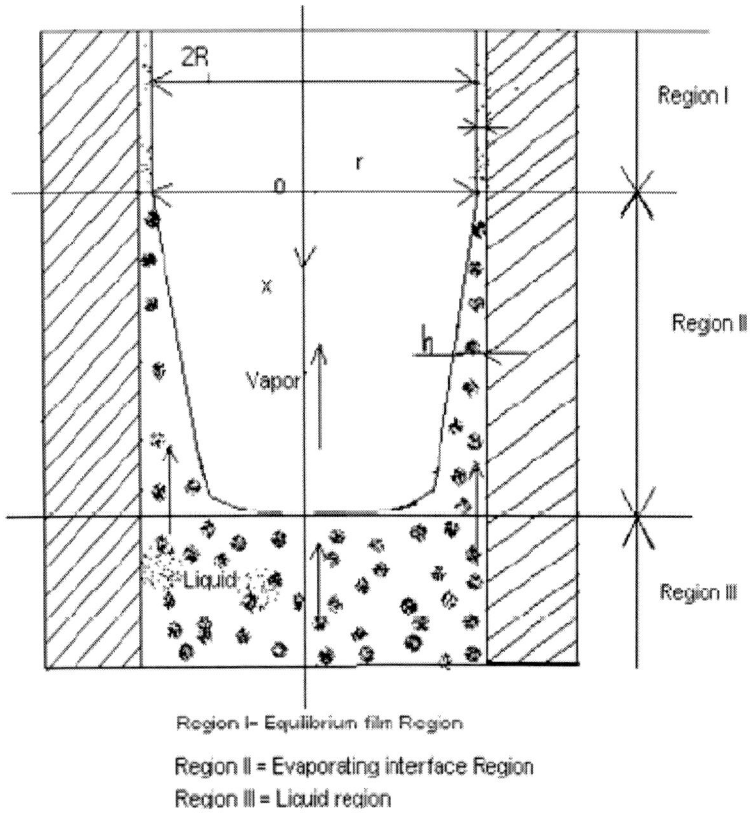

Figure 4. Geometry of Evaporating Film.

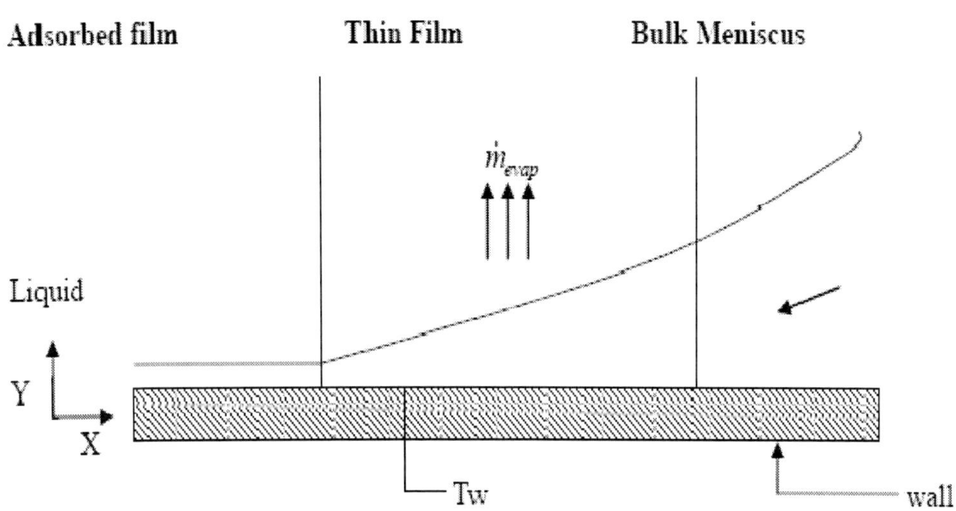

Figure 5. Geometry of Evaporating Film.

DISCUSSION OF RESULTS

An analysis of the entropy generation in a heat transfer device has been presented when heat and mass are being transferred simultaneously under laminar and single phase flow conditions. Control volume analysis has to be used for turbulent flow problem.

The dynamics associated with fluid motion and heat transport in the vicinity of the evaporating meniscus can affect the driving capillary potential by degrading the wetting ability of the working fluid. The change in the wettability is a result of a non-isothermal liquid/vapor interfacial temperatures near the contact line arising from non-uniform wall temperature and evaporation. Either of these influences yield surface tension gradients on the liquid/vapor interface. These surface tension gradients result in thermocapillary stresses acting near the contact line. The degradation leads to a reduction in the capillary pumping potential in capillary heat transfer devices. In order to reduce the thermocapillary stresses, a small amount of a less volatile fluid with a higher surface tension into the working fluid. This produces two effects: firstly, the surface tension of the binary fluid would be larger than the original fluid and secondly, this would result in a concentration gradient in the opposite direction to the temperature gradient. This has been modeled by a binary mixture of n-pentane (surface tension = 0.0155 N/m) and decane (surface tension = 0.0238 N/m). We examined decane mole fraction of 0.0815 for which little degradation in heat transfer was observed in the laboratory. Figure 6 shows experimental data of the wicking height versus temperature for various values of concentration of decane. The addition of decane increases the wicking height due to its higher surface tension but increasing the concentration beyond low levels decreases the wicking height. An examination of the transport properties as a function of conentration, C reveals an increase in density and viscosity and surface tension. This results in competing effects on the wicking height with viscous effects outweighing the increase in surface tension gain when the concentration is increased.

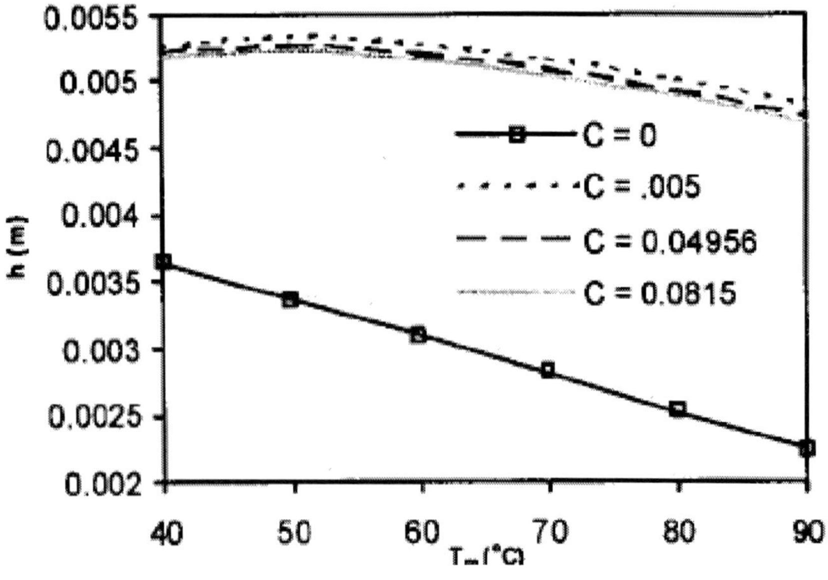

Figure 6. Effect of decane concentration on wicking height.

Figure 7 shows the variation of the entropy generated in the system for different surface heat flux values versus the capillary radius. Figure 8 displays the variation of entropy generated in the system versus the radius of capillary for different values of pressure gradients. As the pressure gradient increases, the optimum radius of the tube for minimum entropy deceases. As the surface heat flux decreases, the optimum radius of the tube decreases. Figures 7 and 8 present data for a binary mixture of n-pentane and decane. The results presented will facilitate in the design of equipment for minimum entropy generation.

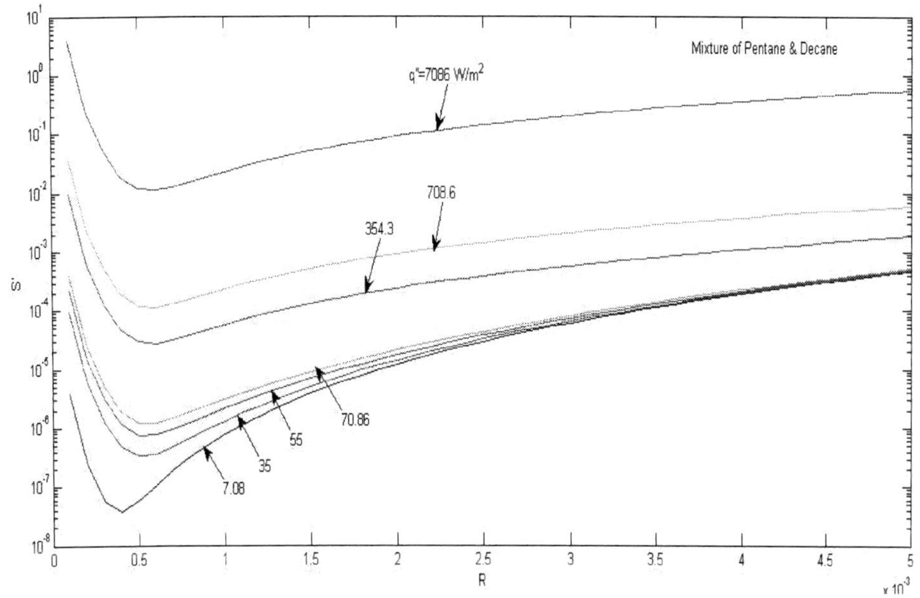

Figure 7. Total Entropy Generated versus Radius of Tube for a binary mixture of n-pentane and decane.

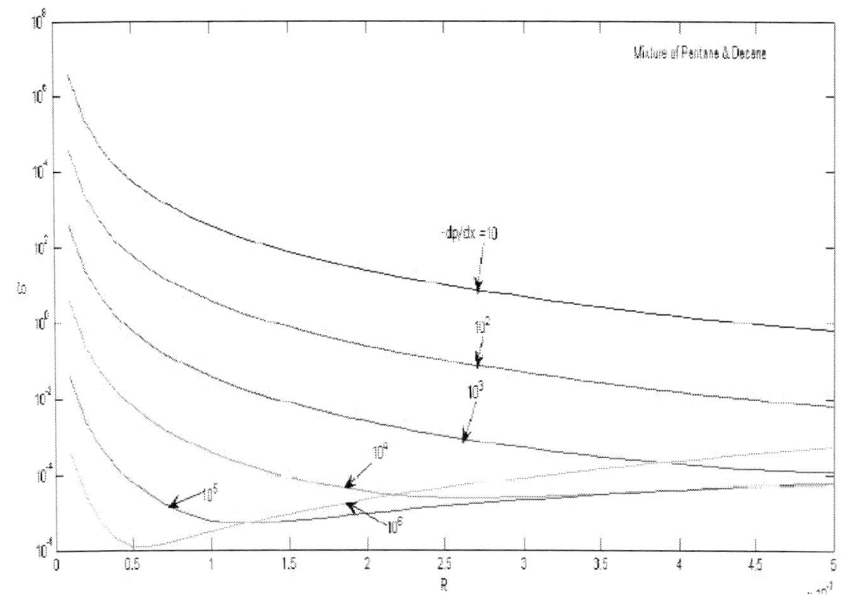

Figure 8. Total Entropy Generated versus Radius of Tube for a binary mixture of n-pentane and decane.

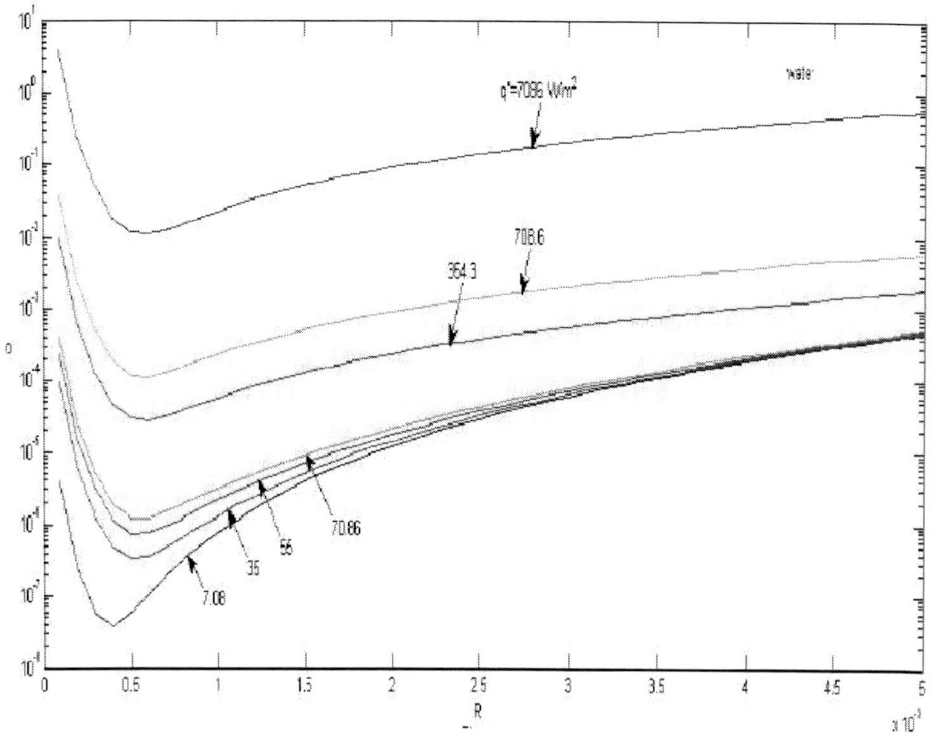

Figure 9. Total Entropy Generated versus Radius of Tube for water.

Figure 9 shows the variation of the entropy generated in the system for different surface heat flux values versus the capillary radius for water. The system has its limitation bearing heat flux input while the pressure difference is low. If the input heat flux is too high then the optimized capillary radius would be large enough that we need to take the gravity effect into account. Furthermore, regardless of the pressure differences, the higher input heat flux the larger optimum radius needed.

Figure 10 displays the variation of entropy generated in the system versus the radius of capillary for different values of pressure gradients for water. As the pressure gradient increases, the optimum radius of the tube for minimum entropy deceases. As the surface heat flux decreases, the optimum radius of the tube decreases.

Figure 11 shows the relationship between the radius of the tube and the capillary. As shown (equation 18), the two radii have a linear relationship and the larger of the radius of capillary is the larger of the radius of the tube. Figure 12 displays the distribution of dimensionless values of the ratio of heat flux component with respect to the total entropy generated (Bejan number).

From Figure 12, we note that the Bejan number decreases with tube radius. The Bejan number shows the importance of heat transfer contribution to the total entropy generation. Bejan number ranges from 0 to 1. Be = 0 is the limit where the irreversibility is dominated by fluid frictional effects and Be = 1 corresponds to the limit where the irreversibility due to heat transfer by virtue of finite temperature differences dominates.

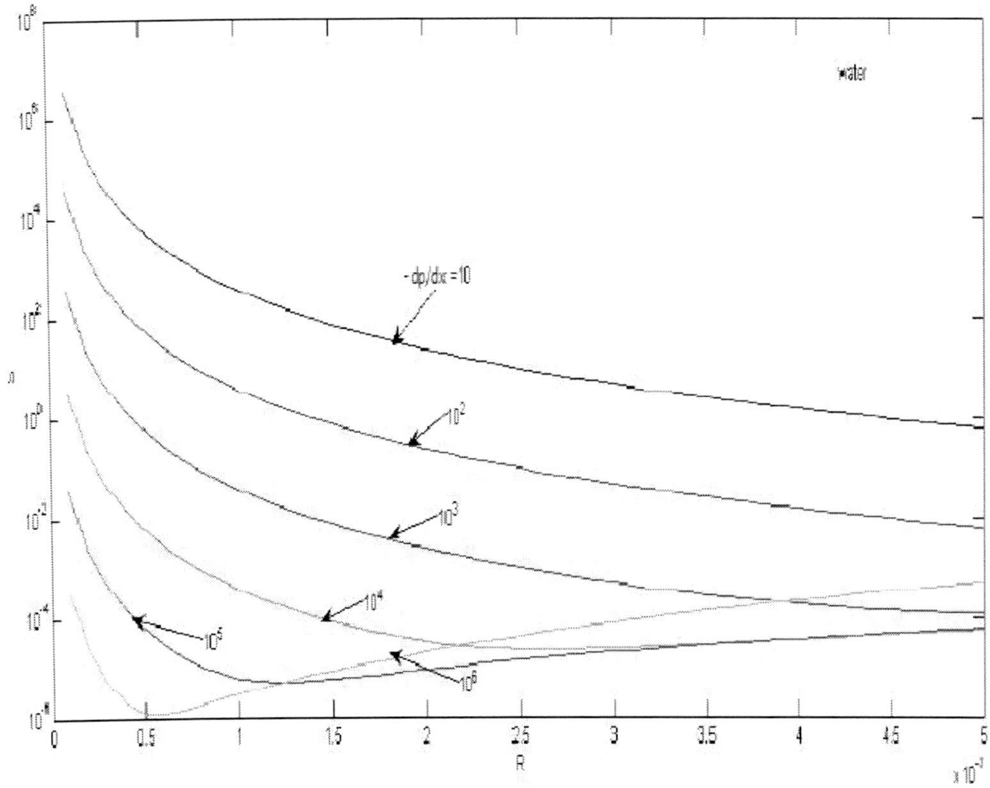

Figure 10. Total Entropy Generated versus Radius of Tube for water.

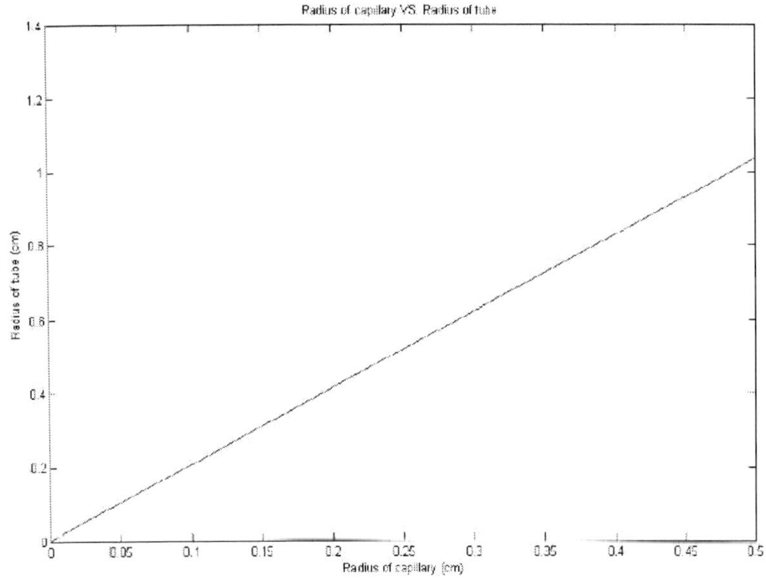

Figure 11. Radius of Capillary versus Tube Radius.

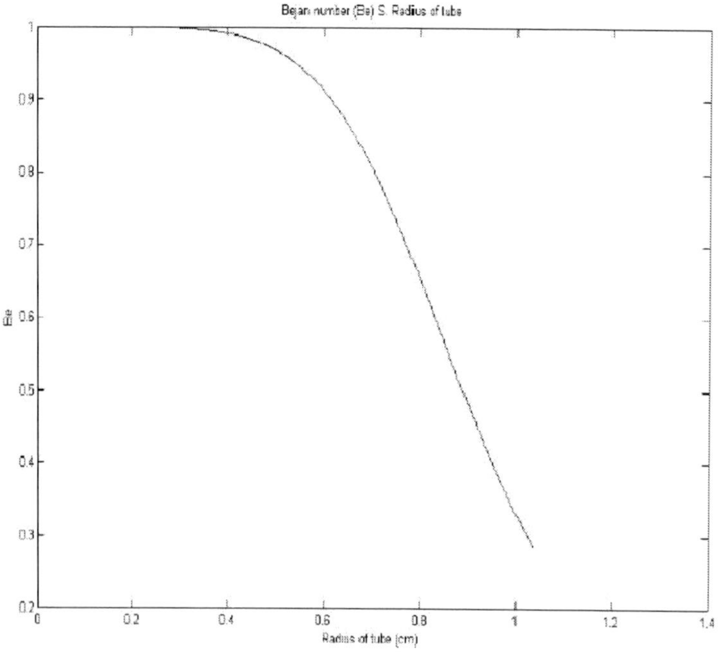

Figure 12. Bejan number versus Radius of Tube.

We have applied the method of thermodynamic optimization to the combined heat and mass transfer in a circular tube and micro scale evaporating thin film energy system. The objective was to identify the geometric configuration that maximized performance subject to a global constraint. The maximization of thermodynamic performance reduced to the minimization of entropy generated when the flow rate is prescribed.

CONCLUSION

The entropy generated in a circular tube and micro scale evaporating thin film due to the combined action of heat and mass transfer has been considered. An optimum radius of the tube was computed for minimum entropy generated. For the capillary radius less than the optimum value, the entropy increases very rapidly. When the capillary radius greater than the optimum value, the entropy generation increases only gradually. Low surface heat flux and high pressure difference also result low entropy generated and changes the optimum radius. However, with high surface heat flux and low pressure difference may increase the optimum radius.

Capillary pumped heat transfer devices do not achieve the maximum predicted heat flux because of interfacial thermocapillary stresses arising from liquid vapor interfacial temperature gradients. This can be counteracted by introducing naturally occurring concentration gradients using binary fluid mixtures.

REFERENCES

[1] B.V. Derjaguin, and Z.M. Zorin, "Optical Study of the Absorption and Surface Condensation of Vapors in the Vicinity of Saturation on a Smooth Surface," *Proc.2nd International Congress on Surface Activity* (London), 2, (1957), 145-152.

[2] X. Xu, and V.P. Carey, "Film Evaporation from a Microgrooved Surface: An Approximate Heat Transfer Model and its Comparison with Experimental Data," *AIAA Journal of Thermophysics*, 4, (1990), 512-520.

[3] J.M. Ha, and G.P. Peterson, "The Interline Heat Transfer of Evaporating Thin Film Along a Micro Grooved Surface," *ASME Journal of Heat Transfer*, 118, (1996), 747-755.

[4] K. Park, K.J. Noh, and K.S. Lee, "Transport Phenomena in the Thin Film Region of a Micro Channel," *International Journal of Heat and Mass Transfer*, 46, (2003), 2381-2388.

[5] S. Das Gupta, I.Y. Kim, and P.C. Wayner, "Use of Kelvin-Clapeyron Equation to Model an Evaporating Curved Microfilm," *J. Heat Transfer, Trans. ASME*, 116, (1994), 1007-1015.

[6] J.P. Longtin, B. Badran, and F.M. Gerner, "A One Dimensional Model of a Micro Heat Pipe During Steady State Operation," *ASME Journal of Heat Transfer*, 116, (1994), 709-715.

[7] V. Sartre, M. Zaghdoudi, and M. Lallemand, "Effect of Interfacial Phenomena on Evaporative Heat Transfer in Micro Heat Pipes," *International Journal of Thermal Sciences*, 39, (2000), 498-504.

[8] W. Qu, T. Ma, J. Miao, and J. Wang, "Effects of Radius and Heat Transfer on the Profile of Evaporating Thin Liquid Film and Meniscus in Capillary Tubes," *International Journal of Heat and Mass Transfer*, 45,(2002),1879-1887.

[9] R.S.R. Gorla, L.W. Byrd and D.M. Pratt , "Entropy Minimization In Phase Change Energy Systems," Technical Report. Thermal Structures Branch, Air Vehicles Directorate, *Air Force Research Laboratory*, Wright Patterson Air Force Base.

[10] R.S.R. Gorla, L.W. Byrd, and D.M. Pratt , "Entropy Minimization In Micro-Scale Evaporating Thin Film Energy Systems," Technical Report. Thermal Structures Branch, Air Vehicles Directorate, Air Force Research Laboratory, Wright Patterson Air Force Base.

[11] R.S.R. Gorla, J.E. Gatica, B. Ghorashi, P. Ineure, and L. Byrd, "Heat Transfer in Thin Liquid Film in the Presence of Electric Field for Non-Isothermal Interfacial Condition," *International Journal of Fluid Mechanics Research*, 29,(2002).

[12] R.S.R. Gorla, and L. W. Byrd, "Effect of Electrostatic Field on Film Rupture," Trans. ASME, *Journal of Fluids Engineering*, 121, (1999),651-657.

[13] R.S.R. Gorla, L.W. Byrd, and D.M. Pratt. Second law analysis for microscale flow and heat transfer. *Applied Thermal Engineering* 27, (2007), 1414-1423.

[14] A. Bejan, "Shape and Structure from Engineering to Nature," Cambridge University Press, Cambridge, UK, (2000).

[15] A. Bejan, "*Entropy Generation through Heat and Fluid Flow*" Wiley, New York,(1982).

[16] A. Bejan, "*Advanced Engineering Thermodynamics*," Wiley, New York, (1997).

[17] A. Bejan, "Entropy Generation Minimization," CRC Press, Boca Raton, Florida (1996).
[18] J.Y. San, C.L. Jan, Second-law analysis of wet cross flow heat exchanger, *Energy* 25(10), (2000), 939-955.
[19] M. Izquierdo, M.de Vega, A. Lecuona, P. Rodriguez, Compressors driven by thermal solar energy: entropy generated, energy destroyed and energetic efficiency. *Solar Energy* 72(4), (2002), 363-375.
[20] S. Mahmud, RA. Fraser, Inherent irreversibility of channel and pipe flows for non-Newtonian fluids. *Int Comm Heat Mass Transfer* 29,(2002), 577-587.
[21] S. Mahmud, RA. Fraser, Thermodynamic analysis of flow and heat transfer inside channel with two parallel plates. *Energy Int J* 2,(2002), 140-146.
[22] SH. Tasnim, S. Mahmud, Entropy generation in a vertical concentric channel with temperature dependent viscosity. *Int Comm Heat Mass Transfer* 29(7),(2002), 907-918.
[23] OM. Haddad, MK. Alkam, MT. Khasawneh, Entropy generation due to laminar forced convection in the entrance region of concentric annulus. *Energy* 29, (2004), 35-55.
[24] PK. Nag, N. Kumar, Second law optimization of convective heat transfer though a duct with constant heat flux. *Int J. Energy Res.*13, (1989), 537-543.
[25] J.Y. San, W.M. Worek and Z. Lavan, Entropy Generation in combined Heat and Mass Transfer. *Int. J. Heat Mass Transfer*, 30,(1987), 1359-1369.
[26] D.M. Pratt, K.D. Kihm, and L.W. Byrd, Binary Fluid Mixture and Thermocapillary Effects on the Wetting Characteristics of a Heated Curved Meniscus. *Proc. ASME/JSME Joint Conference*, Paper TED-AJ03-594, (2003).
[27] D.M. Pratt, J.R. Brown, and K.P. Hallinan. "Thermocapillary Effects on the stability of a Heated, Curved Meniscus." *Trans. ASME,* 120, Feb (1998), 220-226.

In: Frontier Research in Microscale and Nanoscale...
Editor: Lixin Cheng
ISBN 978-1-61324-665-8

Chapter 19

PREDICTION OF THERMO-PHYSICAL PROPERTIES FOR CO_2-OIL MIXTURES AT LOW OIL CONCENTRATION

Rin Yun*
Department of Mechanical Engineering, Hanbat National University, Daejeon 305-719, Korea

ABSTRACT

Due to environmental concerns, CO_2 has been reintroduced as a potential candidate to replace HFCs in refrigeration systems since the 1990s. In a refrigeration cycle, oil is utilized in lubricating a compressor. However, even though oil separators are installed after a compressor, oil is prone to leak to the whole system. In the case of the mixing of CO_2 and even a small amount of oil, the heat transfer performance in the heat exchanger is deteriorated and the pressure drop inside the tube increases. Therefore, to correctly design a CO_2 refrigeration system, it is necessary to precisely estimate the thermodynamic mixture properties of CO_2-lubricant oil. The commonly used method in estimating the mixture properties is the mole-based weighting model. However, the accuracy of the method cannot be assured for the CO_2-lubricant oil under the supercritical region. In the present study, CO_2-lubricant oil mixture properties, including density, viscosity, thermal conductivity, and heat capacity, were estimated by using the mixture models, based on the equation of state (EOS).

Keywords: CO_2-oil mixture, density, viscosity, thermal conductivity, heat capacity, equation of state.

* Tel: 82-42-821-1732, Fax: 82-42-821-1587; E-mail: yunrin@hanbat.ac.kr

NOMENCLATURE

a	Helmholtz free energy, Parameters in the EOS for pure compounds, $kJmol^{-1}$, constant in eq.(23)
a_r	residual function of the Helmholtz free energy, $kJmol^{-1}$
b	parameters in the EOS for pure compounds, constant in eq.(22)
c_p	specific heat, $Jkg^{-1}K^{-1}$
f	temperature-dependent binary interaction factor in eq. (7), scaling parameter in eq. (16)
f_{int}	internal energy factor
h	scaling parameter
h_r	residual function of enthalpy, $kJmol^{-1}$
k, l, m	binary interaction factor
M	molecular mass, $kgkmol^{-1}$
P	pressure, bar
R	gas constant
r	parameter of the eq. (8)
s_r	residual function of entropy, $Jmol^{-1}K^{-1}$
V	volume, m^3
x_i	mole fraction of component i in a mixture

Greeks

α	temperature-dependent function of the energy parameter a
β	experimentally determined constants in eq.(5) and (6)
λ	thermal conductivity, $Wm^{-1}K^{-1}$
μ	dynamic viscosity Nsm^{-2}
τ	binary interaction parameter, corresponding values based on viscosity data of propane
ϕ	corresponding values based on viscosity data of propane
ρ	density, kgm^{-3}
ω	acentric factor

Subscript

α,β	pure components or species in a mixture
c	critical
i, j	index for pure fluids
m	mixture
o	reference fluid value

r reduced

Superscript

R property of reference fluid, propane
* property of the ideal state

1. Introduction

The studies on the natural refrigerant of CO_2 have progressed since the mid-1990s because the existing refrigerants of CFCs or HCFCs are harmful to the environment. The use of the existing refrigerants will gradually fade out, despite their favorable thermo-physical properties. However, when the CO_2 systems are compared with the existing system, they demonstrate lower performance due to the heat rejection to the ambient under the supercritical CO_2 condition and the throttling processes under the large pressure difference. Furthermore, the components of CO_2 systems must be designed to meet the high working pressure. Through the continuous and concentrated efforts to improve the performance of the CO_2 systems to the performance level of the existing system, the state of the art of the CO_2 system has reached a similar standard. The efforts include the development of the compressor and the compact heat exchangers appropriate for the CO_2 system, and the employment of auxiliary components such as an expander. However, there is still much room for improvement of the CO_2 systems. One such improvement is to find and develop the suitable oil for CO_2 systems, for which it is necessary to know the exact thermo-physical properties of the CO_2-oil mixture. According to Dang et al.'s results [1] the effects of oil on gas cooling heat transfer for a mini-channel are more significant compared to those for a large diameter tube. Because the micro- and mini- channels are widely utilized for heat exchangers the thermo-physical properties of the CO_2-oil mixture is crucial when design the microchannel heat exchanger.

When we consider the existing refrigerant oils, the types of oil can be categorized according to the refrigerants. For examples, the series of naphtha and paraffin oils are used for CFC and HCFC refrigerants. Polar synthetic oils of Polyalkylene glycol (PAG), Polyol Ester (POE), and alkyl benzene are utilized for R-134a of HFC. These same polar synthetic oils are considered to be possible refrigerant oils for the CO_2 systems. Table 1 shows the comparison of four lubricants utilized for the CO_2 systems [2]. They have challenging points for each lubricant. Until now, PAG and POE oil are known as the most suitable. For decision of the suitability of the oil to the specific system, miscibility between oil and refrigerant is most important property. Hauk and Weidner [3] investigated the solubility of CO_2 into three oils of PAG, POE, and PAO. At the pressure of 40 bar, the three phases of CO_2-oil mixture, liquid CO_2, and gaseous CO_2 were observed. Over the 50 bar, only two phases of CO_2-oil mixture and liquid CO_2 were found. Pensado et al. [4] measured the densities and dynamic viscosities of CO_2-POE mixtures, whose concentrations of CO_2 are 83.3 wt.% and 92.2 wt.%, by applying a vibrating-wire sensor. The three different types of POE oils were utilized in their studies. Pressure conditions were from 100 to 600 bar, and the temperatures varied from 303.15K to 393.15K by 10K increments. They also suggested the experimental correlations based on their test results for the densities and viscosities of the CO_2-oil mixture. Yokozeki [5] tested the solubility between the CO_2 and POE and PAG oils, and theoretically calculated the solubility of the CO_2-oil mixture by using the Redlich-Kwong type cubic equation of state. The calculations were conducted with the representative critical temperature, critical pressure, and the binary interaction parameters of the oils because there are many different types of POE and PAG oil. The experimental results were well estimated by the theoretical calculations.

Table 1. Comparison of four synthetic lubricants for CO_2 systems [2]

Oil type	Miscibility/Solubility	Lubricity	Suitable system	Challenges of CO_2 systems
Alkyl naphthalene/ alkylbenzne (AN/AB)	Completely immiscible	Good		
Polyalphaolefin (PAO)	Completely immiscible	Normal	Cascade systems	Large oil separator on the compressor discharge is needed
Polyalkylene glycol (PAG)	Partially miscible	Good	Trans-critical system	Oil separation and return systems are needed to avoid long term oil accumulation
Polyol ester (POE)	Miscible	Good	Trans-critical system	Mixture viscosity reduces significantly with temperature increase

Many researchers suggested the calculation methods for estimating the mixture properties of refrigerant-lubricant and pure oils. Zhao and Bansal [2] summarized the method as shown in Table 2. Table 2 also includes the suggested model by Bandarra et al. [6]. Although the estimation methods shown in Table 2 give a good predictability for the conventional refrigerants-oil mixture, it is hard to apply these equations to estimate the properties of CO_2-oil mixture under the working conditions for CO_2. The CO_2 systems work at high pressure and supercritical conditions. Because these equations were developed by using the test results, they have the limits of application for the estimation. The object of this study is to theoretically predict the thermo-physical properties of CO_2-oil mixtures including density, viscosity, and heat capacity, which are important parameters to predict the heat transfer coefficients and pressure drop of CO_2-oil mixtures inside heat exchangers. This theoretical approach to estimate the transport properties for the refrigerant and oil mixture can be practically useful when the experimental data for them aren't available.

2. Prediction Methods of Thermo-Physical Properties of CO_2-Oil Mixtures

The fundamental properties, such as molecular weight, normal boiling point, critical values, and acentric factor, for CO_2 and oil and binary interaction parameters of the mixture are needed for estimating the mixture properties theoretically. Table 3 and 4 show these fundamental properties and the binary interaction parameters for CO_2-POE and CO_2-PAG, respectively. It should be noted that the values shown in Table 4 are selected ones among the suggested values by Yokozeki [5] for the various CO_2-POE and CO_2-PAG oils. This selection is based on how the fundamental properties shown in Table 3 are close between the utilized oils in the present model and the provided ones by Yokoziki [5].

Table 2. Refrigerant-lubricant mixture properties calculation methods ([2], [6])

Nominal oil concentration	$\omega = \dfrac{\dot{m}_{oil}}{\dot{m}_{oil} + \dot{m}_{ref}}$
Local oil concentration	$\omega_{local} = \dfrac{\omega_o}{1-x}$
Mixture dynamic viscosity	$\mu_m = \mu_{ref}^{(1-\omega_{local})} \mu_{oil}^{\omega_{local}}$ $\ln \mu_m = (1-\omega) \ln \mu_{ref} + \omega \ln \mu_{oil}$
Mixture density	$\rho_m = \dfrac{\rho_{oil}}{\left[\left(1 + (1-\omega_{local})(\dfrac{\rho_{oil}}{\rho_{ref}} - 1\right)\right]}$ $\dfrac{1}{\rho_m} = \dfrac{\omega_{local}}{\rho_{oil}} + \dfrac{1-\omega_{local}}{\rho_{ref,l}}$
Specific heat of refrigerant-oil mixture	$c_{pm} = (1-\omega_{local}) c_{p,ref} + \omega_{local} c_{p,oil}$ $c_{pm,l} = \omega c_{p,oil} + (1-\omega) c_{p,ref}$
Mixture conductivity	$k_m = k_{ref}(1-\omega_{local}) + k_{oil}\omega_{local} - 0.72\omega_{local}(1-\omega_{local})(k_{oil} - k_{ref})$
Mixture surface tension	$\sigma_m = \sigma_{ref} + (\sigma_{oil} + \sigma_{ref})\omega_{local}^{1/2}$ $\sigma_m = \sigma_{ref,l} + (\sigma_{oil} - \omega)\sqrt{\omega_{local}}$
Mixture bubble point temperature	$T_{bub} = \dfrac{A}{\ln(p_{sat}) - B}$ $A = a_0 + 182.5\omega_{local} - 724.2\omega_{local}^2 + 3868\omega_{local}^5 - 5269\omega_{local}^7$ $B = b_0 - 0.722\omega_{local} + 2.39\omega_{local}^2 - 13.78\omega_{local}^5 + 17.07\omega_{local}^7$

Table 3. Physical properties of CO_2-oils

Fluids	Mass (g/mol)	T_b (K)	T_c (K)	P_c (kPa)	ω
CO_2	44.01	-	304.1	7377	0.22
PEC9	697.05	537.97	597.1	620	2.07
HEXG	118.18	470.65	623.9	3900	1.15

Table 4. Binary interaction parameters for CO_2-oils

CO_2/OIL	l_{12}	l_{21}	m_{12}	$\tau 0_{12}$	$\tau 1_{12}$ (K)	$\tau 2_{12}$ (K^{-1})
CO_2/PEC9	0.0653	0.0017	0	1	2.55	0
CO_2/HEXG	0.0414	0.0180	0	1	0.439	0

2.1. Density Prediction Model for CO_2-Oil Mixtures

The mixture density can be calculated by solving eq. (1) in terms of V, which was utilized to estimate the solubility of CO_2-oil mixtures by Yokozeki [5]. The parameters of a and b in eq. (1) are given by the Redlich-Kwong type of cubic EOS for pure fluids as eqs. (2) and (3). The binary interaction parameter for a and b can be calculated by using the values shown in Table 2. Eqs. (4), (5), (6), and (7) were utilized to obtain the parameters of a and b.

$$P = \frac{RT}{V-b} - \frac{a(T)}{V(V+b)} \tag{1}$$

$$a = \sum_{i,j=1}^{N} \sqrt{a_i a_j} f_{ij}(T)(1-k_{ij})x_i x_j \tag{2}$$

$$b = \frac{1}{2}\sum_{i,j=1}^{N} (b_i + b_j)(1-k_{ij})(1-m_{ij})x_i x_j \tag{3}$$

$$a_i = 0.427408 \frac{R^2 T_{ci}^2}{P_{ci}} \alpha_i(T), \; b_i = 0.08664 \frac{RT_{ci}}{P_{ci}} \tag{4}$$

$$\alpha(T) = \sum_{k=0}^{\le 3} \beta_k (1/T_r - T_r)^k, T_r \equiv T/T_c \le 1 \tag{5}$$

$$\alpha(T) = \beta_0 + \beta_1[\exp(2(1-T_r)) - 1], \; T_r \ge 1 \tag{6}$$

$$k_{ij} = \frac{l_{ij} l_{ji}(x_i + x_j)}{l_{ji} x_i + l_{ij} x_j}, \; f_{ij}(T) = \tau 0_{ij} + \tau 1_{ij}/T + \tau 2_{ij} T \tag{7}$$

2.2. Viscosity Prediction Model for CO_2-Oil Mixtures

The viscosity of the mixture can be theoretically calculated by using the similarity between the P-V-T and the T- μ -P relation of the mixtures. Because the thermodynamic property of viscosity is the most important parameter in designing the refining facilities of petroleum, the relations of T- μ -P for hydrocarbons and reservoir fluids have been widely investigated. In this study, the dynamic viscosity of CO_2-oil mixtures is estimated by using one of the correlations of the T- μ -P for hydrocarbons and reservoir fluids developed by Guo et al. [7], which shows the minimum mean deviation between the estimated values and the experimental data, when various T- μ -P correlations were tested. Eq. (8) shows the relation of T- μ -P, which has a similar form of the PR EOS. Eqs. (9) and (10) are mixture parameters for a and b in the PR EOS, respectively. Based on the viscosity data of propane, corresponding values of ϕ and τ in eqs. (11) and (12) were determined as functional forms of T_r and P_r. The use of parameter b' in eq.(11) is for describing the temperature effect on the viscosity in the supercritical region. The gas constant R is replaced by r in eq. (8), and it can be calculated by eq. (12). Eqs (13) and (14) are parameters for mixtures of b' and r, respectively. The binary interaction parameters are calculated by using the values shown in Table 4.

$$T = \frac{r_m P}{\mu_m - b'_m} - \frac{a_m}{\mu_m(\mu_m + b_m) + b_m(\mu_m - b_m)} \tag{8}$$

$$a_m = \sum_i x_i a_i \tag{9}$$

$$b_m = \sum_i x_i b_i \tag{10}$$

$$b' = b\phi(T_r, P_r) \tag{11}$$

$$r = r_c \tau(T_r, P_r) \tag{12}$$

$$b'_m = \sum_i \sum_j x_i x_j \sqrt{b'_i b'_j}(1 - k_{ij}) \tag{13}$$

$$r_m = \sum_i x_i r_i \tag{14}$$

2.3. Thermal Conductivity Prediction Model for CO_2-Oil Mixtures

Most theoretical models for the thermal conductivity of gases are based on the idea proposed by Eucken. Eucken proposed for polyatomic gases the separate terms of the translation and internal energy contributions in calculation of the thermal conductivity [8]. In the present study, the thermal conductivity of the mixture under high pressure condition was calculated by the Ely and Hanley model [9] of eq. (15), which shows quite encouraging results for wide range of density. In their model, the thermal conductivity of mixture was calculated by a corresponding states method using methane as the reference component. The Ely and Hanley model contained the terms of density correction and high density contribution, which are represented by eqs. (16) and (17). The thermal conductivity of mixture may be divided into two contributions as shown in eq. (18). One is arising from purely translational effects, and the other from the energy transfer via the internal degree of freedom. The contribution of the internal degree of freedom on the thermal conductivity can be calculated by using eq. (19). The purely translation effects shown in eq. (20) can be divided three contributions, and each are shown in eqs. (21), (22), and (23).

$$\lambda_m - \lambda_m{}^0 = F_{\lambda m} X_{\lambda m} (\lambda^R - \lambda^{R0}) \quad (15)$$

$$F_{\lambda m} = (44.094)^{-1/2} (h_m)^{-2} \sum_i \sum_j x_i x_j (f_{ij} / M_{ij})^{1/2} (h_{ij})^{4/3} \quad (16)$$

$$X_{\lambda m} = \left[1 + \frac{2.1866(\omega_m - \omega^R)}{1 - 0.505(\omega_m - \omega^R)} \right]^{1/2} \quad (17)$$

$$\lambda^R = \lambda_{internal} + \lambda_{translation} \quad (18)$$

$$\lambda_{internal}(T) = \sum_\alpha \sum_\beta x_\alpha x_\beta \lambda''_{\alpha\beta}, \ (\lambda''_{\alpha\beta})^{-1} = 2[(\lambda''_a)^{-1} + (\lambda''_b)^{-1}], \ \frac{\lambda''_\alpha M_\alpha}{\eta^*_\alpha} = f_{int}\left(C^0_{p,\alpha} - \frac{5}{2} R \right) \quad (19)$$

$$\lambda_{translation} = \lambda_o{}^*(T_o) + \lambda^{(1)}(T_o)\rho_o + \Delta\lambda_o(\rho_o, T_o) \quad (20)$$

$$\lambda_o{}^*(T_o) = \frac{15R}{4M_o} \mu_o{}^*(T_o), \ \mu_o{}^*(T_o) = \sum_{n=1}^{9} C_n T_o{}^{(n-4)/3} \quad (21)$$

$$\lambda^{(1)}(T_o) = b_1 + b_2[b_3 - \ln(T_o / b_4)]^2 \quad (22)$$

$$\Delta\lambda_o = \exp[a_1 + a_2 / T_o]\left\{ \exp[(a_3 + a_4 / T_o^{3/2})\rho_o^{0.1} + (\rho_o / \rho_c - 1)\rho_o{}^{0.5}(a_5 + a_6 / T_o + a_7 / T_o^2)] - 1.0 \right\} \quad (23)$$

2.4. Heat Capacity Prediction Model for CO_2-Oil Mixtures

The heat capacity of the CO_2-oil mixture is calculated by using eq. (24) [10]. The molar-based mixing rule can be applied for the heat capacity of the mixture. The residual enthalpy in eq. (25) is obtained by using eqs. (26) and (27). The parameter of V was estimated by the PR EOS.

$$C_{pr} = C_p^* - C_p = \left(\frac{\partial h_r}{\partial T} \right)_p \quad (24)$$

$$h_r = a_r + Ts_r + pV^* - pV \tag{25}$$

$$a_r = a^* - a = RT\ln\frac{|V-b|}{V} - \frac{a(T)}{2\sqrt{2}b}\ln\frac{|V-0.414b|}{V+2.414b} + RT\ln\frac{V}{V^*} \tag{26}$$

$$s_r = s^* - s = R\ln\frac{|V-b|}{V} - R\ln\frac{V}{V^*} \tag{27}$$

3. Results and Discussion

3.1. Density of CO_2-Oil Mixtures

Figure 1 shows the density variation of the CO_2-PEC9 mixture with temperature and pressure. The experimental data shown in the Figure 1 are from Pensado et al. [2]. The mixture density increases with pressure and gradually decreases with increase of temperature. It is also found that the effects of temperature on the density decreases with the increase of pressure. The mean deviation between the estimated values and the experimental data are ±7.93% based on the experimental data.

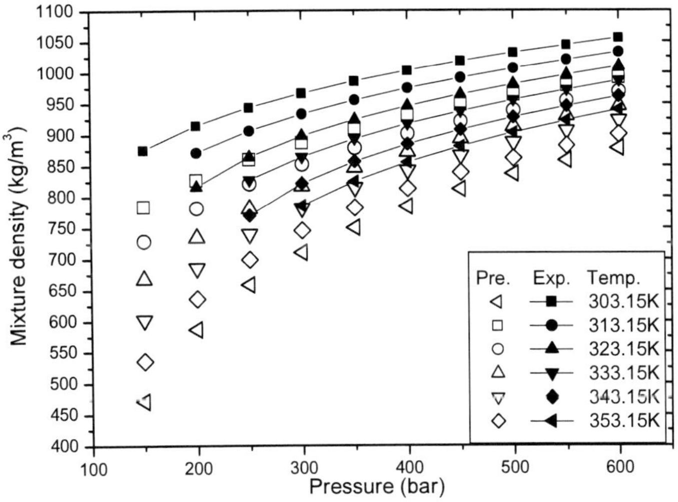

Figure 1. The mixture density of CO_2 (92.2 wt.%) + PEC9 with temperature and pressure variation.

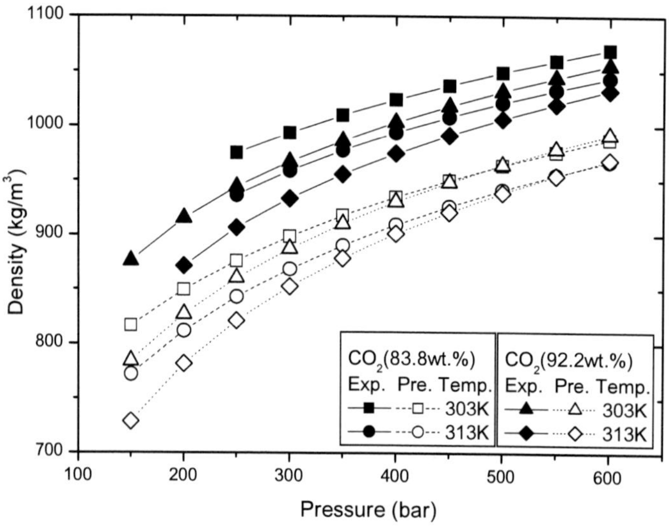

Figure 2. Comparison of mixture densities between CO_2 (92.2 wt.%) and CO_2 (83.8 wt.%).

Figure 2 shows the density of the CO_2-PEC9 mixture with two different weight percents of CO_2. The weight percents of CO_2 are 83.3 and 92.2 wt.%, respectively. The experimental data show that the density of the mixture of 83.3 wt.% of CO_2 is larger than that of 92.2% wt.% of CO_2 by the average of 1.9% at the temperature of 313.15K. As shown in Figure 2, the effects of the concentration of oil to the mixture on the density decreased with the increase of pressure, which is observed in both the experimental and theoretical results.

3.2. Viscosity of CO_2-Oil Mixtures

Figure 3 compares the viscosity of mixtures of CO_2-PEC9 between the calculated and the experimental data [2] with the CO_2 of 92.2 wt.%. The viscosity of the mixture increased with the increase of pressure and the decrease of temperature, which are observed in both the theoretical and experimental results. The mean deviation between the calculated values and the experimental data is ±4% based on the experimental data, and the deviation decreases with the increase of pressure.

Figure 4 shows the dynamic viscosity of the mixture when the concentration of CO_2 is 83.3 wt.%. The variation trends of viscosity observed in Figure 4 are similar to those of Figure 3. However, the mean deviation between the experimental and theoretical data for the mixture of 83.3 wt.% of CO_2 is ±10%. This result shows that the predictability of eq. (8) is getting worse with the increase of oil concentration. This increase of error between the experimental and the calculated data with increase of oil concentration can be partially explained by the limited availability of the binary interaction parameters between the CO_2 and PEC9. In the present study, the binary interaction parameter suggested by Yokozeki [5] is applied to calculate eq. (8). However, when the actual oil concentration in the CO_2 refrigeration system is less than 5% is considered, the present model for the viscosity of CO_2-oil mixture can be applicable. On the other hand, the model shows better predictability at the temperature ranges from 313 to 324K than that of other regions.

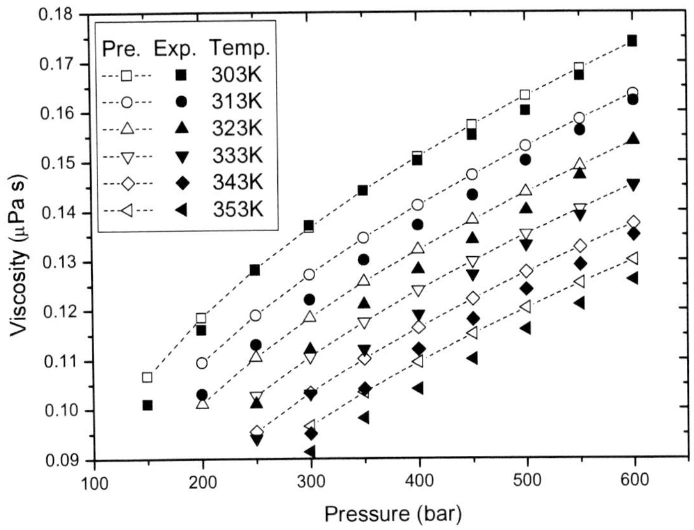

Figure 3. The mixture viscosity of CO_2 (92.2 wt.%) + PEC9 with pressure variation.

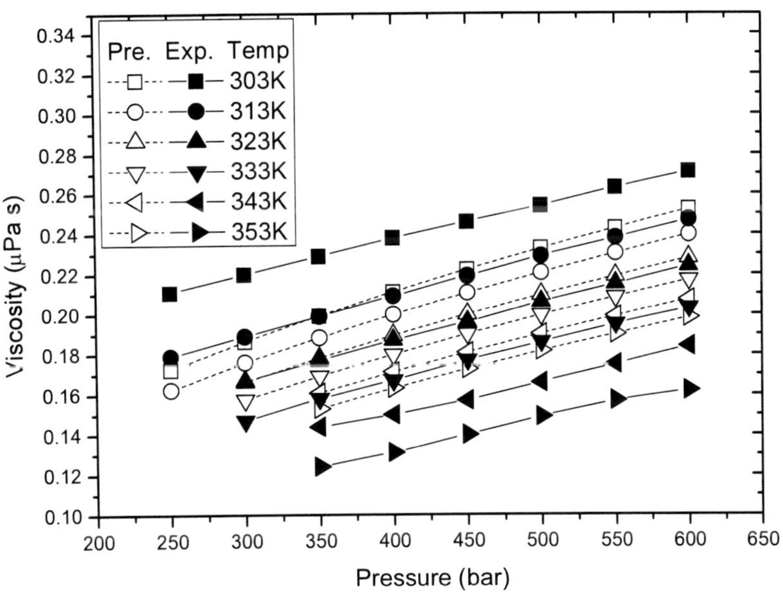

Figure 4. The mixture viscosity of CO_2 (83.3 wt.%) + PEC9 with pressure variation.

3.3. Thermal Conductivity of CO_2-Oil Mixtures

Figure 5 shows the mixture thermal conductivity of CO_2-HEXG. The calculation conditions of pressure are 90 and 150 bar. The temperature varied from 270 to 350K.

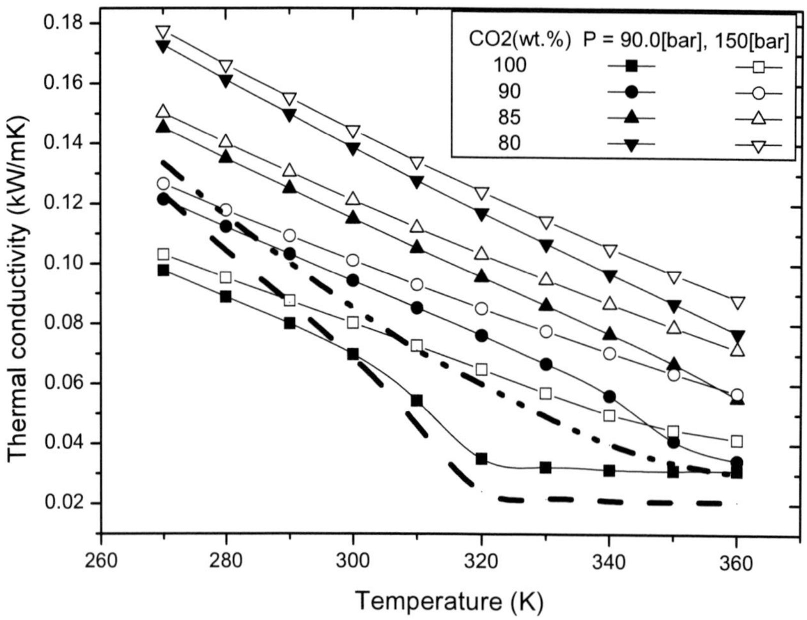

Figure 5. The thermal conductivity for CO_2-oil mixture with the variation of the concentration of CO_2.

Due to the lack of the experimental data of the thermal conductivity for the CO_2-oil mixture under the supercritical condition, the accuracy of the present model of thermal conductivity only depends on the prediction ability of the Ely and Hanley model [9]. The comparison of the present calculations with the result from other correlation for pure CO_2 was conducted. The dotted lines in Figure 5 show the thermal conductivity form the Span et al.'s model [11]. The thermal conductivity of the mixture decreased with the increase of temperature. The thermal conductivity of HEXG at ambient temperature and pressure is suggested as 0.16 W/m-K. When the thermal conductivity of the pure CO_2 is compared with that of CO_2-HEXG mixture, it is getting higher with the increase of oil concentration. When the concentration of CO_2 is 85 wt.%, the thermal conductivity of mixture increases by 99% based on that of the pure CO_2. The thermal conductivity of mixture increased with increase of the pressure. When the pressure increased from 90 to 150 bar, it is increased by 18.6% based on that for 90 bar.

3.4. Heat Capacity of CO_2-Oil Mixtures

Figure 6 shows the variation of the heat capacity of the mixture with increase of weight percentage of HEXG. The estimated heat capacity of pure CO_2 is compared with the data obtained by the other experimental correlation [11]. The pressure is 98 bar, and the temperature ranges from 284K to 394K. The mean deviation between them is ±5.5%. The maximum heat capacity is observed at the pseudo-critical temperature for pure CO_2. However, the characteristics of the heat capacity of pure CO_2 are significantly changed with the increase of the HEXG concentration.

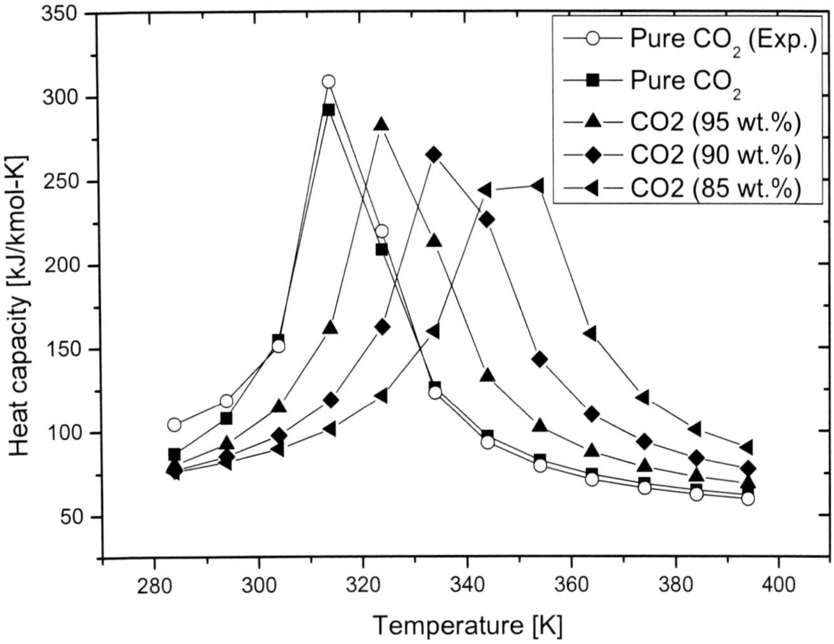

Figure 6. The variation of heat capacity with HEXG concentration and temperature.

The maximum heat capacity is decreased and the temperature at which the maximum value observed shifts to the higher value with the increase of the concentration of HEXG in the mixture. Also, the gradient of heat capacity around the peak value of the heat capacity with temperature is decreased.

CONCLUSION

In the present study, CO_2-lubricant oil mixture properties including density, viscosity, and heat capacity were estimated using the mixture models, based on the equation of state (EOS), and the following conclusions were obtained.

1) The effects of temperature and oil concentration on the mixture density of CO_2-oil decreased with the increase of pressure. The mean deviation of the mixture density with 92.2 wt.% of CO_2 between the experimental data and the estimated values is 7.93%.
2) The mixture viscosity estimated by the T- μ -P relations from the Guo et al. model shows 4% of mean deviation with the experimental data when the concentration of CO_2 in the mixture is 92.2 wt.%.
3) When the thermal conductivity of the pure CO_2 is compared with that of the CO_2-HEXG mixture, it increased with the increase of oil concentration. When the concentration of CO2 is 85 wt.%, the thermal conductivity of mixture increased by 99% based on that of pure CO_2.

4) The heat capacity of CO_2-oil mixtures is estimated by using the residual function of the binary system and the modified Peng-Robinson equation. With the increase of the concentration of HEXG in the mixture, the maximum heat capacity is decreased and the temperature at which the maximum value observed moves to the higher value.

REFERENCES

[1] C. Dang, K. Iino, E. Hihara, Study on two-phase flow pattern of supercritical carbon dioxide with entrained PAG-type lubricating oil in a gas cooler, *Int. J. Refrigeration* 31 (2008) 1265-1272.

[2] X. Zhao, P. Bansal, Critical review of flow boiling heat transfer of CO2-lubricant mixtures. *Int. J. Heat Mass Transfer* 52 (2009) 870-879.

[3] A. Hauk, E. Weidner, Thermodynamic and Fluid-Dynamic Properties of Carbon Dioxide with Different Lubricants in Cooling Circuits for Automobile Application. *Ind. Eng. Chem. Res*. 39 (2000) 4646-4651.

[4] A.S. Pensado, A.A.H. Pádua, J. Comuñas, M.J.P. Fernández, 2008, Viscosity and density measurements for carbon dioxide + pentaerythritol ester lubricant mixtures at low lubricant concentration, *J. of Supercritical Fluids* 44 (2008) 172 - 185.

[5] A. Yokozeki, Solubility correlation and phase behaviors of carbon dioxide and lubricant oil mixtures, *Applied energy* 84 (2007) 159 - 175.

[6] F. Bandarra, L. Cheng, J. Thome, Flow boiling characteristics and flow pattern visualization of refrigerant/lubricant oil mixtures. *Int. J. Refrg*. 32 (2009) 185-202.

[7] X.-Q. Guo, C.-Y. Sun, S. –X. Rong, G. –J. Chen, T. –M. Guo, 2001, Equation of state analog correlations for the viscosity and thermal conductivity of hydrocarbons and reservoir fluids, *Journal of Petroleum Science and Engineering* 30 (2001) 15 - 27.

[8] R.C. Reid, J.M. Prausnitz, B.E. Poling, *The properties of gases and liquids*, 4th ed., Mcgraw-hill, New York, 1988..

[9] J.F. Ely, J.M. Hanley, Prediction of transport properties, *Ind. Eng. Chem*. Fundam 22 (1983) 90-97.

[10] Z. Dalin, Q. Suizheng, S.U. Guanghui, J. Dounan, Estimation of thermodynamic properties of the ternary molten salt system, LiF-NaF-BeF2, by the modified Peng-Robinson equation, *Front. Energy Power Eng*. China 1 (2007) 174-180.

[11] R. Span, W. Wagner, A new equation of state for carbon dioxide covering the fluid region from the triple-point temperature to 1100K at pressures up to 800 MPa, *J. Phys. Chem, Ref.* Data 25(1996) 1509-1596.

In: Frontier Research in Microscale and Nanoscale… ISBN 978-1-61324-665-8
Editor: Lixin Cheng

Chapter 20

Evaluation of Flow Boiling Heat Transfer Correlations with Experimental Data of R134a, R22, R410A and R245fa in Microscale Channels

Lixin Cheng* and Hanfeng Zou

School of Engineering, University of Aberdeen,
King's College, Aberdeen,
AB24 3UE, Scotland, UK

Abstract

Copmarisons of 12 flow boiling heat transfer correlations to experimental data in microscale channels are presented. First, a database containing 1766 experimental data points of flow boiling heat transfer with R134a, R22, R410A and R245fa was compiled from the literature. The database covers a wide range of mass velocities from 20 to 860 kg/m^2s, heat fluxes from 5 to 55 kW/m^2, saturation temperatures from 0 to 50°C and tube diameters from 0.19 to 3 mm. Then, 12 flow boiling heat transfer correlations including 3 macroscale correlations and 9 microscale correlations have been collected from the literature.

The database was compared to these heat transfer prediction methods. Statistical analysis shows that no correlation predict the entire database well, with the best correlation only predicting 60% of the entire data within ±30%. Furthermore, some methods work well for some fluids while they do not work well for others.

In general, predictions for R134a and R410A data are not satisfactory while predictions for R245fa and R22 are acceptable and satisfactory. Effort should be made to develop a general correlation for flow boiling heat transfer in microscale channels according to heat transfer mechanisms and flow regimes which should also be fully understood in future.

* Corresponding author. +44 1224 272795, +44 1224 272497. E-mail addresses: lixincheng@hotmail.com (L. Cheng).

Keywords: Flow boiling; Two-phase flow; Microcale, Micro channels; Correlations; Refrigerants; Evaluation; Prediction; Analysis; Mechanisms.

NOMENCLATURE

Bd	Bond number, $[g(\rho_l-\rho_g)D_h^2/\sigma]$
Bo	boiling number, $[q/(Gi_{fg})]$
C	constant
c_p	specific heat at constant pressure, J/kgK
D_h	internal tube hydraulic diameter, m
d_b	bubble diameter, m
E	convective boiling heat transfer enhancement factor
E'	convective boiling heat transfer enhancement factor
E_2	correct factor defined by Eq. (15)
Fr	Froude number, $[G^2/(\rho_L^2 gD_h)]$
G	total gas and liquid two-phase mass velocity, kg/m^{2}s
g	gravitational acceleration, 9.81 m/s^2
h	heat transfer coefficient, W/m^{2}K
i	data point number
i_{fg}	latent heat of evaporation, J/kg
k	thermal conductivity, W/mK
M	Molercular weight
p	pressure, N/m^2
p_r	redcued pressure $[p/p_{crit}]$
Pr_l	prandtl number, $[\mu_l c_{pl}/k_l]$
q	heat flux, W/m^2
Re_{lo}	Reynolds number considering the total gas-liquid two phase flow as liquid flow $[GD_h/(\mu_L)]$
Re_l	Reynolds number considering only liquid phase flow $[G(1-x)D_h/(\mu_L)]$
Re_{tp}	two phase Reynolds number defined by Eq. (8)
S	nucleate boiling suppression factor
S2	correction factor defined by Eq. (16)
T	temperature, K
T_{sat}	saturation temperature, K
We_l	liquid Weber number considering the total vapor-liquid flow as liquid flow $[G^2D_h/(\rho_l\sigma)]$
We_g	gas phase Weber number $[G^2x^2D_h/(\rho_g\ \sigma)]$
X_{tt}	Martinelli number, $\{[(1-x)/x)]^{0.9}[\rho_g/\rho_l]^{0.5}[\mu_l/\mu_g]^{0.1}\}$
x	vapor quality

Greek Symbols

ϕ^2	two phase friction multiplier
μ	dynamic viscosity, Ns/m^2

ρ	density, kg/m^3
N/m	surface tension
ξ_i	relative error defined by Eq. (21)

Subscripts

b	bubble
cb	convective boiling
crit	critical
fg	latent
g	gas phase
l	liquid phase
lo	considering the total gas-liquid two phase flow as liquid flow
nb	nucleate boiling
p	constant pressure
pool	pool boiling
sat	saturation
sp	single phase
tp	two phase
w	wall

1. Introduction

Applications of microscale thermal and fluid transport phenomena involved in traditional industries and highly specialized fields such as bioengineering, micro-fabricated fluidic systems, microelectronics, aerospace technology, micro heat pipes, chips cooling etc. have been becoming especially important since the late 20th century [1-13]. In recent years, the research in the field of flow boiling heat transfer and two phase flow at a microscale level has been constantly increasing due to the rapid growth of the technology applications that require the transfer of high heat rates in a relatively small space and volume. However, gas liquid two-phase flow and flow boiling heat transfer characteristics in microcale channels are quite different from those in macroscale channels [2-5, 7-8, 6-8, 14-18]. Channel confinement also has a great effect on two-phase flow and flow boiling heat transfer characteristics [16-18]. Furthermore, the available experimental results of flow boiling heat transfer and two phase flow in microscale channels are quite contradictory from one study to another. There are so many issues to be clarified from both theoretical and applied aspects in microscale thermal and fluid transport phenomena. Furthermore, advances in micro-electronics technology continue to develop with surprisingly rapidity and the thermal energy density of electronic devices to be dissipated is becoming much higher. Therefore, it is essential to develop new high heat flux cooling technology to meet the challenging heat dissipation requirements [2-8, 13]. Flow boiling in microscale channels is one of such methods in high heat flux cooling technology. So far, the prediction methods for macroscale flow boiling heat transfer have been well developed [19] while those for microscale flow boiling heat transfer have not yet well established. Thus, it is essential to evaluate the available heat transfer correlations with extensive microscale flow boiling experimental data.

Over the last decade, great effort has been put into identifying the main parameters governing microscale flow boiling heat transfer and also into quantifying their relevance. Different heat transfer mechanisms are dominant according to the vapor quality range, and heat flux and mass velocity levels for macroscale flow boiling [18, 19]. At low vapor qualities, nucleate boiling effects prevail while at high vapor qualities and prior to the liquid dryout, the heat transfer coefficient is mainly controlled by convective effects. The predominance of these mechanisms is commonly considered [14-22] when developing flow-boiling heat transfer coefficient prediction methods. For macroscale flow boiling heat transfer, Chen [20] was one of the pioneers of such an approach according to which the flow boiling heat transfer coefficient is calculated as addition of nucleate-dominate and convective boiling dominate mechanisms. Several authors also developed the Chen type correlations such as Gungor and Winterton [21], Liu and Winterton [22] etc.

For microscale flow boiling, several researchers have applied the Chen type methods in their heat transfer correlations [26-28] while some researchers proposed new empirical correlations based on several dimensionless numbers according to experimental data [29-34]. So far, a number of correlations for flow boiling heat transfer are available for flow boiling in microscale channels. However, most of these correlations were proposed according to limited fluids and test conditions. Therefore, it is necessary to evaluate these available correlations with an independent database.

This paper aims to evaluate and analyse 12 flow boiling heat transfer correlations including macroscale and microscale methods with an extensive of experimental data of R134a, R22, R410A and R236fa in microscale channels obtained from the literature. The selected flow boiling heat transfer correlations include 3 macroscale methods of Chen [20], Gungor and Winterton [21] and Liu and Winterton [22] and 9 microscale methods of Lazarek and Black [29], Kew and Cornwell [30], Tran et al. [31], Warrier et al. [32], Zhang et al. [26], Pamitran et al. [27], Saitoh et al. [28], Sun and Mishima [33] and Li and Wu [34].

2. Macroscale To Microscale Channel Criteria

One important issue should be clarified about the distinction between microscale channels and macroscale channels at first. However, a universal agreement is not yet clearly established in the literature. Instead, there are various definitions on this issue. Cheng and Mewes [3] and Cheng et al. [9] have summarized various definitions of macrocale and microscale channels in the literature. Here just to show one example, based on engineering practice and application areas such as refrigeration industry in the small tonnage units, compact evaporators employed in automotive, aerospace, air separation and cryogenic industries, cooling elements in the field of microelectronics and micro-electro-mechanical-systems (MEMS), Kandlikar [4] defined the following ranges of hydraulic diameters D_h which are attributed to different channels:

- Conventional channels: $D_h > 3$ mm.
- Minichannels: $D_h = 200$ μ m – 3 mm.

- Microchannels: $D_h = 10\ \mu m - 200\ \mu m$.

According to this definition, the distinction between small and conventional size channels is 3 mm. In this study, the distinction between macro- and micro-scale channels by the threshold diameter of 3 mm is adopted due to the lack of a well-established theory, but is in line with those recommended by Kandlikar [4]. Using this threshold diameter enables more relevant studies to be included.

3. Flow Boiling Heat Transfer Database and Correlations

3.1. Description of Flow Boiling Heat Transfer Database

A flow boiling heat transfer database was compiled from the literature, containing 1766 experimental data points of flow boiling heat transfer with R134a, R22, R410A and R245fa under a wide range of test conditions as summarized in Table 1. The database covers mass velocities from 20 to 860 kg/m^{2}s, heat fluxes from 5 to 55 kW/m^2, saturation temperatures from 0 to 50°C and tube diameters from 0.19 to 3 mm. The experimental data were taken from the studies of Pamitran et al. [27], Agostini and Bontemps [36], Choi et al. [37], Bertsch et al. [38], In and Jeong [39], Shiferaw et al. [40], Tibirica and Ribatski [41], Vlasie et al. [42], Wang et al. [43] and Yun et al. [44]. The database will be used to evaluate 12 selected flow boiling heat transfer correlations in the following section.

Table 1. Parametric ranges of the selected flow boiling heat transfer database in microscale channels (total of 1766 data points)

Fluid	Tube diameter D_h (mm)	Saturation temperature T_{sat} (°C)	Heat Flux q (kW/m^2)	Mass flux G (kg/m^{2}s)	Data point number	Data sources
R134a	0.19 ~ 3	5 ~ 50	2.6 ~ 107	20~860	1259	Agostini and Bontemps [36], Choi et al. [37], Bertsch et al. [38], In and Jeong [39]. Shiferaw et al. [40], Tibirica and Ribatski [41], Vlasie et al. [42] and Wang et al. [43].
R22	1.5, 3	10	10 ~ 40	200 ~ 600	113	Choi et al. [37]
R410A	1.36 ~ 3	0 ~10	7.5 ~ 30	200 ~ 600	259	Pamitran et al. [27] and Yun et al. [44].
R245fa	0.54 ~ 2.3	8 ~ 30	2.6 ~ 35	40~700	135	Bertsch et al. [38] and Tibirica and Ribatski [41].

3.2. Selected Flow Boiling Heat Transfer Prediction Methods

12 flow boiling heat transfer correlations have been collected from the literature including 3 macroscale heat transfer correlations of Chen [20], Gungor and Winterton [21] and Liu and Winterton [22] and 9 microscale heat transfer correlations of Zhang et al. [26], Pamitran et al. [27], Saitoh et al. [28], Lazarek and Black [29], Kew and Cornwell [30], Tran et al. [31], Warrier et al. [32], Sun and Mishima [33] and Li and Wu [34].

3.2.1 Flow boiling heat transfer correlations in macroscale channels

It is well known that flow boiling heat transfer is governed mainly by two important mechanisms: nucleate boiling dominated heat transfer and convective boiling dominated heat transfer, which may be expressed as :

$$h_{tp} = h_{nb} + h_{cb} \tag{1}$$

where, the nucleate boiling dominated term hnb and the convective boiling dominated term hcb can be replaced by some restriction factors multiplied by single-phase heat transfer coefficient and nucleate pool boiling heat transfer coefficient in the literature. In what follows, some flow boiling heat transfer correlations with different defined nucleate boiling term and convective boiling term are briefly reviewed.

Different heat transfer mechanisms are dominant in flow boiling according to vapor qualities, heat fluxes and mass velocities. At low vapor qualities, nucleate boiling effects prevail while at high vapor qualities and prior to the liquid dryout, the heat transfer coefficient is mainly controlled by convective boiling. The predominance of these two heat transfer mechanisms is commonly considered in the development of prediction methods for macroscale flow boiling. It is also applied to some micro-scale flow boiling correlations by several researchers.

Chen [20] formulated the first general flow boiling heat transfer correlation by using addition of two components corresponding to the two heat transfer mechanisms with a nucleate boiling suppression factor S and a convective enhancement factor E as:

$$h_{tp} = Eh_l + Sh_{pool} \tag{2}$$

The liquid phase heat transfer coefficient hl is evaluated with the Dittus and Boelter equation [23]:

$$h_l = 0.023\,\mathrm{Re}_l^{0.8}\,\mathrm{Pr}_l^{0.4}\left[\frac{k_l}{D_h}\right] \tag{3}$$

$$\mathrm{Re}_l = \frac{G(1-x)D_h}{\mu_l} \tag{4}$$

$$\Pr_l = \frac{\mu_l c_{pl}}{k_l} \quad (5)$$

The pool boiling heat transfer coefficient hpool is evaluated by the Forster and Zuber correlation [24]:

$$h_{pool} = 0.00122 \frac{k_l^{0.79} c_{pl}^{0.45} \rho_l^{0.49}}{\sigma^{0.5} \mu_l^{0.29} i_{fg}^{0.24} \rho_g^{0.24}} (T_w - T_{sat})^{0.24} [p_{sat}(T_w) - p_l]^{0.75} \quad (6)$$

$$S = \frac{1}{1 + 2.35 \times 10^{-6} \mathrm{Re}_{tp}^{1.17}} \quad (7)$$

$$\mathrm{Re}_{tp} = \frac{G(1-x)D_h}{\mu_l} E^{1.25} \quad (8)$$

$$\text{For } \frac{1}{X_{tt}} \leq 0.1 \qquad E = 1 \quad (9)$$

$$\text{For } \frac{1}{X_{tt}} > 0.1 \quad E = 2.35\left(\frac{1}{X_{tt}} + 0.213\right)^{0.736} \quad (10)$$

S is the nucleate boiling suppression factor that takes into account steeper temperature gradients near the wall due to the fluid motion which tends to suppress the number of bubble active sites. The convective enhancement factor, E, takes into account the increment of convective effects relative to that of the single-phase flow of the liquid, promoted by the flow acceleration due to the evaporation process.

Gungor and Winterton [21] developed a general flow boiling heat transfer correlation which also considers the two heat transfer mechanims in the Chen correlation as:

$$h_{tp} = Eh_l + Sh_{pool} \quad (11)$$

$$E = 1 + 24000 Bo^{1.16} + 1.23(\frac{1}{X_{tt}})^{0.86} \quad (12)$$

$$S = \frac{1}{1 + 1.15 \times 10^{-6} E^2 \mathrm{Re}_l^{1.17}} \quad (13)$$

The cooper pool boiling correlation [25] is used in Eq. (11):

$$h_{pool} = 55 p_r^{0.12} (-\log p_r)^{-0.55} M^{-0.5} q^{0.67} \tag{14}$$

The heat transfer coefficient for the liquid phase hl is evaluated with Eq. (3). If the tube is horizontal and the Froude number is less than 0.05 then E should be multiplied by:

$$E_2 = Fr^{(0.1-2Fr)} \tag{15}$$

and S should be multiplied by:

$$S_2 = \sqrt{Fr} \tag{16}$$

$$Fr \frac{G^2}{\rho_l^2 g D_h} \tag{17}$$

Liu and Winterton [22] developed a general flow boiling heat transfer method in order to predict saturated and subcooled flow boiling heat transfer based on the Chen flow boiling heat transfer mechanisms. In their method, the nucleate boiling and convective boiling contributions are added up according to a power-type asymptotic approach with an exponent of 2 as:

$$h_{tp} = \sqrt{(Eh_{lo})^2 + (Sh_{pool})^2} \tag{18}$$

$$E = \left[1 + x \Pr_l \left(\frac{\rho_l}{\rho_g} - 1\right)\right]^{0.35} \tag{19}$$

$$S = \frac{1}{1 + 0.055 E^{0.1} \mathrm{Re}_{l0}^{0.16}} \tag{20}$$

where hlo is evaluated with the Dittus-Boelter [23] correlation taking total gas liquid two phase flow as single phase liquid flow and nucleate pool boiling heat transfer coefficient hpool is evaluated with the Cooper nucleate pool boiling correlation [25] Eq. (14).

3.2. Flow boilin heat transfer correlations in mcroscale channels

Table 2 summarizes the selected flow boiling heat transfer correlations in microscale channels. These heat transfer prediction methods were developed either by modifying the Chen correlation or correlating their experimental data according to dimensionless numbers by a number of researchers. Various flow boiling mechanisms have been observed in different studies.

Lazarek and Black [29] proposed a new type of correlation in terms of dimensionless Nusselt, Reynolds and boiling numbers. The strong dependence of the local saturated boiling heat transfer coefficient upon heat flux with negligible influence of vapor quality in their study suggested that the wall heat transfer process is controlled by nucleate boiling.

Table 2. Flow boiling heat transfer correlations for micro-scale channels in the literature

Lazarek and Black [29]	$h_{tp} = 30\,\mathrm{Re}_{lo}^{0.857}\,Bo^{0.714}\,\frac{k_l}{D_h}$
Kew and Cornwell [30]	$h_{tp} = 30\,\mathrm{Re}_{lo}^{0.857}\,Bo^{0.714}(1-x)^{-0.143}\,\frac{k_l}{D_h}$
Tran et al. [31]	$h_{tp} = 8.4\times10^{-5}\,Bo^{0.6}We_l^{0.3}\left(\frac{\rho_l}{\rho_g}\right)^{-0.4}$ $We_l = \frac{G^2 D_h}{\sigma\rho_l}$
Warrier et al. [32]	$h_{tp} = \left(1+6Bo^{1/16}+f(Bo)x^{0.65}\right)h_{lo}$ $f(Bo) = -5.3(1-855\,Bo)$
Zhang et al. [26]	$h_{tp} = Sh_{pool} + Fh_{sp}$ h_{sp} may be referred to the paper $S = \frac{1}{1+2.53\times10^{-6}\,\mathrm{Re}_l^{1.17}}$ $E = MAX\left(E',\,1\right)$ $E' = 0.64\left(1+\frac{C}{X_{tt}}+\frac{1}{X_{tt}^2}\right)^{0.5}$
Pamitran et al. [27]	$h_{tp} = Sh_{pool} + Fh_l$ $S = 9.4626\,(\phi^2)^{-0.2747}\,Bo^{0.1285}$ $E = 0.062\phi^2 + 0.938$
Saitoh et al. [28]	$h_{tp} = Sh_{pool} + Eh_l$ $E = 1+\frac{X_{tt}^{-1.05}}{1+We_g^{-0.4}}$ $We_g = \frac{G^2x^2D_h}{\sigma\rho_g}$ $S = \frac{1}{1+0.4\left(\mathrm{Re}_{tp}\times10^{-4}\right)^{1.4}}$ $h_{pool} = 207\frac{k_l}{d_b}\left(\frac{qd_b}{k_lT_l}\right)^{0.745}\left(\frac{\rho_g}{\rho_l}\right)^{0.581}\mathrm{Pr}_l^{0.533}$ $d_b = 0.51\left[\frac{2\sigma}{g\left(\rho_l-\rho_g\right)}\right]^{0.5}$
Sun and Mishima [33]	$h_{tp} = \frac{6\,\mathrm{Re}_{lo}^{1.05}\,Bo^{0.54}}{We_l^{0.191}\left(\rho_l/\rho_g\right)^{0.142}}\,\frac{k_l}{D_h}$
Li and Wu [34]	$h_{tp} = 334Bo^{0.3}\left(Bd\,\mathrm{Re}_l^{0.36}\right)^{0.4}\,\frac{k_l}{D_h}$

Kew and Cornwell [30] found that heat transfer and flow characteristics in microscale channels were significantly different from those observed in macroscale channels. They modified the Lazarek and Black [29] correlation in terms of vapor quality x.

Tran et al. [31] recognized that the dominant heat transfer mechanism is nucleate boiling rather than convective boiling in small diameter channels. They proposed a heat transfer correlation with the Reynolds number being replaced with the liquid Weber number We_l to eliminate the viscous effect in favour of the surface tension effect.

Warrier et al. [32] developed a different saturated flow boiling heat transfer correlation which is only dependent on the boiling number *Bo* and vapour quality *x*.

Zhang et al. [26] modified the Chen [20] correlation to predict flow boiling heat transfer in small diameter tubes. The poop boiling correlation by Foster and Zuber [24] is kept unchanged for the nucleate boiling heat transfer component. The boiling suppression factor *S* in the Chen correlation is also kept unchanged. To determine the convection evaporation enhancement factor *E* and the single-phase heat transfer coefficient, vapor and liquid flow conditions (laminar or turbulent) are taken into account in their method.

Pamitran et al. [27] developed a heat transfer correlation for microscale channels based on the Chen [20] correlation. They proposed new calculation methods for nulceate boiling suppression factor *S* and convection evaporation enhancement factor *E* according to their expeirmental data.

Saitoh et al. [28] proposed a prediction method which covers both macro- and micro-scale flow boiling heat transfer by modifying the Chen correlation [20]. They considered the tube diameter effect on the heat transfer coefficient. Because the fluid flow conditions more strongly affect convective evaporation than nucleate boiling, the convective boiling enhancement factor *S* is expressed as a function of gas phase Weber number We_g and Matinelli number X_{tt}. The Stephen and Abdelsalam [35] nucleate pool boiling correlation is used in their method.

Sun and Mishima [33] found that the heat transfer coefficient is more dependent on the Weber number We_l than the vapor quality *x*, which is similar to the results obtained by Lazarek and Black [29]. They proposed a new flow boiling heat transfer correlation by modifying the Lazarek and Black correlation, taking the effect of Weber number into account in their correlation.

Li and Wu [34] found that the two-phase Nusselt number is greatly dependent on the Bond number *Bd*. Thus, they proposed a new flow boiling heat transfer correlation combining the effects of the Bond number *Bd* and liquid phase Reynolds number Re_l.

4. Comparison of the Database to the Selected Heat Transfer Correlations

The database was compared to the 12 selected flow boiling heat transfer correlation. The statistical analysis is based on relative error ξ_i (the percentage of predicted points within ±30%):

$$\xi_i = \frac{Predicted - Measured}{Measured} \tag{21}$$

Table 3 summarized the statistical analysis of the prediction results by the selected flow boiling heat transfer correlations for each individual fluid and all fluids according to the relative error ξ_i within ±30%. For all the fluids together, the correlations of Saitoh et al. [28], Sun and Mishima [33] and Li and Wu [34] are among the best methods which predict 60%, 56.9% and 55.3% of the database within ±30%, respectively. However, these predictions are still not satisfactory as the predicted percentages are apparently low. In fact, no methods can predict the entire database well according to the comparative results.

Table 3. Statistical analysis of the prediction results by the selected heat transfer correlations for each individual fluid and all fluids (relative error ξ_i within ±30%)

Correlation	R134a	R245fa	R22	R410A	All fluids
Chen [20]	40.5%	72.7%	41.6%	4.6%	37.8%
Gungor and Winterton [21]	64.0%	71.1%	25.7%	10.8%	54.3%
Liu and Winterton [22]	28.5%	39.7%	70.8%	5.8%	28.7%
Lazark and Black [29]	48.6%	0.8%	69.0%	52.5%	46.8%
Kew and Cornwell [30]	55.5%	0.8%	83.2%	56.8%	53.3%
Tran et al. [31]	18.2%	0.8%	24.8%	21.6%	17.8%
Warrier et al. [32]	12.8%	5.8%	31.9%	6.6%	12.6%
Zhang et al. [26]	41.5%	66.1%	42.5%	3.9%	37.9%
Pamitran et al. [27]	48.9%	43.8%	0.0%	48.3%	45.3%
Saitoh et al. [28]	63.0%	65.3%	87.6%	3.1%	60%
Sun and Mishima [33]	67.4%	0.8%	76.1%	27.0%	56.9%
Li andWu [34]	66.0%	76.0%	35.4%	1.2%	55.3%

For R134a data, the best two prediction methods are the correlations of Sun and Mishima [33] and Li and Wu [34], which predict 67.4% and 66% of the experimental data respectively. Figure 1 shows the comparisons of these two heat transfer correlations to the R134a experimental data. The predictions by these two methods are very close but still not satisfactory.

For R245fa, the best two prediction methods are the correlations of Li and Wu [34] and Chen [20], which predict 76% and 72.7% of the experimental data respectively. Figure 2 shows the comparisons of these two methods to the R245a experimental data. The predictions by these two methods are very close and acceptable but may be further improved. It seems that the Chen correlation for macroscale flow boiling can predict the microscale flow boiling of R245fa. This is possibly because the Chen heat transfer mechanisms are also dominant in flow boiling microscale channels.

For R22, the best two prediction methods are the correlations of Saitoh et al. [28] and Kew and Cornwell [30] which predict 87.6% and 83.2% of the experimental data. Figure 3 shows the comparisons of these two methods to the R22a experimental data. These two predictions are actually satisfactory. It should be noticed that the two prediction methods were developed based on different heat transfer mechanisms. The Saitoh et al. correlation is based on the Chen heat transfer mechanisms while the Kew and Cornwell correlation is based on the Lazark and Black heat transfer mechanism. It is not easy to say which the dominant mechanisms are. However, the Saitoh et al. correlation gives better prediction than the Kew and Cornwell. Similar to the prediction for R245fa by the Chen correlation, the possible heat transfer mechanisms underlying flow boiling in microscale channels might be both nucleate boiling heat transfer and convective boiling heat transfer, depending on heat fluxes, mass flow velocities and vapor qualities. However, further research is needed to fully understand the heat transfer mechanisms relating to their corresponding flow regimes.

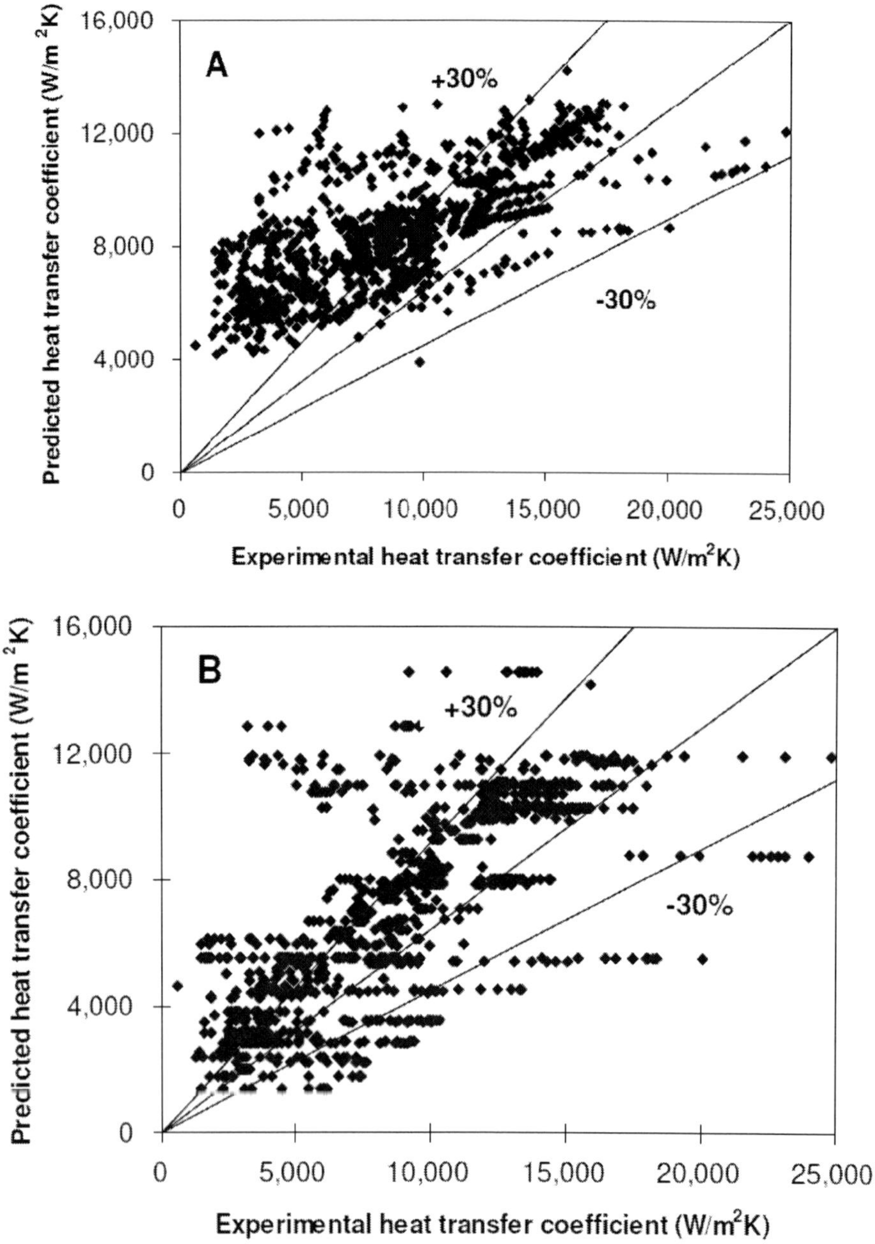

Figure 1. Comparative results of R134a data to the best two correlations: (A) Li and Wu [34]; (B) Sun and Mishima [33].

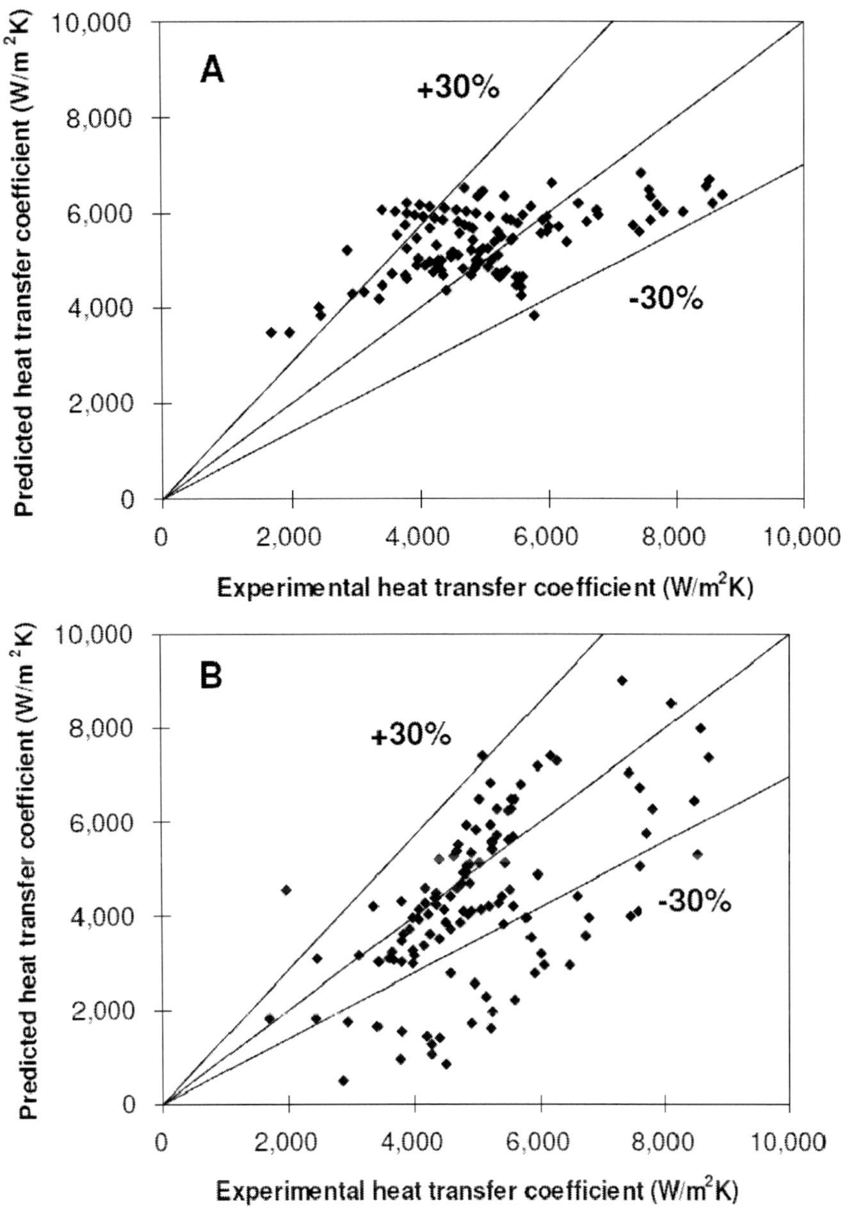

Figure 2. Comparative results of R245fa data to the best two correlations: (A) Li and Wu [34]; (B) Chen [20].

For R410A, the best two prediction methods are the correlation of the Kew and Cronwell [30] and Lazark and Black [29] which predict 56.8% and 52.5% of the experimental data. Figure 4 shows the comparisons of these two methods to the R410A experimental data. The predictions are still not satisfactory. No methods predict the R410A data well. This is possibly due to the effect of non-equilibrium boiling effect as R410A is a binary mixture refrigerant. Further investigation should be done to clarify the heat transfer mechanisms of mixture boiling in microscale channels.

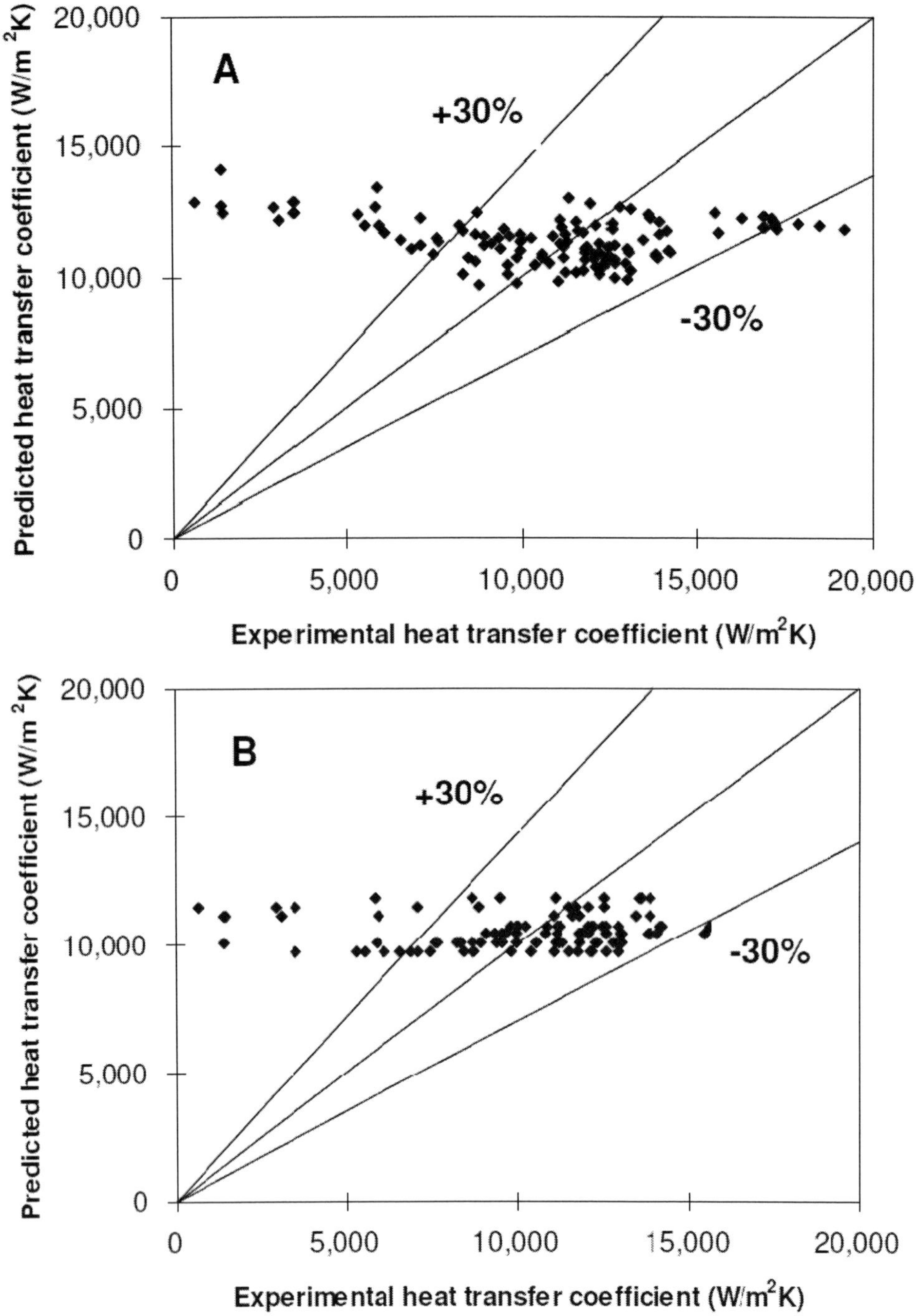

Figure 4. Comparative results of R410A data to the best two correlations: (A) Kew and Cornwell [30]; (B) Lazarek and Black [29].

Overall, no method predicts the entire database well. For R410A and R134a, no method can predict these data over 70% within ±30%. For R245fa, two methods can predict the data over 70% within ±30%. For R22, two methods can predict the data over 80% within ±30%, which is actually quite well.

For macroscale channels, the predictions are generally satisfactory, for example, Cheng and Chen [19] evaluated several correlations including the correlations of Chen [20], Gungor and Winterton [21], Liu and Winterton [22] etc., and found thethese macroscale heat transfer correlations were able to predict their kerosene flow boiling heat transfer data. However, for microscale channels, as it can be seen from the comparisons that even with those methods developed for mciroscale channels, none of the selected prediction methods can predict the entire data of all fluids. Physically speaking, there does not appear to be a simple explanation for this. It is therefore essential to develop a new flow boiling heat transfer model according to the corresponding mechanisms or flow regimes. However, the experimental data in the literature are quite contradictory from one to another study. It is also essential to investigate the heat transfer mechanisms underlying flow boiling in microscale channels which are still not well understood so far. In the available stuides, some researchers adopted the two conventional heat transfer mechanisms in their correlations such as Zhang et al. [26], Pamitran et al. [28] and saitoh et al. [28] while some researchers found different heat transfer mechanisms in their studies and developed their correlations based on these mechanisms such as Lazarek and Black [29], Kew and Cornwell [30], Tran et al. [31], warrier et al. [32], Sun and Mishima [33] and Li and Wu [34]. However, the different heat transfer mechanisms observed by different researchers may be dependent on their test condition range. Flow boiling heat transfer may be strongly affected by heat fluxes, mass velocities, vapor qualities and channel sizes. It is not simple to say which mechanisms are dominant or not generally. Apparently, systematic studies on the heat transfer mechanisms by considering the effects of various parameters are not yet available. Therefore, further efforts should be made to understand the heat transfer mechanisms and more accurate measurements are necessary to assure that heat transfer coefficients are obtained for a wide range of test conditions together with their corresponding flow regimes. Flow regimes observations are necessary so as to understand the physical phenomena and heat transfer mechanisms related to the flow regimes. It is also very important to develop general flow boiling heat transfer methods for microscale channels, which is based on the flow regimes, for better prediction of flow boiling heat transfer coefficients.

CONCLUSION

A database containing 1766 experimental data points of flow boiling heat transfer coefficients of R134a, R22, R410A and R245fa was used to evaluate 12 flow boiling heat transfer correlations including 3 correlations for macroscale channels and 9 correlations for microcale channels. The following conclusions have been obtained through the comaprisons and analysis in the present study:

1) No method of the selected flow boiling heat transfer correlations predicts the entire database well.
2) For R410A and R134a flow boiling heat transfer data, no method can predict these data over 70% within ±30%. The predictions are not satisfactory.

3) For R245fa flow boiling heat transfer data, two methods can predict the data over 70% within ±30%. However, the predictions should be further improved.
4) For R22 flow boiling heat transfer data, two methods can predict the data over 80% within ±30%. The predictions are satisfactory, with the Saitoh et al. correlation predicting 87.6% of the data.
5) Flow boiling heat transfer in microscale channels may be strongly affected by heat fluxes, mass velocities, vapor qualities and channel sizes. It is not simple to say which mechanisms are dominant generally. Apparently, systematic studies on the heat transfer mechanisms by considering the effects of various parameters are not yet available.
6) Further efforts should be made to understand the heat transfer mechanisms in microscale channels and more accurate measurements are necessary to assure that heat transfer coefficients together with their flow regimes are obtained in the future.
7) It is also very important to develop general flow boiling heat transfer methods for microscale channels, with flow regimes considered, for better prediction of flow boiling heat transfer in microscale channels.

REFERENCES

[1] L. Cheng, Microscale and nanoscale thermal and fluid transport phenomena: rapidly developing research fields, *Int. J. Microscale Nanoscale Thermal Fluid Transport Phenomena* 1(1) 2010 (in press).

[2] G.P. Celata, Microscale heat transfer in single- and two-phase flows: scaling, stability and transition, *Int. J. Microscale Nanoscale Thermal Fluid Transport Phenomena* 1(1) 2010 (in press).

[3] L. Cheng, D. Mewes, Review of two-phase flow and flow boiling of mixtures in small and mini channels, *Int. J. Multiphase Flow* 32 (2006) 183-207.

[4] S.G. Kanlikar, Fundamental issues related to flow boiling in minichannels and microchannels, *Exp. Therm. Fluid Sci.* 26 (2002) 389-407.

[5] J.R. Thome, The new frontier in heat transfer: microscale and nanoscale technologies, *Heat Transfer Eng.* 27(9) (2006) 1-3.

[6] G. Hetsroni, A. Mosyak, E. Pogrebnyak, L.P. Yarin, Fluid flow in micro-channels, *Int. J. Heat Mass Transfer* 48 (2005) 1982-1998.

[7] J.R. Thome, Boiling in microchannels: a review of experiment and theory, *Int. J. Heat Fluid flow* 25 (2004) 128-139.

[8] J.R. Thome, State-of-the art overview of boiling and two-phase flows in microschannels, *Heat Transfer Eng.* 27(9) (2006) 4-19.

[9] L. Cheng, G. Ribatski, J.R. Thome, Gas-liquid two-phase flow patterns and flow pattern maps: fundamentals and applications, *ASME Appl. Mech. Rev.* 61 (2008) 050802-1-050802-28.

[10] L. Cheng, D. Mewes, A. Luke, Boiling phenomena with surfactants and polymeric additives: a state-of-the-art review, *Int. J. Heat Mass Transfer* 50 (2007) 2744-2771.

[11] L. Cheng, E. P. Bandarra Filho, J.R. Thome, Nanofluid two-phase flow and thermal physics: a new research frontier of nanotechnology and its challenges, *J. Nanosci. Nanotech.* 8 (2008) 3315-3332.

[12] L. Cheng, Nanofluid heat transfer technologies, Recent Patents on Engineering 3(1) (2009) 1-7.

[13] L. Cheng, J.R. Thome, Cooling of microprocessors using flow boiling of CO_2 in a micro-evaporator: preliminary analysis and performance comparison, *Appl. Therm. Eng.* 29 (2009) 2426-2432.

[14] L. Cheng, G. Ribatski, J. Moreno Quibén, J.R. Thome, New prediction methods for CO_2 evaporation inside tubes: Part I-A two-phase flow pattern map and a flow pattern based phenomenological model for two-phase flow frictional pressure drops, *Int. J. Heat and Mass transfer* 51 (2008) 111-124.

[15] L. Cheng, G. Ribatski, J. Moreno Quibén, J.R. Thome, New prediction methods for CO_2 evaporation inside tubes: Part II-An updated general flow boiling heat transfer model based on flow patterns, *Int. J. Heat and Mass transfer* 51 (2008) 125-135.

[16] J. Moreno Quibén, L. Cheng, R.J. da Silva Lima, J.R. Thome, Flow boiling in horizontal flattened tubes: Part I — Two-phase frictional pressure drop results and model, *Int. J. Heat Mass Transfer*, 52 (2009) 3634-3644.

[17] J. Moreno Quibén, L. Cheng, R.J. da Silva Lima, J.R. Thome, Flow boiling in horizontal flattened tubes: Part II — flow boiling heat transfer results and model, *Int. J. Heat Mass Transfer*, 52 (2009) 3645-3653.

[18] S. Saitoh, H. Daiguji, E. Hihara, Effect of tube diameter on boiling heat transfer of R-134a in horizontal small-diameter tubes, *Int. J. Heat Mass Transfer* 48 (2005) 4973–4984.

[19] L. Cheng and T. Chen, Comparison of six typical correlations for upward flow boiling heat transfer with kerosene in a vertical smooth tube, *Heat Transfer Engineering* 21(5) (2000) 27-34.

[20] J.C. Chen, Correlation for boiling heat transfer to saturated fluids in convective flow, *IandEC Process Des. Dev* 5 (1966) 322–329.

[21] K.E. Gungor and R.H.S. Winterton, A general correlation for flow boiling in tubes and annuli, *Int. J. Heat Mass Transfer* 29 (1986) 351–358.

[22] Z. Liu and R.H.S. Winterton, A general correlation for saturated and subcooled flow boiling in tubes and annuli based on a nucleate pool boiling equation, *Int. J. Heat Mass* Transfer 34 (1991) 2759–2766.

[23] F.W. Dittus, L.M.K. Boelter, Heat transfer in automobile radiators of the tubular type, *Pub. Eng.* vol. 2 p. 443, University of California, Berkeley (1930).

[24] H.K. Forster, N. Zuber, Dynamics of vapor bubbles and boiling heat transfer, *J. AIChE.* 1 (1955) 531-535.

[25] M.G. Cooper, Saturation nucleate pool boiling: A simple correlation, *Int. Chem. Eng. Symp. Ser.* 86 (1984) 785-793.

[26] W. Zhang, T. Hibiki and K. Mishima, Correlation for flow boiling heat transfer in mini-channels, *Int. J. Heat and Mass Transfer* 47 (2004) 5749– 5763.

[27] A.S. Pamitran, K. Choi, J.Oh and H. Oh, Forced convective boiling heat transfer of R-410A in horizontal minichannels, *Int. J. of Refrigeration* 30 (2007) 155-165.

[28] S. Saitoh, H. Daiguji and H. Hihara, Correlation for boiling heat transfer of R-134a in horizontal tubes including effect of tube diameter, *Int. J. Heat Mass Transfer* 50 (2007) 5215–5225.

[29] G.M. Lazarek and S.H. Black, Evaporative heat transfer, pressure drop and critical heat flux in a small vertical tube with R-113, *Int. J. Heat Mass Transfer* 25 (7) (1982) 945-960.

[30] P.A. Kew and K. Cornwell, Correlations for the prediction of boiling heat transfer in small-diameter channels, *Appl. Therm. Eng.*, 17 (1997) 705-715.

[31] T.N. Tran and M.W. Wambsganss, D.M. France, Small circular- and rectangular-channel boiling with two refrigerants, *Int. J. Multiphase Flow* 22 (1996) 485–498.

[32] G.R. Warrier and V.K. Dhir, L.A. Momoda, Heat transfer and pressure drop in marrow rectangular channels, *Exp. Ther. Fluid Sci.* 26 (2002) 53-64.

[33] L. Sun, K. Mishima, An evaluation of prediction methods for saturated methods for saturated flow boiling heat transfer, *Int. J. Heat Mass Transfer* 52 (2009) 5323-5329.

[34] W. Li, Z. Wu, A general correlation for evaporative heat transfer in micro/mini-channels, *Int. J. Heat Mass Transfer* (2010) doi:10.1016/j.ijheatmasstransfer.2010.01.012 C.B.

[35] K. Stephen, M. Abdelsalam, Heat-transfer correlation for natural convection boiling, Int. J. Heat Mass Transfer 23 (1980) 73-87. C. Vlasie, H. Macchi, J. Guilpart, B. Agostini, Flow boiling in small diameter channels, *Int. J. refrigeration* 27 (2004) 191-201.

[36] B. Agostini and A. Bontemps, Vertical flow boiling of refrigerant R134a in small channels, *Int. J. Heat and Fluid Flow* 26 (2005) 296-306.

[37] K. Choi, A.S. Pamitran, C.Y. Oh and J.T. Oh, Boiling heat transfer of R-22, R-134a, and CO_2 in horizontal smooth minichannels, *Int. J. Refrigeration* 30 (2007) 1336-1346.

[38] S.S. Bertsch, E.A. Groll and S.V. Garimella, Effects of heat flux, mass flux, vapor quality, and saturated temperature on flow boiling heat transfer in microchannels, *Int. J. Multiphase Flow* 35 (2009) 142-154.

[39] S. In and S. Jeong, Flow boiling heat transfer characteristics of R123 and R134a in a micro-channel, *Int. J. of Multiphase Flow* 35 (2009) 987-1000.

[40] D. Shiferaw, T.G. Karayiannis and D.B.R. Kenning, Flow boiling in a 1.1 mm tube with R134a: Experimental results and comparison with model, *Int. J. Thermal Sci.* 48 (2009) 331-341.

[41] Tibirica and G. Ribatski, Flow boiling heat transfer of R134a and R245fa in a 2.3 mm tube, *Int. J. Heat Mass Transfer* 53 (2010) 2459-2468.

[42] C. Vlasie, H. Macchi, J. Guilpart, B. Agostini, Flow boiling in small diameter channels, *Int. J. Refrigeration* 27 (2004) 191-201.

[43] L. Wang, M. Chen and M. Groll, Flow boiling heat transfer characteristics of R134a in a horizontal mini tube, *J. Chem. Eng. Data* 54 (2009) 2638-2645.

[44] R. Yun, J.H. Heo and Y. Kim, Evaporative heat transfer and pressure drop of R410A in microchannels, *Int. J. Refrigeration* 29 (2006) 92-100.

In: Frontier Research in Microscale and Nanoscale... ISBN 978-1-61324-665-8
Editor: Lixin Cheng

Chapter 21

MEMS THERMAL CONTROL LABORATORY DESCRIPTION

Ping-Hei Chen*
Mechanical Engineering, National Taiwan University, Taiwan

GENERAL DESCRIPTION

MEMS thermal control Lab. has a research team that is consisted of two post-doctors, one research assistant, seven ph. D. students, and five master students. In this laboratory, the research team focuses on four research topics, a finger size smart sensor, innovative applications of microfluidic devices, surface modification of micro/nano hierarchical structures, and a disposable DNA sensor. Most fabrication process of devices or micro/nano structures are either out sourced to a fabrication company or performed in MEMS laboratory or Nanotechnology center in National Taiwan University. Our laboratory develops the testing instruments to measure the performance of developed sensors, devices, or micro/nano structures.

RESEARCH TOPICS AND FACILITIES IN THE LABORATORY

1. A Finger Size Smart Sensor

1.1 General Description

The smart sensor we try to develop would have the capability to scavenge energy from different energy sources in environment to provide the electricity for a microprocessor, sensors, and radio frequency (RF) module. Since the energy provided from environment is

* Address correspondence to Prof Ping-Hei Chen, No. 1, Sec. 4, Roosevelt Road, Taipei, 10617 Taiwan (R.O.C), Rm. 728, Engineering Bldg., Department of Mechanical Engineering, National Taiwan University, E-mail: phchen@ntu.edu.tw

sparse and random, it becomes a problem to provide stable electricity to the microprocessor, sensors, and RF module. The smart sensor developed in the current stage has a length of 10 cm, a width of 8 cm, and a thickness of 1.5 cm. It requires some major efforts to reduce such a smart sensor into a finger size. In this study, we successfully developed a smart humidity diagnosis system for proton exchange membrane (PEM) fuel cell. This smart humidity sensor will finally be driven by fuel cell energy source without any battery. The proposed smart sensing system integrates the flexible humidity sensing component, fuel cell powering component and RF module including microprocessor, AD/DA converter and antenna. Figure 1 shows the schematic diagram of a smart humidity diagnosis system for PEM fuel cell, and the photographs of the flexible humidity sensor and RF communication module is shown in Figure 1(a) and (b). The dimension of the RF module including microprocessor, AD/DA converter and antenna is around 5 x 3 cm^2; the sensing area and the thickness of the flexible humidity sensor are 210 x 570 μm^2 and 15 μm^2, respectively.

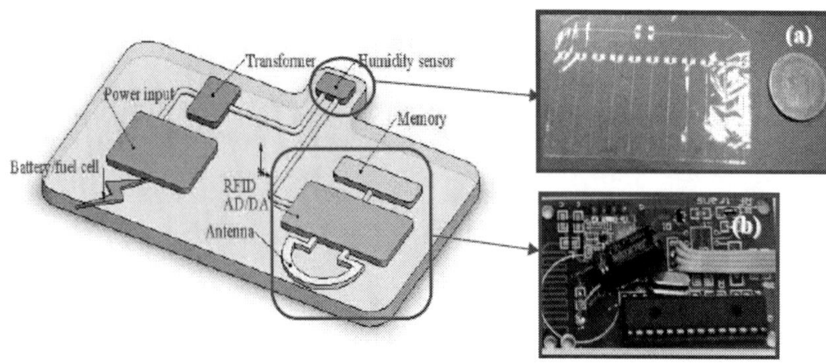

Figure 1. The schematic diagram of a smart humidity diagnosis system for the proton exchange membrane fuel cell: (a) Photograph of the flexible humidity sensors; (b) The diagram of the wireless RF communication module.

1.2. Experimental Facilities

In fuel cell testing, optimization control between electronic load and fuel temperature, humidity, pressure, flow rate is essentially important in order to obtain precise and reliable results. Therefore, our wireless smart sensor is initially applied to the humidity diagnosis of PEM fuel cell. Figure 2 shows the measurement equipment for the wireless humidity sensing device that consists of a flexible humidity sensor and RF communication module.

The proposed flexible humidity sensor is placed into the PEM fuel cell module for in-situ humidity diagnosis, the dimension of PEM fuel cell module is about 4 x 3 cm^2. The PEM fuel cell testing equipment includes humidification system, gas supplied cylinders, gas flow controller, temperature controller and auto humidified liquid supplier. It is used for a single cell and low watts PEM fuel cell testing, and the dimension of the fuel cell testing equipment is 46 x 28 x 48 cm^3.

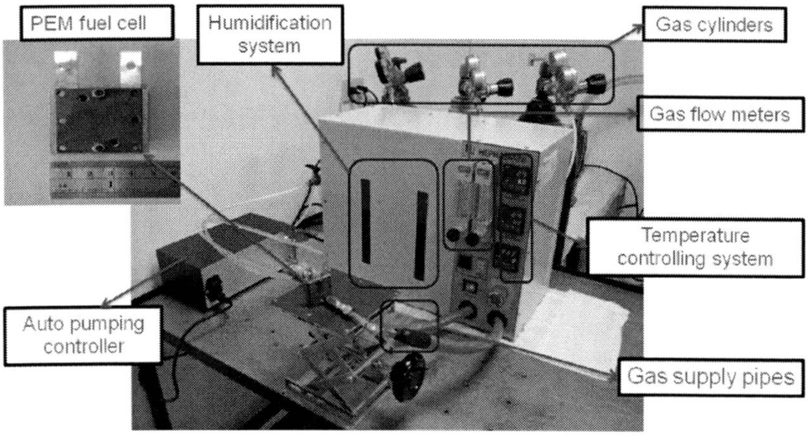

Figure 2. Illustration of the PEM fuel cell testing equipment for the in-situ humidity sensing system.

2. Innovative Applications of Microfluidic Devices

2.1 Introduction

The goal in our laboratory for developing innovative applications of microfluidic devices is to design standard modular components, such as pumps, valves, reservoirs, etc. Different modules can be easily assembled as a complete microfluidic device that can perform specific function, such as separation of cell or particle sieving. At current stage, we have successfully developed the prototypes of the pump, valve, and reservoir. Figure 3 (a) presents the geometry of the valve with the open bottom. We prepare the clay mold first and then use casting method to pour PDMS in the container to fabricate this chamber part.

Figure 3 (b) and (c) show the sizes of the reed and the base plate. Once three individual pieces are ready, we cover the base with a thin layer of wet PDMS before assembling three pieces together. Figure 3 (d) presents the way of assembling and Figure 3 (e) shows the photograph of one sample. The pump consists of two parts: chamber and cover plate, shown in Figure 4 (a) and (b). The clay molds with different volume are prepared before casting. Again, a thin layer of wet PDMS is uniformly wiped on the cover plate to glue two parts together. Figure 4 (c) and (d) shows the assembling of two parts and the photograph of the pump, respectively. The reservoir is also fabricated by the casting method.

A clay mold with several designed channels is prepared before casting. Figure 5 (a) shows the schematic diagram of casting of the reservoir, Figure 5 (b) shows the reservoir photo of the mold, and Figure 5 (c) shows the photo of one sample of the reservoir. Figure 6 shows the schematic diagram of a microfluidic system composed by the modular components. People can confirm their concept of design by using this way to avoid complicated procedures to get a prototype device with unique functions.

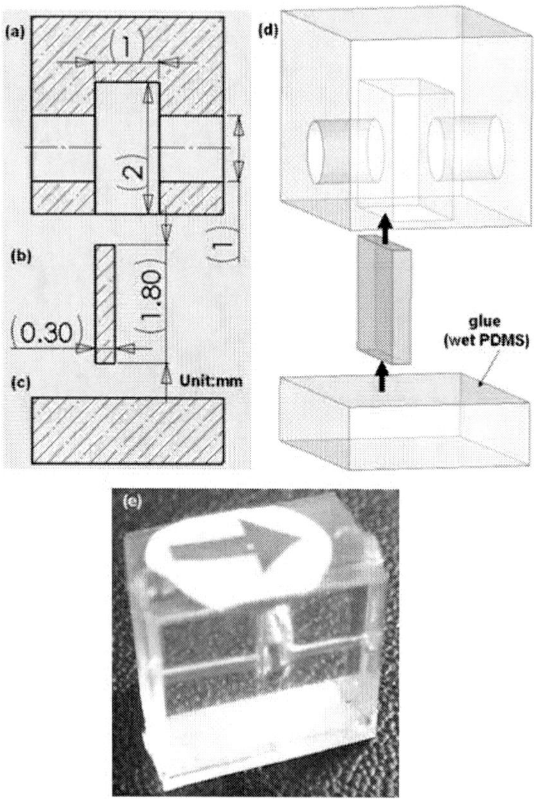

Figure 3. Fabrication of uni-directional valve (a)-(c) Dimensions of three parts (chamber, reed, and base, respectively) before assembling a uni-directional valve. (d) A thin layer of wet PDMS is covered uniformly on the base surface to glue three parts together. (e) The photograph of one sample.

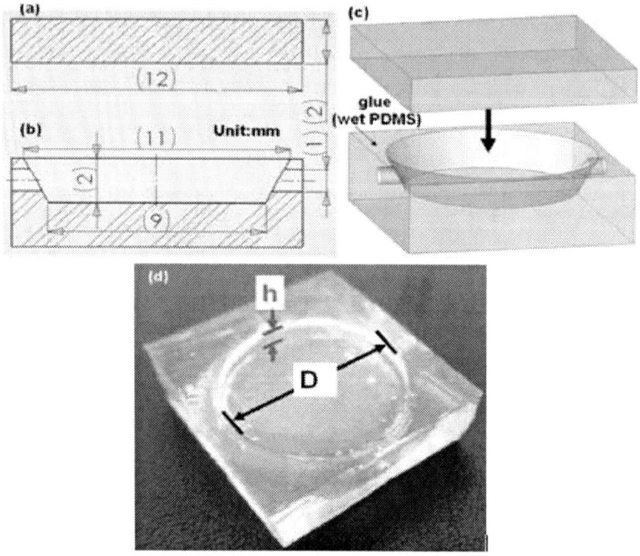

Figure 4. Fabrication of pump (a) and (b) The dimensions of chamber and cover plate. (c) Assembling of two parts (d) Photograph of one sample (D=average diameter, h=height).

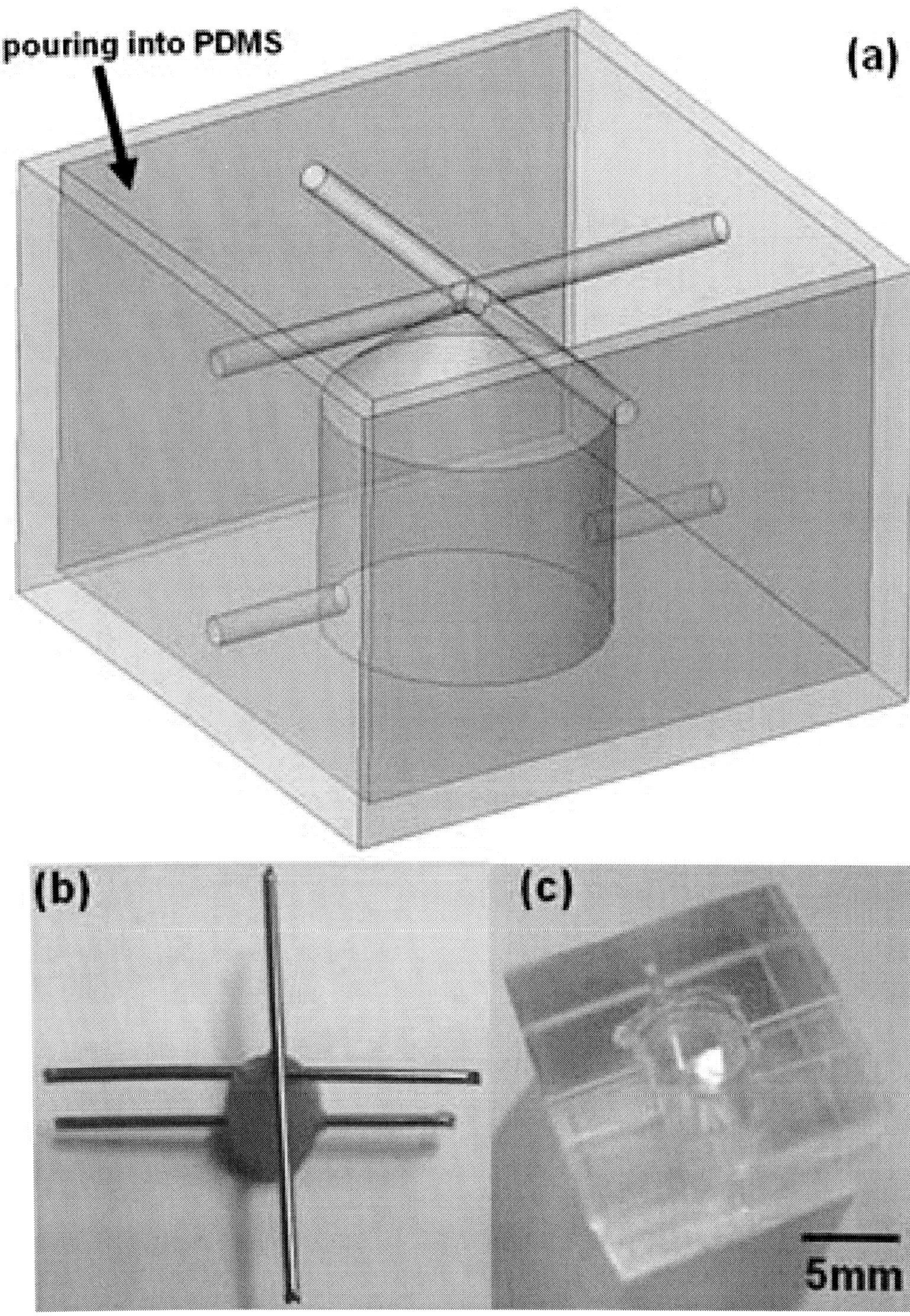

Figure 5. Fabrication of Reservoir (a) Schematic diagram of casting of a reservoir (b) Photograph of the mold. (c) Photograph of one sample of reservoir.

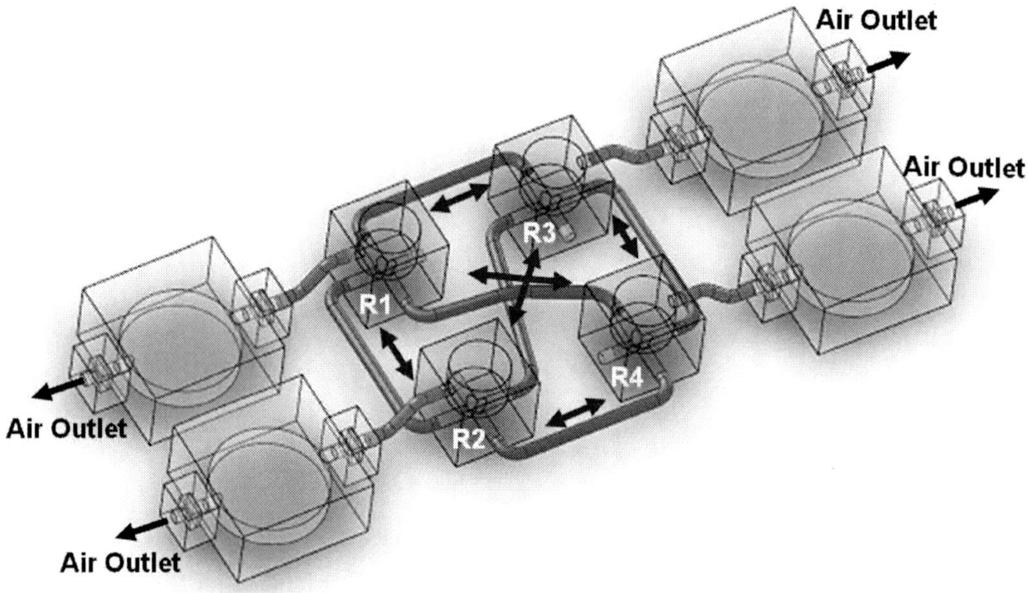

Figure 6. Example of a microfluidic system assembled by the modular components. Each pump (P1-P4) controls one reservoir (R1-R4). By proper operation of the pumps, people can convey the fluids from one reservoir to another for their own purpose.

2.2. Experimental Facilities

The facilities for development of the microfluidic devices include the syringe pumps and the optical system (CCD camera, optical lens, and light source), shown in Figure 7.

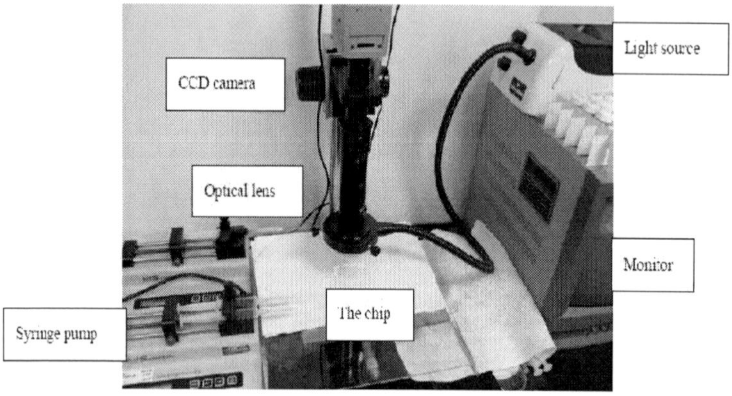

Figure 7. Photograph of the facilities for the operation of the microfluidic devices.

3. Surface Modification with Micro/Nano Structures

3.1. Introduction

Surface characteristic of substrate can be significantly changed by surface modification with micro/nano structures. The aim of this research topic is to obtain a tunable surface wettability by external energy sources, for instance, optical, thermal, electrical excitation. At the current stage, the substrate surface is modified by two different methods, self-assembled approach and laser interference lithography in our laboratory. The self-assembled silica particles method produces superhydrophobic surfaces with micro/nano-silica on glass surface. The superhydrophobic silica thin films are produced by dipping the glass substrate into the mixture of silane couple (METES/DECYS) agents, where silica particles are suspended. The silane couple agent is used for bonding the inorganic substrate such as glass and an organic material such as silica particles. The modified surface can achieve very high contact angle (162.8°). Figure 8 shows the SEM image of the modified surface and the contact angle measurement.

Figure 8. SEM micrographs of modified glass surface and the contact angle measurement (SiO_2/METES, CA 162.8).

The laser interference lithography generates a patterned micro/nano structures on the substrate surface. The contact angle of water droplet on the patterned surface can be significantly varied by the spacing between neighboring nanostructures.

3.2. Experimental Facilities

Figure 9 shows the photograph of the dip coater. To use this equipment, first put the sol-gel solution inside dipping tank, and then fix the sample (glass substrate) on the sample holder. Then, the sample displacement is controlled by control system through a screw. The control system includes a main switch, a speed tuner, and two directional controllers. The advantage of this device is that we can do coating for large glass substrates (35 cm x 35 cm).

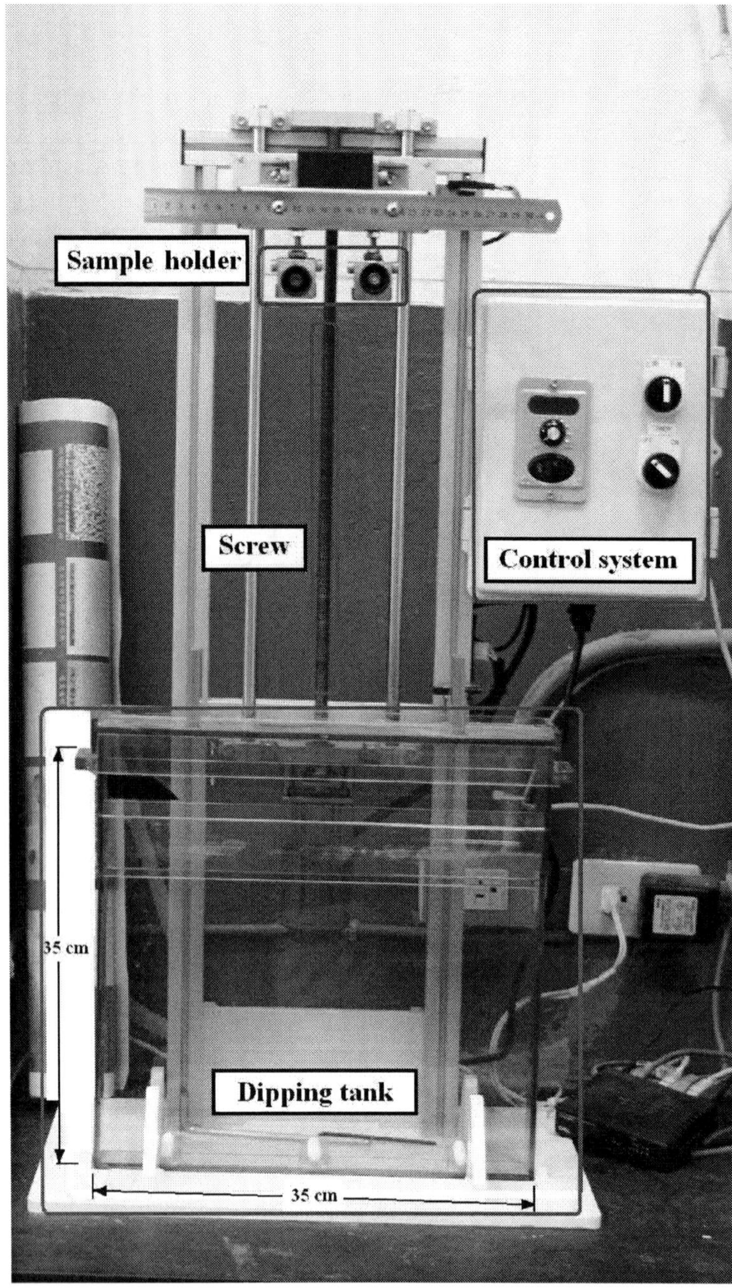

Figure 9. Photograph of dip coater. It consists of a dipping tank on the bottom, a sample holder on the top, and the corresponding control system. The sample holder can move through the screw by the control system.

4. Disposable DNA Detection Platform

4.1. Introduction

To perform DNA detections in a point-of-care device, our laboratory is developing a lab-in-a-capillary-tube for performing following functions, DNA extraction, DNA amplification, and DNA detection. Typical DNA amplification requires thermal cycling roughly from 55°C to 95°C. At the current stage, our laboratory develops a DNA detection platform that is powered with batteries but requires no temperature control system. The capillary tube is heated from below and cooled by the surrounding air freely. Our platform has successfully amplified amplicons with lengths from 87 to 500 bps with different viral genomes. The total time needed from the mixing of DNA and agents to the completion of detection is only 50 minutes (Figure 10). If the battery of our DNA sensor can be further removed, our developed DNA sensor can be disposable.

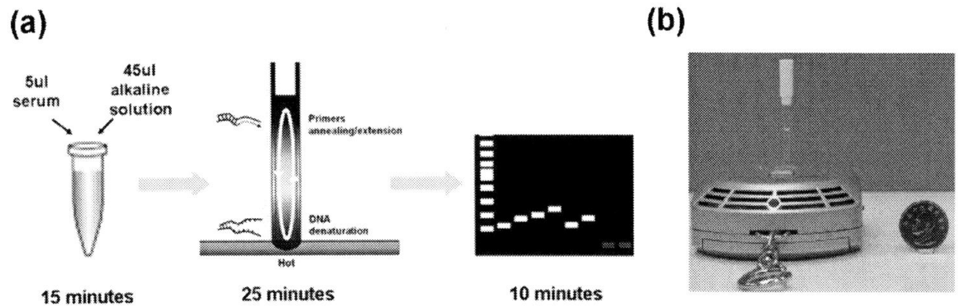

Figure 10. (a) Our disposable DNA sensor can detect target DNA in 50 (15+25+10) minutes. (b) The prototype of disposable DNA sensor. (Dimension: 9.5cm X 7cm X 3cm; Weight: 205g with 3 AAA batteries).

4.2. Experimental Facilities

Figure 11 shows the main traditional facilities used in this topic for validation purposes. It includes an electrophoresis image capture system and a thermal cycler.

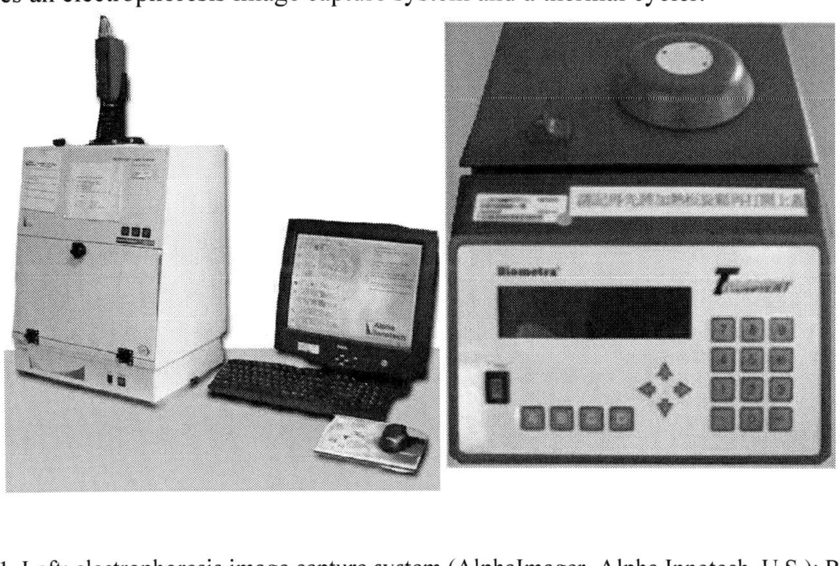

Figure 11. Left: electrophoresis image capture system (AlphaImager, Alpha Innotech, U.S.); Right: thermal cycler (TGradient, Biometra, Germany).

REFERENCES

1.3. Representative Publications

Yang, C. H., Chien, J.-H., Wang, B. Y., Chen, P. H., and Lee, D. S., 2008, "Development of a Flexible Surface Wetness Sensor Using a RFID Technique" *Biomedical Microdevices*, Vol. 10, Feb., pp. 47-54.

Chien, J.-H., Chen, P. H., Kuo, L. S., Lin, C. S., and Wang H., 2007, "Protein Detection Using a Radio Frequency Biosensor with Gold Nanoparticles," *Applied Physics Letters,* Vol. 91, No. 14, Art. No. 143901.

Chang, T. L., Tsai, C. Y., Sun, C. C., Chen, C. C., Kuo, L. S., and Chen, P. H., 2007, "Ultrasensitive Electrical Detection of Protein Using Nanogap Electrodes and Nanoparticle-Based DNA Amplification," *Biosensors and Bioelectronics* Vol. 22, pp. 3139-3145.

2.3. Representative Publications

Tsai, C. H., Liu, D. S., Kuo, L. S., and Chen, P. H., "Rapid mixing between ferro-nanofluid and water in a semi-active Y-Type micromixer," *Sensors and Actuators A,* Vol. 153, pp. 267-273.

Liou, D. S., Kuo, L. S., Chen, P. H., Chang, T. L., and Yang, C. T., 2009, "Axial particle displacements in fluid slugs after passing a simple serpentiform microchannel," *Microfluidic and Nanofluidic,* Vol. 7, pp. 145-148.

Chang, C. W., Cheng, Y. T. Tsai, C. Y., Chien, J. H., Wang, P. Y, and Chen, P. H., 2007, "Periodic Flow Patterns of the Magnetic Fluid in Microchannel," *Journal of Magnetism and Magnetic Materials,* Vol. 310, pp. 2844-2846.

PING-HEI CHEN

Professor Ping-Hei Chen was born in Taichung County, Taiwan on Aug. 31th, 1958. He received bachelor degree in mechanical engineering from National Taiwan University in 1980. After completing his military service, he went to University of Minnesota for his graduate study. He received his master and Ph. D. degree in mechanical engineering from University of Minnesota in 1984 and 1988, respectively.

In 1988, he joined the Dept. of Mechanical Engineering of National Taiwan University as an associate professor. He was promoted to full professor in 1996. After his promotion, he is then chosen to serve as the vice-chairman of the Department of Mechanical Engineering of National Taiwan University. Then, he had served as the chairman of the Department of Mechanical Engineering of National Taiwan University from 1998 to 2001. Due to his excellent achievement in sensors-related researches, he was awarded as a distinguished professor by the National Taiwan University in 2008. He is serving as the program director of thermal, fluid, and energy division of NSC. His major research areas are in MEMS, Biomedical devices, nanotechnology, energy harvesting chips, highly efficient energy systems, and sensors.

Prof. Ping-Hei Chen is a member of ASME, CSME, CIE, CSNM, CSAAE, and CSAT. He was awarded with the excellent engineering professor award by CSME in 2002, the excellent engineering professor award by CIE in 2006, the distinguished research award by NSC in 2008, ASME fellow in 2009. He has published more than 100 journal papers, 130 conference papers, two books, and 14 patents.

EDITORIAL

Dr. Lixin Cheng is the founder and Editor-in-Chief of the *International Journal of Microscale and Nanoscale Thermal and Fluid Transport Phenomena (IJMNTFTP)*. He is lecturer at the University of Aberdeen, UK since 2009. He received his Ph.D. from the State Key Laboratory of Multiphase Flow at Xi'an Jiaotong University, China in 1998. He was scientific collaborator of the Laboratory of Heat and Mass Transfer (LTCM) at the Swiss Federal Institute of Technology in Lausanne (EPFL), Switzerland from 2006 to 2009. He was awarded an Alexander von Humboldt Fellowship and thus worked in the Institute of Process Engineering at the Leibniz University of Hanover, Germany from 2004 to 2006. He was senior research fellow at the London South Bank University, UK from 2001 to 2003, and post-doctoral research fellow at the Eindhoven University of Technology, the Netherlands from 2000 to 2001. His research interests include multiphase flow and heat transfer, enhanced heat transfer, micro-scale heat transfer, nanofluid two-phase flow and heat transfer, compact and micro-heat exchangers, thermal systems, oil-water-gas multiphase flow and chemical transport phenomena. He has published more than 80 papers in journals and conferences, 5 book chapters, and one book (with Dr. Yun Rin) *CO_2 Heat Transfer Fluid Flow and Thermal Systems* coming out in 2010. He also serves as the Editor-in-Chief of e-book series *Advances in Multiphase Flow and Heat Transfer* (2 volumes per year) since 2009.

Index

A

B

C

D

E

F

G

H

I

K

L

M

N

O

P

Q

R

S

T

U

V

W

Y

Z